Interdisciplinary Applied Mathematics

Volume 9

Problems in engineering, computational science, and the physical and biological sciences are using increasingly sophisticated mathematical techniques. Thus, the bridge between the mathematical sciences and other disciplines is heavily traveled. The correspondingly increased dialog between the disciplines has led to the establishment of the series: *Interdisciplinary Applied Mathematics*.

The purpose of this series is to meet the current and future needs for the interaction between various science and technology areas on the one hand and mathematics on the other. This is done, firstly, by encouraging the ways that mathematics may be applied in traditional areas, as well as point towards new and innovative areas of applications; and, secondly, by encouraging other scientific disciplines to engage in a dialog with mathematicians outlining their problems to both access new methods and suggest innovative developments within mathematics itself.

The series will consist of monographs and high-level texts from researchers working on the interplay between mathematics and other fields of science and technology.

Interdisciplinary Applied Mathematics

Volumes published are listed at the end of this book.

Springer
New York
Berlin
Heidelberg
Barcelona
Hong Kong
London
Milan
Paris
Singapore
Tokyo

Weimin Han B. Daya Reddy

Plasticity

Mathematical Theory and Numerical Analysis

With 34 Illustrations

Springer

Weimin Han
Department of Mathematics
University of Iowa
Iowa City, IA 52242
USA
whan@math.uiowa.edu

B. Daya Reddy
Department of Mathematics and
Applied Mathematics
University of Cape Town
7700 Rondebosch
South Africa
bdr@maths.uct.ac.za

Mathematics Subject Classification (1991): 73Exx, 65Nxx, 65Mxx, 73Vxx, 49J40

Library of Congress Cataloging-in-Publication Data
Han, Weimin.
Plasticity : mathematical theory and numerical analysis / Weimin Han, B. Daya Reddy.
p. cm. — (Interdisciplinary applied mathematics ; 9)
Includes bibliographical references.
ISBN 0-387-98704-5 (hc. : alk. paper)
1. Plasticity. 2. Numerical analysis. I. Reddy, B. Dayanand, 1953– . II. Title. III. Series: Interdisciplinary applied mathematics ; v. 9.
QA931.H25 1999
531′.385—dc21 98-51755

Printed on acid-free paper.

Production managed by MaryAnn Cottone; manufacturing supervised by Joe Quatela.
Photocomposed copy produced from the author's LaTeX files.
Printed and bound by Braun-Brumfield, Ann Arbor, MI.
Printed in the United States of America.

9 8 7 6 5 4 3 2 1

ISBN 0-387-98704-5 Springer-Verlag New York Berlin Heidelberg SPIN 10707337

To our wives
HUIDI TANG AND SHAADA
and our children
ELIZABETH AND JORDI

Preface

The basis for the modern theory of elastoplasticity was laid in the nineteenth-century, by TRESCA, ST. VENANT, LÉVY, and BAUSCHINGER. Further major advances followed in the early part of this century, the chief contributors during this period being PRANDTL, VON MISES, and REUSS. This early phase in the history of elastoplasticity was characterized by the introduction and development of the concepts of irreversible behavior, yield criteria, hardening and perfect plasticity, and of rate or incremental constitutive equations for the plastic strain.

Greater clarity in the mathematical framework for elastoplasticity theory came with the contributions of PRAGER, DRUCKER, and HILL, during the period just after the Second World War. Convexity of yield surfaces, and all its ramifications, was a central theme in this phase of the development of the theory.

The mathematical community, meanwhile, witnessed a burst of progress in the theory of partial differential equations and variational inequalities from the early 1960s onwards. The timing of this set of developments was particularly fortuitous for plasticity, given the fairly mature state of the subject, and the realization that the natural framework for the study of initial boundary value problems in elastoplasticity was that of variational inequalities. This confluence of subjects emanating from mechanics and mathematics resulted in yet further theoretical developments, the outstanding examples being the articles by MOREAU, and the monographs by DUVAUT AND J.-L. LIONS, and TEMAM. In this manner the stage was set for comprehensive investigations of the well-posedness of problems in elastoplasticity, while the simultaneous rapid growth in interest in numerical methods ensured that equal attention was given to issues such as the development of solution algorithms, and their convergences.

The interaction between elastoplasticity and mathematics has spawned among many engineering scientists an interest in gaining a better understanding of the modern mathematical developments in the subject. In the same way, given the richness of plasticity in interesting and important mathematical problems, many mathematicians, either students or mature researchers, have developed an interest in understanding the mechanical and engineering basis of the subject, and its connections with the mathematical theory. While there are many textbooks and monographs on plasticity that deal with the mechanics of the subject, they are written mainly for a readership in the engineering sciences; there does not appear to us to have existed an extended account of elastoplasticity which would serve these dual needs of both engineering scientists and mathematicians. It is our hope that this monograph will go some way towards filling that gap.

We present in this work three logically connected aspects of the theory of elastic-plastic solids: the constitutive theory, the variational formulations of the related initial boundary value problems, and the numerical analysis of these problems. These three aspects determine the three parts into which the monograph is divided.

The constitutive theory, which is the subject of Part I, begins with a motivation grounded in physical experience, whereafter the constitutive theory of classical elastoplastic media is developed. This theory is then cast in a convex analytic setting, after some salient results from convex analysis have been reviewed. The term "classical" refers in this work to that theory of elastic-plastic material behavior which is based on the notion of convex yield surfaces, and the normality law. Furthermore, only the small strain, quasi-static theory is treated. Much of what is covered in Part I will be familiar to those working on plasticity, though the greater insights offered by exploiting the tools of convex analysis may be new to some researchers. On the other hand, mathematicians unfamiliar with plasticity theory will find in this first part an introduction that is self-contained and accessible.

Part II of the monograph is concerned with the variational problems in elastoplasticity. Two major problems are identified and treated: the primal problem, of which the displacement and internal variables are the primary unknowns; and the dual problem, of which the main unknowns are the generalized stresses.

Finally, Part III is devoted to a treatment of the approximation of the variational problems presented in the previous part. We focus on finite element approximations in space, and both semi- and fully discrete problems. In addition to deriving error estimates for these approximations, attention is given to the behavior of those solution algorithms that are in common use.

Wherever possible we provide background materials of sufficient depth to make this work as self-contained as possible. Thus, Part I contains a review of topics in continuum mechanics, thermodynamics, linear elasticity, and convex analytic setting of elastoplasticity. In Part II we include a treatment of those topics from functional analysis and function spaces that are relevant to a discussion of the well-posedness of vatriational problems. And Part III begins with an overview of the mathematics of finite elements.

In writing this work we have drawn heavily on the results of our joint collaboration in the past few years. We have also consulted, and made liberal use of the works of many: we mention in particular the major contributions of G. Duvaut and J.-L. Lions, C. Johnson, J.B. Martin, H. Matthies, and J.C. Simo. While we acknowledge this debt with gratitude, the responsibility for any inaccuracies or erroneous interpretations that might exist in this work, rests with its authors.

We thank our many friends, colleagues and family members whose interest, guidance, and encouragement made this work possible.

W.H. B.D.R.
Iowa City Cape Town

Contents

Part I

Continuum Mechanics and Elastoplasticity Theory

1 Preliminaries

1.1 Introduction

The theory of elastoplastic media is now a mature branch of solid and structural mechanics, having experienced significant development during the latter half of this century. In particular, the classical theory, which deals with small-strain elastoplasticity problems, has a firm mathematical basis, and from this basis further developments, both mathematical and computational, have evolved. Small-strain elastoplasticity is well understood, and the understanding of its governing equations can be said to be almost complete. Likewise, theoretical, computational, and algorithmic work on approximations in the spatial and time domains are at a stage at which approximations of desired accuracy can be achieved with confidence.

The finite-strain theory has evolved along parallel lines, although it is considerably more complex and is subject to a number of alternative treatments. The form taken by the governing equations is reasonably settled, though there is as yet no mathematical treatment of existence, uniqueness, and stability analogous to those of the small-strain case. Computationally, great strides have been made in the last two decades, and it is now possible to solve highly complex problems with the aid of the computer.

This monograph focuses on theoretical aspects of the small-strain theory of elastoplasticiy with hardening assumptions. It is intended to provide a reasonably comprehensive and unified treatment of the mathematical theory and numerical analysis, exploiting in particular the great advantages to be gained by placing the theory in a convex-analytic context.

The monograph is divided into three parts. The first part contains the first four chapters and provides a detailed introduction to plasticity, in which the mechanics of elastoplastic behavior is emphasized. The equations describing elastoplastic behavior are subsequently recast in the language and setting of convex analysis. In particular, the flow law can be written in terms of either the dissipation function or the yield function. Thus, it is possible to present the flow law in two alternative yet equivalent forms, which are dual to each other.

The second part of the monograph is taken up with mathematical considerations of the elastoplasticity problem. It begins with some preparations on basic knowledge from functional analysis and weak formulations of boundary value problems. These are the contents of Chapters 5 and 6. Depending on the form of the flow law used, we obtain two formulations for the elastoplasticity problem: the primal variational formulation, which uses the dissipation function to describe the flow law, and the dual variational formulation, which uses the yield function to describe the flow law. The two forms are equivalent. The main task of the second part is a thorough mathematical treatment of the well-posedness of the two alternative formulations of the small-strain problem. The primal variational problem is analyzed in Chapter 7, and the dual variational problem in Chapter 8.

Numerical analysis of the elastoplasticity problem is the topic of the third part. For the convenience of the reader, we introduce the basic ideas of the finite element method and some typical finite element interpolation results in Chapter 9. We then review some standard results in the error analysis for finite element approximations of boundary value problems for differential equations and inequalities. This is followed by error analysis of various semidiscrete and fully discrete approximations for both the primal and dual variational problems. We also discuss convergence properties of a number of solution algorithms commonly used in practice.

Plasticity is a vast research area, and it is impossible to touch on every aspect of this area in a single volume. Thus, several important topics are not included in this monograph, for example, applications of elastoplasticity theory to the analysis of engineering structures, which have been covered in many books on elastoplasticity directed at the engineering community (see, for example, Martin [83] and Chen and Han [22]).

In this book, we deal exclusively with hardening elastoplasticity. The reader will find in Temam [122] a comprehensive mathematical treatment of the elastic, perfectly-plastic problem.

Details of the implementation and behavior of specific algorithms are omitted, as are other topics, such as viscoplasticity, and matters pertaining to the finite-strain problem. These topics are given a comprehensive treatment in the monograph by Simo and Hughes [116] and the extended survey by Simo [114]. Both of these works, and many of the references cited in them, contain a wealth of numerical examples.

The list of the references at the end of the book includes only those that are more relevant to the present exposition, and we do not attempt to make the list complete.

This work summarizes some recent results on mathematical analysis and numerical analysis of the elastoplasticity problem. We hope that it will be useful to those readers who wish to know more about recent developments in the analysis of the elastoplasticity problem and to those who are preparing to carry out research in the area of plasticity. For the convenience of the reader, we include brief introductions to various mathematical materials that should be sufficient for reading the book. In this way, it will not be necessary to have any extensive prior knowledge of advanced mathematical topics, such as functional analysis and convex analysis. Nevertheless, some degree of maturity in mathematics and some knowledge of mechanics are expected from the reader. We hope that the book will also be helpful to those whose main interests lie in the solution of plasticity problems in engineering practice. We are convinced that attempts at solving practical problems in this area—as, indeed, is the case in many other areas—would benefit from a background in the theoretical aspects of the subject.

1.2 Some Historical Remarks

Early works on plasticity. It is generally agreed that the origin of plasticity dates back to a series of papers by Tresca from 1864 to 1872 (see [125]) on the extrusion of metals. In this work the first yield condition was proposed: The condition, known subsequently as the Tresca yield criterion, stated that a metal yields when the maximal shear stress attains a critical value. In the same time period, St. Venant [6] introduced basic constitutive relations for rigid, perfectly plastic materials in plane stress, and suggested that the principal axes of the strain increment coincide with the principal axes of stress. Lévy [76] derived the general equations in three dimensions. In 1886, Bauschinger [8] observed the effect that now carries his name: A previous plastic strain with a certain sign diminishes the resistance of the material with respect to the next plastic strain with the opposite sign. In a landmark paper in 1913, von Mises [92] derived the general equations for plasticity, accompanied by his well-known pressure-insensitive yield criterion (J_2-theory, or octahedral shear stress yield condition).

In 1924, Prandtl [101] extended the St. Venant–Levy–von Mises equations for the plane continuum problem to include the elastic component of strain, and Reuss [111] in 1930 carried out their extension to three dimensions. In 1928, von Mises [93] generalized his previous work for a rigid, perfectly plastic solid to include a general yield function and discussed the relation between the direction of plastic strain increment and the smooth yield surface, thus introducing formally the concept of using the yield function as a plastic potential in the incremental stress–strain relations of the flow theory.

Compared to perfect plasticity, the development of incremental constitutive relations for hardening materials proceeded more slowly. In 1928, Prandtl [102] attempted to formulate general relations for hardening behavior. In 1938, Melan [91] generalized the foregoing concepts of perfect plasticity by giving incremental relations for hardening solids with smooth yield surface and discussing uniqueness results for elastoplastic incremental problems for both perfectly plastic and hardening materials, based on some limiting assumptions.

Since 1940, the theory of plasticity has seen relatively more rapid development. In 1949, Prager [100] obtained a general framework for the plastic constitutive relations for hardening materials with smooth yield functions and recognized the relationship between the convexity of the yield surface plus the normality law and the uniqueness of the associated boundary value problem. Drucker [32], in 1951, proposed his material stability postulate. With this concept, the plastic stress–strain relations together with many related fundamental aspects of the subject may be treated in a unified manner. In 1953, Koiter [72] generalized the plastic stress–strain relations for nonsmooth yield surfaces and obtained some uniqueness and variational results. He introduced the device of using more than one yield function in the stress–strain relations, the plastic strain increment receiving a contribution from each active yield surface and falling within the normal cone to the yield surface. For further details, see [73].

A detailed description of the early development of plasticity theory and a comprehensive list of references on plasticity published before 1980 can be found in Życzkowski [136], which also contains a wealth of discussions on various aspects of plasticity.

Recent mathematical and numerical analysis of problems in plasticity. Mathematical and numerical aspects of the quasistatic problem in elastoplasticity have been the subject of sustained attention since the 1970s. The first systematic mathematical study of the boundary value problems of elastoplasticity is due to Duvaut and Lions [33], who considered the problem for an elastic perfectly plastic material and formulated the problem as a variational inequality. Moreau [95, 96] considered the same issues, but from a more geometric viewpoint. Johnson [64] subsequently extended the analysis in [33] by approaching the problem in two stages; in the first stage the velocity is eliminated and the problem becomes a variational inequality posed on a time-dependent convex set. The second stage involves the solution for the velocity.

The theory for perfectly plastic materials was advanced greatly through the introduction and investigation of the space $BD(\Omega)$ of functions of bounded deformation [88, 90, 123, 124]. This space is essential for a proper study of the perfectly plastic problem, since discontinuity surfaces (sliplines) may be accommodated within this framework; the framework of Sobolev spaces, on the other hand, is not appropriate. A comprehensive summary account of the mathematical theory of perfect plasticity in the framework of the space $BD(\Omega)$ can be found in [122], which is, however, confined to the total strain, or holonomic, problem, an approximate model in which a one-to-one relationship between stress and strain is assumed.

Analysis of the elastoplastic problem with hardening, on the other hand, can be achieved within the framework of Sobolev spaces. There are two alternative formulations of the problem, depending on the form of the flow law. One formulation makes use of the yield function in the plastic flow law and will be called the dual formulation in this work, for reasons that will become clear in Chapter 4. An alternative approach is to express the plastic flow law in terms of the dissipation function, which leads to the primal formulation of the problem. The primal and dual formulations are extensions, respectively, of the displacement and stress problems in linear elasticity.

The first analysis of the dual formulation of the hardening problem is due to Johnson [66], who gave an existence and uniqueness result. A detailed analysis of the primal formulation of the hardening problem was presented by Han, Reddy, and Schroeder [56]. The unknowns are the displacement and internal variables, while the problem takes the form of a variational inequality of the mixed kind: It is an inequality both because of the presence of a nondifferentiable functional in the formulation *and* because the problem is posed on a closed convex cone in a Hilbert space.

Analyses of finite element approximations of the elastoplastic problem have enjoyed limited but steady attention. Johnson [65] considered a formulation of the elastic, perfectly-plastic problem in which stress is the primary variable and derived error estimates for the fully discrete (that is, discrete in both time and space) problem. In a later work, Johnson [67] analyzed fully discrete finite element approximations of the elastoplasticity problem with hardening, in the context of a mixed formulation in which stress and velocity are the variables. Related work can also be found in Hlaváček [60], and summary accounts in Hlaváček, Haslinger, Nečas, and Lovíšek [61], and Korneev and Langer [74].

The dual formulation is a popular approach in practice for the hardening problem; see, for example, the comprehensive treatments of computational aspects of the problem by Simo [114], and by Simo and Hughes [116]. However, while there exist some results on stability, consistency, and convergence for certain numerical approximation schemes, the whole picture is by no means complete.

In comparison, numerical analysis of the primal formulation of the hardening problem did not receive attention until recently. Various schemes for approximating the primal formulation of the hardening problem were analyzed for the first time in [56].

A more classical approach to the analysis and numerical analysis of the hardening plasticity problem is taken by Bonnetier [13], and by Li and Babuška [77, 78]. First, spatial discretization is carried out using finite elements, and the resulting semidiscrete problem is written as a system of highly nonlinear ordinary differential equations. Then it is shown that as the finite element mesh size approaches zero, the solution of the semidiscrete problem converges, and the limit is the solution of the plasticity problem.

The recent work of Han and Reddy [55] provides a comprehensive treatment of the mathematical and numerical analysis of the elastoplasticity problem with hardening. Two alternative variational formulations of the problem are described in a connected way. The formulation based on the dissipation function is defined to be the primal problem, largely because it is a kinematically based formulation: The unknown variables are the displacement, the plastic strain, and internal variables. The formulation based on the yield function is referred to as the dual formulation, the stress being a main unknown. The question of existence and uniqueness of solutions is addressed, taking each of these formulations in turn as a point of departure. Various approximation schemes for each formulation are studied. The schemes considered include semidiscrete approximations in which either the spatial domain or the time domain is discretized, and fully discrete approximations where discretization is carried out with respect to both space and time. Error estimates for these approximations are derived, not only for the approximate stress, but also for the approximate displacement; in comparison, the study in [67] is confined to one involving the stress and velocity, and results on convergence are presented only for the stress.

The resulting discrete systems are nonlinear and large. Various solution algorithms are used in practice to solve these systems. Some popular solution algorithms are discussed in [55], and for the first time convergence of some of the solution algorithms is proved rigorously.

The present monograph is an expanded and updated version of our previous work [55].

1.3 Notation

Throughout this work we will use the popular mathematical symbol $\forall$ to stand for "for any" or "for all." The letter c will denote a generic constant independent of certain quantities (which are clear from the context). The value of c may differ at different places.

Vectors, tensors. Some pertinent results from vector and tensor analysis are summarized here for convenience. More comprehensive sources can be found in the literature (see, for example, Lemaitre and Chaboche [75]).

We will use boldface italic letters to denote vectors and tensors. We adopt the summation convention for repeated indices, unless stated otherwise. Most often, vectors are denoted by lowercase boldface italic letters, and second-order tensors by lowercase boldface Greek letters. Fourth-order tensors are usually denoted by uppercase boldface italic letters.

Our discussion applies to Euclidean space $\mathbb{R}^d$ of any dimension d (in practice, $d = 1, 2, 3$). However, for definiteness of exposition and because of its importance in applications, we will give the presentation in the context of three-dimensional space. Thus, we will make use of a Cartesian coordinate system with an orthonormal basis $\{\boldsymbol{e}_1, \boldsymbol{e}_2, \boldsymbol{e}_3\}$ that is chosen *once and for all*. Where it is necessary to show components of a vector or a tensor, these will always be relative to the orthonormal basis $\{\boldsymbol{e}_1, \boldsymbol{e}_2, \boldsymbol{e}_3\}$.

A second-order tensor $\boldsymbol{\tau}$ is a linear operator mapping vectors to vectors and may be identified with a matrix. For any vector $\boldsymbol{a}$, $\boldsymbol{\tau a}$ represents a vector such that the action of $\boldsymbol{\tau}$ on $\boldsymbol{a}$ is linear; that is, $\boldsymbol{\tau}(\alpha\boldsymbol{a} + \beta\boldsymbol{b}) = \alpha\boldsymbol{\tau a} + \beta\boldsymbol{\tau b}$ for any scalars α, β, and any vectors $\boldsymbol{a}$ and $\boldsymbol{b}$. We will always use a_i, $1 \le i \le 3$, to denote the components of the vector $\boldsymbol{a}$, and τ_{ij}, $1 \le i, j \le 3$, the components of the second-order tensor $\boldsymbol{\tau}$. With the basis defined, the action of the second-order tensor $\boldsymbol{\tau}$ on the vector $\boldsymbol{a}$ may be represented in the form

$$\boldsymbol{\tau a} = \tau_{ij} a_j \boldsymbol{e}_i.$$

The scalar products of two vectors $\boldsymbol{a}$ and $\boldsymbol{b}$, and of two second-order tensors (or matrices) $\boldsymbol{\sigma}$ and $\boldsymbol{\tau}$, are denoted by $\boldsymbol{a} \cdot \boldsymbol{b}$ and $\boldsymbol{\sigma} : \boldsymbol{\tau}$ and have the component representations

$$\boldsymbol{a} \cdot \boldsymbol{b} = a_i b_i, \qquad \boldsymbol{\sigma} : \boldsymbol{\tau} = \sigma_{ij}\tau_{ij}.$$

The magnitudes of a vector $\boldsymbol{a}$ and a second-order tensor $\boldsymbol{\tau}$ are defined by

$$|\boldsymbol{a}| = (\boldsymbol{a} \cdot \boldsymbol{a})^{\frac{1}{2}}, \qquad |\boldsymbol{\tau}| = (\boldsymbol{\tau} : \boldsymbol{\tau})^{\frac{1}{2}}.$$

The vector product $\boldsymbol{c} = \boldsymbol{a} \wedge \boldsymbol{b}$ of two vectors $\boldsymbol{a}$ and $\boldsymbol{b}$ is a vector with components defined by

$$c_i = \epsilon_{ijk} a_j b_k,$$

where ϵ_{ijk} is the permutation symbol: $\epsilon_{ijk} = +1$ for (i, j, k) a cyclic permutation of $(1, 2, 3)$, -1 for (i, j, k) an anticyclic permutation, and is zero otherwise.

The tensor product $\boldsymbol{a} \otimes \boldsymbol{b}$ of two vectors $\boldsymbol{a}$ and $\boldsymbol{b}$ is a second-order tensor defined by the relation

$$(\boldsymbol{a} \otimes \boldsymbol{b})\boldsymbol{c} = (\boldsymbol{b} \cdot \boldsymbol{c})\boldsymbol{a} \quad \forall\, \boldsymbol{c}.$$

Viewed as a matrix, we have the representation

$$\boldsymbol{a} \otimes \boldsymbol{b} = \boldsymbol{a}\boldsymbol{b}^T.$$

Thus the tensor product $\boldsymbol{a}\otimes\boldsymbol{b}$ has the components $a_i b_j$. The nine quantities $\boldsymbol{e}_i \otimes \boldsymbol{e}_j$ form a basis for the space of the second-order tensors, and any such tensor $\boldsymbol{\tau}$ can be represented in the form

$$\boldsymbol{\tau} = \tau_{ij}\boldsymbol{e}_i \otimes \boldsymbol{e}_j.$$

Since we will be working with a fixed basis, there is little point in making a formal distinction between the tensor $\boldsymbol{\tau}$ and the 3×3 matrix of its components, so that unless otherwise stated, $\boldsymbol{\tau}$ will represent the tensor *and* the matrix of its components. With this understanding, it is merely necessary to point out that all the usual matrix operations such as addition, transposition, multiplication, inversion, and so on, apply to tensors, and the standard notation is used for these operations. Thus, for example, $\boldsymbol{\tau}^T$ and $\boldsymbol{\tau}^{-1}$ are, respectively, the transpose and inverse of the tensor (or matrix) $\boldsymbol{\tau}$.

We will use M^3 to denote the space of all the symmetric 3×3 matrices (or second-order symmetric tensors). We will use M_0^3 to denote the subspace of M^3 with vanishing trace; that is,

$$M_0^3 = \{\boldsymbol{\tau} \in M^3 : \operatorname{tr}\boldsymbol{\tau} = 0\},$$

where as usual, $\operatorname{tr}\boldsymbol{\tau} = \tau_{ii}$ is the trace of $\boldsymbol{\tau}$.

One special and important second-order tensor is the *identity* $\boldsymbol{I}$, which is defined by the relation $\boldsymbol{I}\boldsymbol{a} = \boldsymbol{a}$ for any vector $\boldsymbol{a}$. The components of the identity tensor $\boldsymbol{I}$ are the Kronecker delta

$$\delta_{ij} = \begin{cases} 1 & \text{if } j = i, \\ 0 & \text{otherwise.} \end{cases}$$

Every second-order tensor $\boldsymbol{\tau}$ may be additively decomposed into a deviatoric part $\boldsymbol{\tau}^D$ and a spherical part $\boldsymbol{\tau}^S$; these are defined by

$$\boldsymbol{\tau}^S = \tfrac{1}{3}(\operatorname{tr}\boldsymbol{\tau})\boldsymbol{I}, \quad \boldsymbol{\tau}^D = \boldsymbol{\tau} - \tfrac{1}{3}(\operatorname{tr}\boldsymbol{\tau})\boldsymbol{I},$$

so that

$$\boldsymbol{\tau} = \boldsymbol{\tau}^D + \boldsymbol{\tau}^S.$$

For spatial domains of dimension d, a second-order tensor $\boldsymbol{\tau}$ is identified with a $d \times d$ matrix, and the formulae for its deviatoric and spherical parts are modified to

$$\boldsymbol{\tau}^S = \tfrac{1}{d}(\operatorname{tr}\boldsymbol{\tau})\boldsymbol{I}, \quad \boldsymbol{\tau}^D = \boldsymbol{\tau} - \tfrac{1}{d}(\operatorname{tr}\boldsymbol{\tau})\boldsymbol{I}.$$

For planar problems, for example, $d = 2$.

The only higher-order tensors that will occur are those of fourth order, which will appear as tensors of material moduli. These will be denoted by uppercase boldface italic letters. A fourth-order tensor $\boldsymbol{C}$ may be defined as a linear operator mapping the space of second-order tensors into itself. The action of a fourth-order tensor $\boldsymbol{C}$ on a second-order tensor $\boldsymbol{\tau}$ is denoted by $\boldsymbol{C\tau}$ and is the second-order tensor with components $C_{ijkl}\tau_{kl}$, where C_{ijkl} are the components of $\boldsymbol{C}$ relative to the canonical orthonormal basis $\boldsymbol{e}_i \otimes \boldsymbol{e}_j \otimes \boldsymbol{e}_k \otimes \boldsymbol{e}_l$, $1 \leq i, j, k, l \leq 3$. An important special fourth-order tensor is the identity tensor $\boldsymbol{I}$, which satisfies $\boldsymbol{I\tau} = \boldsymbol{\tau}$ for any symmetric second-order tensors $\boldsymbol{\tau}$. This identity tensor has the component representation

$$I_{ijkl} = \tfrac{1}{2}\,(\delta_{ik}\delta_{jl} + \delta_{il}\delta_{jk}).$$

We use the same symbol $\boldsymbol{I}$ for both the second-order and fourth-order identity tensors.

Invariants of second-order tensors (or 3×3 matrices). The problem of finding a scalar λ and a nonzero vector $\boldsymbol{q}$ with

$$\boldsymbol{\tau q} = \lambda \boldsymbol{q}$$

leads to the eigenvalue problem of solving the characteristic equation

$$\det\,(\lambda \boldsymbol{I} - \boldsymbol{\tau}) = 0.$$

This equation can be written equivalently as

$$\lambda^3 - I_1\lambda^2 + I_2\lambda - I_3 = 0,$$

where $I_1(\boldsymbol{\tau})$, $I_2(\boldsymbol{\tau})$, and $I_3(\boldsymbol{\tau})$ are the *principal invariants* of $\boldsymbol{\tau}$. The invariants are defined by

$$\begin{aligned}
I_1 &= \operatorname{tr}\boldsymbol{\tau} = \tau_{ii} = \lambda_1 + \lambda_2 + \lambda_3,\\
I_2 &= \tfrac{1}{2}\{(\operatorname{tr}\boldsymbol{\tau})^2 - \operatorname{tr}\boldsymbol{\tau}^2\} = \tfrac{1}{2}(\tau_{ii}\tau_{jj} - \tau_{ij}\tau_{ji}) = \lambda_1\lambda_2 + \lambda_2\lambda_3 + \lambda_3\lambda_1,\\
I_3 &= \det\boldsymbol{\tau} = \lambda_1\lambda_2\lambda_3.
\end{aligned}$$

Here, λ_1, λ_2, and λ_3, the eigenvalues of $\boldsymbol{\tau}$, are the roots of the characteristic equation (a multiple root is counted repeatedly according to its multiplicity).

We denote by

$$\boldsymbol{\iota}(\boldsymbol{\tau}) = (I_1(\boldsymbol{\tau}),\, I_2(\boldsymbol{\tau}),\, I_3(\boldsymbol{\tau}))$$

the set of three invariants of $\boldsymbol{\tau}$. The eigenvalues λ_i of a matrix $\boldsymbol{\tau}$ are often denoted by τ_i (note the single index) and are called the *principal components* of $\boldsymbol{\tau}$.

Scalar, vector, and tensor fields. The gradient of a scalar field $\phi(\boldsymbol{x})$ is denoted by $\nabla\phi$ and is the vector defined by

$$\nabla\phi = \frac{\partial\phi}{\partial x_i}\boldsymbol{e}_i.$$

The divergence div $\boldsymbol{u}$ and gradient $\nabla\boldsymbol{u}$ of a vector field $\boldsymbol{u}(\boldsymbol{x})$ are respectively a scalar and a second-order tensor field, defined by

$$\operatorname{div}\boldsymbol{u} = \frac{\partial u_i}{\partial x_i},$$
$$\nabla\boldsymbol{u} = \frac{\partial u_i}{\partial x_j}\boldsymbol{e}_i \otimes \boldsymbol{e}_j.$$

Thus the components of $\nabla\boldsymbol{u}$ are $\partial u_i/\partial x_j$. The transpose of $\nabla\boldsymbol{u}$ is denoted by $(\nabla\boldsymbol{u})^T$ and is the second-order tensor with components $\partial u_j/\partial x_i$. The divergence div $\boldsymbol{\tau}$ of a second-order tensor $\boldsymbol{\tau}$ is a vector defined by

$$\operatorname{div}\boldsymbol{\tau} = \frac{\partial \tau_{ij}}{\partial x_j}\boldsymbol{e}_i.$$

For a scalar-valued function $f(\boldsymbol{u})$ of a vector variable $\boldsymbol{u} = (u_1, u_2, u_3)^T$, its derivative with respect to $\boldsymbol{u}$ can be identified with a vector,

$$\frac{\partial f(\boldsymbol{u})}{\partial \boldsymbol{u}} = \frac{\partial f(\boldsymbol{u})}{\partial u_i}\boldsymbol{e}_i.$$

For a scalar-valued function $f(\boldsymbol{\tau})$ of a second-order tensor $\boldsymbol{\tau} = (\tau_{ij})$, the derivative with respect to $\boldsymbol{\tau}$ is a second-order tensor,

$$\frac{\partial f(\boldsymbol{\tau})}{\partial \boldsymbol{\tau}} = \frac{\partial f(\boldsymbol{\tau})}{\partial \tau_{ij}}\boldsymbol{e}_i \otimes \boldsymbol{e}_j.$$

If $\boldsymbol{f}(\boldsymbol{\tau})$ is a matrix-valued function of a second-order tensor variable $\boldsymbol{\tau}$, then its derivative with respect to $\boldsymbol{\tau}$ is a fourth-order tensor with components

$$\frac{\partial \boldsymbol{f}(\boldsymbol{\tau})}{\partial \tau_{ij}} = \frac{\partial f_{kl}(\boldsymbol{\tau})}{\partial \tau_{ij}}\boldsymbol{e}_k \otimes \boldsymbol{e}_l.$$

For a time-dependent quantity z, we will use $\dot{z}$ to denote the partial derivative of z with respect to the temporal variable t.

Landau's notation for orders of magnitude. We will use the "big oh" (O) and "little oh" (o) symbols in the following senses. Given two functions $f(t)$ and $g(t)$ of a real variable t, we say that $f(t)$ is of a lower order of magnitude than $g(t)$ as $t \to 0+$ and write

$$f(t) = o(g(t)), \quad t \to 0+,$$

if

$$\lim_{t\to 0+} \frac{f(t)}{g(t)} = 0.$$

We say that $f(t)$ is dominated by $g(t)$ as $t \to 0+$, and write

$$f(t) = O(g(t)), \quad t \to 0+,$$

if for some positive constants c and δ,

$$|f(t)| \le c\,|g(t)|, \quad t \in (0, \delta).$$

These definitions can be easily adapted to cover other similar expressions, such as

$$f(t) = o(g(t)), \quad t \to 0,$$

or

$$x_n = O(y_n), \quad n \to \infty,$$

for two sequences of numbers $\{x_n\}$ and $\{y_n\}$.

2

Continuum Mechanics and Linear Elasticity

We will be concerned with bodies that at the macroscopic level may be regarded as composed of material that is continuously distributed. By this it is meant, first, that such a body occupies a region of three-dimensional space that may be identified with $\mathbb{R}^3$. The region occupied by the body will of course vary with time as the body deforms; it is convenient, then, for the purpose of keeping track of the evolution of the body's behavior to locate any point in the body by its position vector $\boldsymbol{x}$ with respect to some previously chosen origin $\mathbf{0}$, at a fixed time. For simplicity we will take this to be at the time $t = 0$, and we will assume that the body is undeformed and unstressed in this state, unless stated otherwise. The region occupied by the body at the time $t = 0$ is denoted by Ω, and is called the *reference configuration*. To emphasize the identification between points in the region Ω and points in the undeformed body we will often refer to a point $\boldsymbol{x} \in \Omega$ as a *material point*. If we go further and place a set of Cartesian axes at $\mathbf{0}$, then the position vector $\boldsymbol{x}$ has components x_i $(i = 1, 2, 3)$ with respect to the orthonormal basis $\{\boldsymbol{e}_1, \boldsymbol{e}_2, \boldsymbol{e}_3\}$ associated with this set of axes. The situation is illustrated in Figure 2.1, in which Ω_t is the current configuration, the region occupied by the body at the current time t.

Second, it is assumed that both the properties and the behavior of such a body can be described in terms of functions of position $\boldsymbol{x}$ in the body and time t. Thus, for example, we may associate with the body a scalar temperature distribution θ that varies within the body and with the passage of time, so that the value of the temperature of a material point $\boldsymbol{x}$ at time t is represented by the function $\theta(\boldsymbol{x}, t)$, or equivalently by $\theta(x_1, x_2, x_3, t)$.

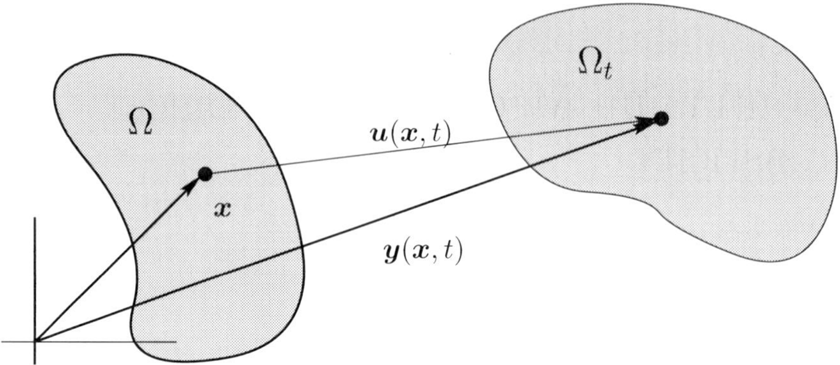

Figure 2.1: Current and undeformed configurations of an arbitrary material body

It will be necessary at some stage to stipulate the properties assumed or expected of functions defined on the body. For the time being there is no need to be too specific about this, except to say that functions will be assumed to possess as many derivatives as are required in order for what follows to make sense. Later we will have to be very careful indeed about the specification of function spaces to which these functions are required to belong.

The study of the behavior of continuous media conveniently begins with a development of a suitable framework within which the motion of the body can be described. This framework is quite independent of any agencies acting on the body, and it is also independent of the constitution of the body. In other words, we are concerned in the first instance solely with the geometry of motion. This is known as kinematics, and we now proceed to set out a framework that will be adequate for future needs.

2.1 Kinematics

As mentioned above, the position of a body in an undeformed state is identified with a region Ω in $\mathbb{R}^3$. With time the body moves and deforms, as a result of the action of various forces (we are not interested in the details of these forces at this point), so that at time t it occupies a new region Ω_t, called the *current configuration* at time t, as is shown in Figure 2.1. This deformation may be described mathematically by introducing a vector-valued function $\boldsymbol{y}$ of position and time, called the *motion*. Thus a material particle initially located at $\boldsymbol{x}$ will have position $\boldsymbol{y}(\boldsymbol{x}, t)$ at time t.

Clearly, we must have $\boldsymbol{y}(\boldsymbol{x}, 0) = \boldsymbol{x}$. For simplicity we denote functions and their values by the same symbol, so that the motion is described by the equation

$$\boldsymbol{y} = \boldsymbol{y}(\boldsymbol{x}, t), \tag{2.1}$$

or in component form,

$$y_i = y_i(x_1, x_2, x_3, t), \quad 1 \leq i \leq 3,$$

for $\boldsymbol{x} \in \Omega$ and $t \in [0, T]$.

The function $\boldsymbol{y}$ will have to satisfy certain conditions if it is used to model adequately the motion of the body. First, we must ensure that no two points get mapped to a single point by $\boldsymbol{y}$; in other words, $\boldsymbol{y}$ must be one-to-one. Second, we must ensure that the motion is orientation-preserving; that is, the *Jacobian* J, defined by

$$J = \det\left(\frac{\partial y_i}{\partial x_j}\right), \tag{2.2}$$

must be positive. Here,

$$\nabla y = \left(\frac{\partial y_i}{\partial x_j}\right)$$

stands for the Jacobian matrix whose (i, j)th element is $\partial y_i / \partial x_j$. Hence, every element of nonzero volume in Ω is mapped to an element of nonzero volume in Ω_t. We are using here the result from calculus that $dy = J dx$, where dx and dy denote the volume elements in Ω and Ω_t.

A sufficient condition for the motion $\boldsymbol{y}$ to be invertible is that there exist a constant $c(\Omega) > 0$, depending only on Ω, such that

$$\sup_{\Omega} |\nabla \boldsymbol{y} - \boldsymbol{I}| < c(\Omega).$$

This result, as well as others on the invertibility of the motion, may be found in [24].

Instead of adopting the function $\boldsymbol{y}$ as the primary unknown variable, it is more convenient to introduce the *displacement* vector $\boldsymbol{u}$ by

$$\boldsymbol{u}(\boldsymbol{x}, t) = \boldsymbol{y}(\boldsymbol{x}, t) - \boldsymbol{x}$$

and to replace the motion by the displacement as the primary unknown. Of course, the displacement alone does not give complete information about the deformation of the body. We need to be able to distinguish, for example, between a simple *rigid body motion*, in which the body is translated and rotated to a new position without deformation (Figure 2.2), and a situation in which the body indeed assumes a new shape. The quantity that we use to

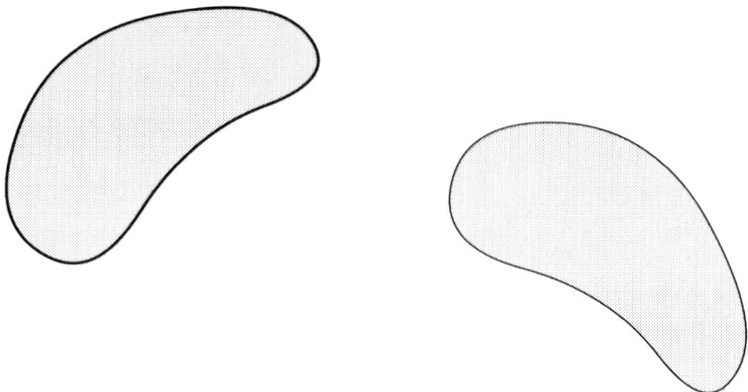

Figure 2.2: An example of rigid body motion

measure deformation is the *strain tensor*. Let us now see how this quantity arises.

Consider a point $\boldsymbol{x}$ in Ω and two fibers of material particles emanating from $\boldsymbol{x}$. These fibers are described by vectors $\Delta\boldsymbol{x}$ and $\delta\boldsymbol{x}$, as is shown in Figure 2.3. The notion of strain emerges naturally if we consider the changes in lengths of these fibers, and the change in the angle between them, under the motion $\boldsymbol{y}$. The fiber $\Delta\boldsymbol{x}$ is mapped to the fiber $\Delta\boldsymbol{y} \equiv \boldsymbol{y}(\boldsymbol{x}+\Delta\boldsymbol{x})-\boldsymbol{y}(\boldsymbol{x})$. Likewise, the fiber $\delta\boldsymbol{x}$ becomes the fiber $\delta\boldsymbol{y} \equiv \boldsymbol{y}(\boldsymbol{x}+\delta\boldsymbol{x})-\boldsymbol{y}(\boldsymbol{x})$. Here, for simplicity in writing, we drop the time variable t in the expression for the motion $\boldsymbol{y}$. We are now in a position to measure changes in lengths and angles.

We assume that the motion is smooth and may be differentiated as many times as required. Then it is possible to expand the term $\boldsymbol{y}(\boldsymbol{x}+\Delta\boldsymbol{x})$ in a Taylor series about $\boldsymbol{x}$ to get

$$\boldsymbol{y}(\boldsymbol{x}+\Delta\boldsymbol{x}) = \boldsymbol{y}(\boldsymbol{x}) + \nabla\boldsymbol{y}\Delta\boldsymbol{x} + o(|\Delta\boldsymbol{x}|),$$

with a similar expression for $\boldsymbol{y}(\boldsymbol{x}+\delta\boldsymbol{x})$. Thus

$$\Delta\boldsymbol{y} \equiv \boldsymbol{y}(\boldsymbol{x}+\Delta\boldsymbol{x}) - \boldsymbol{y}(\boldsymbol{x}) = \nabla\boldsymbol{y}\Delta\boldsymbol{x} + o(|\Delta\boldsymbol{x}|).$$

Since $\nabla\boldsymbol{y}(\boldsymbol{x}) = \boldsymbol{I} + \nabla\boldsymbol{u}(\boldsymbol{x})$, it follows that

$$\Delta\boldsymbol{y} = \Delta\boldsymbol{x} + \nabla\boldsymbol{u}\Delta\boldsymbol{x} + o(|\Delta\boldsymbol{x}|).$$

In exactly the same way we arrive at the expression

$$\delta\boldsymbol{y} = \delta\boldsymbol{x} + \nabla\boldsymbol{u}\,\delta\boldsymbol{x} + o(|\delta\boldsymbol{x}|).$$

We can now consider the expression

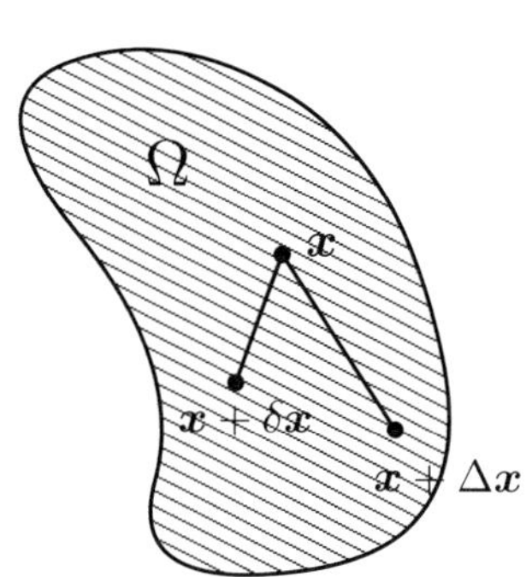

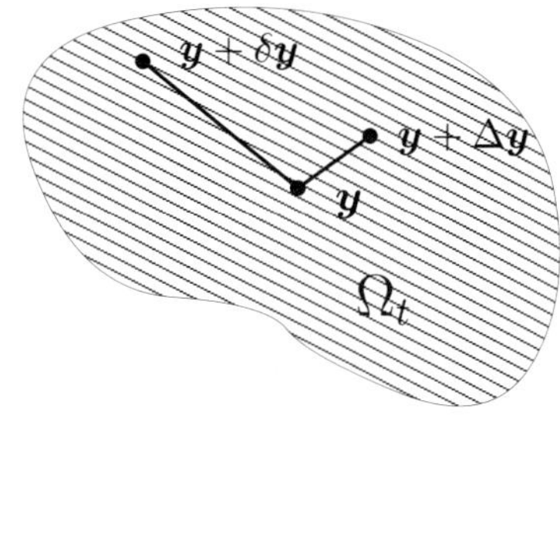

Figure 2.3: Deformed and undeformed configurations of material line elements

$$\begin{aligned}\Delta \boldsymbol{y} \cdot \delta \boldsymbol{y} - \Delta \boldsymbol{x} \cdot \delta \boldsymbol{x} = {} & (\nabla \boldsymbol{u} \Delta \boldsymbol{x}) \cdot \delta \boldsymbol{x} + (\nabla \boldsymbol{u} \delta \boldsymbol{x}) \cdot \Delta \boldsymbol{x} \\ & + (\nabla \boldsymbol{u} \Delta \boldsymbol{x}) \cdot (\nabla \boldsymbol{u} \delta \boldsymbol{x}) + o(|\delta \boldsymbol{x}|^2 + |\Delta \boldsymbol{x}|^2). \quad (2.3)\end{aligned}$$

Though no confusion need arise, it is worth emphasizing that the gradient in (2.3) is with respect to the variable $\boldsymbol{x}$.

The point about the expression (2.3) is that if the body deforms as a rigid body, then obviously we must have $\Delta \boldsymbol{y} \cdot \delta \boldsymbol{y} = \Delta \boldsymbol{x} \cdot \delta \boldsymbol{x}$ for any pair of fibers emanating from any point in the body, since these fibers will not change in length, nor will the angle between them. Thus the right-hand side of (2.3) is identically zero in a rigid body motion. We now go one step further and consider the limit of (2.3) as the lengths of the fibers go to zero. Set $h = \max\{|\Delta \boldsymbol{x}|, |\delta \boldsymbol{x}|\}$, $\boldsymbol{n} = \Delta \boldsymbol{x}/h$, and $\boldsymbol{m} = \delta \boldsymbol{x}/h$; these are assumed to be fixed vectors independent of h. Now divide both sides of (2.3) by h^2, and take the limit as $h \to 0$. This gives

$$\begin{aligned}\lim_{h \to 0} \frac{\Delta \boldsymbol{y} \cdot \delta \boldsymbol{y} - \Delta \boldsymbol{x} \cdot \delta \boldsymbol{x}}{h^2} & = \boldsymbol{n} \cdot \left[\nabla \boldsymbol{u} + (\nabla \boldsymbol{u})^T + (\nabla \boldsymbol{u})^T \nabla \boldsymbol{u}\right] \boldsymbol{m} \\ & \equiv 2\,\boldsymbol{n} \cdot \boldsymbol{\eta}(\boldsymbol{u}) \boldsymbol{m}. \quad (2.4)\end{aligned}$$

We define the *strain tensor* $\boldsymbol{\eta}$ associated with the displacement $\boldsymbol{u}$ by

$$\boldsymbol{\eta}(\boldsymbol{u}) = \tfrac{1}{2}\left[\nabla \boldsymbol{u} + (\nabla \boldsymbol{u})^T + (\nabla \boldsymbol{u})^T \nabla \boldsymbol{u}\right]; \quad (2.5)$$

in component form this expression reads

$$\eta_{ij}(\boldsymbol{u}) = \tfrac{1}{2}(u_{i,j} + u_{j,i} + u_{k,i} u_{k,j}).$$

Though we have been explicit about the fact that the strain is defined for a particular displacement field by writing $\boldsymbol{\eta}(\boldsymbol{u})$, very often we will simply denote the strain by $\boldsymbol{\eta}$ or η_{ij} when there is no danger of confusion.

So we see that the strain tensor is defined in such a way that it is zero if the body undergoes a rigid body motion.

The components of $\boldsymbol{\eta}$ are easily interpreted by referring back to equation (2.4) and by giving the fibers $\Delta\boldsymbol{x}$ and $\delta\boldsymbol{x}$ specific orientations. First, suppose that we identify $\delta\boldsymbol{x}$ with $\Delta\boldsymbol{x}$ at an arbitrary point in the body, and suppose that $\Delta\boldsymbol{x}$ is chosen so that it lies parallel to the x_1-axis. Then (2.4) becomes

$$\lim_{h\to 0}\frac{|\Delta\boldsymbol{y}|^2-|\Delta\boldsymbol{x}|^2}{h^2}=2\,\boldsymbol{e}_1\cdot\boldsymbol{\eta}\boldsymbol{e}_1=2\,\eta_{11},$$

since $\Delta\boldsymbol{x}/h=\boldsymbol{e}_1$ here. Thus we see that in this situation η_{11} equals half the net *change in length* (squared) of a material fiber originally oriented so that it points in the x_1 direction. The other two diagonal components of the strain are interpreted in a similar way.

To see how the off-diagonal components of $\boldsymbol{\eta}$ may be interpreted we return to (2.4) and now choose $\Delta\boldsymbol{x}$ and $\delta\boldsymbol{x}$ at an arbitrary point in the body in such a way that they have equal lengths h and lie parallel to the x_1 and x_2 axes, respectively. Then (2.4) gives

$$\begin{aligned}\lim_{h\to 0}\frac{\Delta\boldsymbol{y}\cdot\delta\boldsymbol{y}-\Delta\boldsymbol{x}\cdot\delta\boldsymbol{x}}{h^2}&=\lim_{h\to 0}\frac{\Delta\boldsymbol{y}\cdot\delta\boldsymbol{y}}{h^2}\\&=2\,\boldsymbol{e}_1\cdot\boldsymbol{\eta}\boldsymbol{e}_2\\&=2\,\eta_{12}.\end{aligned}\tag{2.6}$$

Thus the component η_{12} gives a measure of the *change in angle* between two fibers originally at right angles to each other and oriented so that they were in the x_1 and x_2 directions. The remaining off-diagonal components are interpreted in a similar way.

Because the components of the strain have the interpretations described above, the diagonal components are referred to as *direct strains*, while the off-diagonal components are referred to as *shear strains.*

Earlier we had the result that for a rigid body motion the strain tensor is zero. Now consider a situation in which the strain tensor is zero; then we see from the above interpretation of its components and the observation that the axes may be chosen arbitrarily that no changes in length of fibers take place, nor are there any changes in angles between fibers. Thus the converse is also true: If $\boldsymbol{\eta}=\mathbf{0}$, then the body necessarily undergoes a rigid body motion.

Infinitesimal strain. There are many problems of practical interest for which the deformations can be regarded as "small" in some sense, and under such circumstances it is natural to consider whether the formulation of the problem might be simplified by exploiting this feature. Of course, it is necessary first to formalize and to quantify what is meant by "small," and for the purposes of this work the following definition suffices: A body is said to undergo *infinitesimal deformation* if the displacement gradient $\nabla\boldsymbol{u}$ is sufficiently small so that the nonlinear term in (2.5) can be neglected. When this is the case, we may replace the strain tensor $\boldsymbol{\eta}$ by the *infinitesimal strain tensor* $\boldsymbol{\epsilon}$, which is defined by

$$\boldsymbol{\epsilon}(\boldsymbol{u}) = \tfrac{1}{2}(\nabla \boldsymbol{u} + (\nabla \boldsymbol{u})^T). \tag{2.7}$$

Setting $h = |\nabla \boldsymbol{u}|$, in the case of infinitesimal strains we assume that $h \ll 1$ and that to within an error of $O(h^2)$ as $h \to 0$, $\boldsymbol{\epsilon}$ and $\boldsymbol{\eta}$ coincide.

Characterization of rigid body motions for infinitesimal strain. We have seen earlier that the strain tensor $\boldsymbol{\eta}$ vanishes if and only if the body undergoes a rigid body motion. Since we will study problems in the context of infinitesimal strains, it is necessary to characterize a rigid body motion for situations in which terms of $O(h^2)$ are neglected. Suppose that the body undergoes an infinitesimal rigid body motion, that is, one for which

$$\boldsymbol{\epsilon}(\boldsymbol{u}) = \boldsymbol{0}.$$

Then

$$\nabla \boldsymbol{u} = -(\nabla \boldsymbol{u})^T,$$

so that the displacement gradient is skew. Thus the most general representation of the motion in such a situation is given by

$$\boldsymbol{y}(\boldsymbol{x}) = \boldsymbol{y}_0 + \boldsymbol{\omega}(\boldsymbol{x} - \boldsymbol{x}_0),$$

or, equivalently, by

$$\boldsymbol{u}(\boldsymbol{x}) = \boldsymbol{u}_0 + \boldsymbol{\omega}(\boldsymbol{x} - \boldsymbol{x}_0),$$

where $\boldsymbol{x}_0$ is any point, $\boldsymbol{\omega}$ is a *skew* tensor, and $\boldsymbol{y}_0$ and $\boldsymbol{u}_0$ are either given or arbitrary vectors (for a proof of this result, see [75], Section 3.6). If the motion is a pure translation, then $\boldsymbol{\omega} = \boldsymbol{0}$, while if on the other hand the motion is a pure rotation, then $\boldsymbol{u}_0 = \boldsymbol{0}$. An infinitesimal rigid body motion may be written alternatively as

$$\boldsymbol{u}(\boldsymbol{x}) = \boldsymbol{u}_0 + \boldsymbol{w} \wedge (\boldsymbol{x} - \boldsymbol{x}_0),$$

where $\boldsymbol{w}$ is the unique *axial vector* corresponding to $\boldsymbol{\omega}$; that is, $\boldsymbol{\omega a} = \boldsymbol{w} \wedge \boldsymbol{a}$ for any vector $\boldsymbol{a}$.

Changes in volume; incompressibility. We require a simple measure of the local change in volume accompanying a motion. The volume of the reference configuration is

$$V_0 = \int_\Omega dx,$$

while the volume of the current configuration is

$$V_t = \int_{\Omega_t} dy.$$

Thus the change in volume as a result of the deformation $\boldsymbol{y}$ is simply given by

$$\Delta V \equiv V_t - V_0 = \int_{\Omega_t} dy - \int_{\Omega} dx.$$

Since $\Omega_t = \boldsymbol{y}(\Omega, t)$, we may use the conventional technique for change of variables in an integral to write

$$\int_{\Omega_t} dy = \int_{\Omega} J\, dx,$$

where the Jacobian J has been defined in (2.2). Thus the change in volume is

$$\Delta V = \int_{\Omega} (J(\boldsymbol{x}) - 1)\, dx. \tag{2.8}$$

Once again we are interested in determining the expression for the change in volume for situations in which the underlying deformation can be regarded as infinitesimal. For this purpose we set $h = |\nabla \boldsymbol{u}|$ and write the Jacobian in terms of $\boldsymbol{u}$; thus

$$\begin{aligned} J &= \det(\nabla \boldsymbol{y}) \\ &= \det(\boldsymbol{I} + \nabla \boldsymbol{u}) \\ &= 1 + \operatorname{div} \boldsymbol{u} + O(h^2). \end{aligned}$$

This result follows directly from the definition of the determinant or from the identity (see, for example, [21], page 48)

$$\det(\boldsymbol{A} + \boldsymbol{B}) = (1 + \boldsymbol{B} : \boldsymbol{A}^{-T}) \det \boldsymbol{A} + (1 + \boldsymbol{A} : \boldsymbol{B}^{-T}) \det \boldsymbol{B}$$

for all invertible matrices $\boldsymbol{A}$ and $\boldsymbol{B}$. Substitution in (2.8) yields the result that to within an error of $O(h^2)$,

$$\Delta V = \int_{\Omega} \operatorname{div} \boldsymbol{u}\, dx.$$

In other words, the quantity $\operatorname{div} \boldsymbol{u}$ represents the change in volume per unit volume in an infinitesimal deformation.

A deformation that experiences no change in volume is called *isochoric*; for such a deformation we have

$$J = 1 \quad \forall \boldsymbol{x} \in \Omega,\ t \in [0, T]. \tag{2.9}$$

When an isochoric deformation is infinitesimal, then to within an error of $O(h^2)$ the displacement field satisfies the condition

$$\operatorname{tr} \boldsymbol{\epsilon}(\boldsymbol{u}(\boldsymbol{x}, t)) = \operatorname{div} \boldsymbol{u}(\boldsymbol{x}, t) = 0 \quad \forall \boldsymbol{x} \in \Omega, \ t \in [0, T]. \tag{2.10}$$

It may alternatively happen that a material has the property, possibly idealized, that it is unable to experience a change in volume. This idealization is often made in the case of materials for which, for the range of conditions under which they are being analyzed, the volume change observed is negligible. Such materials are referred to as *incompressible*. Note the difference between isochoric deformations and incompressible materials; in the former case a particular *deformation* is accompanied by no change in volume so that (2.9) and (2.10) are consequences of the deformation, while in the latter case it is a property of the *material* that no matter what the deformation, the body is unable to undergo any change in volume. In this case the conditions (2.9) or (2.10) represent constraints on the possible classes of deformations that are admitted.

2.2 Balance of Momentum; Stress

In this section we move away from the purely geometric nature of kinematics and investigate the consequences for material bodies of the fundamental laws of balance of linear and angular momentum. A further development is the introduction in this context of the notion of stress as a tensorial quantity that characterizes the state of internal forces acting in a body. All variables are assumed to have the requisite degree of smoothness consistent with developments in this section.

It is particularly convenient to develop the notions of momentum and stress in the context of the *reference configuration*; that is, we exploit the fact that field variables are functions of reference position $\boldsymbol{x}$ and time t, so that while the momentum and stress at time t are quantities associated with the configuration of the body at time t, these can easily be expressed, via the mapping (2.1), as functions defined over the reference configuration Ω.

The equations corresponding to local balance of linear and angular momentum are obtained by writing down the expressions that correspond to balance of linear and angular momentum for an arbitrary subset of the body. The local forms of these laws then follow from the arbitrariness of the subset and appropriate smoothness assumptions on the variables.

Now let Ω represent the reference configuration of the body, as before, and Ω_t the current configuration. Furthermore, let Ω' be an arbitrary subset of Ω, which is mapped by the motion to an arbitrary subset Ω'_t of Ω_t. Under these circumstances we may express global quantities associated with the current configuration as integrals over the reference configuration.

The *velocity field* $\dot{\boldsymbol{u}}$ and *acceleration field* $\ddot{\boldsymbol{u}}$ corresponding to a displacement field $\boldsymbol{u}(\boldsymbol{x}, t)$ are defined by

$$\dot{\boldsymbol{u}}(\boldsymbol{x}, t) = \frac{\partial \boldsymbol{u}(\boldsymbol{x}, t)}{\partial t},$$
$$\ddot{\boldsymbol{u}}(\boldsymbol{x}, t) = \frac{\partial^2 \boldsymbol{u}(\boldsymbol{x}, t)}{\partial t^2}.$$

Thus, the *linear momentum* of the subset Ω_t' of Ω_t at time t is defined by

$$\int_{\Omega'} \rho \dot{\boldsymbol{u}} \, dx,$$

and its *angular momentum* by

$$\int_{\Omega'} \boldsymbol{x} \wedge \rho \dot{\boldsymbol{u}} \, dx,$$

in which ρ denotes the mass density of the body, that is, the mass per unit reference volume of the body.

The body is subjected to a system of forces, which are of two kinds. There is the *body force* $\boldsymbol{b}(\boldsymbol{x}, t)$, which represents the force per unit reference volume exerted on the material point $\boldsymbol{x}$ at time t by agencies external to the body; gravity is a canonical example, the body force in this case being $\rho g \boldsymbol{e}$, where g is the gravitational acceleration and $\boldsymbol{e}$ is the unit vector pointing in the downward vertical direction. The second kind of force acting on the body is the *surface traction*. To define this force field it is convenient to begin by introducing, for a given unit vector $\boldsymbol{n}$, the *stress vector* $\boldsymbol{s}_n(\boldsymbol{x}, t)$: If γ is a regular surface in $\bar{\Omega}$ passing through $\boldsymbol{x}$ and having unit normal $\boldsymbol{n}$ at $\boldsymbol{x}$, then $\boldsymbol{s}_n(\boldsymbol{x}, t)$ is the current force per unit reference area exerted by the portion of Ω on the side of γ towards which $\boldsymbol{n}$ points, on the portion of Ω that lies on the other side. Let Γ' denote the boundary of Ω'; then the surface traction at time t is defined to be the stress vector $\boldsymbol{s}_n(\boldsymbol{x}, t)$ $(\boldsymbol{x} \in \Gamma')$ acting on Γ', with $\boldsymbol{n}$ defined to be the outward unit normal on Γ' (see Figure 2.4). While we have chosen to define quantities such as forces and momentum in terms of the reference configuration of the body, there is no difficulty in restating these definitions in terms of the current configuration.

The laws of balance of linear and angular momentum may now be stated.

BALANCE OF LINEAR MOMENTUM. The total force acting on Ω_t' is equal to the rate of change of the linear momentum of Ω_t'; expressed in terms of integrals over the reference configuration,

$$\int_{\Omega'} \rho \ddot{\boldsymbol{u}} \, dx = \int_{\Omega'} \boldsymbol{b} \, dx + \int_{\Gamma'} \boldsymbol{s}_n \, ds. \tag{2.11}$$

Note that in this identity we have used the fact that

$$\frac{\partial}{\partial t} \int_{\Omega'} (\cdot) \, dx = \int_{\Omega'} \frac{\partial}{\partial t} (\cdot) \, dx,$$

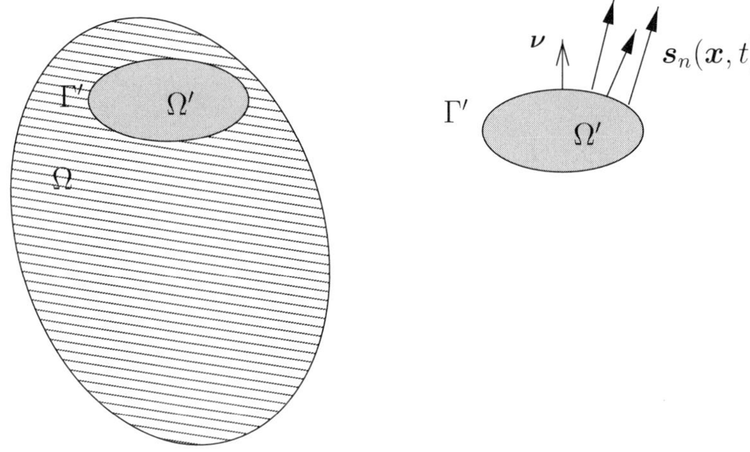

Figure 2.4: The surface traction vector field

since Ω' is chosen independent of time.

Balance of angular momentum. The total moment acting on Ω'_t is equal to the rate of change of the angular momentum of Ω'_t; expressed in terms of integrals over the reference configuration,

$$\int_{\Omega'} \boldsymbol{x} \wedge \rho \ddot{\boldsymbol{u}}\, dx = \int_{\Omega'} \boldsymbol{x} \wedge \boldsymbol{b}\, dx + \int_{\Gamma'} \boldsymbol{x} \wedge \boldsymbol{s}_n\, ds. \tag{2.12}$$

We have the following two important results.

Cauchy's Reciprocal Theorem. Given any unit vector $\boldsymbol{n}$,

$$\boldsymbol{s}_n = -\boldsymbol{s}_{-n}. \tag{2.13}$$

This result is clearly a generalization to deformable bodies of Newton's third law of action and reaction.

Existence of the stress tensor. There exists on $\Omega \times [0, T]$ a second-order tensor field $\boldsymbol{\tau}$, called the first Piola–Kirchhoff stress field, with the property that

$$\boldsymbol{\tau n} = \boldsymbol{s}_n \tag{2.14}$$

for each unit vector $\boldsymbol{n}$.

The derivation of the reciprocal theorem of Cauchy and the proof of the existence of the stress tensor are treated in detail in [2] (page 404), [48] (page 45), and [75] (Section 4.1).

We are now in a position to obtain *local* forms of the two balance laws. In the following we assume that all variables have the degree of differentiability consistent with the manipulations that are carried out.

We begin with the law of balance of linear momentum. From the relationship (2.14) between the surface traction and stress tensor we obtain, using a variant of the Green–Gauss theorem,

$$\int_{\Gamma'} \boldsymbol{s}_n \, ds = \int_{\Gamma'} \boldsymbol{\sigma n} \, ds = \int_{\Omega'} \operatorname{Div} \boldsymbol{\tau} \, dx,$$

so that (2.11) becomes

$$\int_{\Omega'} (\rho \ddot{\boldsymbol{u}} - \boldsymbol{b} - \operatorname{Div} \boldsymbol{\tau}) \, dx = \boldsymbol{0}. \tag{2.15}$$

Here Div is the divergence operator with respect to the reference configuration and expressed in terms of derivatives with respect to x_j. Since the domain Ω' is arbitrary, the integrand in (2.15) must vanish. We thus obtain in local form the *equation of motion*

$$\operatorname{Div} \boldsymbol{\tau} + \boldsymbol{b} = \rho \ddot{\boldsymbol{u}}. \tag{2.16}$$

In component form, the equation of motion reads

$$\frac{\partial \tau_{ij}}{\partial x_j} + b_i = \rho \ddot{u}_i, \quad 1 \le i \le 3.$$

For situations in which all the given data are independent of time, the response of the body will also be independent of time. In this case we have $\boldsymbol{u} = \boldsymbol{u}(\boldsymbol{x})$, $\boldsymbol{\tau} = \boldsymbol{\tau}(\boldsymbol{x})$, and the equation of motion becomes the *equation of equilibrium*

$$\frac{\partial \tau_{ij}}{\partial x_j} + b_i = 0, \quad 1 \le i \le 3. \tag{2.17}$$

We have chosen to present the arguments leading to the equation of motion in the setting of the reference configuration, with $\boldsymbol{x}$ and t as independent variables. Since the motion

$$\boldsymbol{y} = \boldsymbol{y}(\boldsymbol{x}, t)$$

is invertible, it is also possible to treat $\boldsymbol{y}$ as the independent variable and to carry out the development in the *current* configuration. That is, we have $\boldsymbol{x} = \overline{\boldsymbol{x}}(\boldsymbol{y}, t)$ after carrying out the inversion, and so, for example, the velocity $\dot{\boldsymbol{u}}$ has the alternative representation

$$\frac{\partial}{\partial t} \boldsymbol{y}(\boldsymbol{x}, t) = \dot{\boldsymbol{u}}(\overline{\boldsymbol{x}}(\boldsymbol{y}, t), t) \equiv \boldsymbol{v}(\boldsymbol{y}, t).$$

Similar transformations can be carried out with respect to all variables, and the principles of balance of linear and angular momentum are then expressed in terms of integrals over the current configuration Ω_t. As far as the stress goes, an argument identical to that which leads to the existence of the first Piola–Kirchhoff stress tensor gives the existence of a tensor $\boldsymbol{\sigma}$, called the *Cauchy stress*, that has the property that the force per unit *current* area $\boldsymbol{t}_\nu$ on an elemental surface having unit normal $\boldsymbol{\nu}$ is given by

$$\boldsymbol{\sigma}\boldsymbol{\nu} = \boldsymbol{t}_\nu. \tag{2.18}$$

The Cauchy stress therefore has the same relationship to the current configuration as does the first Piola–Kirchhoff stress to the reference configuration.

The use of the principle of balance of linear momentum, when applied in the current configuration, leads to the equation of motion in the form

$$\operatorname{div} \boldsymbol{\sigma} + \boldsymbol{b} = \rho_t \boldsymbol{a}$$

in which $\boldsymbol{a}$ is the acceleration and ρ_t is the mass density per unit current volume. Here div is the divergence operator in the current configuration, so that $\operatorname{div} \boldsymbol{\sigma} = (\partial \sigma_{ij}/\partial y_j)\boldsymbol{e}_i$.

It can be shown that the first Piola–Kirchhoff and Cauchy stresses are related according to

$$\boldsymbol{\sigma} = J^{-1}\boldsymbol{\tau}\,(\boldsymbol{I} + \nabla \boldsymbol{u})^T. \tag{2.19}$$

We have not as yet examined the consequences of the equation (2.12) for balance of angular momentum; by carrying out manipulations similar to those that lead to (2.16), it is possible to show that this balance law implies that

$$\boldsymbol{\tau}\,(\boldsymbol{I} + \nabla \boldsymbol{u})^T$$

is symmetric. Equivalently, we have the classical result that the Cauchy stress is *symmetric*:

$$\boldsymbol{\sigma}^T = \boldsymbol{\sigma}, \quad \text{or} \quad \sigma_{ji} = \sigma_{ij}. \tag{2.20}$$

Stress and the balance laws for infinitesimal deformations. For problems in which deformations are assumed infinitesimal, the distinction between the reference and current configurations may be ignored. To begin with, we may neglect the term $\nabla \boldsymbol{u}$ appearing in (2.19); furthermore, since $J = \det(\boldsymbol{I} + \nabla \boldsymbol{u}) = 1 + \operatorname{div} \boldsymbol{u} + O(h^2)$, we may set $J \approx 1$. Likewise, $\rho_t = J^{-1}\rho \approx \rho$, to within an error $O(h)$. Thus the distinction between the first Piola–Kirchhoff and Cauchy stresses disappears. In addition, since

$$\frac{\partial}{\partial x_j} = \frac{\partial y_i}{\partial x_j}\frac{\partial}{\partial y_i} = \left(\delta_{ij} + \frac{\partial u_i}{\partial x_j}\right)\frac{\partial}{\partial y_i},$$

it follows that when $\nabla \boldsymbol{u}$ is small, we may replace derivatives with respect to y_j by derivatives with respect to x_j. In summary, then, the principles of balance of linear and angular momentum are, in local form, and for infinitesimal deformations,

$$\operatorname{div} \boldsymbol{\sigma} + \boldsymbol{b} = \rho \ddot{\boldsymbol{u}}, \tag{2.21}$$

$$\boldsymbol{\sigma}^T = \boldsymbol{\sigma}. \tag{2.22}$$

2.3 Linearly Elastic Materials

We are moving towards a situation in which the behavior of a material body is described by a system of partial differential equations. So far, we have the equation of motion (2.16) and the strain–displacement relation (2.5); equivalently, if we assume that the deformation is infinitesimal, we will deal with equations (2.21) and (2.7). In either case these represent, when written out in component form, a total of nine equations: three from the equation of motion and six from the strain–displacement relation (taking into account the symmetry of $\boldsymbol{\epsilon}$). The total number of unknowns is fifteen: three components of displacement, six components of the strain and six components of the stress (again accounting for the symmetry of $\boldsymbol{\epsilon}$ and $\boldsymbol{\sigma}$). Thus it is clear that six additional equations are required if we are to have a problem that is at least in principle solvable.

Physical considerations also dictate that the description of the problem so far is incomplete: The kinematics have been described, and the balance laws are accounted for, but as yet there is no description of the particular material behavior. This information, embodied in the *constitutive equations* of the material, will provide the remaining equations of the problem.

Later on, we will embark on a detailed study of the constitutive equations that describe elastoplastic behavior. An essential precursor to such a study is an understanding of the equations governing elastic behavior. We review in this section the salient ideas, confining attention to linearly elastic materials.

A body is *linearly elastic* if the stress depends linearly on the infinitesimal strain, that is, if the stress and strain are related to each other through an equation of the form

$$\boldsymbol{\sigma} = \boldsymbol{C}\boldsymbol{\epsilon}, \tag{2.23}$$

where $\boldsymbol{C}$, called the *elasticity tensor*, is a linear map from the space of symmetric matrices or second-order tensors into itself. Like $\boldsymbol{\sigma}$, $\boldsymbol{\epsilon}$, $\boldsymbol{u}$, and other variables, the elasticity tensor is a function of position in the body. It does not, however, depend on time. If the density ρ and the elasticity tensor $\boldsymbol{C}$ are independent of position, the body is said to be *homogeneous*.

The map $\boldsymbol{C}$ may be represented as a fourth-order tensor as follows: Relative to the orthonormal basis $\{\boldsymbol{e}_i\}$ we have

$$\begin{aligned}\sigma_{ij} &= \boldsymbol{e}_i \cdot \boldsymbol{\sigma} \boldsymbol{e}_j \\ &= \boldsymbol{e}_i \cdot (\boldsymbol{C}\boldsymbol{\epsilon})\boldsymbol{e}_j \\ &= \boldsymbol{e}_i \cdot (\boldsymbol{C}(\epsilon_{kl}\boldsymbol{e}_k \otimes \boldsymbol{e}_l))\, \boldsymbol{e}_j \\ &= \boldsymbol{e}_i \cdot (\boldsymbol{C}(\boldsymbol{e}_k \otimes \boldsymbol{e}_l))\, \boldsymbol{e}_j \epsilon_{kl} \\ &= C_{ijkl}\epsilon_{kl},\end{aligned}$$

where C_{ijkl}, the components of $\boldsymbol{C}$, are defined by

$$C_{ijkl} = \boldsymbol{e}_i \cdot (\boldsymbol{C}(\boldsymbol{e}_k \otimes \boldsymbol{e}_l))\, \boldsymbol{e}_j.$$

It follows that the constitutive equation (2.23) has the component form

$$\sigma_{ij} = C_{ijkl}\epsilon_{kl}. \tag{2.24}$$

Properties of the elasticity tensor. Without loss of generality, we may assume the elasticity tensor to have the symmetry properties

$$C_{ijkl} = C_{jikl} = C_{ijlk}. \tag{2.25}$$

This is argued as follows. Since $\boldsymbol{\epsilon}$ is symmetric, we have, from (2.24),

$$\sigma_{ij} = C_{ijkl}\epsilon_{lk} = C_{ijlk}\epsilon_{kl}.$$

Hence

$$\sigma_{ij} = \frac{1}{2}\left(C_{ijkl} + C_{ijlk}\right)\epsilon_{kl}.$$

Similarly, using the symmetry of $\boldsymbol{\sigma}$, we have

$$\sigma_{ij} = \frac{1}{2}\left(C_{ijkl} + C_{jikl}\right)\epsilon_{kl}.$$

Therefore, the relation (2.24) can be equivalently expressed as

$$\sigma_{ij} = \frac{1}{4}\left(C_{ijkl} + C_{ijlk} + C_{jikl} + C_{jilk}\right)\epsilon_{kl}.$$

In other words, if necessary, we may redefine the tensor $\boldsymbol{C}$ for the relation (2.24) such that the symmetry properties (2.25) hold.

Later, when we consider elastic constitutive equations that are derived from a strain energy or free energy function, it will be seen that the elasticity tensor possesses the additional symmetry property

$$C_{ijkl} = C_{klij}. \tag{2.26}$$

The elasticity tensor is *positive definite* if

$$\boldsymbol{\epsilon} : \boldsymbol{C}\boldsymbol{\epsilon} > 0 \quad \text{for all nonzero symmetric second-order tensors } \boldsymbol{\epsilon}. \tag{2.27}$$

Furthermore, $\boldsymbol{C}$ is said to be *strongly elliptic* (see [82, 127]) if

$$(\boldsymbol{a} \otimes \boldsymbol{b}) : \boldsymbol{C}(\boldsymbol{a} \otimes \boldsymbol{b}) > 0 \quad \text{for all nonzero vectors } \boldsymbol{a} \text{ and } \boldsymbol{b}. \tag{2.28}$$

In component form, (2.28) reads

$$C_{ijkl} a_i a_k b_j b_l > 0 \quad \text{if } a_i a_i > 0 \text{ and } b_i b_i > 0.$$

Finally, $\boldsymbol{C}$ is said to be *pointwise stable* ([82], page 321) if there exists a constant $\alpha > 0$ such that

$$\boldsymbol{\epsilon} : \boldsymbol{C}\boldsymbol{\epsilon} \geq \alpha \, |\boldsymbol{\epsilon}|^2 \quad \text{for all symmetric second-order tensors } \boldsymbol{\epsilon}. \tag{2.29}$$

It should be clear from these definitions that pointwise stability implies, but is not implied by, strong ellipticity. It is also clear that pointwise stability is equivalent to pointwise positive definiteness, under the assumption that $\boldsymbol{C}$ is continuous on $\overline{\Omega}$.

Sometimes it is convenient to work not with stress as a function of strain, but the other way around. If the relationship (2.23) is invertible (and this will always be the case for real materials) then we may write

$$\boldsymbol{\epsilon} = \boldsymbol{A}\boldsymbol{\sigma}, \tag{2.30}$$

where the fourth-order tensor $\boldsymbol{A}$ is known as the *compliance tensor*; it is the inverse of $\boldsymbol{C}$ and therefore has the property that

$$\boldsymbol{A}(\boldsymbol{C}\boldsymbol{\epsilon}) = \boldsymbol{\epsilon} \quad \forall \, \boldsymbol{\epsilon}, \ \boldsymbol{\epsilon}^T = \boldsymbol{\epsilon},$$

and

$$\boldsymbol{C}(\boldsymbol{A}\boldsymbol{\sigma}) = \boldsymbol{\sigma} \quad \forall \, \boldsymbol{\sigma}, \ \boldsymbol{\sigma}^T = \boldsymbol{\sigma}.$$

2.4 Isotropic Elasticity

It is often the case that materials possess preferred directions or symmetries. For example, timber can be regarded as an orthotropic material, in the sense that it possesses particular constitutive properties along the grain and at right angles to the grain of the wood. The greatest degree of symmetry is possessed by a material that has no preferred directions; that is, say, its response to a force is independent of its orientation. This property is known as isotropy, and a material with such a property is called *isotropic*.

Isotropic linearly elastic materials occur in abundance, and so form an important subclass of materials whose properties we need to model mathematically. The most striking mathematical effect of isotropy is that it reduces the twenty-one independent components C_{ijkl} of $\boldsymbol{C}$ (taking account of the symmetry properties (2.25) and (2.26)) to *two*. Of course, these two material coefficients are not unique, and a new pair may be generated by combining a given pair in different ways. The most appropriate choice of material coefficients for isotropic elastic materials will depend on the application in mind. We will discuss some of the more common variants.

First, for an isotropic linearly elastic material we have the result that the components of the elasticity tensor are given by

$$C_{ijkl} = \lambda\delta_{ij}\delta_{kl} + \mu(\delta_{ik}\delta_{jl} + \delta_{il}\delta_{jk}), \tag{2.31}$$

where δ_{ij} is the Kronecker delta. In coordinate-free form the elasticity tensor is defined to be the fourth-order tensor $\boldsymbol{C}$ that satisfies

$$(\boldsymbol{a} \otimes \boldsymbol{b}) : \boldsymbol{C}(\boldsymbol{c} \otimes \boldsymbol{d}) = \lambda\,(\boldsymbol{a} \cdot \boldsymbol{b})(\boldsymbol{c} \cdot \boldsymbol{d}) + \mu\,[(\boldsymbol{a} \cdot \boldsymbol{c})(\boldsymbol{b} \cdot \boldsymbol{d}) + (\boldsymbol{a} \cdot \boldsymbol{d})(\boldsymbol{b} \cdot \boldsymbol{c})] \tag{2.32}$$

for all vectors $\boldsymbol{a}, \boldsymbol{b}, \boldsymbol{c}$, and $\boldsymbol{d}$. The scalars λ and μ are called *Lamé moduli*. The stress-strain relation (2.23) in this case is easily found to be given by

$$\boldsymbol{\sigma} = \lambda\,(\mathrm{tr}\,\boldsymbol{\epsilon})\boldsymbol{I} + 2\,\mu\,\boldsymbol{\epsilon}. \tag{2.33}$$

For the purpose of interpreting the moduli, and of defining alternative pairs of moduli for isotropic elastic materials, it is convenient to carry out an orthogonal decomposition of both the stress and the strain into what are known as spherical and deviatoric components; the first is associated solely with volumetric changes, while the latter is associated with shearing stresses and deformations. To achieve this decomposition we recall that any second-order tensor $\boldsymbol{\tau}$ may be written in the form

$$\boldsymbol{\tau} = \boldsymbol{\tau}^D + \boldsymbol{\tau}^S, \tag{2.34}$$

where the deviatoric and spherical parts $\boldsymbol{\tau}^D$ and $\boldsymbol{\tau}^S$ of $\boldsymbol{\tau}$ are defined, respectively, by

$$\boldsymbol{\tau}^D = \boldsymbol{\tau} - \tfrac{1}{3}(\mathrm{tr}\boldsymbol{\tau})\boldsymbol{I}, \quad \boldsymbol{\tau}^S = \tfrac{1}{3}(\mathrm{tr}\,\boldsymbol{\tau})\boldsymbol{I}. \tag{2.35}$$

The maps $\boldsymbol{\tau} \mapsto \boldsymbol{\tau}^D$ and $\boldsymbol{\tau} \mapsto \boldsymbol{\tau}^S$ can be regarded as orthogonal projections on the space of second-order tensors when this space is equipped with the inner product $\boldsymbol{\tau} : \boldsymbol{\sigma} = \tau_{ij}\sigma_{ij}$. Indeed, we have $(\boldsymbol{\tau}^D)^S = (\boldsymbol{\tau}^S)^D = \boldsymbol{0}$, and

$$\begin{aligned}
\boldsymbol{\tau}^D : \boldsymbol{\tau}^S &= (\boldsymbol{\tau} - \boldsymbol{\tau}^S) : \boldsymbol{\tau}^S \\
&= \boldsymbol{\tau} : \boldsymbol{\tau}^S - |\boldsymbol{\tau}^S|^2 \\
&= \tau_{ij}\tfrac{1}{3}\tau_{kk}\delta_{ij} - |\boldsymbol{\tau}^S|^2 \\
&= \tfrac{1}{3}\tau_{ii}\tau_{kk} - |\boldsymbol{\tau}^S|^2 \\
&= 0,
\end{aligned}$$

since $|\boldsymbol{\tau}^S|^2 = \frac{1}{3}(\tau_{ii})^2$. The constitutive equation can thus be written in the *uncoupled form* (by applying the operators $(\cdot)^D$ and $(\cdot)^S$ successively to (2.33))

$$\boldsymbol{\sigma}^D = 2\,\mu\,\boldsymbol{\epsilon}^D, \tag{2.36}$$

$$\boldsymbol{\sigma}^S = \lambda\,(\operatorname{tr}\boldsymbol{\epsilon})\boldsymbol{I}^S + 2\,\mu\,\boldsymbol{\epsilon}^S = 3\,(\lambda + \tfrac{2}{3}\mu)\,\boldsymbol{\epsilon}^S. \tag{2.37}$$

The scalar μ is also known as the *shear modulus* (for reasons that are evident from (2.36)), while the material coefficient $K \equiv \lambda + \frac{2}{3}\mu$ is known as the *bulk modulus* because it measures the ratio between the spherical stress and volume change. Thus an alternative pair of elastic coefficients to the Lamé moduli is $\{\mu, K\}$. Note that the shear modulus is often denoted by G, especially in the engineering literature.

Yet another important alternative pair of material coefficients arises from direct consideration of the behavior of the length of an elastic rod when it is subjected to a uniaxial stress. Suppose that the Cartesian axes are aligned in such a way that an isotropic elastic rod lies parallel to the x_1-axis (see Figure 2.5) and is subjected to a uniform stress with $\sigma_{11} \neq 0$ and all other components being zero. The effect will be that the rod experiences only direct strains, on account of its isotropy. We are interested here first in the ratio $\sigma_{11}/\epsilon_{11}$ and second in the ratio $\epsilon_{22}/\epsilon_{11}$, or, equivalently, $\epsilon_{33}/\epsilon_{11}$. The associated material coefficients are known, respectively, as *Young's modulus* and *Poisson's ratio*:

$$\text{Young's modulus} \quad E = \frac{\sigma_{11}}{\epsilon_{11}},$$

$$\text{Poisson's ratio} \quad \nu = -\frac{\epsilon_{22}}{\epsilon_{11}}.$$

Thus Young's modulus measures the slope of the stress–strain curve and is analogous to the stiffness of a spring, while Poisson's ratio measures lateral contraction. Since we expect a tensile stress to be accompanied by an extension of the material and since we also know from experience that most common materials would respond to an extension in one direction with a contraction in the transverse direction (think of what happens when a rubber band is extended), it follows that one expects both E and ν to be positive quantities. We will see later that further restrictions are placed on the ranges of E and ν by thermodynamic or mathematical considerations.

From (2.31) it is a straightforward task to obtain a relationship between the pairs $\{\lambda, \mu\}$ and $\{E, \nu\}$. Since for the case of pure tension we have

$$\boldsymbol{\sigma} = \begin{pmatrix} \sigma_{11} & 0 & 0 \\ 0 & 0 & 0 \\ 0 & 0 & 0 \end{pmatrix} \quad \text{and} \quad \boldsymbol{\epsilon} = \begin{pmatrix} \epsilon_{11} & 0 & 0 \\ 0 & \epsilon_{22} & 0 \\ 0 & 0 & \epsilon_{22} \end{pmatrix},$$

it follows that

$$E = \frac{\mu(2\mu + 3\lambda)}{\mu + \lambda} \tag{2.38}$$

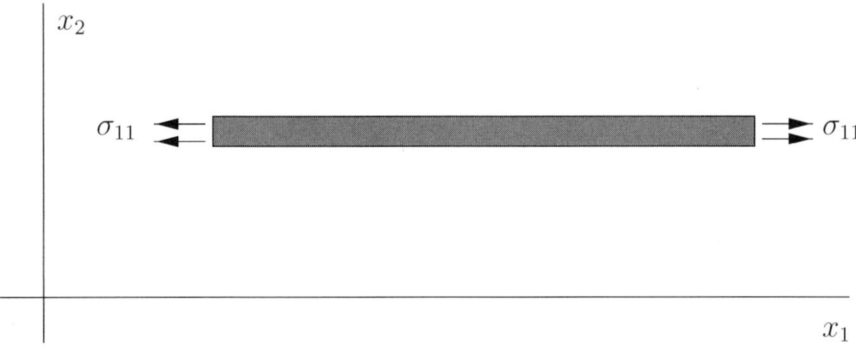

Figure 2.5: A rod in a state of uniaxial stress

and

$$\nu = \frac{\lambda}{2(\mu + \lambda)} . \tag{2.39}$$

The constitutive relation (2.33) can be put in an alternative useful form involving E and ν by inverting it and making use of (2.38) and (2.39); this gives

$$\boldsymbol{\epsilon} = E^{-1}[(1 + \nu)\boldsymbol{\sigma} - \nu(\operatorname{tr} \boldsymbol{\sigma})\boldsymbol{I}]. \tag{2.40}$$

The conditions of pointwise stability and strong ellipticity introduced earlier both lead to constraints on admissible ranges for the material constants. Indeed, it is possible to show ([82], page 241) that an isotropic linearly elastic material is

(a) pointwise stable if and only if $\mu > 0$ and $3\lambda + 2\mu > 0$ (or, in terms of Young's modulus and Poisson's ratio, if and only if $E > 0$ and $-1 < \nu < \frac{1}{2}$);

(b) strongly elliptic if and only if $\mu > 0$ and $\lambda + 2\mu > 0$ (or if and only if $E > 0$, and $\nu < \frac{1}{2}$ or $\nu > 1$).

2.5 A Thermodynamic Framework for Elasticity

The developments in the preceding sections were described in a purely mechanical framework, without bringing into play any thermodynamic considerations. Since it is our intention in this monograph to deal only with processes that take place under isothermal conditions, it would appear that there is indeed no need to take account of thermodynamics. This is, however, not quite the case. Since the primary goal is to present a theory of elastoplasticity and since plasticity as a constitutive theory can be conveniently developed within a thermodynamic framework, it will be necessary to bring thermodynamics into play, albeit in the context of isothermal processes. Plasticity is most conveniently described in the framework of thermodynamics with *internal variables*. We postpone discussion of internal variable theories to Section 2.7, while in this section we sketch the basic thermodynamic theory within which linear elasticity can be described.

Suppose that a material body is subjected to a body force $\boldsymbol{b}$ in its interior and a surface traction $\boldsymbol{s}$ on the boundary. Suppose also, for now, that the body is subjected to thermal equivalents of these mechanical sources: In its interior the *heat source* r per unit volume, and across its boundary the *heat flux* $\boldsymbol{q}$ per unit area.

We begin with the *first law of thermodynamics*, which is essentially a statement of balance of energy. This law states that for any part Ω' of the body Ω, the rate of change of total internal energy plus kinetic energy is equal to the rate of work done on that part of the body by the mechanical forces, plus the heat supply. Mathematically the law may be expressed in the form

$$\frac{d}{dt}\int_{\Omega'}(e + \tfrac{1}{2}\rho|\dot{\boldsymbol{u}}|^2)\,dx = \int_{\Omega'}\boldsymbol{b}\cdot\dot{\boldsymbol{u}}\,dx + \int_{\Gamma'}\boldsymbol{s}\cdot\dot{\boldsymbol{u}}\,ds + \int_{\Omega'}r\,dx - \int_{\Gamma'}\boldsymbol{q}\cdot\boldsymbol{n}\,ds. \tag{2.41}$$

Here e represents the internal energy per unit volume, $\dot{\boldsymbol{u}}$ is the velocity vector, and $\Gamma' = \partial\Omega'$ is the boundary of Ω'. The minus sign in front of the term involving the heat flux appears because $\boldsymbol{n}$ is the outward unit normal vector to the surface, while $\boldsymbol{q}$ is the heat flux per unit area in the direction of $\boldsymbol{n}$, so that $-\int_{\Gamma'}\boldsymbol{q}\cdot\boldsymbol{n}\,ds$ is the total flow of heat across Γ' *into* the body. This law may be simplified by the use of the divergence theorem: Indeed, observe that

$$\begin{aligned}\int_{\Gamma'}\boldsymbol{s}\cdot\dot{\boldsymbol{u}}\,ds &= \int_{\Gamma'}\boldsymbol{\sigma n}\cdot\dot{\boldsymbol{u}}\,ds \\ &= \int_{\Omega'}\boldsymbol{\sigma}:\nabla\dot{\boldsymbol{u}}\,dx + \int_{\Omega'}\operatorname{div}\boldsymbol{\sigma}\cdot\dot{\boldsymbol{u}}\,dx \\ &= \int_{\Omega'}\boldsymbol{\sigma}:\dot{\boldsymbol{\epsilon}}\,dx + \int_{\Omega'}\operatorname{div}\boldsymbol{\sigma}\cdot\dot{\boldsymbol{u}}\,dx,\end{aligned}$$

where in the last step we invoked the symmetry of $\boldsymbol{\sigma}$. Substituting this result in (2.41) and making use of equation (2.16) of balance of momentum, we obtain the first law in the form

$$\frac{d}{dt}\int_{\Omega'}e\,dx = \int_{\Omega'}\boldsymbol{\sigma}:\dot{\boldsymbol{\epsilon}}\,dx + \int_{\Omega'}r\,dx - \int_{\Gamma'}\boldsymbol{q}\cdot\boldsymbol{n}\,ds.$$

Here and below we use the notation $\dot{\boldsymbol{\epsilon}} = \boldsymbol{\epsilon}(\dot{\boldsymbol{u}})$. The *local* form of this law may be obtained by assuming first that all variables in the above relation are sufficiently smooth, and then by converting the surface integral involving the heat flux to a volume integral with the use of the divergence theorem. This gives

$$\int_{\Omega'} (\dot{e} - \boldsymbol{\sigma} : \dot{\boldsymbol{\epsilon}} - r + \operatorname{div} \boldsymbol{q})\, dx = 0,$$

which in turn leads to the local form

$$\dot{e} = \boldsymbol{\sigma} : \dot{\boldsymbol{\epsilon}} + r - \operatorname{div} \boldsymbol{q}. \tag{2.42}$$

The second essential postulate of thermodynamics is the *second law*. For this we require first the notion of the *entropy* η per unit volume, and the *absolute temperature* $\theta > 0$. The entropy flux across the bounding surface Γ' into the body Ω' is given by $-\int_{\Gamma'} \theta^{-1} \boldsymbol{q} \cdot \boldsymbol{n}\, ds$, while the entropy supplied by the exterior is given by $\int_{\Omega'} \theta^{-1} r\, dx$. The second law states that the rate of increase in entropy in the body is not less than the total entropy supplied to the body by the heat sources. That is,

$$\frac{d}{dt} \int_{\Omega'} \eta\, dx \geq \int_{\Omega'} \theta^{-1} r\, dx - \int_{\Gamma'} \theta^{-1} \boldsymbol{q} \cdot \boldsymbol{n}\, ds. \tag{2.43}$$

By the same process used to obtain the local form (2.42) of the first law from (2.41) we may obtain the local form of the second law, which reads

$$\dot{\eta} \geq -\operatorname{div} (\theta^{-1} \boldsymbol{q}) + \theta^{-1} r. \tag{2.44}$$

The inequalities (2.43) and (2.44) are known as the *Clausius–Duhem form* of the second law of thermodynamics.

It is customary in elasticity and elastoplasticity to work with the *Helmholtz free energy* ψ, defined by

$$\psi = e - \eta\theta,$$

rather than with the internal energy. With this substitution and the use of (2.42), the local form of the second law becomes

$$\dot{\psi} + \eta\dot{\theta} - \boldsymbol{\sigma} : \dot{\boldsymbol{\epsilon}} + \theta^{-1} \boldsymbol{q} \cdot \nabla\theta \leq 0. \tag{2.45}$$

The inequality (2.45) is known as the *local dissipation inequality*.

Now we specialize to the situation in which subsequent developments will take place, namely, that of isothermal processes. Thus the temperature distribution in a body is assumed to be uniform and equal to the ambient temperature. Furthermore, it is assumed that there is no flow of heat, and also that there is no heat supply from the exterior. Under these circumstances the local dissipation inequality takes the simpler form

$$\dot{\psi} - \boldsymbol{\sigma} : \dot{\boldsymbol{\epsilon}} \leq 0. \tag{2.46}$$

Henceforth we will at all times make the assumptions just described, so that temperature will not appear as a variable. Furthermore, both the heat flux vector and heat supply will be assumed zero in what follows.

Elastic constitutive equations. We are now in a position to obtain the equations describing elastic material behavior. We define an elastic material to be one for which the constitutive equations take the form

$$\psi = \psi(\boldsymbol{\epsilon}), \tag{2.47}$$

$$\boldsymbol{\sigma} = \boldsymbol{\sigma}(\boldsymbol{\epsilon}). \tag{2.48}$$

That is, the free energy and stress depend only on the current strain; there is no dependence on the history of behavior, for example. It should be remarked that the more general point of departure is to take the free energy and stress to be functions of the *displacement gradient* $\nabla \boldsymbol{u}$ rather than the strain. That these variables in fact depend on $\nabla \boldsymbol{u}$ through its symmetric part, the strain $\boldsymbol{\epsilon}$, is a consequence of the principle of material frame indifference (see [82]). We circumvent these considerations by assuming from the outset a dependence on $\boldsymbol{\epsilon}$ rather than on $\nabla \boldsymbol{u}$.

The functions appearing in (2.47) and (2.48) are assumed to be sufficiently smooth with respect of their arguments that as many derivatives as required may be taken.

It is an immediate consequence of the local dissipation inequality that the stress is determined by ψ through the relation

$$\boldsymbol{\sigma} = \frac{\partial \psi}{\partial \boldsymbol{\epsilon}}. \tag{2.49}$$

To see this, we substitute (2.47) in the local dissipation inequality (2.46) to obtain

$$\left(\frac{\partial \psi}{\partial \boldsymbol{\epsilon}} - \boldsymbol{\sigma} \right) : \dot{\boldsymbol{\epsilon}} \leq 0. \tag{2.50}$$

Then (2.49) follows from the fact that (2.50) holds for all $\dot{\boldsymbol{\epsilon}}$. The *linearly elastic material* is recovered from (2.49) by assuming that the free energy is a quadratic function of the strain; that is,

$$\psi(\boldsymbol{\epsilon}) = \tfrac{1}{2} \boldsymbol{\epsilon} : \boldsymbol{C} \boldsymbol{\epsilon}, \tag{2.51}$$

or

$$\psi(\boldsymbol{\epsilon}) = \tfrac{1}{2} C_{ijkl} \epsilon_{ij} \epsilon_{kl}.$$

Then the constitutive equation (2.23) is immediately recovered from (2.49) by substitution of (2.51). The thermodynamic framework is not entirely equivalent to the mechanical framework adopted earlier, though. One distinction lies in the symmetries of $\boldsymbol{C}$. From (2.49) and (2.51) we find that

$$\boldsymbol{\sigma} = \tfrac{1}{2} \left(\boldsymbol{C} + \boldsymbol{C}^T \right) \boldsymbol{\epsilon}.$$

Here $\frac{1}{2}(\boldsymbol{C}+\boldsymbol{C}^T)$ is the symmetric part of $\boldsymbol{C}$. Replacing $\boldsymbol{C}$ by $\frac{1}{2}(\boldsymbol{C}+\boldsymbol{C}^T)$ in (2.51) does not change the value of $\psi(\boldsymbol{\epsilon})$. Hence in the definition (2.51) we will replace $\boldsymbol{C}$ by its symmetric part, though for convenience we continue to denote this symmetrized tensor by $\boldsymbol{C}$. We then have

$$\boldsymbol{C} = \frac{\partial^2 \psi}{\partial \boldsymbol{\epsilon} \partial \boldsymbol{\epsilon}}, \tag{2.52}$$

and in addition to the symmetries given in (2.25) (these still hold, in view of the symmetry of the stress and strain), we must have the additional symmetry

$$C_{ijkl} = C_{klij}. \tag{2.53}$$

We will henceforth take as a basis for the description of linearly elastic material behavior the thermodynamic framework, so that in particular, the symmetry (2.53) will be assumed valid. Note that this symmetry is satisfied with the coefficients (2.31) for *isotropic* elastic materials.

2.6 Initial–Boundary and Boundary Value Problems for Linear Elasticity

The stage has now been reached where it is possible to give a clear and complete formulation of the problems that need to be solved in order to obtain a complete description of the deformation of a linearly elastic body. Suppose such a body initially occupies a domain $\Omega \subset \mathbb{R}^3$ and that the body has boundary Γ, which comprises nonoverlapping parts Γ_u and Γ_t with $\Gamma = \bar{\Gamma}_u \cup \bar{\Gamma}_t$. Suppose that the body force $\boldsymbol{b}(\boldsymbol{x}, t)$ is given in Ω, the displacement $\bar{\boldsymbol{u}}(\boldsymbol{x}, t)$ is given on the part Γ_u of the boundary, and the surface traction $\bar{\boldsymbol{s}}(\boldsymbol{x}, t)$ is given on the remainder Γ_t of the boundary, for $t \in [0, T]$. The initial values of the displacement and velocity are given by $\boldsymbol{u}(\boldsymbol{x}, 0) = \boldsymbol{u}_0(\boldsymbol{x})$ and $\dot{\boldsymbol{u}}(\boldsymbol{x}, 0) = \dot{\boldsymbol{u}}_0(\boldsymbol{x})$. Then the *initial–boundary value problem of linear elasticity* is the following: Find the displacement field $\boldsymbol{u}(\boldsymbol{x}, t)$ that satisfies, in Ω and for $t \in [0, T]$,

the *equation of motion*

$$\operatorname{div} \boldsymbol{\sigma} + \boldsymbol{b} = \rho \ddot{\boldsymbol{u}}, \tag{2.54}$$

the *strain–displacement relation*

$$\boldsymbol{\epsilon}(\boldsymbol{u}) = \tfrac{1}{2}\left(\nabla \boldsymbol{u} + (\nabla \boldsymbol{u})^T\right), \tag{2.55}$$

the *elastic constitutive relation*

$$\boldsymbol{\sigma} = \boldsymbol{C}\boldsymbol{\epsilon}, \tag{2.56}$$

the *boundary conditions*

$$\boldsymbol{u} = \bar{\boldsymbol{u}} \text{ on } \Gamma_u \quad \text{and} \quad \boldsymbol{\sigma n} = \bar{\boldsymbol{s}} \quad \text{on } \Gamma_t, \tag{2.57}$$

and the *initial conditions*

$$\boldsymbol{u}(\boldsymbol{x}, 0) = \boldsymbol{u}_0(\boldsymbol{x}) \quad \text{and} \quad \dot{\boldsymbol{u}}(\boldsymbol{x}, 0) = \dot{\boldsymbol{u}}_0(\boldsymbol{x}), \quad \boldsymbol{x} \in \Omega. \tag{2.58}$$

We may take the displacement vector field as the primary unknown, and eliminate the stress and strain from the governing equations by substitution; this gives the equation of motion in the form

$$\operatorname{div}(\boldsymbol{C}\boldsymbol{\epsilon}(\boldsymbol{u})) + \boldsymbol{b} = \rho \ddot{\boldsymbol{u}}. \tag{2.59}$$

Similarly, the second boundary condition in (2.57) becomes

$$(\boldsymbol{C}\boldsymbol{\epsilon}(\boldsymbol{u}))\,\boldsymbol{n} = \bar{\boldsymbol{s}} \quad \text{on } \Gamma_t. \tag{2.60}$$

When the data are independent of the time, or when the data can be reasonably approximated as being time-independent, the initial–boundary value problem becomes a *boundary value problem*. In this case the body force $\boldsymbol{b}(\boldsymbol{x})$ is given in Ω, the displacement $\bar{\boldsymbol{u}}(\boldsymbol{x})$ is given on Γ_u and the surface traction $\bar{\boldsymbol{s}}$ is given on Γ_t. The problem is now to find the displacement field $\boldsymbol{u}(\boldsymbol{x})$ that satisfies the *equation of equilibrium*

$$\operatorname{div} \boldsymbol{\sigma} + \boldsymbol{b} = \boldsymbol{0} \quad \text{in } \Omega \tag{2.61}$$

together with (2.55)–(2.57). As before, the stress can be eliminated from this problem to give (2.59) with the right-hand side equal to zero.

The variational formulation of the boundary value problem for linear elasticity, (2.61) and (2.55)–(2.57), will be discussed in Chapter 6, as well as the question of well-posedness of this problem.

2.7 Thermodynamics with Internal Variables

The thermodynamic theory presented in Section 2.5 is not entirely adequate for modeling the behavior of a wide range of phenomena. There are situations involving chemically reacting continuous media, for example, in which it is necessary to account for the individual reactions taking place. This may be accomplished by adding to the conventional variables (temperature, strain, and so on) a number of *internal variables* that represent the degree of advancement of the various reactions.

A similar situation obtains in the case of elastoplastic media, the focus of attention of this monograph. Whereas the theory of continuum thermodynamics in its standard form, as presented in Section 2.5, is quite adequate as a framework for the discussion of elasticity, and even of thermoelasticity, it is essential that hidden or internal variables be introduced in order that the theory may serve as a basis for the mathematical description of elastoplastic material behavior. The characteristic features of plasticity will be discussed at length in Chapter 3 and subsequent chapters. In this concluding section of Chapter 2 we extend the thermodynamic theory of Section 2.5 by presenting the theory of thermodynamics with internal variables in a form that will suffice as a basis for the theory of plasticity later. The fundamental references here are those of Coleman and Gurtin [27] and Halphen and Nguyen [49]; in addition, the survey article of Gurtin [47] is a good source for further details, as is the text by Lemaitre and Chaboche [75].

The first and second laws of thermodynamics remain valid in their earlier forms (2.42) and (2.45); here we are concerned with a constitutive theory that will be an extension of that for elastic materials presented earlier. As in that situation we specialize from the outset to isothermal processes in which the temperature is constant and there is no heat flux.

Then we consider materials for which the Helmholtz free energy and stress are given as functions of the strain *and* a set of m internal variables $\boldsymbol{\xi}_1, \boldsymbol{\xi}_2, \ldots, \boldsymbol{\xi}_m$. Some of these may be scalars and some tensors, depending on the application.

The constitutive equations are thus of the form

$$\psi = \psi(\boldsymbol{\epsilon}, \boldsymbol{\xi}_1, \ldots, \boldsymbol{\xi}_m), \tag{2.62}$$

$$\boldsymbol{\sigma} = \boldsymbol{\sigma}(\boldsymbol{\epsilon}, \boldsymbol{\xi}_1, \ldots, \boldsymbol{\xi}_m). \tag{2.63}$$

Unlike the case of elasticity, in which historical effects are irrelevant, the above representations do not suffice for the case in which internal variables are present, and it is necessary to add to this pair of equations an *evolution equation* in which the rate of change of each of the $\boldsymbol{\xi}_i$ is given by an equation of the form

$$\dot{\boldsymbol{\xi}}_i = \boldsymbol{\beta}_i(\boldsymbol{\epsilon}, \boldsymbol{\xi}_1, \ldots, \boldsymbol{\xi}_m), \quad 1 \leq i \leq m. \tag{2.64}$$

Later we will adopt a specialized form of (2.64), but for now it is important merely to note that such an equation is necessary to complete the description of constitutive behavior.

As in Section 2.5 we assume that all functions appearing in (2.62)–(2.64) are sufficiently smooth with respect to their arguments that as many derivatives as required may be taken.

By introducing (2.62) and (2.64) in the reduced dissipation inequality (2.45) we find that

$$\left(\frac{\partial \psi}{\partial \boldsymbol{\epsilon}} - \boldsymbol{\sigma}\right) : \dot{\boldsymbol{\epsilon}} + \frac{\partial \psi}{\partial \boldsymbol{\xi}_i} : \dot{\boldsymbol{\xi}}_i \leq 0. \tag{2.65}$$

In view of the arbitrariness of the rate of change $\dot{\boldsymbol{\epsilon}}$ appearing in (2.65) we conclude that

$$\boldsymbol{\sigma} = \frac{\partial \psi}{\partial \boldsymbol{\epsilon}}. \tag{2.66}$$

We now introduce the *thermodynamic forces* $\boldsymbol{\chi}_i$ conjugate to $\boldsymbol{\xi}_i$; these are defined by

$$\boldsymbol{\chi}_i = -\frac{\partial \psi}{\partial \boldsymbol{\xi}_i}, \quad 1 \leq i \leq m. \tag{2.67}$$

Then, taking account of (2.66) we see that

$$\boldsymbol{\chi}_i : \dot{\boldsymbol{\xi}}_i \geq 0. \tag{2.68}$$

The inequality (2.68) will play a major role later in the construction of a constitutive theory for plastic materials. The left-hand side may be interpreted as a rate of dissipation due to those internal agencies modeled by the internal variables; indeed, we have here a quantity that is a scalar product of force-like variables ($\boldsymbol{\chi}_i$) with the rate of change of strain-like variables ($\boldsymbol{\xi}_i$). Under these circumstances (2.68) declares that the dissipation rate due to internal agencies is nonnegative.

3
Elastoplastic Media

In this chapter we begin to look at the features that characterize elastoplastic materials and at how these physical features are translated into a mathematical theory. The theory has grown slowly during this century, with the impetus for development coming alternately from physical understanding of such materials and from insight into how the physical attributes might be modeled mathematically. We will eventually arrive, towards the end of this chapter, at a theory that is now regarded as classical and that incorporates all the main features of elastoplasticity. This theory may be further generalized, and placed in a unifying framework, if the ideas and techniques of convex analysis are employed. While the entire theory could easily have been developed *ab initio* in such a framework, we have chosen instead to focus first on giving a clear outline of the main features of the mathematical theory, without introducing any sophisticated ideas from convex analysis. In this way, we hope that the connection between physical behavior and its mathematical idealization may be more readily seen. Once such a theory is in place, the business of abstraction and generalization, using the tools of convex analysis, may begin.

3.1 Physical Background and Motivation

It is perhaps useful to begin by summarizing briefly what is understood by linearly elastic behavior. The details were discussed at some length in Chapter 2; briefly, one might say that elastic materials are those for which the stress is completely determined by the strain, and vice versa. For linear elasticity, furthermore, the relation between the stress and the strain is linear.

To illustrate the fundamental features of elastoplastic materials, we consider for simplicity a situation of uniaxial stress in a body; that is, $\sigma \equiv \sigma_{11}$ is nonzero, while all other components of the stress are zero. Such a situation would apply in the case of a thin rod to which is applied at each end, and acting in opposite directions, a force per unit area of intensity σ (Figure 3.1(a)). In this idealized situation the stress does not depend on the position. Alternatively, one might prefer to consider a situation of uniaxial stress in which stress *is* a function of position, and for the purpose of developing a constitutive theory we then consider the relationship between stress and strain at a fixed but arbitrary point in the body.

Suppose that the graph of stress σ versus strain $\epsilon \equiv \epsilon_{11}$ is plotted in order to record the history of behavior during a program of loading. For example, if the force acting on the rod is gradually increased, we will have a corresponding change of length in the rod, and therefore a corresponding increase in strain. For an elastoplastic material it will be observed that up to a value σ_0, say, of stress, the material behaves in a linearly elastic fashion. If the applied force, and hence also the stress, is increased further, behavior deviates from the linear relation in the manner shown in Figure 3.1(b); various possibilities may arise here, but a common feature, particularly of metals, is that there is a decrease in the slope of the curve of stress versus strain. This slope will continue to decrease, and eventually a variety of phenomena may take place. For example, the material may rupture, at which point the experiment will necessarily be regarded as concluded. Alternatively, the slope may reach a value of zero, after which it becomes negative (Figure 3.1(c)). Yet another alternative is that the curve has a point of inflection, after which the slope begins to rise again (Figure 3.1(d)). All of these features, and others yet, are important, the importance of any particular feature depending on the application in question and on the range of stress that is expected to be experienced. The situation in Figure 3.1(b), in which the curve continues to rise, albeit at a slope less than that when $\sigma < \sigma_0$, is known as *hardening* behavior. The situation shown in Figure 3.1(c) for strains greater than $\bar{\epsilon}$, in which the slope is negative, is known as *softening*. This behavior is encountered in materials such as soil and concrete, both of which may be modeled adequately as elastoplastic materials. The kind of *stiffening* behavior shown in Figure 3.1(d) is seen in the stress–strain curves of some metals. The stress σ_0 is known as the *initial yield stress*; it is the threshold of elastic behavior reached from a state of zero stress and zero strain.

To some extent the curves in Figure 3.1 are idealizations of what one would actually encounter in practice, in the sense that some local features may be absent. Consider, for example, the stress–strain curve of mild steel, shown in Figure 3.2(a). This curve exhibits the following features. First,

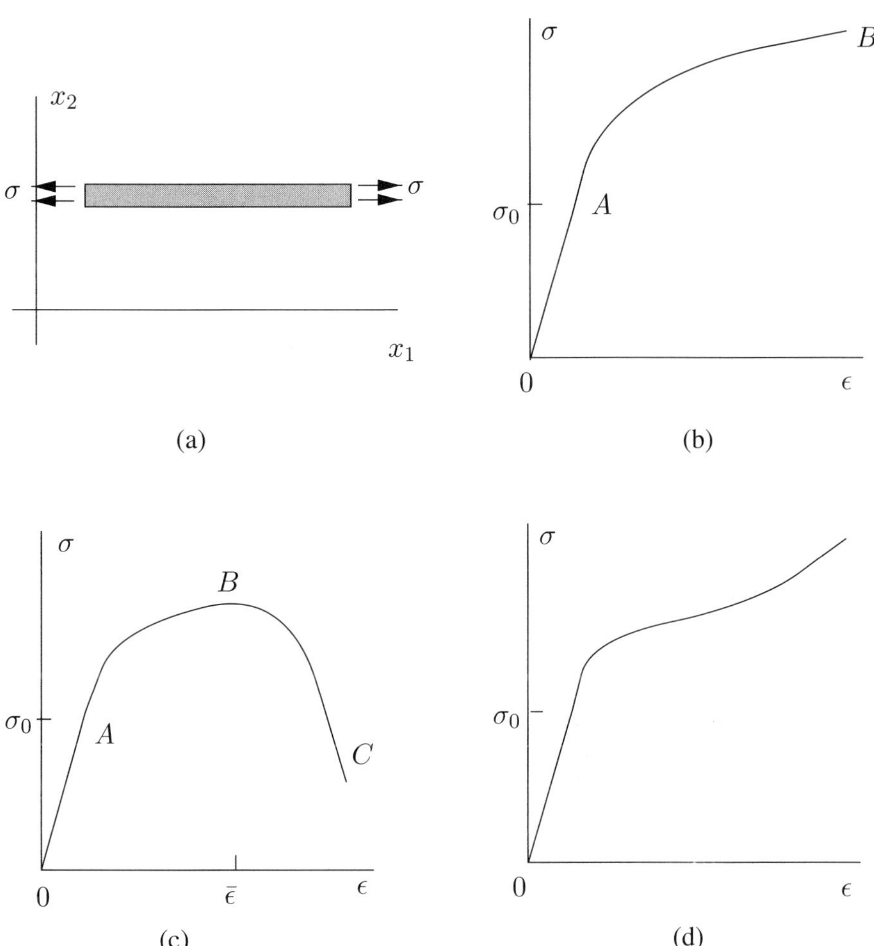

Figure 3.1: (a) An elastoplastic rod in uniaxial tension; (b) stress–strain behavior showing hardening; (c) stress–strain behavior showing hardening and softening; (d) stiffening in the plastic range

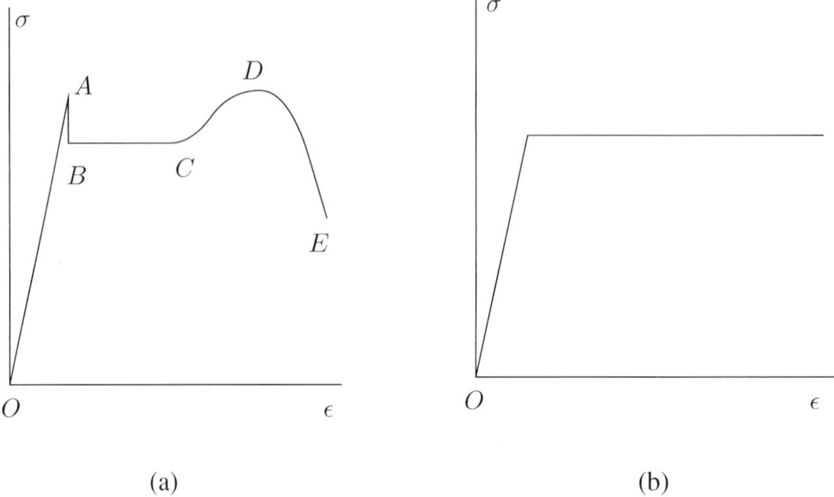

Figure 3.2: (a) The stress–strain curve for mild steel, and (b) its perfectly plastic idealization

there is the linearly elastic region, represented by the section OA of the curve. There is then a sharp and sudden drop in stress, after that the curve has a slope that is barely discernible from zero. This is the section BC of the curve. Hardening behavior then ensues (CD), followed by softening (DE) and, eventually, rupture. A theory that attempts to incorporate all of these features will necessarily be complex, with little consequent reward. Certainly, a feature such as section AB of the curve may be omitted from a model without impairing to any significant degree the viability of the resulting theory. Furthermore, in some applications of practical interest the observed behavior would not include that corresponding to the point C and beyond of the stress–strain curve. For such applications, it is natural to replace the curve of Figure 3.2(a) by the idealized curve of Figure 3.2(b). This is the case of *perfect plasticity*, in which zero hardening occurs or is assumed to occur.

Similar behavior will be observed if the sense of the applied forces is changed so that the stress is compressive ($\sigma < 0$). In this case the strain is also negative, and a typical response would be that shown in Figure 3.3(a). Variants, corresponding to the different features shown in Figures 3.1(b), (c), and (d), also occur. The response in compression does not necessarily mirror that in tension; the initial compressive yield stress $(-\sigma_0')$ may differ in magnitude from the tensile yield value σ_0, and the nature of the curve for $\sigma < -\sigma_0'$ may also differ from the corresponding part for $\sigma > \sigma_0$ of the

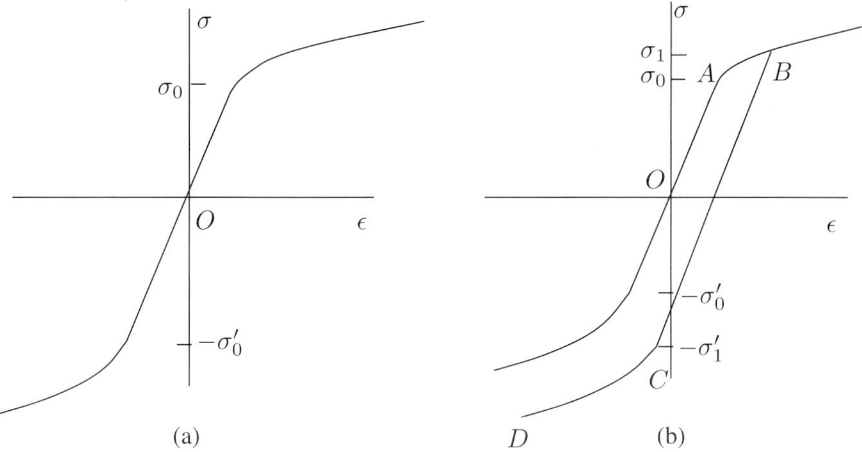

Figure 3.3: (a) Behavior in tension and compression; (b) the path-dependence of plastic behavior

curve in tension. In other words, the function $\sigma(\epsilon)$ is not necessarily an odd one.

The above considerations illustrate clearly the *nonlinearity* inherent in plastic behavior. The next feature that we introduce is that of *irreversibility*, or *path-dependence*. By this it is meant that unlike the case of elasticity, the state of stress does not revert to its original state upon removal of applied forces. Instead, it is observed that a reversal in the stress takes place *elastically*. This is illustrated in Figure 3.3(b). If the direction of loading is reversed at $\sigma_1 > \sigma_0$, the path followed is not the original curve (if this were the case, we would in fact merely have nonlinearly elastic behavior); rather, the material behaves elastically, and the path followed is the straight line BC having slope E, the same slope as that of the line segment OA. This phenomenon is known as *elastic unloading*. Elastic behavior continues until the *new* yield stress $-\sigma_1'$ is reached, after which the curve CD would be followed if the stress were to be decreased further. We thus have an *initial elastic range*, that is, $\sigma \in (-\sigma_0', \sigma_0)$, which includes the unstressed, undeformed state (the origin). We also have subsequent elastic ranges, such as the interval $(-\sigma_1', \sigma_1)$, that are reached only as a result of plastic deformation having taken place.

It is the feature of irreversibility that sets an elastoplastic material apart from an elastic one; the nonlinear behavior described before is not a feature peculiar to plastic materials, since nonlinearly elastic behavior is possible, and indeed common. But the feature of irreversibility implies that we no longer have a one-to-one relationship between stress and strain. In order to

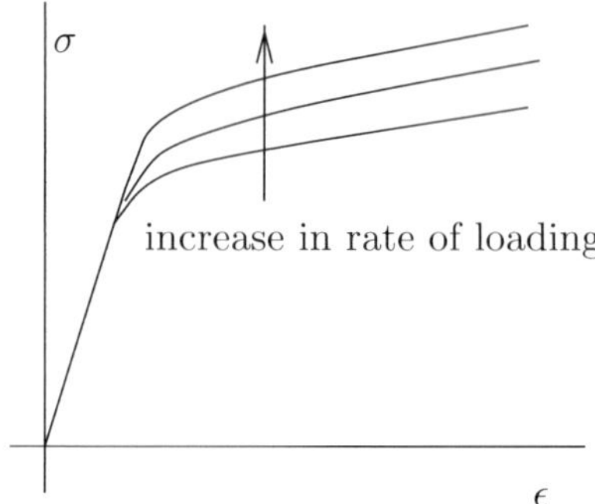

Figure 3.4: Dependence of the stress–strain response on the rate of loading

know the state of stress corresponding to a given strain it is necessary to know the *history* of loading, as Figure 3.3(b) illustrates.

A further feature of elastoplastic behavior that we will incorporate in the general theory is that of *rate-independence*. To see what this implies, consider once again the situation that gives rise to the stress–strain curve of Figure 3.1(b), but suppose this time that the experiment is repeated a few times, the force being applied at a different rate each time. It is found that the elastic response is unchanged, while the response in the plastic range (when $\sigma > \sigma_0$) differs, in a manner shown in Figure 3.4. We will neglect this feature of rate-dependence, thereby restricting the theory either to those materials in which rate-dependence is not a significant phenomenon, or to those situations in which processes occur at rates sufficiently low that rate-dependent effects can be neglected.

The mechanical behavior that we characterize as plastic is very different, at the microstructural or crystalline level, from elastic behavior. It is not appropriate to go into the details here except to point out that at such a level, elastic behavior arises from the deformation of crystal lattices, whereas plastic behavior is typically characterized by irreversible slipping occurring along preferred glide planes. In the plastic range both of these kinds of deformation take place, so that the total strain is made up of an elastic and a plastic component. In other words, we may decompose the strain additively, as shown in Figure 3.5, into an elastic component e and a plastic component p: $\epsilon = e + p$. The elastic part of the strain is given, as before, by Hooke's law; that is, $e = \sigma/E$. The matter of finding the plastic component is an issue that we still have to address. It is clear, of course, from the irreversible nature of plastic behavior that it is unrealistic to expect to have a relationship of the form $\sigma = \sigma(p)$ or $p = p(\sigma)$, since such a relation takes no account of the stress history. Instead, we resolve the question of the plastic constitutive relation by seeking an expression for the *plastic strain rate*. In particular, we pose the question in the following

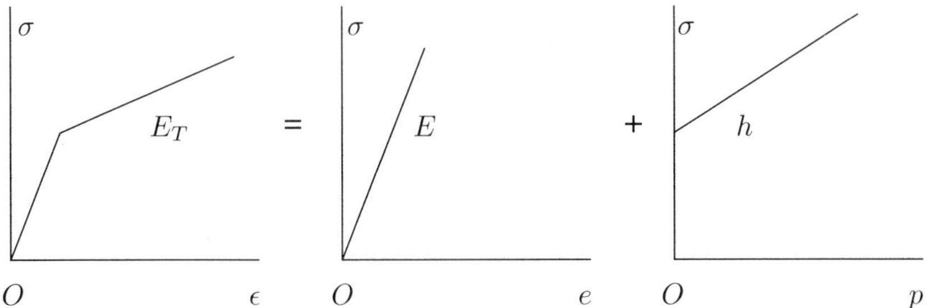

Figure 3.5: Decomposition of the strain into elastic and plastic parts

way: Given the state of stress and the history of behavior of the material point, express the plastic strain rate as a function of the stress and of the history. The motivation for such an approach may be seen from Figure 3.3(b): At $\sigma = \sigma_1$ the plastic strain rate will be nonzero only if the stress increases. If the stress decreases, so that we have elastic unloading, then elastic behavior takes place, and the plastic strain rate is consequently zero.

Suppose then that we follow the curve OAB in Figure 3.3(b). At the point B the region of elastic behavior is the interval $(-\sigma_1', \sigma_1)$. Thus a decrease in stress at B will lead to elastic behavior along the straight line BC. This is known as *elastic unloading*. On the other hand, if the stress is increased at B or decreased at C, then plastic deformation will take place. This behavior is known as *plastic loading* or *plastic hardening*. In other words,

$$\dot{p} = 0 \ \text{ if } \left\{ \begin{array}{ll} & \sigma \in (-\sigma_1', \sigma_1) \\ \text{or} & \sigma = \sigma_1 \text{ and } \dot{\sigma} < 0 \\ \text{or} & \sigma = -\sigma_1' \text{ and } \dot{\sigma} > 0, \end{array} \right. \tag{3.1}$$

and

$$\dot{p} = \frac{\dot{\sigma}}{h} \ \text{ if } \left\{ \begin{array}{ll} & \sigma = \sigma_1 \text{ and } \dot{\sigma} > 0 \\ \text{or} & \sigma = -\sigma_1' \text{ and } \dot{\sigma} < 0. \end{array} \right. \tag{3.2}$$

Here h, a measure of the degree of hardening, is a positive scalar that depends on the history. In the simple example considered here the hardening constant is defined by

$$\frac{1}{h} = \frac{1}{E_T} - \frac{1}{E},$$

where E is Young's modulus, the slope of the elastic curve, and E_T is the slope of the stress–strain curve at $\sigma = \sigma_1$ in Figure 3.3(b).

Equation (3.2) does not hold for the limiting case $E_T \to 0$ of a perfectly plastic material. For this case, the plastic strain rate is nonzero, but its value has to be determined by other considerations.

3.2 Three-Dimensional Elastoplastic Behavior

We now embark on the task of constructing a theory of elastoplasticity for arbitrary three-dimensional behavior, by generalizing in an appropriate way those features of plasticity that were discussed in the last section in the context of the one-dimensional situation. There will be other features as well that are absent when a one-dimensional problem is considered and that will have to be incorporated here.

Isothermal behavior. While thermal effects are important in certain practical situations, the theory of elastoplastic behavior can be developed fully in an isothermal context, that is, one in which it is assumed that no temperature changes take place and no flow of heat occurs. We will assume throughout for convenience that isothermal conditions obtain, so that temperature will not be a variable in the theory that is being developed.

Rate-independence. As mentioned in the previous section, it is generally the case that plastic behavior is rate-dependent: The response of the material depends on the rate at which the process takes place. There is a wide range of materials, however, that respond in an essentially rate-independent fashion for slow processes, and there is likewise a wide range of practical situations in which such slowly varying processes occur. This type of behavior is called quasistatic: In addition to the rate-independence of the material, the rate at which deformation takes place is sufficiently low for the inertial term in the equation of motion (2.16) to be neglected. We will develop a theory of plasticity for quasistatic situations in which the material is assumed to be rate-independent. The appropriate extension to rate-dependent behavior is the theory of viscoplasticity (see, for example, [75, 80, 114] for accounts of viscoplasticity).

The primary variables. We begin by deciding on the primary variables in terms of which the theory will be constructed. The variables required for a complete description of material behavior are essentially of two kinds: kinematic or geometric variables, and force- or stress-like variables.

The first kinematic variable of interest is the strain $\boldsymbol{\epsilon}$, which characterizes the local deformation. We will show shortly that the total strain can be decomposed into two parts: the elastic strain $\boldsymbol{e}$, due to the elastic behavior of the material point, and the plastic strain $\boldsymbol{p}$, which characterizes the irreversible part of the deformation. The elastic and plastic strains, like the total strain, are symmetric second-order tensors.

In addition to these variables we need kinematic quantities that will account for the internal restructuring that takes place during plastic behavior. For this purpose it is convenient to introduce a set of *internal variables*, denoted collectively by $\boldsymbol{\xi} = (\boldsymbol{\xi}_i)_{i=1}^m$. These internal variables characterize features such as hardening, and may be scalars or tensors. The theory of thermodynamics rules out internal variables that are vectors (see [27], page

610), and we will therefore exclude right from the outset vectorial internal variables. Certainly, this does not cause any problems with regard to adequate modeling of physical behavior, since there are no vectorial internal variables that come to mind.

The precise role of the plastic strain in the internal variable theory will become clear shortly; in particular, it will be seen that it would be premature to assume the plastic strain to be one of the internal variables, since the additive decomposition of the strain, and the existence of the plastic strain, follow as *consequences* of the thermodynamics of internal variables. There are instances, though, in which it happens that the plastic strain can be identified with one of the internal variables; one such example is that of linear kinematic hardening (see Section 3.4).

Whether or not internal variables are required will depend on the particular features that one would wish to incorporate in the theory. An obvious example is hardening behavior, which is conveniently characterized through an appropriate choice of internal variables. Other possibilities also exist (see [75], page 60).

The stress-like variables are of two kinds: the stress $\boldsymbol{\sigma}$, and a set $\boldsymbol{\chi} = (\boldsymbol{\chi}_i)_{i=1}^m$ of *internal forces* that are generated as a result of the internal restructuring that occurs during plastic deformation. Clearly, the intention is that these internal forces be conjugate to the internal variables $\boldsymbol{\xi} = (\boldsymbol{\xi}_i)_{i=1}^m$ in the same way in which the stress is a quantity conjugate to the strain, in the sense that the quantity (internal force) $\times$ (rate of change of internal variable) gives a rate of work done, or one of dissipation. The precise relationship between the conjugate forces and the kinematic variables will become clear when we carry out this development in the framework of thermodynamics with internal variables, as set out in Section 2.7.

For convenience we will set $\boldsymbol{\Sigma} = (\boldsymbol{\sigma}, \boldsymbol{\chi})$, and this $(m+1)$-tuple will be known as the *generalized stress*, while we will refer to the $(m+1)$-tuple $\boldsymbol{P} \equiv (\boldsymbol{p}, \boldsymbol{\xi})$ as the *generalized plastic strain*. Thus $\boldsymbol{\Sigma}$ and $\boldsymbol{P}$ are conjugate in the sense that the product $\boldsymbol{\Sigma} : \dot{\boldsymbol{P}} \equiv \boldsymbol{\sigma} : \dot{\boldsymbol{p}} + \boldsymbol{\chi}_i : \dot{\boldsymbol{\xi}}_i$ represents the rate of dissipation due to plastic deformation; this product will be of particular significance later.

Thermodynamic considerations. As was discussed in Section 2.7, for elastoplastic materials it is sufficiently general to consider the free energy and the stress to be given as functions of the total strain and the set of internal variables. The constitutive equations are thus of the form

$$\psi = \psi(\boldsymbol{\epsilon}, \boldsymbol{\xi}), \tag{3.3}$$

$$\boldsymbol{\sigma} = \boldsymbol{\sigma}(\boldsymbol{\epsilon}, \boldsymbol{\xi}). \tag{3.4}$$

We will in fact show later that the decomposition of strain into elastic and plastic parts may be *deduced* from the theory presented here, and that the free energy can equivalently be written as a function of *elastic strain* and internal variables.

To complete the specification of the constitutive equations it is also necessary to express the evolution of the internal variables as functions of the dependent variables, that is,

$$\dot{\boldsymbol{\xi}} = \boldsymbol{\beta}(\boldsymbol{\epsilon}, \boldsymbol{\xi}). \tag{3.5}$$

Returning once again to Section 2.7 we see that as a consequence of the second law of thermodynamics, the stress is given by

$$\boldsymbol{\sigma} = \frac{\partial \psi}{\partial \boldsymbol{\epsilon}}. \tag{3.6}$$

Furthermore, the internal forces $\boldsymbol{\chi}$ introduced earlier are now *defined* to be those quantities conjugate to the internal variables in the sense that

$$\boldsymbol{\chi}_i = -\frac{\partial \psi}{\partial \boldsymbol{\xi}_i}, \quad i = 1, \dots, m. \tag{3.7}$$

Alternatively, the internal forces may enter the second law directly by recognizing that the scalar quantity $\boldsymbol{\chi}_i : \dot{\boldsymbol{\xi}}_i$ represents internal dissipation. Either way, the reduced dissipation inequality now becomes

$$\boldsymbol{\chi}_i : \dot{\boldsymbol{\xi}}_i \geq 0. \tag{3.8}$$

Additive decomposition of strain. We show next how it is possible, in the present framework, to *deduce* from thermodynamic considerations the decomposition of the strain into its elastic and inelastic parts. The approach followed is similar to that in [80].

We begin by temporarily replacing the strain by the stress as an independent variable in the constitutive description. This is achieved by introducing the *Gibbs free energy* h, which is defined through a Legendre transformation by the formula

$$h(\boldsymbol{\sigma}, \boldsymbol{\xi}) = \boldsymbol{\sigma} : \boldsymbol{\epsilon} - \psi.$$

Then the relation conjugate to (3.6) is

$$\boldsymbol{\epsilon} = \frac{\partial h}{\partial \boldsymbol{\sigma}}. \tag{3.9}$$

If we set $\boldsymbol{A} = \partial \boldsymbol{\epsilon}/\partial \boldsymbol{\sigma}$ and $\boldsymbol{B}_i = \partial \boldsymbol{\epsilon}/\partial \boldsymbol{\xi}_i$, then the strain rate is given by

$$\dot{\boldsymbol{\epsilon}} = \boldsymbol{A}\dot{\boldsymbol{\sigma}} + \boldsymbol{B}_i \dot{\boldsymbol{\xi}}_i. \tag{3.10}$$

Now, it is known (see, for example, [80]) that for crystalline solids the elastic compliance $\boldsymbol{A}$ is insensitive to irreversible processes, so that its dependence on $\boldsymbol{\xi}$ may be neglected. That is, $\boldsymbol{A} = \boldsymbol{A}(\boldsymbol{\sigma})$, and it follows that $\boldsymbol{B}_i = \boldsymbol{B}_i(\boldsymbol{\xi})$, $1 \leq i \leq m$, since

$$\mathbf{0} = \frac{\partial}{\partial \boldsymbol{\xi}_i} \frac{\partial \boldsymbol{\epsilon}}{\partial \boldsymbol{\sigma}} = \frac{\partial}{\partial \boldsymbol{\sigma}} \frac{\partial \boldsymbol{\epsilon}}{\partial \boldsymbol{\xi}_i} = \frac{\partial \boldsymbol{B}_i}{\partial \boldsymbol{\sigma}}.$$

Thus the strain may be decomposed additively in the form

$$\boldsymbol{\epsilon} = \boldsymbol{e}(\boldsymbol{\sigma}) + \boldsymbol{p}(\boldsymbol{\xi}), \tag{3.11}$$

where the *elastic strain* $\boldsymbol{e}$ depends only on the stress, while the *plastic strain* $\boldsymbol{p}$ is a function only of the internal variables. These strains are given by the formulae

$$\boldsymbol{e}(\boldsymbol{\sigma})(t) = \int_0^t \boldsymbol{A}(\boldsymbol{\sigma}(s))\dot{\boldsymbol{\sigma}}(s)\,ds \equiv \int_{\mathbf{0}}^{\boldsymbol{\sigma}(t)} \boldsymbol{A}(\boldsymbol{\sigma})\,d\boldsymbol{\sigma}$$

and

$$\boldsymbol{p}(\boldsymbol{\xi})(t) = \int_0^t \boldsymbol{B}(\boldsymbol{\xi}(s))\dot{\boldsymbol{\xi}}(s)\,ds \equiv \int_{\mathbf{0}}^{\boldsymbol{\xi}(t)} \boldsymbol{B}(\boldsymbol{\xi})\,d\boldsymbol{\xi}.$$

In the case where $\boldsymbol{A}$ is independent of $\boldsymbol{\sigma}$, the elastic strain is given as a function of stress by (2.30), that is,

$$\boldsymbol{e} = \boldsymbol{A}\boldsymbol{\sigma} \quad \text{or} \quad \boldsymbol{\sigma} = \boldsymbol{C}\boldsymbol{e}, \tag{3.12}$$

and we see that the fourth-order *compliance* tensor $\boldsymbol{A}$ is the inverse of the elasticity tensor $\boldsymbol{C}$.

Free energy as a function of elastic strain and internal variables. Since (3.9) and (3.11) imply that

$$\frac{\partial e_{ij}}{\partial \sigma_{kl}} = \frac{\partial \epsilon_{ij}}{\partial \sigma_{kl}} = \frac{\partial \epsilon_{kl}}{\partial \sigma_{ij}} = \frac{\partial e_{kl}}{\partial \sigma_{ij}},$$

it follows from a theorem of multivariable calculus that a potential function $h^e(\boldsymbol{\sigma})$ exists such that

$$\boldsymbol{e} = \frac{\partial h^e}{\partial \boldsymbol{\sigma}}. \tag{3.13}$$

This potential function in turn has the Legendre transform $\psi^e(\boldsymbol{e})$ defined by

$$\psi^e = \boldsymbol{\sigma} : \boldsymbol{e} - h^e$$

and with the property that

$$\boldsymbol{\sigma} = \frac{\partial \psi^e}{\partial \boldsymbol{e}}. \tag{3.14}$$

Since $\boldsymbol{\sigma} : \boldsymbol{e} = \psi^e(\boldsymbol{e}) + h^e(\boldsymbol{\sigma})$, it now follows by the properties of Legendre transformations that

$$\begin{aligned} h(\boldsymbol{\sigma}, \boldsymbol{\xi}) &= \boldsymbol{\sigma} : (\boldsymbol{e} + \boldsymbol{p}) - \psi(\boldsymbol{\epsilon}, \boldsymbol{\xi}) \\ &= h^e(\boldsymbol{\sigma}) + \boldsymbol{\sigma} : \boldsymbol{p}(\boldsymbol{\xi}) - \psi^p(\boldsymbol{\xi}), \end{aligned}$$

where the inelastic part ψ^p of the free energy is given by

$$\psi^p(\boldsymbol{\xi}) = \psi(\boldsymbol{\epsilon}, \boldsymbol{\xi}) - \psi^e(\boldsymbol{e}).$$

The function ψ^p indeed depends only on the internal variables, since

$$\boldsymbol{\epsilon} = \frac{\partial h}{\partial \boldsymbol{\sigma}} = \frac{\partial h^e}{\partial \boldsymbol{\sigma}} + \boldsymbol{p}(\boldsymbol{\xi}) - \frac{\partial \psi^p}{\partial \boldsymbol{\sigma}},$$

whence it is seen that $\partial \psi^p / \partial \boldsymbol{\sigma} = \mathbf{0}$.

Summarizing, the Helmholtz free energy ψ and Gibbs free energy h may be decomposed additively into elastic and plastic parts according to

$$\psi(\boldsymbol{\epsilon}, \boldsymbol{\xi}) = \psi^e(\boldsymbol{e}) + \psi^p(\boldsymbol{\xi}) \equiv \hat{\psi}(\boldsymbol{e}, \boldsymbol{\xi}), \tag{3.15}$$

$$h(\boldsymbol{\sigma}, \boldsymbol{\xi}) = h^e(\boldsymbol{\sigma}) + h^p(\boldsymbol{\xi}), \tag{3.16}$$

where $\boldsymbol{e} = \boldsymbol{\epsilon} - \boldsymbol{p}(\boldsymbol{\xi})$.

If we return to the second law (2.46) and this time use (3.15), then we obtain (3.14) and are left with the reduced dissipation inequality in the form

$$\boldsymbol{\sigma} : \dot{\boldsymbol{p}} + \boldsymbol{\chi}_i : \dot{\boldsymbol{\xi}}_i \geq 0, \tag{3.17}$$

or, more concisely,

$$\boldsymbol{\Sigma} : \dot{\boldsymbol{P}} \geq 0, \tag{3.18}$$

where the conjugate forces are now defined by

$$\boldsymbol{\chi}_i = -\frac{\partial \hat{\psi}}{\partial \boldsymbol{\xi}_i} = -\frac{\partial \psi^p}{\partial \boldsymbol{\xi}_i}, \quad 1 \leq i \leq m. \tag{3.19}$$

It is worth pointing out that (3.17) is not in conflict with (3.8), since the internal forces $\boldsymbol{\chi}_i$ are defined in two different ways: either by (3.7) or by (3.19).

EXAMPLE 3.1. To fix ideas, and to provide a concrete working example for later developments, we give an example on what the free energy function looks like for a very common case, namely, that corresponding to coupled linear kinematic and linear isotropic hardening. Hardening has been discussed briefly in the previous section in the one-dimensional context and will be treated in a little more detail later. Suffice to say for now that for this case there are two internal variables, a tensor $\boldsymbol{\alpha}$ corresponding to the back stress in kinematic hardening and a nonnegative scalar γ that determines expansion of the yield surface in isotropic hardening (the notion of the yield surface will be introduced below, while kinematic and isotropic hardening laws will be discussed in detail in Section 3.4). Thus we set $\boldsymbol{\xi} = (\boldsymbol{\xi}_1, \xi_2)$, $\boldsymbol{\xi}_1 = \boldsymbol{\alpha}$, and $\xi_2 = \gamma$, while the conjugate forces are denoted by $\boldsymbol{\chi} = (\boldsymbol{a}, g)$.

For the case in which the elastic behavior of the material is linear, the elastic part of the Helmholtz free energy ψ^e necessarily has the form

$$\psi^e(\boldsymbol{e}) = \tfrac{1}{2}\boldsymbol{e} : \boldsymbol{C}\boldsymbol{e}, \tag{3.20}$$

so that (3.14) gives

$$\boldsymbol{\sigma} = \boldsymbol{C}\boldsymbol{e} = \boldsymbol{C}(\boldsymbol{\epsilon} - \boldsymbol{p}), \tag{3.21}$$

the generalized Hooke's law.

For linear hardening behavior the plastic part of the Helmholtz free energy takes the form

$$\psi^p(\boldsymbol{\alpha}, \gamma) = \tfrac{1}{2}k_1|\boldsymbol{\alpha}|^2 + \tfrac{1}{2}k_2\gamma^2, \tag{3.22}$$

where k_1 and k_2 are nonnegative scalars associated with kinematic and isotropic hardening, respectively. The conjugate forces are immediately obtained from (3.19) and are

$$\boldsymbol{a} = -k_1\boldsymbol{\alpha}, \tag{3.23}$$

$$g = -k_2\gamma. \tag{3.24}$$

□

Plastic incompressibility. The next postulate is one that has not been discussed previously, since it is not one that manifests itself in an obvious way in a one-dimensional situation. In metal plasticity it is observed that changes in volume are almost exclusively of an elastic nature. That is, there is no change in volume accompanying plastic deformation, so that following the discussion leading to (2.10) in Section 2.1, we assume

$$\operatorname{tr}\boldsymbol{p} = p_{ii} = 0. \tag{3.25}$$

The elastic region and yield surface. We have seen earlier in the discussion of one-dimensional behavior that at any stage there is a well defined (open) region of elastic behavior, and that values of stress outside this range cannot be reached without plastic behavior taking place. In order to see how to characterize this behavior in a multidimensional context, consider the situation in which the stress–strain graph takes the form shown in Figure 3.6(a); stress is an odd function of strain, and hardening takes place at a constant rate, determined by the scalar h. Thus the initial elastic range is the interval $(-\sigma_0, \sigma_0)$. Furthermore, when elastic unloading takes place at a stress σ_1 ($\sigma_1 > \sigma_0$), the new elastic range is assumed to be the interval $(-\sigma_0 + hp, \sigma_0 + hp)$. This is an example of *linear kinematic hardening*, which we will revisit in a more general context in Section 3.4. In this situation there is a single internal variable ξ, so that the generalized strain is $\boldsymbol{P} = (p, \xi)$, while the corresponding generalized stress is taken to be

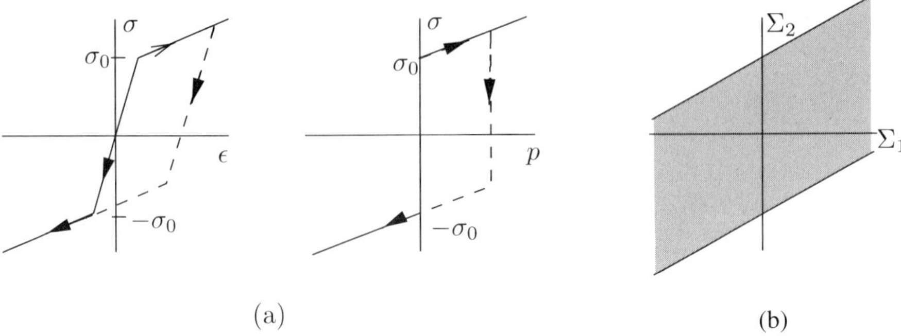

Figure 3.6: (a) Stress–strain curves for kinematic hardening; (b) the situation in generalized stress space

$\mathbf{\Sigma} = (\sigma, \chi)$, where χ, known as the back-stress, represents the translation of the initial elastic region in stress space. For this simple case of linear kinematic hardening the back-stress is related to the plastic strain through $\chi = -k_1 p$, as we saw in the example earlier.

Now it is clear from Figure 3.6 that for any given value of plastic strain p, the region of elastic behavior is the range of values of σ satisfying the inequality $|\sigma + \chi| < \sigma_0$. Rewriting this in terms of components of the generalized stress, we find that the elastic range is given by

$$|\Sigma_1 + \Sigma_2| < \sigma_0.$$

This is the shaded area illustrated in Figure 3.6(b). The above discussion may be summarized by saying that at all times, the generalized stress lies in the closed set $\mathcal{S}$ (that is, the interior region plus its boundary) defined by

$$\mathcal{S} = \{\mathbf{\Sigma} = (\Sigma_1, \Sigma_2) : |\Sigma_1 + \Sigma_2| \leq \sigma_0\}.$$

Purely elastic behavior takes place when $\mathbf{\Sigma}$ lies in the interior of $\mathcal{S}$, while plastic loading may take place only when $\mathbf{\Sigma}$ lies on the boundary of $\mathcal{S}$. Finally, the region exterior to the set $\mathcal{S}$ is not attainable.

The concept of a fixed region of admissible generalized stresses, though illustrated in a rather simple context, is in fact a key ingredient of the theory of plasticity. We now proceed to formalize this postulate. It is assumed that at all times, the generalized stress lies in a closed, connected set $\mathcal{S}$ of *admissible generalized stresses*. The interior of this set is called the *elastic region* and is denoted by $\mathcal{E}$. The boundary of $\mathcal{S}$ is denoted by $\mathcal{B}$ and is known as the *yield surface*. The region $\mathcal{S}^c$ (the complement of the region of admissible stresses) is not attainable.

The significance of an elastic region is that *purely elastic behavior* takes place when $\boldsymbol{\Sigma} \in \mathcal{E}$ or when the generalized stress moves from $\mathcal{B}$ to the interior of $\mathcal{S}$. The latter behavior is known as *elastic unloading*. Plastic behavior takes place only if $\boldsymbol{\Sigma}$ lies on the yield surface and continues to lie on the yield surface; this is known as *plastic loading*.

The above discussion may be made easier if we assume momentarily that we can describe the surface $\mathcal{B}$ and the elastic region $\mathcal{E}$ with the use of a function ϕ: $\mathcal{B} = \{\boldsymbol{\Sigma} : \phi(\boldsymbol{\Sigma}) = 0\}$ and $\mathcal{E} = \{\boldsymbol{\Sigma} : \phi(\boldsymbol{\Sigma}) < 0\}$. Then we make the following assumptions about the rate of change of generalized plastic strains:

$$\dot{\boldsymbol{P}} = \boldsymbol{0} \;\text{ if } \begin{cases} \phi(\boldsymbol{\Sigma}) < 0 \\ \text{or} \\ \phi(\boldsymbol{\Sigma}) = 0 \text{ and } \dot{\phi} < 0; \end{cases} \tag{3.26}$$

$$\dot{\boldsymbol{P}} \text{ may be nonzero only if } \phi(\boldsymbol{\Sigma}) = 0 \text{ and } \dot{\phi} = 0.$$

The requirement that

$$\phi = \dot{\phi} = 0 \tag{3.27}$$

during plastic loading is known as the *consistency condition*.

It is convenient, particularly in practice, to consider the nature of the projection of the yield surface onto the space of stresses. That is, we consider the function $\overline{\phi}(\boldsymbol{\sigma}) \equiv \phi(\boldsymbol{\sigma}, \boldsymbol{\chi})$ for fixed $\boldsymbol{\chi}$, as shown in Figure 3.7.

Now suppose that $\phi(\boldsymbol{\sigma}, \boldsymbol{\chi}) = 0$ and the stress rate is such that

$$\frac{\partial \phi}{\partial \boldsymbol{\sigma}} : \dot{\boldsymbol{\sigma}} > 0. \tag{3.28}$$

Since for plastic loading we have

$$0 = \dot{\phi} = \frac{\partial \phi}{\partial \boldsymbol{\sigma}} : \dot{\boldsymbol{\sigma}} + \frac{\partial \phi}{\partial \boldsymbol{\chi}_i} : \dot{\boldsymbol{\chi}}_i, \tag{3.29}$$

it is clear that there has to be a corresponding change in the forces $\boldsymbol{\chi}$, and consequently also in the projection of the yield surface in the stress space. This therefore defines a new, or current, yield surface in the stress space (see Figure 3.7 for the case where $\boldsymbol{\chi}$ is a scalar). On the other hand, if $\phi(\boldsymbol{\sigma}, \boldsymbol{\chi}) = 0$ and elastic unloading takes place, then by the definition of elastic behavior there will be no change in the internal variables, nor in the forces conjugate to these variables. Consequently, we will have, from (3.26),

$$0 > \dot{\phi} = \frac{\partial \phi}{\partial \boldsymbol{\sigma}} : \dot{\boldsymbol{\sigma}}, \tag{3.30}$$

all $\dot{\boldsymbol{\chi}}$ being zero. The yield surface in operation remains unchanged. In Section 3.4, when we consider specific forms of hardening behavior, we will

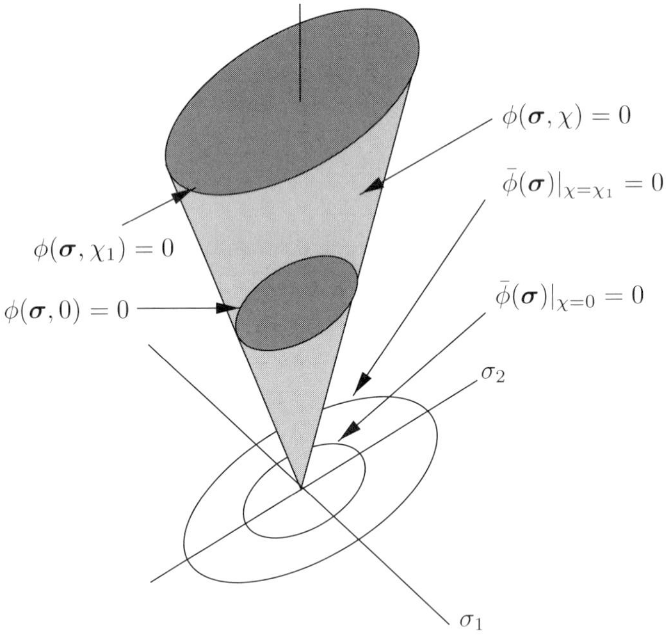

Figure 3.7: The yield surface $\phi(\boldsymbol{\sigma}, \chi) = 0$ and its projection onto stress space

be able to give simple geometric interpretations to the manner in which the surface $\overline{\phi} = 0$ changes under conditions of plastic loading. For completeness it should be added that the defining feature of a perfectly plastic material is that the elastic region, and hence also the yield surface, depends only on the stress; we thus have $\phi(\boldsymbol{\sigma}) = \overline{\phi}(\boldsymbol{\sigma}) = 0$. Plastic behavior takes place when

$$0 = \dot{\phi} = \frac{\partial \phi}{\partial \boldsymbol{\sigma}} : \dot{\boldsymbol{\sigma}}; \tag{3.31}$$

geometrically, the stress moves around the yield surface during plastic deformation, and the yield surface in the stress space remains unchanged by this behavior. Such a situation is known as *neutral loading* to distinguish it from those situations, such as when hardening takes place, in which the surface $\overline{\phi} = 0$ changes during the course of plastic deformation.

These ideas are also readily illustrated for the one-dimensional example discussed earlier. Indeed, in Figure 3.3(b) the initial elastic region in the stress space is simply $(-\sigma_0', \sigma_0)$. But upon plastic loading it is seen that the subsequent elastic region expands and becomes the interval $(-\sigma_1', \sigma_1)$. We will find it useful in subsequent developments to view elastic regions and their associated yield surfaces in these two alternative forms: one, as a fixed region in the space of generalized stresses, and two, as a region in the space of stresses that can change.

Maximum plastic work. The final postulate that we require for a complete theory of plasticity has its origins in the work of von Mises, Taylor, and Bishop and Hill (see [80] for further details). The postulate of maximum plastic work, which can be justified on physical grounds by appealing to the behavior of crystals undergoing plastic deformation, can be stated as follows in its original form: Given a state of stress $\boldsymbol{\sigma}$ for which $\overline{\phi}(\boldsymbol{\sigma}) = 0$ and a plastic strain rate $\dot{\boldsymbol{p}}$ associated with $\boldsymbol{\sigma}$, then

$$\boldsymbol{\sigma} : \dot{\boldsymbol{p}} \geq \boldsymbol{\tau} : \dot{\boldsymbol{p}} \tag{3.32}$$

for all admissible stresses $\boldsymbol{\tau}$, that is, stresses $\boldsymbol{\tau}$ satisfying $\overline{\phi}(\boldsymbol{\tau}) \leq 0$.

This postulate may be stated in an alternative form by introducing the *rate of plastic work* $W(\dot{\boldsymbol{p}}) = \boldsymbol{\sigma} : \dot{\boldsymbol{p}}$ associated with a plastic strain rate $\dot{\boldsymbol{p}}$. Then the postulate of maximum plastic work states that

$$W(\dot{\boldsymbol{p}}) = \max\{\boldsymbol{\tau} : \dot{\boldsymbol{p}} : \ \overline{\phi}(\boldsymbol{\tau}) \leq 0\}.$$

The principle of maximum plastic work is a vital constituent of the theory of plasticity. Depending on the viewpoint adopted, it may be treated as a postulate, as has been the case here, or it may be obtained as a consequence of an alternative postulate. The latter is a popular route, in which it is common to take as a fundamental assumption the postulate on stability due to Drucker; this is the route followed, for example, in [83] for the case of nonsoftening materials. It can then be shown that maximum plastic work follows from the stability postulate.

We will adopt the postulate of maximum plastic work in a more general form, which incorporates the dissipation, or rate of plastic work, due to the internal variables. First, we will assume that the zero generalized stress $\mathbf{\Sigma} = \mathbf{0}$ always belongs to $\mathcal{S}$, the space of admissible or achievable generalized stresses. Secondly, we extend the classical form of the principle of maximum plastic work by postulating the following: Given a generalized stress $\mathbf{\Sigma} \in \mathcal{S}$ and an associated generalized strain rate $\dot{\boldsymbol{P}}$, the inequality

$$\mathbf{\Sigma} : \dot{\boldsymbol{P}} \geq \boldsymbol{T} : \dot{\boldsymbol{P}} \tag{3.33}$$

holds for all generalized stresses $\boldsymbol{T} \in \mathcal{S}$.

In particular, we see that (3.33) is in accord with the reduced dissipation inequality (3.18), since we have, choosing $\boldsymbol{T} = \mathbf{0}$ (recall that $\mathbf{0} \in \mathcal{S}$ by assumption),

$$\mathbf{\Sigma} : \dot{\boldsymbol{P}} \geq 0.$$

Consequences of the maximum plastic work inequality. There are two major consequences of the maximum plastic work inequality. First, it can be shown that the generalized plastic strain rate $\dot{\boldsymbol{P}}$ associated with a generalized stress $\mathbf{\Sigma}$ on the yield surface $\mathcal{B}$ is normal to the tangent hyperplane at the point $\mathbf{\Sigma}$ to the yield surface $\mathcal{B}$. This result is generally referred to as the *normality law*. In the event that the yield surface is not smooth, the normality law states that $\dot{\boldsymbol{P}}$ lies in the cone of normals at $\mathbf{\Sigma}$. Second, it can be shown that the region $\mathcal{E}$ (or $\mathcal{S}$) is *convex*.

We will not attempt to give a rigorous proof of these two assertions here; this will be left for Chapter 4, where the role of the maximum plastic work inequality and its consequences can be seen more clearly with the theory placed in a convex-analytic framework.

To obtain the normality law, here we consider the case that the yield surface is smooth. Let $\mathbf{\Sigma}'$ be a unit tangent vector lying on the tangent hyperplane of the yield surface at $\mathbf{\Sigma}$. Consider a sequence of generalized stresses $\boldsymbol{T} = \mathbf{\Sigma} + \mathbf{\Sigma}'_\epsilon$ that lie on the yield surface and have the property that $\mathbf{\Sigma}'_\epsilon \to \mathbf{0}$ and

$$\frac{\mathbf{\Sigma}'_\epsilon}{\|\mathbf{\Sigma}'_\epsilon\|} \to \mathbf{\Sigma}' \quad \text{as } \epsilon \to 0.$$

Then from (3.33) we have

$$\mathbf{\Sigma}'_\epsilon : \dot{\boldsymbol{P}} \leq 0,$$

and thus,

$$\frac{\mathbf{\Sigma}'_\epsilon}{\|\mathbf{\Sigma}'_\epsilon\|} : \dot{\boldsymbol{P}} \leq 0.$$

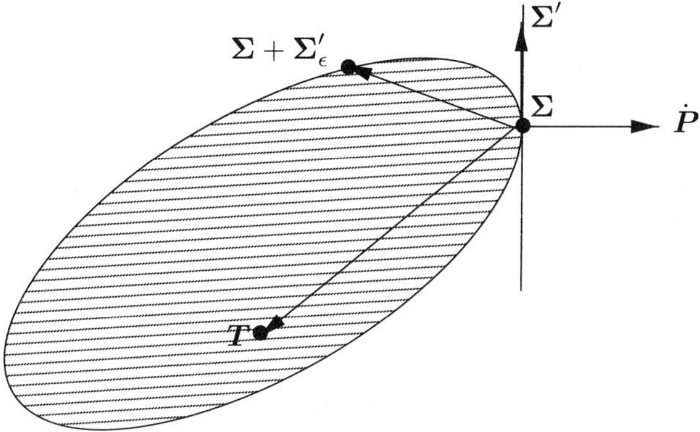

Figure 3.8: Convexity of the yield region and the normality law

Taking the limit $\epsilon \to 0$, we get

$$\boldsymbol{\Sigma}' : \dot{\boldsymbol{P}} \leq 0.$$

Since $(-\boldsymbol{\Sigma}')$ is also a unit tangent vector, we have

$$-\boldsymbol{\Sigma}' : \dot{\boldsymbol{P}} \leq 0.$$

Therefore,

$$\boldsymbol{\Sigma}' : \dot{\boldsymbol{P}} = 0$$

for any vector $\boldsymbol{\Sigma}'$ tangent to the yield surface at $\boldsymbol{\Sigma}$. Hence, $\dot{\boldsymbol{P}}$ is normal to the yield surface at $\boldsymbol{\Sigma}$. To show that $\mathcal{S}$ is a convex set we must show that $\mathcal{S}$ lies to one side of the tangent plane at any point $\boldsymbol{\Sigma} \in \mathcal{B}$, as shown in Figure 3.8. Since $\dot{\boldsymbol{P}}$ lies in the normal to the surface at $\boldsymbol{\Sigma}$, we must show equivalently that the scalar product $(\boldsymbol{\Sigma} - \boldsymbol{T}) : \dot{\boldsymbol{P}}$ for any $\boldsymbol{T} \in \mathcal{S}$ is always nonnegative; this is precisely (3.33).

With the normality law established, it is possible to express the generalized plastic strain rate in a useful form if the yield surface is *smooth*, that is, if it has a well-defined gradient at each point. Since $\dot{\boldsymbol{P}}$ lies parallel to the normal to $\mathcal{B}$ at $\boldsymbol{\Sigma}$, we may write

$$\dot{\boldsymbol{P}} = \lambda \nabla \phi(\boldsymbol{\Sigma}), \tag{3.34}$$

where λ is a nonnegative scalar, called the *plastic multiplier*. This equation may be further reduced by writing out separately the components corresponding to $\boldsymbol{p}$ and $\boldsymbol{\xi}_i$, in the form

$$\dot{\boldsymbol{p}} = \lambda \frac{\partial \phi}{\partial \boldsymbol{\sigma}}, \tag{3.35}$$

$$\dot{\boldsymbol{\xi}}_i = \lambda \frac{\partial \phi}{\partial \boldsymbol{\chi}_i}, \quad 1 \leq i \leq m. \tag{3.36}$$

The conditions on λ may be given in succinct form through a *complementarity condition*; that is,

$$\lambda \geq 0, \quad \phi \leq 0, \quad \lambda\phi = 0. \tag{3.37}$$

The last of conditions (3.37) expresses the fact that λ and ϕ are not simultaneously nonzero, while the first two conditions constrain the signs of λ and ϕ. The last condition then implies that positive λ is possible only when $\phi = 0$, in which case plastic deformation takes place, while negative ϕ implies that λ must be zero, in which case the plastic deformation rate is zero.

Assume that at a certain time t_0, the condition $\phi(t_0) = 0$ is satisfied. Here, we use the notation $\phi(t) \equiv \phi(\boldsymbol{\Sigma}(t))$. Since the generalized stress is constrained by the condition $\phi \leq 0$ at all times, we must have $\dot{\phi}(t_0) \leq 0$, assuming the derivative to exist. Furthermore, if $\dot{\phi}(t_0) < 0$, then the generalized stress has the tendency to move towards the interior of $\mathcal{S}$, and we have elastic unloading, so that $\lambda = 0$. Plastic loading, for which $\lambda > 0$, may take place only if $\dot{\phi}(t_0) = 0$. Summarizing, we have the *consistency condition*:

$$\text{When } \phi = 0, \quad \lambda \geq 0, \ \dot{\phi} \leq 0, \ \lambda\dot{\phi} = 0. \tag{3.38}$$

EXAMPLE 3.1 (CONTINUED). Returning to the example introduced earlier for the case of perfect plasticity, if the yield function is given by

$$\phi(\boldsymbol{\sigma}) = \Phi(\boldsymbol{\sigma}) - c_0 \leq 0,$$

where c_0 is a constant, then the extension to kinematic and isotropic hardening entails the introduction of terms that describe translation and expansion of the yield surface (see Section 3.4). That is, the yield function now becomes, with $\boldsymbol{\Sigma} = (\boldsymbol{\sigma}, \boldsymbol{a}, g)$,

$$\phi(\boldsymbol{\Sigma}) = \Phi(\boldsymbol{\sigma} + \boldsymbol{a}) + g - c_0 \leq 0, \tag{3.39}$$

so that the yield surface translates by an amount $(-\boldsymbol{a})$ and expands by an amount $(-g)$. The flow law (3.35)–(3.36) becomes

$$(\dot{\boldsymbol{p}}, \dot{\boldsymbol{\alpha}}, \dot{\gamma}) = \lambda(\boldsymbol{n}, \boldsymbol{n}, 1), \tag{3.40}$$

where $\boldsymbol{n} = \nabla\Phi(\boldsymbol{\sigma} + \boldsymbol{a})$. Thus we see that the kinematic hardening variable $\boldsymbol{\alpha}$ may be identified with the plastic strain $\boldsymbol{p}$ and the multiplier $\lambda \geq 0$ with the rate of change of the internal variable γ characterizing isotropic hardening. □

Summary of the equations of elastoplasticity. We are now in a position to summarize in one place the equations for the initial–boundary value problem for elastoplastic media. Since this exposition is confined to rate-independent plasticity, which is a valid approximation in the case of processes occurring relatively slowly, it follows that the inertial term in the equations of motion (2.16) will be negligible in such processes. This term is therefore omitted. The resulting equations nevertheless describe processes that are not static, and for this reason we refer to the behavior that is modeled here as *quasistatic*, to emphasize that on the one hand it is not dynamic in the conventional sense (so that the inertial term may be omitted), yet time rates do appear in the governing equations. Thus the problem that emerges is indeed an initial–boundary value problem. We now summarize the variables and the equations of this problem:

Kinematic variables

displacement	$\boldsymbol{u}$
strain	$\boldsymbol{\epsilon} = \frac{1}{2}\left(\nabla \boldsymbol{u} + (\nabla \boldsymbol{u})^T\right)$
plastic strain	$\boldsymbol{p}$
elastic strain	$\boldsymbol{e} = \boldsymbol{\epsilon} - \boldsymbol{p}$
internal variables	$\boldsymbol{\xi} = (\boldsymbol{\xi}_1, \boldsymbol{\xi}_2, \cdots, \boldsymbol{\xi}_m)$
generalized plastic strain	$\boldsymbol{P} = (\boldsymbol{p}, \boldsymbol{\xi}_1, \boldsymbol{\xi}_2, \cdots, \boldsymbol{\xi}_m)$

Dynamic variables

stress	$\boldsymbol{\sigma}$
conjugate forces	$\boldsymbol{\chi} = (\boldsymbol{\chi}_1, \boldsymbol{\chi}_2, \cdots, \boldsymbol{\chi}_m)$
generalized stress	$\boldsymbol{\Sigma} = (\boldsymbol{\sigma}, \boldsymbol{\chi}_1, \boldsymbol{\chi}_2, \cdots, \boldsymbol{\chi}_m)$

Scalar functions

free energy	$\psi(\boldsymbol{\epsilon}, \boldsymbol{\xi}) = \hat{\psi}(\boldsymbol{e}, \boldsymbol{\xi}) = \psi^e(\boldsymbol{e}) + \psi^p(\boldsymbol{\xi})$
yield function	$\phi(\boldsymbol{\Sigma})$

Equilibrium equation

$$\operatorname{div} \boldsymbol{\sigma} + \boldsymbol{b} = \boldsymbol{0}$$

Constitutive equations

$$\boldsymbol{\sigma} = \partial \psi^e / \partial \boldsymbol{e}$$
$$\boldsymbol{\chi}_i = -\partial \psi^p / \partial \boldsymbol{\xi}_i, \ 1 \leq i \leq m$$
$$\dot{\boldsymbol{P}} = \lambda \, \partial \phi / \partial \boldsymbol{\Sigma}$$
$$\lambda \geq 0, \ \phi \leq 0, \ \lambda \phi = 0$$
$$\text{when } \phi = 0, \ \lambda \geq 0, \ \dot{\phi} \leq 0, \ \lambda \dot{\phi} = 0$$

3.3 Examples of Yield Criteria

The theory developed thus far gives very little indication of the nature of yield criteria or, equivalently, of the elastic region. The aim of this section is to introduce a few concrete examples of yield criteria that are typical, as well as important in practice.

The intention is to provide examples of the yield function $\phi(\boldsymbol{\Sigma})$. Since most yield criteria for hardening materials are constructed by extending the criteria that pertain to perfect plasticity, we begin in this section by looking at yield criteria for this special case. The following section is devoted to the extensions that are necessary in order to model hardening effects.

Now, in the case of perfect plasticity there are no internal variables $\boldsymbol{\xi}$, and hence no conjugate forces $\boldsymbol{\chi}$. There is thus no loss of generality in replacing $\boldsymbol{\Sigma}$ by $\boldsymbol{\sigma}$ as the argument for the yield function ϕ.

The simplest yield criteria make use of the assumption of plastic isotropy. When this is the case, it follows that the dependence of ϕ on $\boldsymbol{\sigma}$ is necessarily through the scalar invariants of $\boldsymbol{\sigma}$; for the definition of the invariants, see Section 1.3. The assumption of plastic incompressibility permits a further simplification: Since there is no change in volume accompanying plastic deformation, the spherical part of the stress is assumed to have no influence on the response in the plastic range. Thus instead of expressing ϕ as a function of the stress invariants, it is possible to go one step further and write ϕ as a function of the invariants of the *stress deviator* $\boldsymbol{\sigma}^D$, defined in (2.35) by

$$\boldsymbol{\sigma}^D = \boldsymbol{\sigma} - \tfrac{1}{3}(\operatorname{tr}\boldsymbol{\sigma})\boldsymbol{I}. \tag{3.41}$$

The first invariant, or trace, of any deviatoric tensor vanishes, so we may now write

$$\phi = \phi(I_2(\boldsymbol{\sigma}^D), I_3(\boldsymbol{\sigma}^D)), \tag{3.42}$$

where again the function ϕ and its value are denoted by the same symbol, there being no danger of confusion.

The von Mises yield criterion. This is the simplest yield criterion. It is based on the assumption that the threshold of elastic behavior is determined by the elastic shear energy density

$$\frac{1}{4\mu}\boldsymbol{\sigma}^D : \boldsymbol{\sigma}^D = \frac{1}{4\mu}\operatorname{tr}(\boldsymbol{\sigma}^D)^2,$$

which is an alternative second invariant and is therefore denoted by

$$\frac{1}{2\mu}I_2'(\boldsymbol{\sigma}^D),$$

where $I_2'(\boldsymbol{\sigma}^D) = \frac{1}{2}\operatorname{tr}(\boldsymbol{\sigma}^D)^2$. The threshold value at which plastic deformation occurs is determined by appealing to the one-dimensional situation: In this case, with the axes aligned so that σ_{11} is the only nonzero component of the stress, yielding will occur when $\sigma_{11} = \sigma_0$, the initial yield stress in tension, which is an easily determined quantity. It is readily found that in the one-dimensional case, $I_2'(\boldsymbol{\sigma}^D) = \frac{2}{3}\sigma_0^2$. The von Mises yield criterion is therefore given by

$$\phi(\boldsymbol{\sigma}) = I_2'(\boldsymbol{\sigma}^D) - \tfrac{2}{3}\sigma_0^2 = 0. \tag{3.43}$$

In component form the expression reads

$$\tfrac{1}{2}\left[(\sigma_{11}-\sigma_{22})^2 + (\sigma_{22}-\sigma_{33})^2 + (\sigma_{33}-\sigma_{11})^2 + 6(\sigma_{12}^2 + \sigma_{23}^2 + \sigma_{13}^2)\right] - \sigma_0^2 = 0. \tag{3.44}$$

This yield surface may be readily visualized by considering various special cases in which the stress state is three-dimensional or less. For example, if the axes are aligned locally with the *principal axes* of $\boldsymbol{\sigma}$, then (3.44) becomes

$$\tfrac{1}{2}\left[(\sigma_1-\sigma_2)^2 + (\sigma_2-\sigma_3)^2 + (\sigma_3-\sigma_1)^2\right] - \sigma_0^2 = 0; \tag{3.45}$$

this is the equation of a right circular cylinder of radius $\sqrt{\frac{2}{3}}\sigma_0$ that is inclined at equal angles to the three coordinate axes $(\sigma_1, \sigma_2, \sigma_3)$, as is shown in Figure 3.9.

Another important special case is that of *plane stress*, for which $\sigma_{i3} = 0$. There are again three independent components of the stress, and (3.44) becomes in this case

$$(\sigma_{11}^2 - \sigma_{11}\sigma_{22} + \sigma_{22}^2 + 3\sigma_{12}^2) - \sigma_0^2 = 0. \tag{3.46}$$

This is the equation of an ellipsoid relative to the axes $(\sigma_{11}, \sigma_{22}, \sigma_{12})$.

The Tresca yield criterion. A significant feature of the von Mises yield criterion is that the yield surface is smooth, so that all of the results in Section 3.2, including for example the normality law, apply to this case. However, yield surfaces need not be smooth, and there are in fact important examples of yield surfaces that are piecewise smooth, in the sense that they are made up of smooth manifolds whose intersections form corners, or vertices. The Tresca yield criterion is one such standard example. This criterion is based on the assumption that the elastic threshold is reached when the maximum shear stress reaches a particular value. The assumptions of isotropy and plastic incompressibility are also made here. In terms of the principal stresses, the maximum shear stress is given by

$$\tfrac{1}{2}\max_{i \neq j} |\sigma_i - \sigma_j|.$$

As with the von Mises yield condition, the threshold value is obtained in terms of the yield stress in one-dimensional tension σ_0; the maximum shear stress in this case is obviously $\frac{1}{2}\sigma_0$, so that the Tresca yield surface is defined by

$$\phi(\boldsymbol{\sigma}) = \max_{i \neq j} |\sigma_i - \sigma_j| - \sigma_0 = 0. \tag{3.47}$$

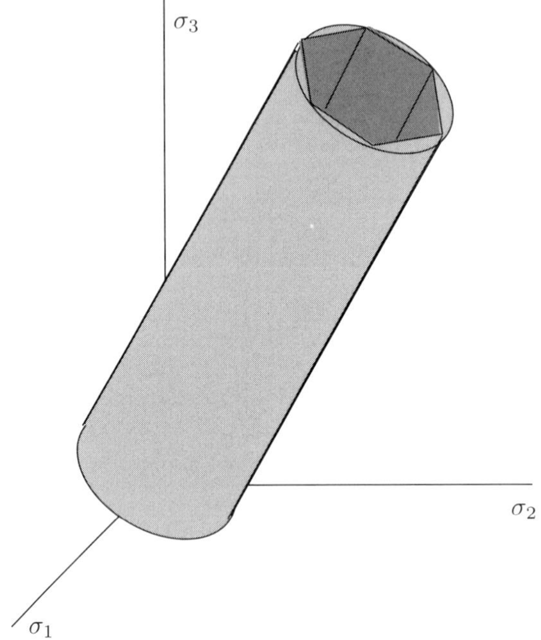

Figure 3.9: The von Mises and Tresca cylinders in principal stress space

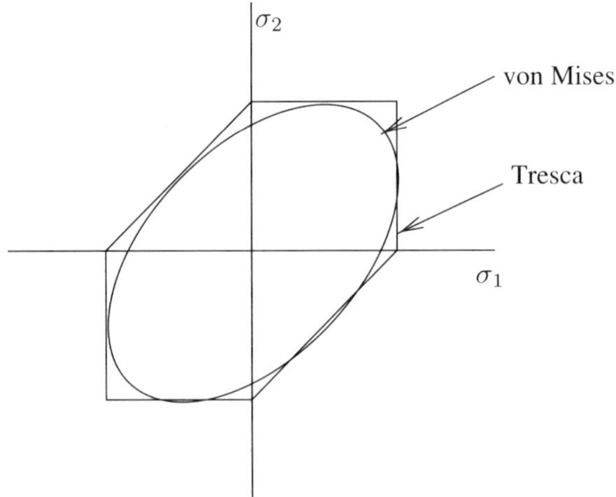

Figure 3.10: The von Mises and Tresca yield surfaces in biaxial stress space

Relative to the axes defined by the principal stress directions the yield surface corresponding to the Tresca criterion is a right hexagonal prism with its axis equally inclined to the three principal axes. The axis of the prism thus coincides with that of the von Mises cylinder, and in fact the cylinder circumscribes this hexagonal prism, as is shown in Figure 3.9. This function can be expressed in terms of the second and third invariants of $\boldsymbol{\sigma}^D$:

$$\begin{aligned}\phi(\boldsymbol{\sigma}) &= \tilde{\phi}(I_2'(\boldsymbol{\sigma}^D), I_3(\boldsymbol{\sigma}^D)) \\ &= 4I_2'(\boldsymbol{\sigma}^D)^3 - 27I_3(\boldsymbol{\sigma}^D)^2 - 9\sigma_0^2 I_2'(\boldsymbol{\sigma}^D)^2 + 6\sigma_0^4 I_2(\boldsymbol{\sigma}^D) - \sigma_0^6. \quad (3.48)\end{aligned}$$

A special case worth considering is that of biaxial stress, in which the only nonzero principal stresses are σ_1 and σ_2; the yield surface (or rather, curve) is obtained from the intersection of the surface in Figure 3.9 with the σ_1-σ_2 plane, and is shown in Figure 3.10. The circumscribing von Mises curve is also shown in the figure.

Anisotropic yield criteria. In the absence of conditions of isotropy, yield criteria obviously have a more complex form than those that have been examined hitherto. In particular, it is no longer possible to express such criteria as functions solely of scalar invariants. Here we give a brief indication of the form taken by yield criteria for the general case in which the material is anisotropic in the plastic range.

A typical example is a generalization of the von Mises yield criterion (3.43), which assumes the form

$$\phi(\boldsymbol{\sigma}) = \tfrac{1}{2}\boldsymbol{\sigma} : \boldsymbol{D}\boldsymbol{\sigma} - k^2 = 0, \quad (3.49)$$

in which $\boldsymbol{D}$ is a fourth-order tensor possessing the symmetries

$$D_{ijkl} = D_{klij} = D_{jikl}. \quad (3.50)$$

If the assumption of independence of mean stress is retained, then the yield function must be invariant under additions of spherical tensors to the stress. That is, we require that

$$\phi(\boldsymbol{\sigma} + \alpha\boldsymbol{I}) = \phi(\boldsymbol{\sigma}) \quad \forall\alpha \in \mathbb{R},$$

or

$$(\boldsymbol{\sigma} + \alpha\boldsymbol{I}) : \boldsymbol{D}(\boldsymbol{\sigma} + \alpha\boldsymbol{I}) - \boldsymbol{\sigma} : \boldsymbol{D}\boldsymbol{\sigma} = 0 \quad \forall\alpha \in \mathbb{R}.$$

Expansion and simplification of the left-hand side yields the condition

$$D_{ijkk} = D_{iikl} = 0, \quad (3.51)$$

which must be satisfied by the components of $\boldsymbol{D}$.

A special case of the criterion (3.49) is that due to Hill, in which there exist three mutually perpendicular planes of symmetry. If a basis is chosen such that the three basis vectors are normal to the three planes of symmetry, then the tensor $\boldsymbol{D}$ has only nine independent components (as opposed to fifteen if it satisfies only (3.50) and (3.51)), and

$$\begin{aligned}\boldsymbol{\sigma} : \boldsymbol{D}\boldsymbol{\sigma} &= D(\sigma_{22} - \sigma_{33})^2 + B(\sigma_{11} - \sigma_{33})^2 + C(\sigma_{11} - \sigma_{22})^2 \\ &\quad + D\sigma_{23}^2 + E\sigma_{13}^2 + F\sigma_{12}^2,\end{aligned}$$

where $D_{1111} = B + C$, $D_{1122} = -C$, $D_{1133} = -B$, and so on.

3.4 Hardening Laws

With some concrete examples of yield criteria available for perfectly plastic materials, it is now a relatively straightforward matter to generalize to the case of materials that experience hardening. In order to do this it becomes necessary to include internal variables in the description of the yield criteria, as has been indicated before.

Isotropic hardening. This type of hardening is characterized by a single *scalar* internal variable, which is denoted here by γ. The corresponding scalar conjugate force is denoted by g. It is generally the case that γ is chosen to be some measure of accumulated plastic deformation. Two typical choices are the *total plastic dissipation* W_p, defined by

$$W_p(t) = \int_0^t \boldsymbol{\sigma}(\tau) : \dot{\boldsymbol{p}}(\tau)\, d\tau,$$

and the *equivalent plastic strain* $p(t)$, a scalar measure of the total plastic strain, defined by

$$p(t) = \sqrt{\tfrac{2}{3}} \int_0^t |\dot{\boldsymbol{p}}(\tau)|\, d\tau.$$

The reason for the inclusion of the factor $\sqrt{\frac{2}{3}}$ in the definition of equivalent plastic strain may be seen by appealing to the special case of one-dimensional strain: When this is the case, symmetry and the fact that $\operatorname{tr} \boldsymbol{p} = 0$ leads to

$$\dot{\boldsymbol{p}} = \begin{bmatrix} \dot{p}_1 & 0 & 0 \\ 0 & -\frac{1}{2}\dot{p}_1 & 0 \\ 0 & 0 & -\frac{1}{2}\dot{p}_1 \end{bmatrix}.$$

It follows that $\sqrt{\frac{2}{3}}\, |\dot{\boldsymbol{p}}| = |\dot{p}_1|$.

An isotropic hardening yield criterion is one in which the yield function takes the form

$$\phi(\boldsymbol{\Sigma}) = \phi(\boldsymbol{\sigma}, g) = \Phi(\boldsymbol{\sigma}) + G(g) - \sqrt{\tfrac{2}{3}}\,\sigma_0, \tag{3.52}$$

where $G(\cdot)$ is a monotone increasing function satisfying $G(0) = 0$. Therefore, as can be seen from (3.52), isotropic hardening amounts to the projection of the yield surface in the stress space expanding isotropically (in the sense that it retains its original shape) by an amount that is determined by the function value $G(g)$, and therefore by the amount of plastic deformation that has already taken place, as shown earlier in Figure 3.7.

Since the free energy now has the form

$$\hat{\psi}(\boldsymbol{e}, \gamma) = \tfrac{1}{2}\boldsymbol{e} : \boldsymbol{C}\boldsymbol{e} + \psi^p(\gamma),$$

the conjugate force g is determined as a function of γ from

$$g = -(\psi^p)'(\gamma). \tag{3.53}$$

As an example, consider the case of the von Mises yield criterion with isotropic hardening. In this situation the yield function becomes

$$\phi(\boldsymbol{\sigma}, g) = I_2'(\boldsymbol{\sigma}^D) - \sqrt{\tfrac{2}{3}}\sigma_0(1 - g).$$

If we set

$$\psi^p(\gamma) = \tfrac{1}{2}k_2\gamma^2$$

as before, then $g = -k_2\gamma$, and the region of admissible stresses comprises those pairs $(\boldsymbol{\sigma}, g)$ for which

$$I_2'(\boldsymbol{\sigma}^D) \le \sqrt{\tfrac{2}{3}}\sigma_0(1 - g).$$

Thus, for example, in a situation of biaxial stresses the yield surface at any time is an ellipse that is similar to the initial ellipse, or initial yield surface. This is shown in Figure 3.11.

It is possible to accommodate more general forms of isotropic hardening, in which the expansion of the yield surface depends nonlinearly on γ. For example, suppose that ψ^p is given by

$$\psi^p(\gamma) = \tfrac{1}{2}k_2\gamma^2 + (k_\infty - k_3)(\gamma + \beta^{-1}e^{-\beta\gamma}) - k_3\gamma,$$

where $k_2 > 0$ as before and $k_\infty \ge k_3 > 0$. Then

$$g(\gamma) = -k_2\gamma + (k_3 - k_\infty)(1 - e^{-\beta\gamma}) + k_3.$$

The meaning of the constants in these functions can be gleaned from the one-dimensional situation, illustrated in Figure 3.12.

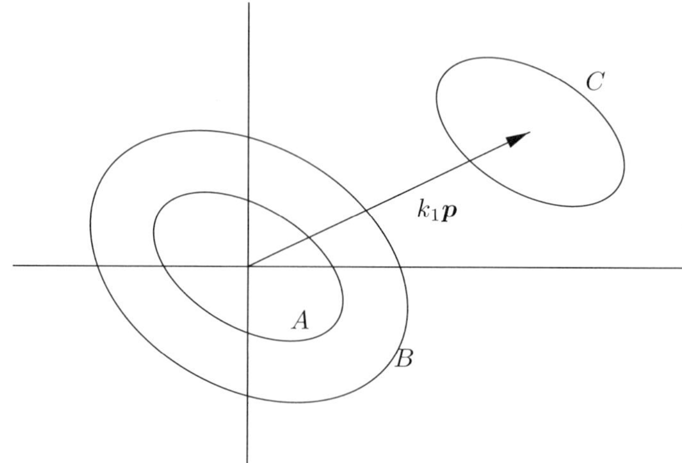

Figure 3.11: Isotropic and kinematic hardening behavior: A is the initial yield surface ($k_1 = k_2 = 0$), B is a subsequent yield surface after isotropic hardening ($k_1 = 0$, $k_2 \neq 0$), and C is a subsequent yield surface with kinematic hardening ($k_1 \neq 0$, $k_2 = 0$)

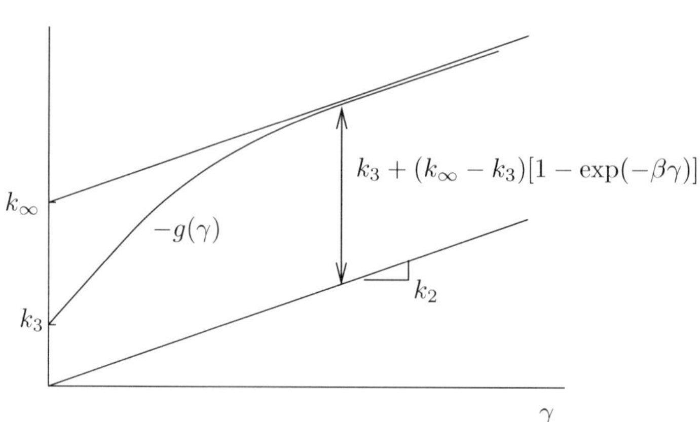

Figure 3.12: A nonlinear isotropic hardening function

Kinematic hardening. The nature of kinematic hardening has been described briefly in Section 3.2; whereas isotropic hardening causes the initial yield surface to undergo a homogeneous expansion, kinematic hardening is characterized by a *translation* of the initial yield surface. The most common form is linear kinematic hardening, with which the name of Prager is generally associated. In this case there is again a single internal variable $\boldsymbol{\alpha}$, which is generally taken to be the plastic strain tensor $\boldsymbol{p}$. The free energy takes the form

$$\psi(\boldsymbol{e}, \boldsymbol{p}) = \tfrac{1}{2}\boldsymbol{e} : \boldsymbol{C}\boldsymbol{e} + \tfrac{1}{2}k_1|\boldsymbol{p}|^2,$$

in which $k_1 > 0$ is the hardening constant. The corresponding conjugate force, denoted by $\boldsymbol{a}$, is found to be

$$\boldsymbol{a} = -k_1\boldsymbol{p}.$$

The yield function is obtained by introducing a translation into the standard or initial function. Thus

$$\phi(\boldsymbol{\Sigma}) = \phi(\boldsymbol{\sigma}, \boldsymbol{a}) = \Phi(\boldsymbol{\sigma} + \boldsymbol{a}).$$

For the von Mises yield condition, for example, the elastic region comprises the set of generalized stresses for which

$$\phi(\boldsymbol{\sigma}, \boldsymbol{a}) < 0 \iff |\boldsymbol{\sigma}^D + \boldsymbol{a}^D| < \sqrt{\tfrac{2}{3}}\sigma_0.$$

The translation of the yield surface due to kinematic hardening is illustrated in Figure 3.11.

It is not a simple matter to extend these ideas to the case of nonlinear kinematic hardening. In fact, it is not possible to do this within the framework constructed here. Instead, it would be necessary to turn to a model encompassing a *nonassociated* flow law, in which there is an additional scalar function $\varphi(\boldsymbol{\Sigma})$, distinct from the yield function $\phi(\boldsymbol{\Sigma})$, which serves as a potential for the rate of change of internal variables in the sense that

$$\dot{\boldsymbol{P}} = \lambda\frac{\partial\varphi}{\partial\boldsymbol{\Sigma}}.$$

The form of the yield criterion remains unchanged, though, and is still given by $\phi(\boldsymbol{\Sigma}) = 0$.

The extension of the standard theory to incorporate nonlinear kinematic hardening in this way is not pursued here; a good reference on the topic is the work by Lemaitre and Chaboche ([75], Sections 5.4.3 and 5.4.4).

Combined kinematic and isotropic hardening. There is no difficulty in combining isotropic and kinematic hardening in a single model. When this is the case, there are two internal variables, the plastic strain $\boldsymbol{p}$ (assuming that we deal with linear kinematic hardening) and the scalar internal variable γ, and the free energy takes the uncoupled form

$$\psi(\boldsymbol{e}, \boldsymbol{p}, \gamma) = \tfrac{1}{2}\boldsymbol{e} : \boldsymbol{C}\boldsymbol{e} + \tfrac{1}{2}k_1|\boldsymbol{p}|^2 + \psi^p(\gamma).$$

The conjugate forces are determined as functions of the internal variables in the usual way as before, and the yield function now becomes

$$\phi(\boldsymbol{\sigma}, \boldsymbol{a}, g) = \Phi(\boldsymbol{\sigma} + \boldsymbol{a}) + G(g) - \sqrt{\tfrac{2}{3}}\,\sigma_0 \leq 0.$$

The yield function may also contain parameters that determine the relative influence of kinematic and isotropic hardening. For example, one may employ a function of the form

$$\phi(\boldsymbol{\sigma}, \boldsymbol{a}, g) = \Phi(\boldsymbol{\sigma} + \theta\boldsymbol{a}) + (1 - \theta)G(g) - \sqrt{\tfrac{2}{3}}\,\sigma_0 \leq 0$$

in which $0 \leq \theta \leq 1$. The value of θ then determines the extent of each of the two forms of hardening and includes the limiting cases $\theta = 0$ for isotropic hardening only, and $\theta = 1$ for kinematic hardening only.

4

The Plastic Flow Law in a Convex-Analytic Setting

The previous chapter has been devoted to the presentation of the basic theory of elastoplasticity in a fairly classical manner. While this theory is adequate in its own right, it is highly advantageous from the point of view of carrying out a mathematical and numerical analysis of the ensuing initial–boundary value problem to recast the constitutive theory in a convex-analytic framework. That will be the aim of this chapter.

We begin in Section 4.1 by collecting together some definitions and results from convex analysis that are of relevance to later developments. Further details, including proofs of results, may be found in Rockafellar [113] and Ekeland and Temam [34]; the former monograph is concerned entirely with convex analysis on finite-dimensional spaces, while the latter develops the theory in an infinite-dimensional (topological vector) space context.

The first place in which we will need to draw on results from convex analysis will be in Section 4.2, in which the elastoplastic initial–boundary value problem is formulated. These results will be required in particular in the formulation of the constitutive relations, which are idealized as pointwise relations involving scalars, vectors, and tensors, so that the setting is finite-dimensional. In later chapters, on the other hand, where variational problems are discussed, results from infinite-dimensional convex analysis will be required. Thus the review of basic results from convex analysis will be carried out in an infinite-dimensional setting, with particular examples from $\mathbb{R}^n$ being given where appropriate.

Basic results from functional analysis will be reviewed in the following chapter. It will be necessary in this chapter to refer occasionally to some concepts introduced there, but we will keep such references to a minimum.

The reader interested only in the forms of the variational problems for elastoplasticity and the analysis and discrete approximations of the variational problems may skip most of the material in this chapter. To be able to follow later developments in this work, the reader needs to be familiar with the three equivalent formulations of the flow law and the contents of Examples 4.8 and 4.9.

4.1 Some Results from Convex Analysis

Let X be a normed vector space, with topological dual X', the space of the linear continuous functionals on X. For $x \in X$ and $x^* \in X'$ the action of x^* on x is denoted by $\langle x^*, x\rangle$. In the finite-dimensional case, X' is isomorphic to, and hence may be identified with, X. For example, the dual space of the Euclidean space $\mathbb{R}^d$ may be identified with $\mathbb{R}^d$ itself, and the action of a vector $\boldsymbol{v} \in (\mathbb{R}^d)' = \mathbb{R}^d$ on $\boldsymbol{u} \in \mathbb{R}^d$ is usually defined to be the scalar product of the two vectors:

$$\langle \boldsymbol{v}, \boldsymbol{u}\rangle = \boldsymbol{v} \cdot \boldsymbol{u}.$$

Examples of infinite-dimensional normed spaces and their duals—in particular, function spaces—will be given in Section 5.2.

Convex sets. Let Y be a subset of X. The interior and boundary of Y are denoted respectively by $\operatorname{int}(Y)$ and $\operatorname{bdy}(Y)$. The subset Y is said to be *convex* if

$$\text{for any } x, y \in Y \text{ and } 0 < \theta < 1, \quad \theta x + (1-\theta)y \in Y. \tag{4.1}$$

In other words, the subset Y is convex if and only if the line segment joining any two points of Y lies entirely in Y.

The *normal cone* to a convex set Y at x, denoted by $N_Y(x)$, is a set in X' defined by

$$N_Y(x) = \{x^* \in X' : \langle x^*, y - x\rangle \le 0 \ \ \forall y \in Y\}. \tag{4.2}$$

The set $N_Y(x)$ is indeed a cone, since for any $x^* \in N_Y(x)$ and any $\lambda > 0$, $\lambda x^* \in N_Y(x)$. When $x \in \operatorname{int}(Y)$, we clearly have $N_Y(x) = \{0\}$, while at least in a finite-dimensional context, for $x \in \operatorname{bdy}(Y)$, $N_Y(x)$ can be identified with the cone of normals at x in the space X. These notions are illustrated in Figure 4.1; in the figure, the boundary $\operatorname{bdy}(Y)$ is smooth at $\boldsymbol{x}$, and the normal cone $N_Y(\boldsymbol{x})$ degenerates to the one-dimensional set spanned by the outward normals at $\boldsymbol{x}$. At the nonsmooth boundary point $\boldsymbol{y}$, on the other hand, $N_Y(\boldsymbol{y})$ is a nontrivial cone.

Convex functions. It will be convenient to allow functions to take values of $\pm\infty$. Let f be a function defined on X, with values in $\overline{\mathbb{R}} \equiv \mathbb{R} \cup \{\pm\infty\}$,

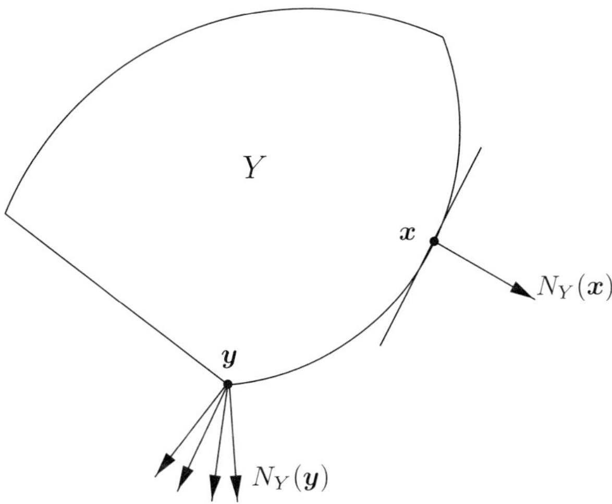

Figure 4.1: The normal cone to a convex set

the extended real line. The *effective domain* of f, denoted by $\mathrm{dom}\,(f)$, is defined by

$$\mathrm{dom}\,(f) = \{x \in X: \ f(x) < \infty\}.$$

The *epigraph* of f, denoted by $\mathrm{epi}\,(f)$, is the set of ordered pairs in $X \times \mathbb{R}$ defined by

$$\mathrm{epi}\,(f) = \{(x, \alpha) \in X \times \mathbb{R}: \ f(x) \leq \alpha\}.$$

That is, the epigraph of f consists of the set of points that lie above the graph of f.

The function f is said to be *convex* if

$$f(\theta x + (1-\theta)y) \leq \theta f(x) + (1-\theta) f(y) \ \ \forall\, x, y \in X, \ \forall\, \theta \in (0,1). \quad (4.3)$$

The function f is said to be *strictly convex* if the strict inequality in (4.3) holds whenever $x \neq y$. Here we follow the convention that $\infty + \infty = \infty$ and $(-\infty) + (-\infty) = -\infty$, while an expression of the form $\infty + (-\infty)$ is undefined. We note that a function is convex if and only if its epigraph is a convex set.

Some other properties of functions on a normed vector space, not particularly related to convexity, will also be of relevance later and are summarized here. The function f is said to be *positively homogeneous* if

$$f(\alpha x) = \alpha f(x) \quad \forall\, x \in X, \ \forall\, \alpha > 0,$$

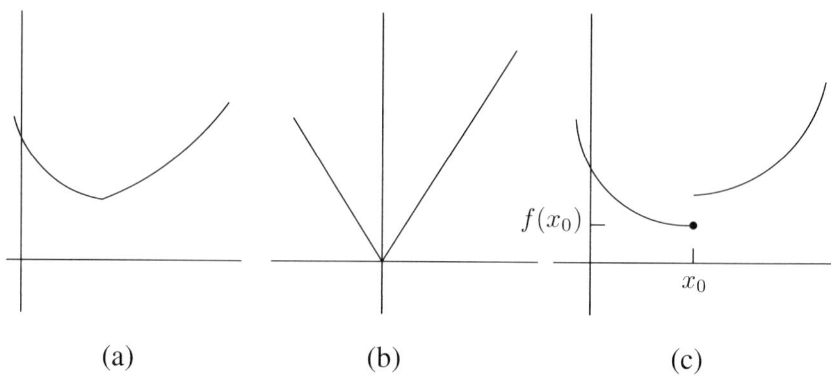

Figure 4.2: Illustrations in one dimension of (a) a strictly convex function; (b) a positively homogeneous function; and (c) a lower semicontinuous function

proper if

$$f(x) < +\infty \text{ for at least one } x \in X \text{ and } f(x) > -\infty \ \forall x \in X,$$

and *lower semicontinuous* (l.s.c.) if

$$\liminf_{n \to \infty} f(x_n) \geq f(x) \tag{4.4}$$

for any sequence $\{x_n\}$ converging to x. These definitions are illustrated in Figure 4.2 for the case in which $X = \mathbb{R}$. A continuous function is lower semicontinuous, but the converse is not true.

It can be shown that f is l.s.c. if and only if its epigraph is closed and that every proper convex function in a finite-dimensional space is continuous on the interior of its effective domain.

A sequence $\{x_n\}$ in a normed space X converges *weakly* to an element $x \in X$ if and only if

$$\lim_{n \to \infty} \langle x^*, x_n \rangle = \langle x^*, x \rangle \quad \forall x^* \in X'.$$

If X is a finite-dimensional space, then the concepts of weak and strong convergence coincide. In Chapter 5 we will encounter the notion of *weak convergence* in a normed space, and in the context of function spaces will see the connection between this type of convergence and the convergence of Fourier series (see Section 5.2). For now, what is of particular interest is the notion of *weak lower semicontinuity* of a function. A function f is weakly lower semicontinuous (abbreviated weakly l.s.c.) if the inequality (4.4) holds for any sequence $\{x_n\}$ converging *weakly* to x. Obviously, a weakly l.s.c. function is l.s.c. Conversely, we have a very useful result: *If f is convex and l.s.c., then it is weakly l.s.c.*

A function $g : X \to [0, \infty]$ is called a *gauge* if

$$\begin{aligned} & g(x) \geq 0 \quad \forall x \in X, \\ & g(0) = 0, \\ & g \text{ is convex, positively homogeneous, and l.s.c.} \end{aligned} \tag{4.5}$$

We remark that the conventional definition of a gauge (see [113]) is that of a function with all the above properties, *excluding* the lower semicontinuity.

For any set $S \subset X$, the *indicator function* I_S of S is defined by

$$I_S(x) = \begin{cases} 0 & x \in S, \\ +\infty & x \notin S, \end{cases} \tag{4.6}$$

and the *support function* σ_S is defined on X' by

$$\sigma_S(x^*) = \sup\{\langle x^*, x\rangle : \ x \in S\}. \tag{4.7}$$

If f is a function on X with values in $\overline{\mathbb{R}}$, the conjugate (often referred to as the Legendre–Fenchel conjugate) function f^* of f is defined by

$$f^*(x^*) = \sup_{x \in X} \{\langle x^*, x\rangle - f(x)\}, \quad x^* \in X'. \tag{4.8}$$

From this definition it is easily seen that the support function is conjugate to the indicator function:

$$I_S^* = \sigma_S. \tag{4.9}$$

Furthermore, if f is proper, convex, and l.s.c., then so is f^*, and in fact,

$$(f^*)^* \equiv f^{**} = f. \tag{4.10}$$

In particular, if S is nonempty, convex, and closed, its indicator function I_S is proper, convex, and l.s.c. So for such a set S,

$$I_S = \sigma_S^* = I_S^{**}. \tag{4.11}$$

Given a convex function f on X, for any $x \in X$ the *subdifferential* $\partial f(x)$ of f at x is the (possibly empty) subset of X' defined by

$$\partial f(x) = \{x^* \in X' : f(y) \geq f(x) + \langle x^*, y - x\rangle \ \ \forall\, y \in X\}. \tag{4.12}$$

A member of $\partial f(x)$ is called a *subgradient* of f at x. According to the definition, when $f(x) = +\infty$, $\partial f(x) = \varnothing$. In the context of functions on $\mathbb{R}^d$, if f is differentiable at x, then

$$\partial f(x) = \{\nabla f(x)\}.$$

At a corner point $(x_0, f(x_0))$, the subdifferential $\partial f(x_0)$ is the set of the slopes of all the lines lying below the graph of f and passing through the

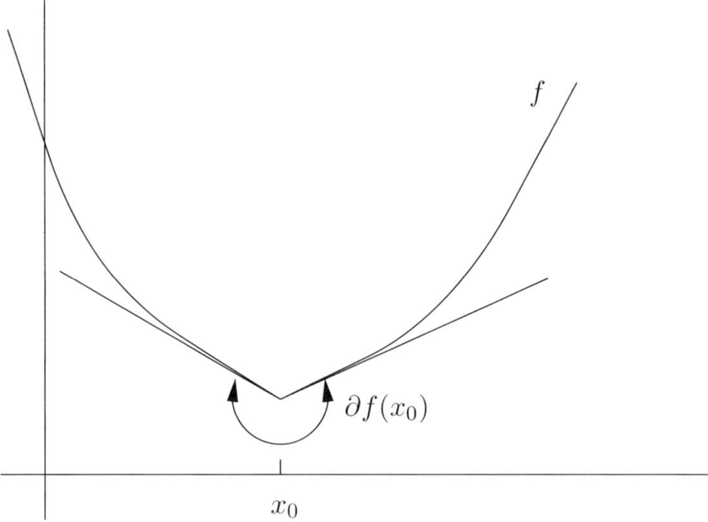

Figure 4.3: Subgradient of a nonsmooth, convex function of a single variable

point $(x_0, f(x_0))$. This is illustrated in Figure 4.3. For the special case of the indicator function it is evident from (4.2) that

$$\partial I_S(x) = N_S(x) \quad \text{for } x \in S. \tag{4.13}$$

A result of fundamental importance in later developments is that for a proper, convex, and l.s.c. function f,

$$x^* \in \partial f(x) \text{ iff } x \in \partial f^*(x^*). \tag{4.14}$$

We have the following results.

LEMMA 4.1. *Let f be a proper, convex, l.s.c. function on a normed space X, and define*

$$\text{dom}\,(\partial f) \equiv \{x \in X : \ \partial f(x) \neq \varnothing\}.$$

Then

(a) $\text{dom}\,(\partial f) \neq \varnothing$ *and* $\text{dom}\,(\partial f)$ *is dense in* $\text{dom}\,(f)$;

(b) *for any proper, convex, l.s.c. functions f and g on X,*

$$\partial f(x) = \partial g(x) \quad \forall\, x \in X$$

if and only if

$$f = g + \text{constant}.$$

LEMMA 4.2. *Let g be a gauge on a reflexive Banach space X. Define a closed convex set K in X' by*

$$K = \{x^* \in X' : \langle x^*, x\rangle \le g(x) \quad \forall x \in X\}. \tag{4.15}$$

Then

(a) *g is the support function of K;*

(b) *the function g^* conjugate to g is the indicator function of K:*

$$g^*(x^*) = \begin{cases} 0 & x^* \in K, \\ +\infty & \text{otherwise;} \end{cases}$$

(c) $K = \partial g(0)$;

(d) $x^* \in \partial g(x) \Longleftrightarrow x \in \partial g^*(x^*) = N_K(x^*)$.

Maximal responsive relations. To set the stage for a complete specification later of the flow law in three equivalent alternative forms, we introduce and explore the properties of a particular multivalued map. Consider a correspondence $G : x \mapsto G(x)$ that associates with each $x \in X$ a (possibly empty) *set* in X'.

DEFINITION 4.3. (a) Let G be a map that associates with each $x \in X$ a *set* $G(x) \subset X'$. The map G is said to be *responsive* if

$$0 \in G(0) \tag{4.16}$$

and if for any x_1, $x_2 \in X$,

$$\langle y_1 - y_2, x_1\rangle \ge 0 \quad \text{and} \quad \langle y_2 - y_1, x_2\rangle \ge 0 \tag{4.17}$$

whenever $y_1 \in G(x_1)$ and $y_2 \in G(x_2)$.

(b) A responsive map G is said to be *maximal responsive* if there is no other responsive map whose graph properly includes the graph of G. Equivalently, G is maximal responsive if the condition

$$\langle y_1 - y_2, x_1\rangle \ge 0 \;\text{ and }\; \langle y_2 - y_1, x_2\rangle \ge 0 \;\; \forall\, y_2 \in G(x_2),\; \forall\, x_2 \in X \tag{4.18}$$

implies that $y_1 \in G(x_1)$.

Figure 4.4 illustrates a simple one-dimensional maximal responsive map. The following theorem makes plain the significance of maximal responsive maps in this context.

THEOREM 4.4. [38] *Let G be a multivalued map on X, with $G(x) \subset X'$ for any $x \in X$. Then the following statements are equivalent:*

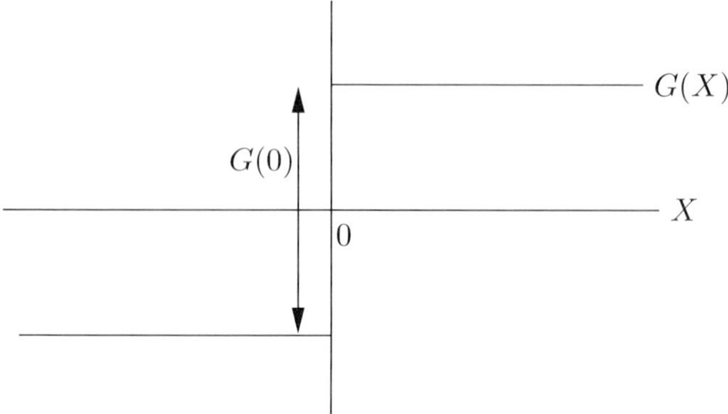

Figure 4.4: An example of a maximal responsive map

(a) *the mapping G is maximal responsive;*

(b) *there exists a gauge g on X such that*

$$G(x) = \partial g(x) \quad \forall\, x \in X.$$

Furthermore, when G is maximal responsive, it determines g uniquely, and with the set $\operatorname{dom}(G) = \{x \in X : G(x) \neq \varnothing\}$, *we have*

$$g(x) = \begin{cases} \langle x^*, x\rangle & \forall\, x^* \in G(x),\ x \in \operatorname{dom}(G); \\ +\infty & \forall\, x \notin \overline{\operatorname{dom}(G)}. \end{cases} \tag{4.19}$$

REMARK. The above theorem is similar in nature to results that connect maximal cyclic monotone maps with the gradients of convex l.s.c. functions (see, for example, [113], [130]).

PROOF OF THEOREM 4.4. Assume first that the condition (b) holds. It follows from Lemma 4.2(d) that

$$G(x) = \{x^* \in K : \langle x^*, x\rangle \geq \langle y^*, x\rangle \ \ \forall\, y^* \in K\}, \tag{4.20}$$

since $G = \partial g$, where K is defined by (4.15). In particular, (4.17) holds. Property (4.16) follows from the observation that $0 \in K = \partial g(0)$ (Lemma 4.2(c)). To show that G is maximal, consider any pair $(\bar{x}, \bar{x}^*)$ such that

$$\langle x^* - \bar{x}^*, x\rangle \geq 0 \ \text{ and } \ \langle \bar{x}^* - x^*, \bar{x}\rangle \geq 0 \tag{4.21}$$

for all $x \in X$, $x^* \in G(x)$. We must verify that $\bar{x}^* \in G(\bar{x})$. We have, from the first part of (4.21) and the definition $G(x) = \partial g(x)$, that

$$\langle \bar{x}^*, x\rangle \leq g(x) \quad \forall\, x \in X,$$

so that $\bar{x}^* \in K$. The second part of (4.21) then implies by (4.20) that $\bar{x}^* \in G(\bar{x})$, as required.

Conversely, suppose that (a) holds. We show that

$$G(0) \supset G(x) \quad \forall\, x \in X. \tag{4.22}$$

Consider any $\overline{x}$ and $\overline{x}^* \in G(x)$. From (4.17),

$$\langle x^* - \overline{x}^*, x\rangle \geq 0 \quad \text{and} \quad \langle \overline{x}^* - x^*, \overline{x}\rangle \geq 0$$

whenever $x^* \in G(x)$. Hence the pair $(0, \overline{x}^*)$ has the property that

$$\langle x^* - \overline{x}^*, x\rangle \geq 0 \quad \text{and} \quad \langle \overline{x}^* - x^*, 0\rangle \geq 0$$

whenever $x^* \in G(x)$, and so $(0, \overline{x}^*)$ could be added to the graph of G without violating (4.17). Since G is maximal responsive, we must have $x^* \in G(0)$, whence (4.22).

The above argument actually establishes that $G(0)$ coincides with the set

$$K = \{\overline{x}^* \in X' : \ \langle x^* - \overline{x}^*, x\rangle \geq 0 \ \ \forall\, x \in X, \ x^* \in G(x)\}. \tag{4.23}$$

This set is closed and convex, and contains 0, by property (4.16). From (4.22) and (4.23),

$$\overline{x}^* \in G(\overline{x}) \ \Rightarrow \ \overline{x}^* \in G(0) = K \ \Rightarrow \overline{x} \in N_K(\overline{x}^*). \tag{4.24}$$

Let g be the support function of K. Since g is the support function of a closed convex set containing 0, it is a gauge ([113], Chapter 15), and

$$\overline{x}^* \in \partial g(\overline{x}) \iff \overline{x} \in N_K(\overline{x}^*).$$

Moreover, ∂g is a responsive map. Furthermore, (4.24) implies that the graph of G is included in the graph of ∂g. Inasmuch as G is maximal responsive, we may conclude that $G = \partial g$, whence part (b).

To establish the uniqueness of g, we recall (Lemma 4.1) that two l.s.c. proper convex functions have the same subdifferential if and only if they differ by an additive constant. We fix this constant by the requirement that $g(0) = 0$, thereby defining g uniquely.

To establish (4.19), we note that g is the support function of K, defined by (4.15), so that from (4.20), $g(x) = \langle x^*, x\rangle$ when $x^* \in G(x)$. Since $\text{dom}\,(G) = \text{dom}\,\partial(G) = \text{dom}\,(D)$ (Lemma 4.1), we also have $g(x) = +\infty$ when $x \notin \text{dom}\,(G)$, whence (4.19). □

Polar functions. It will be useful in discussing the admissible region in the context of convex analysis to consider this region as a closed convex set K whose boundary (the yield surface) is the level set of a convex function g; that is, for some constant $c_0 > 0$,

$$K = \{x^* \in X' : \ g(x^*) \leq c_0\}.$$

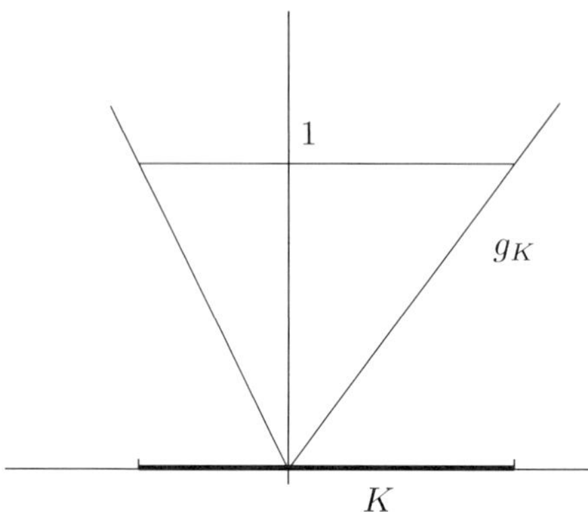

Figure 4.5: An illustration of the gauge g_K corresponding to a set $K \subset \mathbb{R}$

Given a closed convex set K, it is in fact possible to define g in such a way that it is a gauge, so that epi (g) is a closed convex cone containing the origin. Furthermore, this function g has a special relation to the support function σ_K.

We define the gauge g_K of the set K by

$$g_K(x^*) = \inf\{\mu > 0 : \ x^* \in \mu K\}, \tag{4.25}$$

where $\mu K = \{\mu y : \ y \in K\}$. From Lemma 4.2(a) and (4.25), we see that an alternative form for g_K is

$$g_K(x^*) = \inf\{\mu > 0 : \ \langle x^*, x\rangle \leq \mu\sigma_K(x) \ \ \forall\, x \in X\}. \tag{4.26}$$

Note that $g_K(x^*)$ can take on the value $+\infty$ (when $x^* \notin \mu K$ for any $\mu > 0$). An illustration of the function g_K is given in Figure 4.5. Now assume that $\sigma_K(x) = 0$ if and only if $x = 0$. Then g_K and σ_K are related by

$$g_K(x^*) = \sup_{0 \neq x \in \operatorname{dom} \sigma_K} \frac{\langle x^*, x\rangle}{\sigma_K(x)}. \tag{4.27}$$

In other words, we have the inequality

$$\langle x^*, x\rangle \leq g_K(x^*)\sigma_K(x) \ \ \forall\, x \in \operatorname{dom}(\sigma_K), \ x^* \in \operatorname{dom}(g_K). \tag{4.28}$$

Let $x^* \in \operatorname{bdy}(K)$. Then

$$\sup_{0 \neq y \in \operatorname{dom}(\sigma_K)} \frac{\langle x^*, y\rangle}{\sigma_K(y)} = 1, \tag{4.29}$$

and the supremum is achieved when $y = x$, the conjugate to x^* in the sense of Lemma 4.2(d). Thus for $x^* \in K$ and $x^* \in \partial\sigma_K(x)$, $x \neq 0$, we have

$$\langle x^*, x\rangle = g_K(x^*)\sigma_K(x). \tag{4.30}$$

Hence whereas I_K and σ_K are conjugate in the sense of (4.9), the relationships (4.28) and (4.30) define g_K and σ_K as *polar conjugates* of each other; g_K is the *polar function* of σ_K, and we write $g_K = \sigma_K^\circ$. Furthermore, just as $\sigma_K^{**} = \sigma_K$, it can be shown that $\sigma_K^{\circ\circ} = \sigma_K$ (for this the lower semicontinuity of σ_K is required).

This type of relationship between pairs of functions has also been investigated by Hill [59], who refers to such pairs as dual potentials.

The following result will allow us later to establish the normality law in a form involving the yield function.

LEMMA 4.5. *Let g be nonnegative and convex, with $g(0) = 0$ and x a point in the interior of* dom(g) *such that $g(x) > 0$. Set $C = \{z : \; g(z) \leq g(x)\}$. Then $y \in N_C(x)$ if and only if there exists $\lambda \geq 0$ such that $y \in \lambda\partial g(x)$.*

REMARK. Lemma 4.5 appears in [113] as Corollary 23.7.1. Here we include a simplified proof of the result following [38].

PROOF OF LEMMA 4.5. First we assume that $y \in \lambda\partial g(x)$. From the definition of the subdifferential, we have

$$\lambda\, g(z) \geq \lambda\, g(x) + \langle y, z - x\rangle \quad \forall\, z \in X,$$

from which

$$\langle y, z - x\rangle \leq \lambda\, [g(z) - g(x)] \leq 0 \quad \forall\, z \in C,$$

since $\lambda \geq 0$. Thus, $y \in N_C(x)$.

Conversely, assume that $y \in N_C(x)$. We need to show that there are $\lambda \geq 0$ and $y_0 \in X'$ such that

$$g(z) \geq g(x) + \langle y_0, z - x\rangle \quad \forall\, z \in X. \tag{4.31}$$

If $y \neq 0$ and $\langle y, z - x\rangle = 0$, then $x \in \text{bdy}\,(C)$, and since C is convex, $z \in (\text{int}\, C)^c$, the complement of int C. Hence, $g(z) \geq g(x)$ and (4.31) holds. Assume then that $\langle y, z - x\rangle < 0$. For $z \in X\backslash C$, we have $g(z) - g(x) \geq 0$. Thus (4.31) holds for any $\lambda \geq 0$. Finally, suppose that $z \in C$. We see that

$$g(z) - g(x) - \mu\,\langle y, z - x\rangle = \left\{\frac{g(z) - g(x)}{\langle y, z - x\rangle} - \mu\right\}\, \langle y, z - x\rangle \geq 0,$$

provided that

$$\mu > \max\{(g(z) - g(x))/\langle y, z - x\rangle : z \in C\}.$$

The result follows with $\lambda = \mu^{-1}$. $\square$

Duality theory and a posteriori error analysis. Duality theory in convex analysis is a powerful tool for the purpose of deriving a posteriori error estimates for some problems and numerical procedures, including regularization methods. We briefly review the general framework for a posteriori error analysis presented in Han [50]. In Section 12.4 we will use regularization methods to solve some variational problems arising in plasticity; we will show there how the general framework discussed below can be used to derive a posteriori error estimates for the solutions of the regularized problems.

Let Z and S be two normed spaces, and Z' and S' their dual spaces. Assume that there exists a linear continuous operator $F \in \mathcal{L}(Z, S)$, with dual $F^* \in \mathcal{L}(S', Z')$. Let J be a function mapping $Z \times S$ into $\overline{\mathbb{R}}$, and consider the minimization problem

$$\inf_{z \in Z} J(z, Fz). \tag{4.32}$$

Recalling the definition (4.8), we see that the conjugate function of J is

$$J^*(z^*, s^*) = \sup_{z \in Z, s \in S} \left[\langle z^*, z \rangle + \langle s^*, s \rangle - J(z, s) \right],$$

for $z^* \in Z'$, $s^* \in S'$.

We have the following result ([50]).

THEOREM 4.6. *Assume that:*

(a) *Z is a reflexive Banach space, and S a normed space;*

(b) *the functional $J : Z \times S \to \overline{\mathbb{R}}$ is proper, l.s.c., and strictly convex;*

(c) *there exists $z_0 \in Z$ such that $J(z_0, Fz_0) < \infty$ and $s \mapsto J(z_0, s)$ is continuous at Fz_0;*

(d) *$J(z, Fz) \to \infty$ for any $z \in Z$ such that $\|z\|_Z \to \infty$.*

Then the problem (4.32) has a unique solution $y \in Z$. If we define

$$D(y, z) = J(z, Fz) - J(y, Fy) \tag{4.33}$$

for any $z \in Z$ with $J(z, Fz) < \infty$, then

$$D(y, z) \leq J(z, Fz) + J^*(F^* s^*, -s^*) \quad \forall\, s^* \in S'. \tag{4.34}$$

In our application of the theorem later in Section 12.4, the problem (4.32) is a variational inequality of the second kind (cf. Section 6.2, which provides a brief introduction to elliptic variational inequalities), y is the solution to the variational inequality, and we take z in (4.34) to be the solution of a regularized problem. The auxiliary variable (the dual variable) s^* appearing in (4.34) will be suitably chosen from the relation satisfied by the solution of the regularized problem. A positive lower bound, in terms of the "error" $\|y - z\|$, will be established for the quantity $D(y, z)$ defined in (4.33). Then (4.34) will provide an a posteriori error estimate for the solution of the regularized problem.

4.2 Basic Plastic Flow Relations of Elastoplasticity

We now return to the subject of Chapter 3, where the equations governing elastoplastic behavior were introduced, and show how these equations can be recast in the framework of convex analysis. Naturally, the focus will be on the yield criterion and flow law as summarized, for example, in (3.34)–(3.38).

Recall from Section 3.2 that the convexity of the yield surface and the normality law (3.34) were both obtained as consequences of the maximum plastic work inequality. The argument used to arrive at these properties was somewhat heuristic, though, but the results presented in the last section, particularly Lemma 4.2 and Theorem 4.4, will not only provide a means of establishing those same results in a more rigorous fashion, but will also allow us to derive results of greater generality. Specifically, the restriction to smooth yield surfaces will be dropped, and furthermore, we will be able to derive various alternatives to the standard form of the plastic flow law.

In this section it will be convenient to equate X with the set of generalized plastic strain rates. Recall that a generalized plastic strain is of the form $\boldsymbol{P} = (\boldsymbol{p}, \boldsymbol{\xi})$, where $\boldsymbol{p}$ is the plastic strain and $\boldsymbol{\xi}$ the set of internal variables. Thus, in referring to the results in Section 4.1, it is useful to make the substitution

$$x \longleftrightarrow \dot{\boldsymbol{P}}.$$

Likewise, we equate X' with the set of generalized stresses $\boldsymbol{\Sigma} = (\boldsymbol{\sigma}, \boldsymbol{\chi})$, where $\boldsymbol{\chi}$ is the set of thermodynamic forces, and make the association

$$x^* \longleftrightarrow \boldsymbol{\Sigma}.$$

Suppose, then, that the yield surface constitutes the *boundary* of a closed region K. We refer to the interior of this set as the *elastic region*. Likewise, as in Section 3.2, the set K may be represented in the form

$$K = \{\boldsymbol{\Sigma} : \ \phi(\boldsymbol{\Sigma}) \leq 0\},$$

so that the yield surface is the set of points $\boldsymbol{\Sigma}$ that satisfy $\phi(\boldsymbol{\Sigma}) = 0$, while the elastic region is given by the set of points $\boldsymbol{\Sigma}$ for which $\phi(\boldsymbol{\Sigma}) < 0$.

Against this background, we now return to Theorem 4.4 and to the notion of maximal responsiveness. Let G be a responsive map having as its values subsets of X'. Then the condition (4.16) implies that the set of thermodynamic forces corresponding to zero generalized plastic strain rate contains the zero generalized stress. Relation (4.17), on the other hand, is a generalization of the *maximum plastic work inequality* (3.33); indeed, for the case of perfect plasticity, (4.17) becomes, by an obvious change of notation,

$$(\boldsymbol{\sigma}_0 - \boldsymbol{\sigma}_1) : \dot{\boldsymbol{p}}_0 \geq 0 \quad \text{and} \quad (\boldsymbol{\sigma}_1 - \boldsymbol{\sigma}_0) : \dot{\boldsymbol{p}}_1 \geq 0$$

whenever $\boldsymbol{\sigma}_0 \in G(\dot{\boldsymbol{p}}_0)$ and $\boldsymbol{\sigma}_1 \in G(\dot{\boldsymbol{p}}_1)$, and comparison with (3.32) establishes the relationship between these two sets of conditions.

Figure 4.4 in the last section shows a simple example of a maximal responsive map, of the kind that occurs in perfect plasticity.

Given the close relationship between the notion of responsiveness and the maximum plastic work inequality, it is natural to enquire as to the conditions under which one may deduce from responsiveness properties the convexity of the yield surface and the normality rule, in the same way as these properties follow from the maximum plastic work inequality. The answer lies in Theorem 4.4 read together with Lemma 4.2: We see there that *maximal responsiveness* implies the existence of a closed convex set K of achievable values of the generalized stress, and furthermore, it implies also a generalized normality rule in the sense that

$$\dot{\boldsymbol{P}} \in N_K(\boldsymbol{\Sigma}). \tag{4.35}$$

But Theorem 4.4 goes much further than this, in that it implies the *equivalence* between the formulation (4.35) and the formulation

$$\boldsymbol{\Sigma} \in G(\dot{\boldsymbol{P}}). \tag{4.36}$$

Since

$$\boldsymbol{\Sigma} \in \operatorname{int}(K) \Longrightarrow N_K(\boldsymbol{\Sigma}) = \{\boldsymbol{0}\},$$

we see from (4.35) that in the elastic region, $\dot{\boldsymbol{P}} = \boldsymbol{0}$.

To summarize, the existence of a maximal responsive map on X is equivalent to the existence of a convex elastic region and the normality law.

We turn next to the function g that appears in Lemma 4.2. In the context of plasticity this function, the support function of K, is known as the *dissipation function*, and is denoted by D. Thus we have the association

$$g \longleftrightarrow D$$

in Lemma 4.2. The nomenclature arises from the use of the definition (4.7), which gives

$$D(\dot{\boldsymbol{P}}) = \sup\{\boldsymbol{T} : \dot{\boldsymbol{P}} : \ \boldsymbol{T} \in K\} = \boldsymbol{\Sigma} : \dot{\boldsymbol{P}}, \tag{4.37}$$

say, in which $\boldsymbol{\Sigma}$ is the point at which the supremum is achieved, and the term on the extreme right-hand side is the rate of plastic work, or dissipation.

The Legendre–Fenchel conjugate D^* is the indicator function of the set K of admissible generalized stresses, and part (d) of Lemma 4.2 implies that (4.35) is equivalent to the condition

$$\boldsymbol{\Sigma} \in \partial D(\dot{\boldsymbol{P}}). \tag{4.38}$$

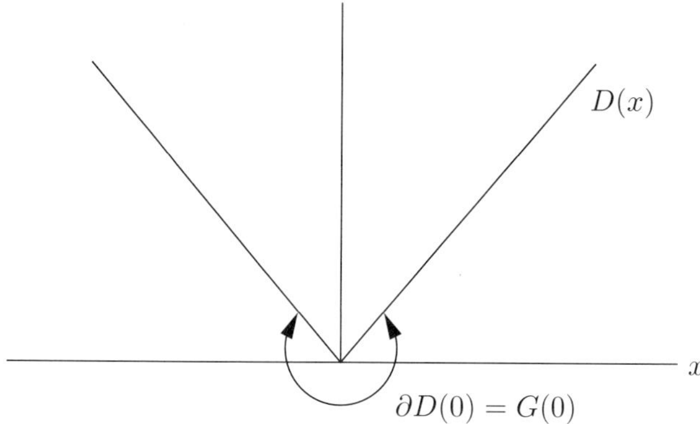

Figure 4.6: The support function D corresponding to the map G in Figure 4.4

The connection between the support function D and the maximal responsive map is then made via Theorem 4.4.

Returning to the example shown in Figure 4.4, Figure 4.6 shows the support function corresponding to G, as well as the relationship

$$G = \partial D. \tag{4.39}$$

This equation identifies D as a *pseudopotential* for $\boldsymbol{\Sigma}$ (see [94, 95]).

The rich structure embodied in the theory of Section 4.1 therefore permits *three equivalent* formulations of the flow law in plasticity:

G maximal responsive, (I)
$\boldsymbol{\Sigma} \in G(\dot{\boldsymbol{P}})$

$$\Updownarrow$$

D convex, positively homogeneous, l.s.c.,
$D(\dot{\boldsymbol{P}}) \geq 0, \quad D(\mathbf{0}) = 0,$ (II)
$\boldsymbol{\Sigma} \in \partial D(\dot{\boldsymbol{P}})$

$$\Updownarrow$$

K closed, convex, contains $\mathbf{0}$,
$D^* =$ indicator function of K, (III)
$\dot{\boldsymbol{P}} \in \partial D^*(\boldsymbol{\Sigma}) = N_K(\boldsymbol{\Sigma}).$

Here $G = \partial D$.

The formulation (III) is well known, and goes back to Moreau [94]. Formulation (II) is sometimes mentioned as a consequence of (III). Formulation

(I) was presented for the first time in [38], though it has some connection with the works of Rice [112] and Hill [59].

These three formulations show clearly the *minimal* assumptions that need to be made if an acceptable classical theory of plasticity is to emerge. In particular, we see that (I) and (II) do not require the assumption of an elastic region and a yield surface: These are *consequences.*

Practical considerations would dictate which of these formulations would be most appropriate for the problem at hand. For example, (III) is the most often used, and in fact has been the goal of the theory developed in Chapter 3. Formulation (II) has been used in [19, 36, 84, 86]; we will see later that it leads to a variational formulation of the initial–boundary value problem, which can be regarded as the natural extension of the corresponding displacement problem for linear elasticity. Formulation (I) has limitations in that it is not simple or natural to formulate evolution equations in this form, except perhaps for problems posed in one dimension. The major benefit of (I), though, is that it resolves the issue of how much information needs to be added to the assumption of the maximum plastic work inequality in order for this inequality to form the basis of an internal variable theory of plasticity.

Polar functions: the relationship between the yield and dissipation functions. We turn now to the discussion of polar functions in Section 4.1 and to the consequences of polar relationships in plasticity. From (4.25) we see that it is possible to define a gauge g (the subscript K is omitted without any ambiguity) corresponding to which the region of admissible stresses may be expressed in the form

$$K = \{\boldsymbol{\Sigma} : \; g(\boldsymbol{\Sigma}) \leq c_0\}. \tag{4.40}$$

The function g thus defined is just one possible representation of the *yield function*, albeit an important one, in that it is the representation corresponding to which the yield function is a gauge. To distinguish this from other representations, we refer to the gauge g as the *canonical yield function.* It is defined by

$$g(\boldsymbol{\Sigma}) = \inf\{\mu > 0 : \; \boldsymbol{\Sigma} \in \mu K\}, \tag{4.41}$$

which also makes it clear that *every* yield surface can be represented by a gauge. We use distinct symbols here and henceforth to distinguish between an arbitrary representation (f) of a yield surface and its representation by means of the canonical function (g).

Assume that $D(\boldsymbol{\Lambda}) = 0$ if and only if $\boldsymbol{\Lambda} = \mathbf{0}$. We see from (4.27) that g and D are related by

$$g(\boldsymbol{\Sigma}) = \sup_{\mathbf{0} \neq \boldsymbol{\Lambda} \in \operatorname{dom}(D)} \frac{\boldsymbol{\Sigma} : \boldsymbol{\Lambda}}{D(\boldsymbol{\Lambda})}.$$

Now consider the case in which $\boldsymbol{\Sigma} \in \partial K$, the boundary of K; then

$$\sup_{\mathbf{0} \neq \boldsymbol{\Lambda} \in \operatorname{dom}(D)} \frac{\boldsymbol{\Sigma} : \boldsymbol{\Lambda}}{D(\boldsymbol{\Lambda})} = 1,$$

and the supremum is achieved when $\boldsymbol{\Lambda} = \dot{\boldsymbol{P}}$, where $\dot{\boldsymbol{P}}$ is conjugate to $\boldsymbol{\Sigma}$ in the sense of an equality in (4.37). Thus for $\boldsymbol{\Sigma} \in K \cap \partial D(\dot{\boldsymbol{P}})$ and $\dot{\boldsymbol{P}} \neq \mathbf{0}$,

$$\boldsymbol{\Sigma} : \dot{\boldsymbol{P}} = g(\boldsymbol{\Sigma}) D(\dot{\boldsymbol{P}}). \tag{4.42}$$

Hence, whereas D and D^* are conjugate in the Legendre–Fenchel sense, D and g are polar in the sense of (4.42).

With the concept of the canonical yield function and its properties at our disposal, it is now possible, with the aid of Lemma 4.5, to give a generalized version of the classical form (3.34) of the flow law.

LEMMA 4.7. *Let g be nonnegative and convex, with $g(\mathbf{0}) = 0$ and $\boldsymbol{\Sigma}$ a point in the interior of* dom (g) *such that $g(\boldsymbol{\Sigma}) > 0$. Set $K = \{\boldsymbol{T} : \ g(\boldsymbol{T}) \leq g(\boldsymbol{\Sigma})\}$. Then $\dot{\boldsymbol{P}} \in N_K(\boldsymbol{\Sigma})$ if and only if there exists $\lambda \geq 0$ such that*

$$\dot{\boldsymbol{P}} \in \lambda \, \partial g(\boldsymbol{\Sigma}).$$

Various results follow from the lemma. Firstly, the reduction to (3.34) is obvious in the case of smooth functions g. Secondly, it is possible to characterize the multiplier λ. Indeed, we have

$$\dot{\boldsymbol{P}} \in \lambda \partial g(\boldsymbol{\Sigma}) \Longleftrightarrow \lambda \, g(\boldsymbol{T}) \geq \lambda \, g(\boldsymbol{\Sigma}) + \dot{\boldsymbol{P}} : (\boldsymbol{T} - \boldsymbol{\Sigma}) \quad \forall \boldsymbol{T}.$$

By setting first $\boldsymbol{T} = \mathbf{0}$, then $\boldsymbol{T} = 2\boldsymbol{\Sigma}$, and by using the properties of g, we arrive at the identity

$$\lambda = D(\dot{\boldsymbol{P}}). \tag{4.43}$$

Thus the scalar multiplier has a simple interpretation as the dissipation associated with a particular internal variable rate.

Lemma 4.7 may also be applied to the dissipation function. Setting $g = D$ and defining

$$C = \{\boldsymbol{Q} : \ D(\boldsymbol{Q}) \leq D(\dot{\boldsymbol{P}})\}$$

for given $\dot{\boldsymbol{P}} \neq \mathbf{0}$, we have immediately, for $\boldsymbol{\Sigma}$ related to $\dot{\boldsymbol{P}}$ through (4.38),

$$\boldsymbol{\Sigma} \in N_C(\dot{\boldsymbol{P}}) \quad \text{or} \quad \boldsymbol{\Sigma} \in \lambda \, \partial D(\dot{\boldsymbol{P}})$$

for some $\lambda > 0$ (we exclude the possibility $\lambda = 0$, since $\boldsymbol{\Sigma} \neq \mathbf{0}$). The situation is illustrated in Figure 4.7: In X' the conjugate pair $(\boldsymbol{\Sigma}, \dot{\boldsymbol{P}})$ is such that $\dot{\boldsymbol{P}}$ lies in the normal cone to K (the level set $g(\boldsymbol{\Sigma}) \leq 1$) at $\boldsymbol{\Sigma}$, while in X we find that $\boldsymbol{\Sigma}$ lies in the normal cone to C (the level set $D(\boldsymbol{Q}) \leq D(\dot{\boldsymbol{P}})$ at $\dot{\boldsymbol{P}}$).

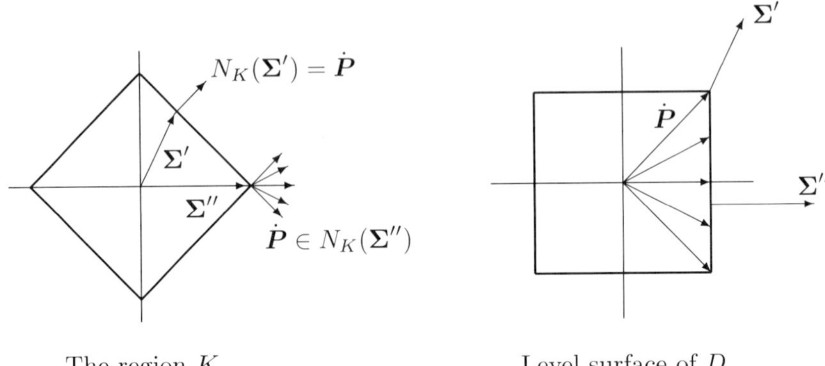

Figure 4.7: The relationship between the admissible region K and its support function D

We conclude this section with several concrete examples that illustrate the theory presented here.

EXAMPLE 4.8. COMBINED LINEAR KINEMATIC AND ISOTROPIC HARDENING. We continue the discussion of Example 3.1 in Section 3.2 for the case of coupled linear kinematic and isotropic hardening with an arbitrary yield function. For this case there are two internal variables, a symmetric tensor $\boldsymbol{\alpha}$ corresponding to the back-stress in kinematic hardening and a scalar γ that determines the expansion of the yield surface in isotropic hardening. Thus we set $\boldsymbol{\xi} = (\boldsymbol{\alpha}, \gamma) \in M^3 \times \mathbb{R}_+ = X$, while the conjugate force is denoted by $\boldsymbol{\chi} = (\boldsymbol{a}, g)$. Recall that M^3 is the set of all the symmetric matrices of order 3, and $\mathbb{R}_+$ denotes the half real line of nonnegative numbers. The part of the free energy function associated with hardening is now

$$\psi^p(\boldsymbol{\alpha}, \gamma) = \tfrac{1}{2} k_1 |\boldsymbol{\alpha}|^2 + \tfrac{1}{2} k_2 \gamma^2, \tag{4.44}$$

where k_1 and k_2 are nonnegative scalars associated with kinematic and isotropic hardening, respectively. As is shown in Example 3.1, the stress and conjugate forces are

$$\boldsymbol{\sigma} = \boldsymbol{C}\boldsymbol{e} = \boldsymbol{C}(\boldsymbol{\epsilon} - \boldsymbol{p}), \tag{4.45}$$

$$\boldsymbol{a} = -k_1 \boldsymbol{\alpha}, \tag{4.46}$$

$$g = -k_2 \gamma. \tag{4.47}$$

For the case of kinematic and isotropic hardening the yield function is, with $\boldsymbol{\Sigma} = (\boldsymbol{\sigma}, \boldsymbol{a}, g)$, of the form

$$\phi(\boldsymbol{\Sigma}) = \Phi(\boldsymbol{\sigma} + \boldsymbol{a}) + g - c_0 \leq 0. \tag{4.48}$$

From the preceding theory we see that it is always possible to express ϕ as a gauge g in $\boldsymbol{\Sigma}$.

With the yield function given by (4.48), the flow law (4.35) becomes

$$(\dot{\boldsymbol{p}}, \dot{\boldsymbol{\alpha}}, \dot{\gamma}) = \lambda(\boldsymbol{n}, \boldsymbol{n}, 1), \tag{4.49}$$

where $\boldsymbol{n} = \nabla\Phi(\boldsymbol{\sigma} + \boldsymbol{a})$. It is also seen that the kinematic hardening variable $\boldsymbol{\alpha}$ may be identified with $\boldsymbol{p}$, and the multiplier $\lambda \geq 0$ with the rate of change of the internal variable γ characterizing isotropic hardening. In particular, then, $\boldsymbol{\alpha} \in M_0^3$.

A simple example of a dissipation function is the function corresponding to the von Mises yield condition. For this case we have

$$\Phi(\boldsymbol{\sigma}) = |\boldsymbol{\sigma}^D| \equiv \sqrt{\sigma_{ij}^D \sigma_{ij}^D},$$

where $\boldsymbol{\sigma}^D = \boldsymbol{\sigma} - \frac{1}{3}(\operatorname{tr}\boldsymbol{\sigma})\boldsymbol{I}$ is the deviatoric part of $\boldsymbol{\sigma}$. The tensor $\boldsymbol{n} = (\boldsymbol{\sigma}^D + \boldsymbol{a}^D)/|\boldsymbol{\sigma}^D + \boldsymbol{a}^D|$ on the right-hand side of (4.49) is a unit tensor. It follows that $\dot{\gamma} = |\dot{\boldsymbol{p}}|$, and so γ can be interpreted as the equivalent plastic strain. It is now convenient to identify the kinematic hardening variable $\boldsymbol{\alpha}$ with $\boldsymbol{p}$; then the Helmholtz free energy function may be expressed in the form

$$\psi(\boldsymbol{e}, \boldsymbol{p}, \gamma) = \tfrac{1}{2}\,\boldsymbol{e} : \boldsymbol{C}\,\boldsymbol{e} + \tfrac{1}{2}\,k_1|\boldsymbol{p}|^2 + \tfrac{1}{2}\,k_2\gamma^2.$$

We can rewrite the flow law (4.49) in the equivalent form

$$(\dot{\boldsymbol{p}}, \dot{\gamma}) = \lambda(\boldsymbol{n}, 1),$$

or

$$(\dot{\boldsymbol{p}}, \dot{\gamma}) \in N_K(\tilde{\boldsymbol{a}}, g),$$

where

$$\tilde{\boldsymbol{a}} = \boldsymbol{\sigma} + \boldsymbol{a} = \boldsymbol{\sigma} - k_1\boldsymbol{p}$$

and

$$K = \{(\tilde{\boldsymbol{a}}, g) : \ |\tilde{\boldsymbol{a}}^D| + g - c_0 \leq 0, \ g \leq 0\}.$$

Equivalently, the flow law can be rewritten as

$$(\tilde{\boldsymbol{a}}, g) \in \partial D(\dot{\boldsymbol{p}}, \dot{\gamma}). \tag{4.50}$$

Using the definition (4.37), the dissipation function can be computed, for $(\boldsymbol{q}, \mu) \in M_0^3 \times \mathbb{R}_+$, according to

$$
\begin{aligned}
D(\boldsymbol{q}, \mu) &= \sup \left\{ \tilde{\boldsymbol{a}} : \boldsymbol{q} + g\,\mu : \ |\tilde{\boldsymbol{a}}^D| + g \le c_0, \ g \le 0 \right\} \\
&= \sup \left\{ \tilde{\boldsymbol{a}}^D : \boldsymbol{q} + g\,\mu : \ |\tilde{\boldsymbol{a}}^D| + g \le c_0, \ g \le 0 \right\} \\
&= \sup \left\{ |\tilde{\boldsymbol{a}}^D|\,|\boldsymbol{q}| + g\,\mu : \ |\tilde{\boldsymbol{a}}^D| + g \le c_0, \ g \le 0 \right\} \\
&= \sup \left\{ (c_0 - g)\,|\boldsymbol{q}| + g\,\mu : \ g \le 0 \right\} \\
&= \sup \left\{ c_0\,|\boldsymbol{q}| + g\,(\mu - |\boldsymbol{q}|) : \ g \le 0 \right\}.
\end{aligned}
$$

Therefore, for $(\boldsymbol{q}, \mu) \in M_0^3 \times \mathbb{R}_+$,

$$
D(\boldsymbol{q}, \mu) = \begin{cases} c_0\,|\boldsymbol{q}| & \text{if} \quad |\boldsymbol{q}| \le \mu, \\ +\infty & \text{if} \quad |\boldsymbol{q}| > \mu. \end{cases} \tag{4.51}
$$

We observe also that (4.50) can be rewritten in the form

$$
\begin{gathered}
D(\boldsymbol{q}, \mu) \ge D(\dot{\boldsymbol{p}}, \dot{\gamma}) + (\boldsymbol{\sigma} - k_1 \boldsymbol{p}) : (\boldsymbol{q} - \dot{\boldsymbol{p}}) - k_2 \gamma\,(\mu - \dot{\gamma}) \\
\forall\, \boldsymbol{q} \in M_0^3, \ \mu \in \mathbb{R}.
\end{gathered} \tag{4.52}
$$

□

EXAMPLE 4.9. LINEAR KINEMATIC HARDENING. We now consider a special case of the previous example. In the relations of Example 4.8, we let $k_2 = 0$ and drop the variables γ and g. As a result, we obtain corresponding relations for the elastoplastic deformation of a material undergoing the linear kinematic hardening with the von Mises yield condition. We need only one internal variable to describe the evolution of the yield surface in the kinematic hardening, and this variable is identified with the plastic strain $\boldsymbol{p}$. The plastic flow law is

$$
\dot{\boldsymbol{p}} \in N_K(\boldsymbol{\sigma} - k_1 \boldsymbol{p}),
$$

where

$$
K = \{ \tilde{\boldsymbol{a}} : \ |\tilde{\boldsymbol{a}}^D| - c_0 \le 0 \}.
$$

The plastic flow law can be equivalently written as

$$
\boldsymbol{\sigma} - k_1 \boldsymbol{p} \in \partial D(\dot{\boldsymbol{p}}),
$$

where the dissipation function

$$
D(\boldsymbol{q}) = c_0\,|\boldsymbol{q}|
$$

is now finite and Lipschitz continuous. □

EXAMPLE 4.10. BENDING AND EXTENSION OF A BEAM. The theory presented earlier applies in general situations and is not dictated by any physical assumptions other than those embodied in the properties possessed by

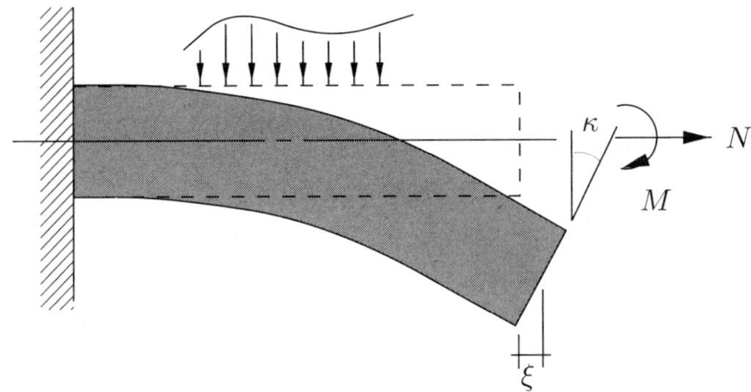

Figure 4.8: Generalized stresses and plastic strains acting on a rigid-plastic beam

the sets and functions appearing there. In particular, while our prime motivation has been evolution laws for continuous media, there is no restriction in applying the results to more specialized situations, such as those arising from theories for particular structural types such as beams, plates, and shells.

Consider, for example, a rigid-perfectly-plastic beam of arbitrary cross-section, subject to bending and extension. The Kirchhoff assumption is imposed; that is, sections initially plane and normal to the axis of the beam remain plane and normal after deformation. The two generalized plastic strains are the axial extension ξ at the centroidal axis and the cross-sectional rotation κ (see Figure 4.8), so we write $\boldsymbol{P} = (\xi, \kappa)$. Under these circumstances it is a straightforward matter to show that the region K of admissible forces consists of those values of bending moment M and axial force N satisfying

$$\pm aM \geq b^2 N^2 - 1, \tag{4.53}$$

where a and b are constants depending on the cross-sectional geometry and yield stress (see Figure 4.9(a)). The forces conjugate to κ and ξ are thus M and N. According to (4.40) it is possible to express the region K as a level set $\{\boldsymbol{\Sigma} : \ g(\boldsymbol{\Sigma}) \leq 1\}$, where g is a gauge and $\boldsymbol{\Sigma} = (M, N)$. To do this we rewrite (4.53) as

$$\pm aM + 1 \geq b^2 N^2 \tag{4.54}$$

and complete the square to get

$$(1 \pm aM)^2 \geq b^2 N^2 + \frac{a^2}{4} M^2, \tag{4.55}$$

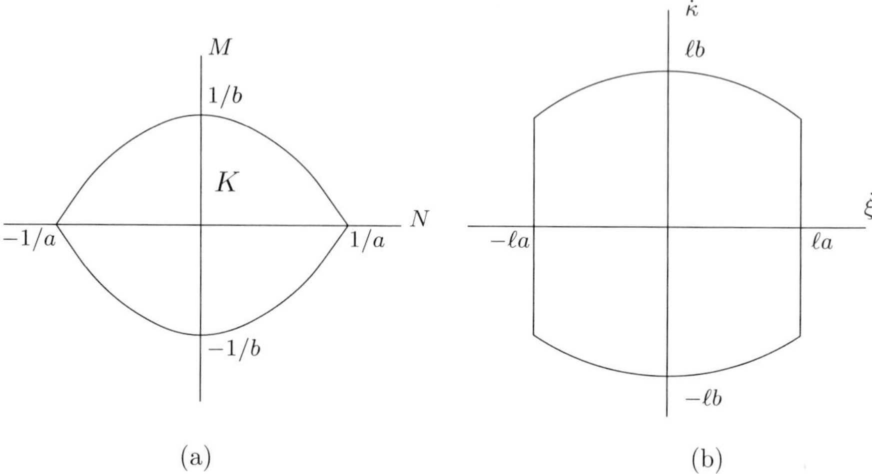

Figure 4.9: The yield surface and ℓ-level set of the dissipation function, for the rigid-plastic beam

whence the canonical yield function is

$$g(\boldsymbol{\Sigma}) \equiv \mp\frac{a}{2}M + \left(b^2N^2 + \frac{a^2}{4}M^2\right)^{1/2} \leq 1. \tag{4.56}$$

The dissipation function is then found from

$$\begin{aligned} D(\dot{\boldsymbol{P}}) &= \sup_{\boldsymbol{\Sigma}\in K} \boldsymbol{\Sigma} : \dot{\boldsymbol{P}} \\ &= \sup_{g(M,N)\leq 1} (M\dot{P}_1 + N\dot{P}_2) \\ &= \sup_{g(M,N)=1} (M\dot{P}_1 + N\dot{P}_2). \end{aligned}$$

Consider any $\dot{\boldsymbol{P}}$ satisfying $\dot{\boldsymbol{P}} \in \lambda\partial g(\boldsymbol{\Sigma})$ for $\boldsymbol{\Sigma}$ such that g is differentiable, that is, for the set

$$\left\{(M,N) : \, |N| < \frac{1}{b}, \; M = \pm\frac{1}{a}(1 - b^2N^2)\right\}.$$

For these values we have $\dot{\boldsymbol{P}} = \lambda\,\partial g/\partial\boldsymbol{\Sigma}$, from which we find, after some manipulation, that

$$D(\dot{\kappa}, \dot{\xi}) = \frac{a}{4b}\frac{\dot{\xi}^2}{\dot{\kappa}} + \frac{1}{a}\dot{\kappa}.$$

When $g(\mathbf{\Sigma}) = 1$, $M = 0$, and $N = \pm 1/b$, g is not differentiable. At these points

$$D(\dot{\kappa}, \dot{\xi}) = \pm \frac{1}{b} \dot{\xi}.$$

The level set $\{\boldsymbol{Q} : D(\boldsymbol{Q}) \leq D(\dot{\boldsymbol{P}})\}$ is shown in Figure 4.9(b). □

Part II

The Variational Problems of Elastoplasticity

5

Results from Functional Analysis and Function Spaces

The focus of Part II of this monograph will be, firstly, on the construction of variational formulations of the initial–boundary value problem of elastoplasticity, and, secondly, on the well-posedness of these variational problems. There are a number of tools from functional analysis that are called upon in the course of such analyses, and naturally the variational problems themselves are posed on particular function spaces. For these reasons we begin Part II by reviewing, in this chapter, those results from functional analysis that are pertinent to subsequent developments. We also collect in one place a number of results pertaining to function spaces, especially Sobolev spaces.

The overviews are not intended to be comprehensive, and full details may be found in monographs devoted to functional analysis and function spaces. The text [106] by Reddy may be consulted for an introduction to functional analysis that is aimed at those interested in variational problems and finite elements. Extended summary accounts of the relevant subject matter may also be found in Zeidler [128, 131, 132] and in Dautray and Lions [30]. There exist a number of accessible accounts of function spaces and, in particular, the theory of Sobolev spaces, some examples of which are Adams [1], Dautray and Lions [30], Reddy [106], Renardy and Rogers [109], and Zeidler [129, 131].

The reader who is familiar with the basics of functional analysis, Sobolev spaces, variational formulations of boundary value problems, and variational inequalities may skip Chapters 5 and 6 and go on directly to Chapter 7.

5.1 Results from Functional Analysis

We assume that the notion of a vector, or linear, space and the standard properties of vector spaces are familiar to the reader. All vector spaces are assumed to be defined over the field of real numbers.

Normed spaces and Banach spaces. Let V be a vector space. A *seminorm* on V is a map $|\cdot| : V \to \mathbb{R}^+$ that is subadditive and positively homogeneous; that is,

$$|u+v| \le |u| + |v|, \quad |\alpha v| = |\alpha|\,|v| \quad \forall\, u, v \in V,\ \forall\, \alpha \in \mathbb{R}. \tag{5.1}$$

It can be shown that the properties (5.1) imply that $|0| = 0$ and $|v| \ge 0$ for all $v \in V$.

A *norm* $\|\cdot\|$ on V is a seminorm that has the additional property of positive definiteness:

$$\|v\| = 0 \ \text{ iff } \ v = 0. \tag{5.2}$$

If $\|\cdot\|$ is a norm on V, then the pair $(V, \|\cdot\|)$ is called a *normed space*. Usually, the norm $\|\cdot\|$ defined over the space V is conventional or is clear from the context, and we simply denote the normed space by V. The notion of norm is a generalization of the absolute value for real numbers. The quantity $\|v\|$ is used to measure the length of a vector $v \in V$, while $\|u - v\|$ is used to measure the distance between two vectors u and v in V.

Two norms $\|\cdot\|$ and $|\!|\!|\cdot|\!|\!|$ on a normed space V are said to be *equivalent* if there are positive constants c_1 and c_2 such that

$$c_1\|v\| \le |\!|\!|v|\!|\!| \le c_2\|v\| \quad \forall\, v \in V. \tag{5.3}$$

It is a well-known result that on a finite-dimensional space, any two norms are equivalent. On the other hand, an infinite-dimensional space can be endowed with different norms that are not equivalent. For instance, both $\|v\|^{(1)} = \max_{0\le x\le 1} |v(x)|$ and $\|v\|^{(2)} = \int_0^1 |v(x)|\,dx$ are norms on the space of continuous functions $C([0,1])$, but these norms are not equivalent. In the study of boundary value problems, it is sometimes more convenient to use a norm different from, yet equivalent to, the conventional norm of the function space for the problem.

A sequence $\{v_n\}$ in a normed space V *converges* (strongly) to $v \in V$ if and only if $\lim_{n\to\infty} \|v_n - v\| = 0$. When this is the case, we write $v_n \to v$ and say that v is the (strong) limit of the sequence $\{v_n\}$. If $\|\cdot\|$ and $|\!|\!|\cdot|\!|\!|$ are equivalent norms on V, then a sequence $\{v_n\} \subset V$ converges to v in the norm $\|\cdot\|$ if and only if it converges to v in $|\!|\!|\cdot|\!|\!|$.

Let A be a subset of a normed space V. The set A is said to be *closed* in V if and only if $v_n \in A$ and $v_n \to v$ imply that $v \in A$. The *closure* $\overline{A}$ of A is the smallest closed set in V containing A. Loosely speaking, the closure $\overline{A}$ is obtained from the set A by adding the "boundary points" of A. The set A is *dense* in V if for every $v \in V$ there exists a sequence $\{v_n\}$ in A such that $v_n \to v$. A canonical example is that the set of rational numbers is dense in the space of the real numbers, with the absolute value as the norm. Finally, A is said to be *bounded* if for some constant M, $\|v\| \le M$ for every $v \in A$.

Closely related to the study of the convergence of sequences is the notion of a Cauchy sequence. A *Cauchy sequence* $\{v_n\}_{n=1}^{\infty}$ in V is a sequence that has the property that for any $\epsilon > 0$ there exists a number $N(\epsilon)$ such that $\|v_n - v_m\| < \epsilon$ for all $n, m > N(\epsilon)$. Certainly, all convergent sequences are Cauchy sequences, though the converse is not true. A subset A of a normed space V is *complete* if and only if every Cauchy sequence in A has a strong limit in A.

A complete normed space is called a *Banach space*. Hence a Banach space is characterized by the property that any Cauchy sequence converges in the space. The following statement establishes a relationship between completeness and closedness in normed spaces.

PROPOSITION 5.1. *A subset of a Banach space is complete if and only if it is closed.*

Linear operators and linear functionals. Let V and W be vector spaces. A map $L : V \to W$ is also called an *operator*. The operator L is *linear* from V to W if it is additive and homogeneous, that is, if

$$\begin{aligned} L(u+v) &= L(u) + L(v), \\ L(\alpha v) &= \alpha L(v), \end{aligned}$$

for all $u, v \in V$ and $\alpha \in \mathbb{R}$. For a linear operator L, we often write $L(v)$ as Lv. A linear operator is called a *linear functional* if $W = \mathbb{R}$.

The *range* $\mathcal{R}(L)$ and *kernel*, or *null, space* $\mathcal{N}(L)$ of L are subspaces of W and V, defined respectively by

$$\begin{aligned} \mathcal{R}(L) &= \{w \in W : \; w = L(v) \text{ for some } v \in V\}, \\ \mathcal{N}(L) &= \{v \in V : \; L(v) = 0\}. \end{aligned}$$

The range $\mathcal{R}(L)$ is the set of the images under the mapping L, while the null space $\mathcal{N}(L)$ consists of the solutions of the equation $L(v) = 0$. Obviously, $0 \in \mathcal{N}(L)$.

A special operator worth mentioning explicitly is the *projection operator* $P : V \to V$, from a vector space V into itself, which is defined to have the property

$$P^2 = P, \quad \text{or} \quad P^2 v = Pv, \quad \forall\, v \in V.$$

A simple example is the (orthogonal) projection onto the closed ball $\overline{B}(\mathbf{0}, r) = \{\boldsymbol{x} \in \mathbb{R}^d : \; \|\boldsymbol{x}\| \le r\}$ in the Euclidean space $\mathbb{R}^d$:

$$P(\boldsymbol{x}) = \begin{cases} \boldsymbol{x} & \text{if } \boldsymbol{x} \in \overline{B}(\mathbf{0}, r), \\ r\, \boldsymbol{x}/\|\boldsymbol{x}\| & \text{otherwise.} \end{cases}$$

If V and W are normed spaces and L is a map from V into W, then L is said to be *continuous* if $v_n \to v$ in V implies that $L(v_n) \to L(v)$ in W. Furthermore, the map L is said to be *bounded* if for any $r > 0$, there is a constant $R \geq 0$ such that

$$\|L(v)\| \leq R \quad \forall\, v \in V,\ \|v\| \leq r.$$

When L is a linear operator, the boundedness of L is characterized by the existence of a constant $M \geq 0$ such that

$$\|L(v)\| \leq M\|v\| \quad \forall\, v \in V. \tag{5.4}$$

The properties of continuity and boundedness are equivalent in the case of linear operators: *A linear operator is continuous if and only if it is bounded.*

An operator L from V to W is said to be *Lipschitz continuous* if there exists a constant $c > 0$ such that

$$\|L(v_1) - L(v_2)\| \leq c\,\|v_1 - v_2\| \quad \forall\, v_1, v_2 \in V.$$

Lipschitz continuous operators are continuous, but the converse is not true in general. On the other hand, a linear operator is Lipschitz continuous if and only if it is continuous.

Bilinear forms. Let V and W be vector spaces. A map $b : V \times W \to \mathbb{R}$ is called a *bilinear form* if it is linear in each slot, that is, for any $v_1, v_2, v \in V$, $w_1, w_2, w \in W$, and $\alpha, \beta \in \mathbb{R}$,

$$\begin{aligned} b(\alpha v_1 + \beta v_2, w) &= \alpha\, b(v_1, w) + \beta\, b(v_2, w), \\ b(v, \alpha w_1 + \beta w_2) &= \alpha\, b(v, w_1) + \beta\, b(v, w_2). \end{aligned}$$

A bilinear form $b : V \times W \to \mathbb{R}$ is *continuous* (or *bounded*) if there exists a constant $M > 0$ such that

$$b(v, w) \leq M\|v\|_V \|w\|_W \quad \forall\, v \in V,\ w \in W. \tag{5.5}$$

For the case in which $W = V$, we say that the bilinear form is *symmetric* if

$$b(v_1, v_2) = b(v_2, v_1) \quad \forall\, v_1, v_2 \in V, \tag{5.6}$$

and V-*elliptic* if there exists a constant $\alpha > 0$ such that

$$b(v, v) \geq \alpha\|v\|^2 \quad \forall\, v \in V. \tag{5.7}$$

Isomorphisms; completions. A linear continuous map L from V to W is an isomorphism if and only if L is both *injective*, that is, one-to-one, and *surjective*, that is, $\mathcal{R}(L) = W$.

If V is a normed space, then its *completion* is a Banach space $\hat{V}$, which has the property that there exists an isomorphism from V onto a dense subspace of $\hat{V}$. A standard example is that the completion of the continuous function space $C([0,1])$ in the norm $\|v\|_0 = (\int_0^1 |v(x)|^2 dx)^{1/2}$ is the Lebesgue space $L^2(0,1)$.

The space $\mathcal{L}(V,W)$; dual space. Let V and W be normed spaces. We denote by $\mathcal{L}(V,W)$ the space of all bounded linear operators from V to W. For $L \in \mathcal{L}(V,W)$, from (5.4) the quantity

$$\|L\| = \sup_{0 \neq v \in V} \frac{\|Lv\|}{\|v\|} = \sup_{\|v\| \leq 1} \|Lv\| \tag{5.8}$$

is well-defined; furthermore, it can be shown that $\|L\|$ thus defined is a norm on $\mathcal{L}(V,W)$. The space $\mathcal{L}(V,W)$ endowed with the norm (5.8) is a Banach space if W is a Banach space.

The space $\mathcal{L}(V,\mathbb{R})$ of bounded linear functionals on V is known as the *dual space* of V and is denoted by V'. Clearly, then, since $\mathbb{R}$ is complete, V' is a Banach space with the norm

$$\|L\| = \sup_{\|v\| \leq 1} |Lv|. \tag{5.9}$$

Often, we will use ℓ for a bounded linear functional on a normed space V and denote the action of ℓ on a member $v \in V$ by $\langle \ell, v \rangle$ rather than $\ell(v)$. Here, $\langle \cdot, \cdot \rangle$ is the duality pairing between V' and V. In Section 5.2 we will see examples of duality in the context of the function space $L^p(\Omega)$.

Monotone and strongly monotone operators. The notion of monotonicity is an extension of the concept of an increasing function of one real variable. An operator $T : V \to V'$ is said to be monotone if

$$\langle T(u) - T(v), u - v \rangle \geq 0 \quad \forall\, u, v \in V.$$

Furthermore, if there exists a constant $c > 0$ such that

$$\langle T(u) - T(v), u - v \rangle \geq c\, \|u - v\|^2 \quad \forall\, u, v \in V,$$

then T is called a *strongly monotone* operator.

Biduals and reflexivity. The dual V' of a normed space V is a Banach space, and V' itself also has a dual $V'' \equiv (V')'$, called the *bidual* of V. The bidual is, of course, a Banach space. It is possible to show that there exists a bounded linear map J from V to its bidual that is *one-to-one*, and that furthermore is an *isometry*: $\|Jv\| = \|v\|$ for all $v \in V$. Thus it is possible to *identify* V with a subspace $J(V)$ of V''. The normed space V is said to be *reflexive* if we in fact have

$$J(V) = V''. \tag{5.10}$$

We then write loosely $V'' = V$ to indicate the identification between V and its bidual. Obviously, a reflexive normed space must be a Banach space.

A canonical example of a reflexive Banach space is $L^p(\Omega)$ for $p \in (1, \infty)$, in which $\Omega \subset \mathbb{R}^d$ is an open set. The spaces $L^1(\Omega)$ and $L^\infty(\Omega)$, on the other hand, are not reflexive, as will be seen in the next section.

Weak and weak* convergence. In modern analysis, the typical strategy employed in showing that a minimization problem has a solution involves several steps. Under appropriate assumptions one chooses a minimizing sequence, shows that the minimizing sequence is bounded, extracts from the bounded minimizing sequence a subsequence that converges in some sense, and finally proves that the limit is a minimizer. Now, over a finite-dimensional space, any bounded sequence contains a convergent subsequence. But infinite-dimensional spaces do not enjoy this property; for example, the sequence

$$\{1, \sin \pi x, \cos \pi x, \ldots, \sin n\pi x, \cos n\pi x, \ldots\}$$

is bounded in the space $L^2(0,1)$ with the norm $\|v\| = (\int_0^1 |v(x)|^2 dx)^{1/2}$ and yet has no convergent subsequence. Fortunately, for many commonly used spaces, and in particular for reflexive Banach spaces, any bounded sequence does contain what is known as a *weakly* convergent subsequence. Thus, by the device of weak convergence we are able to make use of the standard strategy when studying minimization problems in infinite-dimensional spaces.

Let V be a normed space and V' its dual. A sequence $\{v_n\}$ in V is said to converge *weakly* in V to v if

$$\lim_{n \to \infty} \langle \ell, v_n \rangle = \langle \ell, v \rangle \quad \forall\, \ell \in V'. \tag{5.11}$$

The notation

$$v_n \rightharpoonup v$$

is used to indicate weak convergence. Strong (norm) convergence implies weak convergence, but the converse does not hold, with the exception of *finite-dimensional spaces*, for which the two forms of convergence coincide.

From the theory of Fourier series, we know that for any $v \in (L^2(0,1))' = L^2(0,1)$,

$$\langle v(x), \sin n\pi x \rangle = \int_0^1 v(x) \sin n\pi x \, dx \to 0,$$
$$\langle v(x), \cos n\pi x \rangle = \int_0^1 v(x) \cos n\pi x \, dx \to 0,$$

as $n \to \infty$. Therefore, the bounded sequence

$$\{1, \sin \pi x, \cos \pi x, \ldots, \sin n\pi x, \cos n\pi x, \ldots\}$$

converges weakly to 0 in the space $L^2(0,1)$.

Let V be a normed space and V' its dual. A sequence $\{\ell_n\}$ in V' is said to converge *weakly** in V' to ℓ if

$$\lim_{n\to\infty} \langle \ell_n, v\rangle = \langle \ell, v\rangle \quad \forall\, v \in V. \tag{5.12}$$

Weak* convergence is denoted by

$$\ell_n \overset{*}{\rightharpoonup} \ell.$$

We note that $\{\ell_n\}$ *converges weakly in* V' to ℓ if and only if

$$\lim_{n\to\infty} \langle \ell_n, v\rangle = \langle \ell, v\rangle \quad \forall\, v \in V''.$$

Thus weak* is a novel concept only if V is not reflexive.

Compactness and weak compactness. A subset V_1 of a normed space V is said to be (sequentially) *compact* if every bounded sequence in V_1 has a subsequence that converges in V_1. Likewise, V_1 is *weakly compact* if every bounded sequence in V_1 has a subsequence that converges weakly in V_1.

A linear operator $L : V \to W$ is said to be *compact* if the image under L of a bounded sequence in V contains a subsequence converging in W; that is, if $\{v_n\} \subset V$ is bounded, then there exists a subsequence $\{v_{n_j}\}$ and $w \in W$ such that $Lv_{n_j} \to w$ in W. If in the above definition convergence is changed from strong to weak, $Lv_{n_j} \rightharpoonup w$ in W, then L is said to be *weakly compact*. Evidently, L is compact if and only if it maps bounded sets to compact sets.

The following result will be important in later developments.

THEOREM 5.2 (EBERLEIN–SMULYAN). *A reflexive Banach space V is weakly compact; that is, if $\{v_n\}$ is a bounded sequence in V, then it is possible to extract from $\{v_n\}$ a subsequence that converges weakly in V. If, furthermore, the limit v is independent of the subsequence extracted, then the whole sequence $\{v_n\}$ converges weakly to v.*

Embeddings. Embedding results are especially important when we compare Sobolev spaces with different indices; details of Sobolev spaces are given in the next section. Let V and W be normed spaces with $V \subset W$. If there is a constant $c > 0$ such that

$$\|v\|_W \le c\,\|v\|_V \quad \forall\, v \in V, \tag{5.13}$$

we say V *is continuously embedded in* W, and write

$$V \hookrightarrow W.$$

This property can be interpreted in various ways; for example, (5.13) states that the identity operator $I : V \to W$ is bounded, or equivalently, continuous. Thus the continuous embedding of V in W implies also that if $v_n \to v$ in V, then $v_n \to v$ in W.

The subspace V is said to be *compactly embedded in* W if

$$v_n \rightharpoonup v \text{ in } V \text{ implies that } v_n \to v \text{ in } W.$$

This property is expressed in the form

$$V \hookrightarrow\hookrightarrow W,$$

and is equivalent to the statement that the identity operator from V into W is compact.

Dual operators. The generalization to normed spaces of the notion of the transpose of a matrix has many applications in functional analysis. To carry out such a generalization we begin with normed spaces V and W and their duals V' and W'. Let A be a linear operator with domain $\mathcal{D}(A) \subset V$ and range in W. Given $w' \in W'$ we pose the question, under what conditions does there exist $v' \in V'$ such that

$$\langle w', Av \rangle = \langle v', v \rangle \quad \forall v \in \mathcal{D}(A)? \tag{5.14}$$

It can be shown that a necessary and sufficient condition for (5.14) to hold is that $\mathcal{D}(A)$ be dense in V; when this is the case, v' is determined uniquely by w'. When $\mathcal{D}(A)$ is the whole space V, then this procedure defines a linear operator A' from W' to V' such that $A'w' = v'$. The operator A' is called the *dual* operator of A, and we may write

$$\langle w', Av \rangle = \langle A'w', v \rangle \quad \forall v \in V,\ w' \in W'.$$

If $\mathcal{D}(A) = V$ and A is bounded, then A' is also bounded, and

$$\|A'\| = \|A\|.$$

The following important theorem records further relationships between a bounded linear operator and its dual.

THEOREM 5.3 (CLOSED RANGE THEOREM). *Let V and W be two Banach spaces, and let A be a bounded linear operator from V to W with dual operator A'. Then the following statements are equivalent:*

(a) *$\mathcal{R}(A)$ is closed in W.*

(b) *$\mathcal{R}(A')$ is closed in V'.*

(c) $\mathcal{R}(A) = [\operatorname{Ker} A']^\circ \equiv \{w \in W : \ \langle \ell, w \rangle = 0 \ \ \forall \ell \in \operatorname{Ker} A'\}$.

(d) $\mathcal{R}(A') = [\operatorname{Ker} A]^\circ \equiv \{\ell \in V' : \ \langle \ell, v \rangle = 0 \ \ \forall v \in \operatorname{Ker} A\}$.

The next result follows directly from the closed range theorem.

COROLLARY 5.4. *The following results hold:*

$$\begin{aligned} \mathcal{R}(A) = W \ &\text{ iff } \|A'\ell\| \ge c_1 \|\ell\| \ \ \forall \ell \in W', \\ \mathcal{R}(A') = V' \ &\text{ iff } \|Av\| \ge c_2 \|v\| \ \ \forall v \in V. \end{aligned}$$

The Babuška–Brezzi condition for bilinear forms [4, 17]. Suppose that $b : V \times W \to \mathbb{R}$ is a continuous bilinear form; that is,

$$|b(v,w)| \le \alpha_b \|v\|_V \|w\|_W \quad \forall\, v \in V,\ w \in W. \tag{5.15}$$

The bilinear form $b(\cdot,\cdot)$ is said to satisfy the Babuška–Brezzi condition if there exists a constant $\beta_b > 0$ such that

$$\sup_{0 \ne w \in W} \frac{|b(v,w)|}{\|w\|_W} \ge \beta_b \|v\|_V \quad \forall\, v \in V. \tag{5.16}$$

The following theorem relates the Babuška–Brezzi condition to the closed range theorem.

THEOREM 5.5. *Let $b : V \times W \to \mathbb{R}$ be a continuous bilinear form, and define bounded linear operators $B : V \to W'$ and $B' : W \to V'$ according to*

$$b(v,w) = \langle Bv, w\rangle = \langle B'w, v\rangle \quad \forall\, v \in V,\ w \in W.$$

Then the following are equivalent:

(a) *The bilinear form $b(\cdot,\cdot)$ satisfies the Babuška–Brezzi condition* (5.16).

(b) *The operator B is an isomorphism from* $(\operatorname{Ker} B)^\perp$ *onto W', where*

$$\operatorname{Ker} B = \{v \in V :\ b(v,w) = 0 \quad \forall\, w \in W\}.$$

(c) *The operator B' is an isomorphism from W onto* $(\operatorname{Ker} B)^\circ$, *where*

$$(\operatorname{Ker} B)^\circ = \{\ell \in V' :\ \langle \ell, v\rangle = 0 \ \ \forall\, v \in \operatorname{Ker} B\}.$$

The Babuška–Brezzi condition plays a crucial role in the analysis of mixed, or saddle-point, variational problems.

Inner products and Hilbert spaces. An inner product is a generalization of the ordinary vector scalar product in $\mathbb{R}^d$ to an arbitrary vector space. Let V be a vector space. An *inner product* on V is a symmetric bilinear form $(\cdot,\cdot) : V \times V \to \mathbb{R}$ that is also positive definite; that is, $(\cdot,\cdot)$ has the following properties:

$$\begin{aligned}
(u,v) &= (v,u) \quad \forall\, u,v \in V,\\
(\alpha u_1 + \beta u_2, v) &= \alpha(u_1,v) + \beta(u_2,v) \quad \forall\, u_1,u_2,v \in V,\ \alpha,\beta \in \mathbb{R},\\
(v,v) &\ge 0 \quad \forall\, v \in V, \quad \text{and} \quad (v,v) = 0 \Longleftrightarrow v = 0.
\end{aligned}$$

A space V endowed with an inner product $(\cdot,\cdot)$ is called an *inner product space*. When the definition of $(\cdot,\cdot)$ is clear, we will simply denote the inner product space by V.

Every inner product generates a norm according to

$$\|v\| = (v, v)^{1/2},$$

so that every inner product space is a normed space.

A *complete inner product space* is called a *Hilbert space*. Hence a Hilbert space is a Banach space whose norm is induced by an inner product.

The following theorems summarize well-known and fundamental properties of inner products and Hilbert spaces.

THEOREM 5.6 (CAUCHY–SCHWARZ INEQUALITY). *Let V be an inner product space. Then*

$$|(u, v)| \le \|u\| \, \|v\| \quad \forall\, u, v \in V.$$

THEOREM 5.7 (RIESZ REPRESENTATION THEOREM). *There exists an isometric isomorphism from a Hilbert space V onto its dual V'. More precisely, for any $\ell \in V'$ there exists a unique $u \in V$ such that*

$$\langle \ell, v \rangle = (u, v) \quad \forall\, v \in V.$$

Conversely, for any $u \in V$, the mapping $v \mapsto (u, v)$ determines an $\ell \in V'$. Furthermore, $\|\ell\| = \|u\|$.

By the Riesz representation theorem, we may identify a Hilbert space with its dual. Consequently the Hilbert space can also be identified with its bidual. Therefore, we have the following result.

COROLLARY 5.8. *Every Hilbert space is reflexive.*

Combining the results of Corollary 5.8 and Theorem 5.2, we see that in a Hilbert space, every bounded sequence has a weakly convergent subsequence.

On an inner product space V,

$$v_n \rightharpoonup v \Longrightarrow \|v\| \le \liminf_{n \to \infty} \|v_n\|.$$

In other words, in an inner product space, the norm function $\|\cdot\|$ is weakly l.s.c. This result is easily proved by noting that for a fixed $w \in V$, the mapping $v \mapsto (w, v)$ defines a linear continuous functional on V, and therefore

$$\|v\|^2 = (v, v) = \lim_{n \to \infty} (v, v_n) \le \|v\| \, \liminf_{n \to \infty} \|v_n\|.$$

We will use this result in Section 8.2.

The next well-known result is useful in proving the unique solvability of elliptic variational problems.

THEOREM 5.9 (LAX–MILGRAM LEMMA). *Let V be a Hilbert space, b :*

$V \times V \to \mathbb{R}$ a bilinear form that is both continuous and V-elliptic, and $\ell : V \to \mathbb{R}$ a bounded linear functional. Then the problem

$$b(u, v) = \langle \ell, v \rangle \quad \forall\, v \in V$$

has a unique solution $u \in V$, and for some constant $c > 0$ independent of ℓ,

$$\|u\| \le c\, \|\ell\|.$$

Later, in Section 8.2, we will use the following result from the theory of monotone operators.

THEOREM 5.10. *Assume that V is a Hilbert space and that $T : V \to V'$ is strongly monotone and Lipschitz continuous. Then for any $\ell \in V'$, the equation $T(u) = \ell$ has a unique solution $u \in V$.*

THEOREM 5.11 (PROJECTION THEOREM). *Let K be a nonempty closed convex subset in a Hilbert space V and let $u \in V$. Then there exists a unique element $u_0 \in K$ such that*

$$\|u - u_0\| = \inf_{v \in K} \|u - v\|.$$

The element u_0 is called the projection $P(u)$ of u on K and is characterized by the inequality

$$(u - u_0, v - u_0) \le 0 \quad \forall\, v \in V.$$

Using the inequality characterization of the projection, it is easy to verify that the projection operator is nonexpansive, that is,

$$\|P(u) - P(v)\| \le \|u - v\| \quad \forall\, u, v \in V,$$

and monotone, that is,

$$(P(u) - P(v), u - v) \ge 0 \quad \forall\, u, v \in V.$$

COROLLARY 5.12. *If K is a closed subspace of a Hilbert space V, then for any $u \in V$ there exists a unique element $u_0 \in K$ such that*

$$(u - u_0, v) = 0 \quad \forall\, v \in V.$$

The map $u \mapsto Pu = u_0$ is linear and defines an orthogonal projection onto K.

5.2 Function Spaces

We introduce in this section some function spaces that will be relevant to the subsequent developments in this monograph. The function spaces to be discussed include the spaces $C^m(\Omega)$ and $C^m(\overline{\Omega})$ of m-times continuously differentiable functions on Ω and $\overline{\Omega}$, the Lebesgue spaces $L^p(\Omega)$, the Sobolev spaces $W^{m,p}(\Omega)$, and their Hilbert space specializations $H^m(\Omega) = W^{m,2}(\Omega)$. The spaces will be defined on an open bounded domain $\Omega \subset \mathbb{R}^d$ that will be assumed to possess certain prescribed condition of smoothness. In order to give a proper treatment of time-dependent problems, we will later introduce vector-valued function spaces, which permit functions of space and time to be interpreted as maps from a time interval into a Banach or Hilbert space of functions.

5.2.1 *The Spaces $C^m(\Omega)$, $C^m(\overline{\Omega})$, and $L^p(\Omega)$*

Let Ω be a bounded domain in $\mathbb{R}^d$ ($d \leq 3$ for most applications). Before going on to discuss function spaces, we introduce the useful multi-index notation.

Multi-index notation. Let $\mathbb{Z}_+^d$ denote the set of all ordered d-tuples of nonnegative integers. A member of $\mathbb{Z}_+^d$ will usually be denoted by α or β, where, for example,

$$\alpha = (\alpha_1, \alpha_2, \dots, \alpha_d),$$

each component α_i being a nonnegative integer.

We denote by $|\alpha|$ the sum $|\alpha| = \alpha_1 + \alpha_2 + \cdots + \alpha_d$, called the length of α, and by $D^\alpha v$ the partial derivative

$$D^\alpha v = \frac{\partial^{|\alpha|} v}{\partial x_1^{\alpha_1} \, \partial x_2^{\alpha_2} \cdots \partial x_d^{\alpha_d}}.$$

Thus if $|\alpha| = m$, then $D^\alpha v$ will denote one of the mth partial derivatives of v. For example, $\alpha = (1, 0, 3)$ belongs to $\mathbb{Z}_+^3$, with $|\alpha| = \alpha_1 + \alpha_2 + \alpha_3 = 1 + 0 + 3 = 4$, and in this case the partial derivative $D^\alpha v$ is the fourth derivative defined by

$$D^\alpha v = \frac{\partial^4 v}{\partial x_1^{\alpha_1} \, \partial x_2^{\alpha_2} \, \partial x_3^{\alpha_3}} = \frac{\partial^4 v}{\partial x_1^1 \, \partial x_2^0 \, \partial x_3^3} = \frac{\partial^4 v}{\partial x_1 \, \partial x_3^3}.$$

Spaces of continuous and continuously differentiable functions. We denote by $C(\Omega)$ the space of all real-valued functions that are continuous on Ω. Since Ω is open, a function from the space $C(\Omega)$ is not necessarily bounded; consider, for example, the continuous function $v(x) = \ln x$ on $(0, 1)$. We denote further by $C(\overline{\Omega})$ the space of functions that are *bounded and uniformly continuous* on Ω. The notation $C(\overline{\Omega})$ is consistent with the fact that a bounded and uniformly continuous function on Ω has a unique continuous extension to $\overline{\Omega}$. The space $C(\overline{\Omega})$ is a *Banach space* with the norm

$$\|v\|_{C(\overline{\Omega})} = \sup\{|v(\boldsymbol{x})| : \boldsymbol{x} \in \Omega\} \equiv \max\{|v(\boldsymbol{x})| : \boldsymbol{x} \in \overline{\Omega}\}.$$

For any nonnegative integer m, $C^m(\Omega)$ is defined to be the space of functions that together with their derivatives of order less than or equal to m are continuous; that is,

$$C^m(\Omega) = \{v \in C(\Omega) : \ D^\alpha v \in C(\Omega) \text{ for } |\alpha| \le m\}.$$

We likewise set

$$C^m(\overline{\Omega}) = \{v \in C(\overline{\Omega}) : \ D^\alpha v \in C(\overline{\Omega}) \text{ for } |\alpha| \le m\}.$$

It is common practice to write $C(\Omega)$ and $C(\overline{\Omega})$ instead of $C^0(\Omega)$ and $C^0(\overline{\Omega})$. The space $C^m(\overline{\Omega})$ can be endowed with the seminorm

$$|v|_{C^m(\overline{\Omega})} = \sum_{|\alpha|=m} \|D^\alpha v\|_{C(\overline{\Omega})},$$

and it becomes a Banach space when endowed with the norm

$$\|v\|_{C^m(\overline{\Omega})} = \sum_{j=0}^{m} |v|_{C^j(\overline{\Omega})} = \sum_{|\alpha|\le m} \|D^\alpha v\|_{C(\overline{\Omega})}.$$

Finally, we set

$$C^\infty(\Omega) = \{v \in C(\Omega) : \ v \in C^m(\Omega) \ \ \forall\, m \in \mathbb{Z}_+\}$$

and

$$C^\infty(\overline{\Omega}) = \{v \in C(\overline{\Omega}) : \ v \in C^m(\overline{\Omega}) \ \ \forall\, m \in \mathbb{Z}_+\}.$$

These are spaces of infinitely differentiable functions.

Hölder spaces. A function v defined on Ω is said to be Lipschitz continuous if for some constant c,

$$|v(\boldsymbol{x}) - v(\boldsymbol{y})| \le c\,|\boldsymbol{x} - \boldsymbol{y}| \quad \forall\, \boldsymbol{x}, \boldsymbol{y} \in \Omega.$$

More generally, v is said to be a Hölder continuous function with exponent $\beta \in (0, 1]$ if for some constant c,

$$|v(\boldsymbol{x}) - v(\boldsymbol{y})| \le c\,|\boldsymbol{x} - \boldsymbol{y}|^\beta \quad \text{for } \boldsymbol{x}, \boldsymbol{y} \in \Omega.$$

We define $C^{0,\beta}(\overline{\Omega})$ to be the Hölder space of functions in $C(\overline{\Omega})$ that are Hölder continuous with the exponent β. With the norm

$$\|v\|_{C^{0,\beta}(\overline{\Omega})} = \|v\|_{C(\overline{\Omega})} + \sup\Big\{ \frac{|v(\boldsymbol{x}) - v(\boldsymbol{y})|}{|\boldsymbol{x} - \boldsymbol{y}|^\beta} : \ \boldsymbol{x}, \boldsymbol{y} \in \Omega, \ \boldsymbol{x} \ne \boldsymbol{y} \Big\},$$

the space $C^{0,\beta}(\overline{\Omega})$ becomes a Banach space.

For a nonnegative integer m and $\beta \in (0,1]$, we define the Hölder space

$$C^{m,\beta}(\overline{\Omega}) = \left\{ v \in C^m(\overline{\Omega}) : \ D^\alpha v \in C^{0,\beta}(\overline{\Omega}) \text{ for all } \alpha \text{ with } |\alpha| = m \right\},$$

which is a Banach space when it is endowed with the norm

$$\|v\|_{C^{m,\beta}(\overline{\Omega})} = \|v\|_{C^m(\overline{\Omega})} + \sum_{|\alpha|=m} \sup\Big\{ \frac{|D^\alpha v(\boldsymbol{x}) - D^\alpha v(\boldsymbol{y})|}{|\boldsymbol{x}-\boldsymbol{y}|^\beta} : \ \boldsymbol{x},\boldsymbol{y} \in \Omega, \ \boldsymbol{x} \neq \boldsymbol{y} \Big\}.$$

The spaces $L^p(\Omega)$. For any number $p \in [1,\infty)$, we denote by $L^p(\Omega)$ the space of (equivalence classes of) measurable functions v for which

$$\int_\Omega |v(\boldsymbol{x})|^p \, dx < \infty,$$

where integration is understood to be in the sense of Lebesgue. The space $L^p(\Omega)$ is a *Banach space* when endowed with the norm $\|\cdot\|_{0,p,\Omega}$ defined by

$$\|v\|_{0,p,\Omega} = \left(\int_\Omega |v(\boldsymbol{x})|^p \, dx \right)^{1/p}.$$

The reason for including the zero in the subscript of the notation $\|\cdot\|_{0,p,\Omega}$ will become clear when Sobolev spaces are introduced. When there is no danger of confusion, reference to the domain Ω will be omitted in the symbol for norms. It will also be convenient to write $\|\cdot\|_{0,\Omega}$ or even $\|\cdot\|_0$ for the norm on $L^2(\Omega)$ when this is unlikely to be ambiguous.

The quantity $\|\cdot\|_{0,p}$ is a norm only when it is understood that u represents an equivalence class of functions, two functions being equivalent if they are equal almost everywhere (a.e.), that is, equal everywhere except on a subset of Ω of Lebesgue measure zero.

The definition of the spaces $L^p(\Omega)$ can be extended to include the case $p = \infty$ in the following manner. We define the essential supremum (denoted by ess sup) of any measurable function v by

$$\operatorname{ess\,sup}_\Omega v = \inf\{M \in (-\infty,\infty] : \ v(\boldsymbol{x}) \leq M \text{ a.e. in } \Omega\}.$$

Then v is said to be *essentially bounded above* if $\operatorname{ess\,sup}_\Omega v < \infty$. A similar definition of essential infimum may be given, leading to the notion of a function that is essentially bounded below. We say that v is *essentially bounded* if both $\operatorname{ess\,sup}_\Omega v$ and $\operatorname{ess\,inf}_\Omega v$ are finite.

Then we may define

$$L^\infty(\Omega) = \{v : \ v \text{ is essentially bounded on } \Omega\}.$$

This space is a *Banach space* when endowed with the norm

$$\|v\|_{0,\infty,\Omega} = \operatorname{ess\,sup}_\Omega |v|.$$

Since all continuous functions on a bounded closed set are bounded, we have

$$C(\overline{\Omega}) \hookrightarrow L^\infty(\Omega).$$

The case $p = 2$ is special, in that $L^2(\Omega)$ is an inner product space (indeed, a Hilbert space) when endowed with the inner product

$$(u, v)_{0,\Omega} = \int_\Omega u(\boldsymbol{x})\, v(\boldsymbol{x})\, dx.$$

This inner product, in turn, generates the norm $\|\cdot\|_{0,2,\Omega}$.

Let v be a function defined on Ω. We say that $v \in L^p_{\text{loc}}(\Omega)$ if for any proper subset $\Omega' \subset\subset \Omega$, $v \in L^p(\Omega')$.

We summarize in the following theorem two important inequalities.

THEOREM 5.13.

(a) (THE MINKOWSKI INEQUALITY) *If $u, v \in L^p(\Omega)$, $1 \le p \le \infty$, then $u \pm v \in L^p(\Omega)$ and*

$$\|u \pm v\|_{0,p} \le \|u\|_{0,p} + \|v\|_{0,p}.$$

(b) (THE HÖLDER INEQUALITY) *Let p, q, and r be numbers satisfying $p, q, r \ge 1$ and $p^{-1} + q^{-1} = r^{-1}$. Suppose that $u \in L^p(\Omega)$ and $v \in L^q(\Omega)$. Then $uv \in L^r(\Omega)$ and*

$$\|uv\|_{0,r} \le \|u\|_{0,p}\|v\|_{0,q}.$$

For the special case $p = q = 2$, the Hölder inequality reduces to the Cauchy–Schwarz inequality

$$\|uv\|_{0,2} \le \|u\|_{0,2}\|v\|_{0,2} \quad \forall\, u, v \in L^2(\Omega).$$

Dual spaces and reflexivity. We define the dual exponent q of $p \in [1, \infty)$ by $1/p + 1/q = 1$ (with the usual convention that $q = \infty$ when $p = 1$). Then the topological dual $[L^p(\Omega)]'$ of $L^p(\Omega)$ may be identified with $L^q(\Omega)$. In particular, $L^2(\Omega)$ may be identified with its dual space. For $1 < p < \infty$ the roles of p and q are symmetric, and so it is clear that

$$L^p(\Omega) = (L^q(\Omega))' = (L^p(\Omega))''.$$

Thus the spaces $L^p(\Omega)$ are reflexive for $1 < p < \infty$.

The spaces $L^1(\Omega)$ and $L^\infty(\Omega)$ are *not* reflexive, though it is possible to identify $L^\infty(\Omega)$ with the dual of $L^1(\Omega)$; this identification is expressed in the form

$$L^\infty(\Omega) = (L^1(\Omega))'.$$

On the other hand, $L^1(\Omega)$ can be identified only with a proper subspace of $(L^\infty(\Omega))'$.

5.2.2 Sobolev Spaces

Assumptions about domains. We introduce a definition that will suffice for most purposes when smoothness assumptions about the boundary of a domain need to be made.

For any point $\boldsymbol{x} = (x_1, x_2, \ldots, x_d) \in \mathbb{R}^d$, set

$$y = x_d \text{ and } \hat{\boldsymbol{x}} = (x_1, x_2, \ldots, x_{d-1}) \in \mathbb{R}^{d-1}.$$

An open set Ω in $\mathbb{R}^d$ is said to have a *Lipschitz-continuous boundary* Γ if there exist constants $\alpha > 0$ and $\beta > 0$, a finite number of *local* coordinate systems $(\hat{\boldsymbol{x}}^m, y^m)$, and local maps f^m, $m = 1, \ldots, M$, that are Lipschitz-continuous on their respective domains of definition $\{\hat{\boldsymbol{x}}^m : |\hat{\boldsymbol{x}}^m| \leq \alpha\}$ such that

$$\Gamma = \cup_{m=1}^{M} \{(\hat{\boldsymbol{x}}^m, y^m) : y^m = f^m(\hat{\boldsymbol{x}}^m), |\hat{\boldsymbol{x}}^m| \leq \alpha\},$$

and for $m = 1, \ldots, M$,

$$\{(\hat{\boldsymbol{x}}^m, y^m) : f^m(\hat{\boldsymbol{x}}^m) < y^m < f^m(\hat{\boldsymbol{x}}^m) + \beta, |\hat{\boldsymbol{x}}^m| \leq \alpha\} \subset \Omega,$$
$$\{(\hat{\boldsymbol{x}}^m, y^m) : f^m(\hat{\boldsymbol{x}}^m) - \beta < y^m < f^m(\hat{\boldsymbol{x}}^m), |\hat{\boldsymbol{x}}^m| \leq \alpha\} \subset \mathbb{R}^d \backslash \overline{\Omega}.$$

The situation is depicted in Figure 5.1 for the two-dimensional case. More generally, we say that the boundary is of class X if the functions f^m are of class X, and that it is *smooth* if $X = C^\infty$.

With a slight abuse of terminology, a domain with a Lipschitz boundary is also referred to as a Lipschitz domain, with obvious modifications in nomenclature for boundaries of other classes. *In the following, we always assume that Ω is a Lipschitz domain, unless stated otherwise.* We note, though, that such an assumption is, in fact, not needed for some of the results stated below, for example, in Theorem 5.14 (b) and (c).

Distributions. As a prelude to introducing the Sobolev spaces we review the definition and properties of distributions, which permit the notion of differentiation to be extended to functions that are not differentiable in the classical sense.

We first introduce the space $C_0^\infty(\Omega)$ of *smooth functions with compact support*, defined to be functions in $C^\infty(\Omega)$ that vanish outside a compact subset of Ω. In particular, then, for any $\phi \in C_0^\infty(\Omega)$ we have $\phi = 0$ in a neighborhood of the boundary Γ of Ω.

The space of functions with compact support does not have a "standard" norm topology, but it suffices for our purposes to define convergence in this space as follows: A sequence $\{\phi_k\}$ in $C_0^\infty(\Omega)$ is said to converge to ϕ in $C_0^\infty(\Omega)$ if

(a) there exists a compact set K in Ω such that ϕ_k vanishes outside K for any k, and

(b) for each multi-index α, $D^\alpha \phi_k \to D^\alpha \phi$ uniformly in Ω.

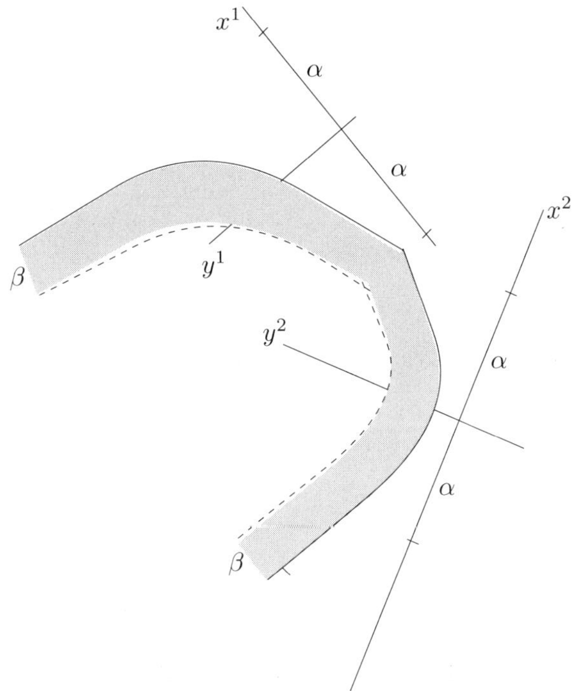

Figure 5.1: An illustration of the definition of a domain with Lipschitz-continuous boundary

The space $C_0^\infty(\Omega)$ endowed with this notion of convergence is called the *space of test functions* and is denoted by $\mathcal{D}(\Omega)$.

A *distribution* on Ω is a *continuous linear functional* on $\mathcal{D}(\Omega)$. That is, a linear functional ℓ on $\mathcal{D}(\Omega)$ is a distribution if and only if

$$\phi_k \to \phi \text{ in } \mathcal{D}(\Omega) \text{ implies } \langle \ell, \phi_k \rangle \to \langle \ell, \phi \rangle.$$

The space of distributions is denoted by $\mathcal{D}'(\Omega)$.

Any *locally integrable* function $u \in L^1_{\text{loc}}(\Omega)$ may be identified with a distribution, in the sense that there exists a unique distribution ℓ_u for which

$$\langle \ell_u, \phi \rangle = \int_\Omega u\, \phi\, dx \quad \forall\, \phi \in \mathcal{D}(\Omega).$$

In this case we write u for both the function and the associated distribution.

By an appropriate extension of the classical Green's formula it is possible to define derivatives of any order for distributions. Indeed, given $u \in \mathcal{D}'(\Omega)$ and a multi-index α with $|\alpha| = m$, the (distributional) derivative $D^\alpha u$ of u

is a distribution defined by

$$\langle D^\alpha u, \phi \rangle = (-1)^m \langle u, D^\alpha \phi \rangle \quad \forall \phi \in \mathcal{D}(\Omega).$$

For the case in which u is m-times continuously differentiable, $D^\alpha u$ coincides with the corresponding mth-order derivative in the classical sense.

The Sobolev spaces $W^{m,p}(\Omega)$. For any nonnegative integer m and real number $p \geq 1$ or $p = \infty$, we define

$$W^{m,p}(\Omega) = \{v \in L^p(\Omega) : \; D^\alpha v \in L^p(\Omega) \text{ for any } \alpha \in \mathbb{Z}_+^d \text{ with } |\alpha| \leq m\},$$

where derivatives are taken in the distributional sense. Norms in the spaces $W^{m,p}(\Omega)$ are defined by

$$\|v\|_{m,p,\Omega} = \Big(\sum_{|\alpha| \leq m} \|D^\alpha v\|_{0,p,\Omega}^p \Big)^{1/p}, \quad 1 \leq p < \infty, \tag{5.17}$$

and

$$\|v\|_{m,\infty,\Omega} = \max_{|\alpha| \leq m} \|D^\alpha v\|_{0,\infty,\Omega}. \tag{5.18}$$

With the norm defined above, the space $W^{m,p}(\Omega)$ becomes a Banach space. We also introduce here the seminorms on the spaces $W^{m,p}(\Omega)$:

$$|v|_{m,p,\Omega} = \Big(\sum_{|\alpha| = m} \|D^\alpha v\|_{0,p,\Omega}^p \Big)^{1/p}, \quad 1 \leq p < \infty,$$

$$|v|_{m,\infty,\Omega} = \max_{|\alpha| = m} \|D^\alpha v\|_{0,\infty,\Omega}.$$

The space $W^{m,p}(\Omega)$ is *reflexive* if and only if $1 < p < \infty$. We note here that $W^{0,p}(\Omega) = L^p(\Omega)$.

The case $p = 2$ is special, in that $W^{m,2}(\Omega)$ may be assigned an inner product. We set $W^{m,2}(\Omega) \equiv H^m(\Omega)$ and define the inner product on this space by

$$(u, v)_{m,\Omega} = \sum_{|\alpha| \leq m} (D^\alpha u, D^\alpha v)_{0,\Omega},$$

where as before, $(\cdot,\cdot)_{0,\Omega}$ denotes the $L^2(\Omega)$ inner product. With this inner product, $H^m(\Omega)$ is a *Hilbert space*. The corresponding norm will be denoted by $\|\cdot\|_{m,2,\Omega}$ or simply by $\|\cdot\|_{m,\Omega}$, or even $\|\cdot\|_m$, depending on the particular context.

Given the definition of the Sobolev norm (5.17) or (5.18), it follows that convergence in $W^{m,p}(\Omega)$ of a sequence $\{v_l\}_{l=1}^\infty$ to a function v is equivalent to the requirement that

$$D^\alpha v_l \to D^\alpha v \;\text{ in } L^p(\Omega), \quad \text{for } |\alpha| \leq m.$$

A similar result holds for weak convergence in $W^{m,p}(\Omega)$.

Some properties regarding embeddings and inclusions are summarized in the following theorem.

THEOREM 5.14. *The following statements are valid:*

(a) $W^{m,p}(\Omega) \hookrightarrow\hookrightarrow W^{k,p}(\Omega)$ *if* $m > k$.

(b) $\mathcal{D}(\Omega) \subset W^{m,p}(\Omega)$.

(c) $C^m(\overline{\Omega}) \hookrightarrow W^{m,p}(\Omega)$.

(d) $C^\infty(\overline{\Omega}) \cap W^{m,p}(\Omega)$ *is dense in* $W^{m,p}(\Omega)$*; in other words, a function in* $W^{m,p}(\Omega)$ *can be approximated by a sequence of functions smooth up to the boundary.*

(d) (Sobolev compact embedding) *If* $k < m - d/p$ *with* $1 \le p \le \infty$*, then* $W^{m,p}(\Omega) \hookrightarrow\hookrightarrow C^k(\overline{\Omega})$*; in particular,* $W^{m,p}(\Omega) \hookrightarrow\hookrightarrow C^k(\overline{\Omega})$.

It is possible also to define Sobolev spaces $W^{m,p}(\Omega)$ for noninteger values of m. We will require such spaces only for the case of functions defined on the boundary Γ of the domain Ω, so we give here a brief review for this situation.

The space $W^{s,p}(\Gamma)$ for noninteger s. Let Ω be a bounded domain in $\mathbb{R}^d$ ($d \ge 2$) with Lipschitz boundary Γ. Suppose that $1 \le p < \infty$ and $\sigma \in (0,1)$. For a function $v \in L^p(\Gamma)$, set

$$F_\sigma(v) = \int_{\Gamma\times\Gamma} \frac{|v(\boldsymbol{x}) - v(\boldsymbol{y})|^p}{|\boldsymbol{x} - \boldsymbol{y}|^{d-1+\sigma p}}\, ds(\boldsymbol{x})\, ds(\boldsymbol{y})$$

and

$$\|v\|_{\sigma,p,\Gamma} = \left(\int_\Gamma |v|^p\, ds + F_\sigma(v) \right)^{1/p}.$$

Then $W^{\sigma,p}(\Gamma)$ is defined to be the space of functions $v \in L^p(\Gamma)$ for which $\|v\|_{\sigma,p,\Gamma} < \infty$. This is a Banach space with the norm $\|\cdot\|_{\sigma,p,\Gamma}$; furthermore, it is reflexive for $1 < p < \infty$.

More generally, for $s = m + \sigma$ with $m \in \mathbb{Z}_+$ and $\sigma \in (0,1)$, the space $W^{s,p}(\Gamma)$ is defined in a similar way: It consists of all the functions v such that any tangential derivatives of order less than or equal to m of the function v belong to $L^p(\Gamma)$, and any tangential derivative $D^\alpha v$ of order $|\alpha| = m$ satisfies $F_\sigma(D^\alpha v) < \infty$.

Trace theorems. A uniformly continuous function v on a bounded domain Ω with boundary Γ has a well-defined boundary value, usually denoted by $v|_\Gamma$. This property may be expressed in an alternative manner by the introduction of a map γ called the *trace* operator, which associates with each $v \in C(\overline{\Omega})$ its boundary value $\gamma v = v|_\Gamma$, a function belonging to $C(\Gamma)$.

For a function $v \in W^{m,p}(\Omega)$ the issue of its boundary value is less straightforward: the restriction of v to Γ need not make sense, since Γ is a set of measure zero, and two functions in $W^{m,p}(\Omega)$ are identified if they are equal a.e. Fortunately, it is possible to extend the notion of the trace operator for continuous functions in $C(\overline{\Omega})$ to functions in $W^{m,p}(\Omega)$ for certain ranges of the indices m and p. This result is summarized in the following.

THEOREM 5.15 (TRACE THEOREM). *Assume that* $1 \leq p \leq \infty$ *and* $m > 1/p$. *Then there exists a unique bounded linear surjective mapping* $\gamma : W^{m,p}(\Omega) \to W^{m-1/p,p}(\Gamma)$ *such that* $\gamma v = v|_\Gamma$ *when* $v \in W^{m,p}(\Omega) \cap C(\overline{\Omega})$.

In future, when the trace γv of a Sobolev function v on the boundary is defined, we will simply write v for the trace γv.

The trace theorem can be extended to higher-order derivatives on the boundary. In order to avoid complications arising from compatibility conditions we confine attention to higher-order *normal derivatives*, since, for example, the tangential derivative of a function is completely defined if the function itself is known along a boundary.

Let $\boldsymbol{n} = (n_1, \ldots, n_d)^T$ denote the outward unit normal to the boundary Γ of Ω, assumed here to be smooth. The kth normal derivative of a function $v \in C^k(\overline{\Omega})$ is then defined by

$$\frac{\partial^k v}{\partial n^k} \equiv n_{i_1} \cdots n_{i_k} \frac{\partial^k v}{\partial x_{i_1} \cdots \partial x_{i_k}}.$$

The following theorem states the fact that this definition can be extended to functions in certain Sobolev spaces.

THEOREM 5.16 (SECOND TRACE THEOREM). *Assume that* Ω *is a bounded open set with a* $C^{k,1}$ *boundary* Γ. *Assume that* $1 \leq p \leq \infty$ *and* $m > k+1/p$. *Then there exist unique bounded linear and surjective mappings* $\gamma_j : W^{m,p}(\Omega) \to W^{m-j-1/p,p}(\Gamma)$ $(j = 0, 1, \ldots, k)$ *such that* $\gamma_j v = (\partial^j v/\partial n^j)|_\Gamma$ *when* $v \in W^{m,p}(\Omega) \cap C^{k,1}(\overline{\Omega})$.

It is important to note that the ranges of the trace operators are proper subsets of $L^p(\Gamma)$. On the other hand, it can be shown that $W^{m-j-1/p,p}(\Gamma)$ is *dense* in $L^p(\Gamma)$, for $j = 0, 1, \ldots, k$.

The space $W_0^{m,p}(\Omega)$. With the definition of traces at our disposal, it is now possible to consider those subspaces of Sobolev spaces characterized by the fact that the functions vanish on the boundary. To this end we define

$$W_0^{m,p}(\Omega) = \{v \in W^{m,p}(\Omega) : \ \gamma_j v = 0 \text{ for } j < m - 1/p\}.$$

This space may be *equivalently defined by*

$$W_0^{m,p}(\Omega) = \text{the closure of } C_0^\infty(\Omega) \text{ in } W^{m,p}(\Omega).$$

Immediately we see that any function in $W_0^{m,p}(\Omega)$ can be approximated by a sequence of $C_0^\infty(\Omega)$ functions with respect to the norm of $W^{m,p}(\Omega)$.

From the definition and the second trace theorem, $W_0^{m,p}(\Omega)$ is a *closed subspace* of $W^{m,p}(\Omega)$. When $p = 2$, we write $H_0^m(\Omega)$ to replace $W_0^{m,2}(\Omega)$. In particular, we will frequently use the space

$$H_0^1(\Omega) = \{v \in H^1(\Omega) : \ v = 0 \text{ a.e. on } \Gamma\}.$$

Equivalent norms. The following result can be used to generate various equivalent norms (cf. the definition (5.3)) on Sobolev spaces. Recall that over the Sobolev space $W^{k,p}(\Omega)$, $|v|_{k,p,\Omega}$ is the seminorm defined by

$$|v|_{k,p,\Omega} = \Big(\int_\Omega \sum_{|\alpha|=k} |D^\alpha v|^p \, dx\Big)^{1/p}.$$

THEOREM 5.17 (EQUIVALENT NORM THEOREM). *Let Ω be an open, bounded, connected set in $\mathbb{R}^d$ with a Lipschitz boundary, $k \geq 1$, $1 \leq p < \infty$. Assume that $f_j : W^{k,p}(\Omega) \to \mathbb{R}$, $1 \leq j \leq J$, are seminorms on $W^{k,p}(\Omega)$ satisfying two conditions:*

(H_1) $0 \leq f_j(v) \leq c\,\|v\|_{k,p,\Omega}$ $\forall\, v \in W^{k,p}(\Omega)$, $1 \leq j \leq J$.

(H_2) If v is a polynomial of degree less than or equal to $k-1$ and $f_j(v) = 0$, $1 \leq j \leq J$, then $v = 0$.

Then, the quantity

$$\|v\| = |v|_{k,p,\Omega} + \sum_{j=1}^J f_j(v)$$

or

$$\|v\| = \Big(|v|_{k,p,\Omega}^p + \sum_{j=1}^J f_j(v)^p\Big)^{1/p}$$

defines a norm on $W^{k,p}(\Omega)$, which is equivalent to the norm $\|v\|_{k,p,\Omega}$.

PROOF. We will prove that the quantity

$$\|v\| = |v|_{k,p,\Omega} + \sum_{j=1}^J f_j(v)$$

is a norm on $W^{k,p}(\Omega)$ equivalent to the norm $\|v\|_{k,p,\Omega}$. That

$$\|v\| = \Big(|v|_{k,p,\Omega}^p + \sum_{j=1}^J f_j(v)^p\Big)^{1/p}$$

is also an equivalent norm can be proved similarly.

By the condition (H_1), we see that for some constant $c > 0$,

$$\|v\| \le c\,\|v\|_{k,p,\Omega} \quad \forall\, v \in W^{k,p}(\Omega).$$

So we need only to show that there is another constant $c > 0$ such that

$$\|v\|_{k,p,\Omega} \le c\,\|v\| \quad \forall\, v \in W^{k,p}(\Omega).$$

We argue by contradiction. Suppose that this inequality is false; then we can find a sequence $\{v_l\} \subset W^{k,p}(\Omega)$ with the properties

(a) $\|v_l\|_{k,p,\Omega} = 1$,

(b) $\|v_l\| \le 1/l$

for $l = 1, 2, \ldots$. From Property (b), we see that as $l \to \infty$,

$$|v_l|_{k,p,\Omega} \to 0$$

and

$$f_j(v_l) \to 0, \quad 1 \le j \le J.$$

Since $\{v_l\}$ is a bounded sequence in $W^{k,p}(\Omega)$, and since

$$W^{k,p}(\Omega) \hookrightarrow\hookrightarrow W^{k-1,p}(\Omega),$$

there is a subsequence of the sequence $\{v_l\}$, still denoted by $\{v_l\}$, and a function $v \in W^{k-1,p}(\Omega)$ such that

$$v_l \to v \quad \text{in } W^{k-1,p}(\Omega), \text{ as } l \to \infty.$$

This property and $|v_l|_{k,p,\Omega} \to 0$ as $l \to \infty$, together with the uniqueness of a limit, imply that

$$v_l \to v \quad \text{in } W^{k,p}(\Omega), \text{ as } l \to \infty$$

and

$$|v|_{k,p,\Omega} = \lim_{l\to\infty} |v_l|_{k,p,\Omega} = 0.$$

We then conclude that v is a polynomial of degree less than or equal to $k-1$. On the other hand, from the continuity of the functionals f_j, $1 \le j \le J$, we find that

$$f_j(v) = \lim_{l\to\infty} f_j(v_l) = 0, \quad 1 \le j \le J.$$

Using the assumption (H_2), we see that $v = 0$, which contradicts the fact that

$$\|v\|_{k,p,\Omega} = \lim_{l\to\infty} \|v_l\|_{k,p,\Omega} = 1.$$

The proof of the result is now completed. □

Many useful inequalities can be derived as consequences of Theorem 5.17. For example, let us apply the theorem to the special case $k = 1$, $p = 2$, $J = 1$, and

$$f_1(v) = \int_{\partial\Omega} |v|\, ds.$$

We can then conclude that there exists a constant $c > 0$, depending only on Ω, such that the inequality

$$\|v\|_{1,\Omega} \le c\,|v|_{1,\Omega} \quad \forall\, v \in H_0^1(\Omega) \tag{5.19}$$

holds. This result is known as the Poincaré–Friedrichs inequality. It follows from (5.19) that the seminorm $|\cdot|_1$ is a norm on $H_0^1(\Omega)$, equivalent to the usual $H^1(\Omega)$-norm.

More generally, if Γ_0 is an open, nonempty subset of the boundary Γ, then there is a constant $c > 0$, depending only on Ω, such that

$$\|v\|_{1,\Omega} \le c\,|v|_{1,\Omega} \quad \forall\, v \in H_{\Gamma_0}^1(\Omega). \tag{5.20}$$

Here,

$$H_{\Gamma_0}^1(\Omega) = \{v \in H^1(\Omega) : v = 0 \text{ a.e. on } \Gamma_0\}.$$

This inequality can be derived by applying Theorem 5.17 with $k = 1$, $p = 2$, $J = 1$, and

$$f_1(v) - \int_{\Gamma_0} |v|\, ds.$$

Korn's first inequality. We now give details of an inequality that is of central importance in elasticity and elastoplasticity. Let Ω be a nonempty, open, bounded, and connected set in $\mathbb{R}^3$ with a Lipschitz boundary. Given a function $\boldsymbol{u} \in [H^1(\Omega)]^3$, the linearized strain tensor is defined by (2.7). Then Korn's inequality states that there exists a constant $c > 0$ depending only on Ω such that

$$\|\boldsymbol{u}\|^2_{[H^1(\Omega)]^3} \le c \int_\Omega |\boldsymbol{\epsilon}(\boldsymbol{u})|^2 dx \quad \forall\, \boldsymbol{u} \in [H_0^1(\Omega)]^3. \tag{5.21}$$

This inequality can be proved first for $C_0^\infty(\Omega)$ functions by an integration by parts technique (if we use the equivalent norm $\|\nabla \cdot\|_{[L^2(\Omega)]^3}$ in the space $[H_0^1(\Omega)]^3$), and then extended to $[H_0^1(\Omega)]^3$ by a density argument. The inequality (5.21) is a special case of the more general Korn's first inequality (cf. [24])

$$\|\boldsymbol{u}\|^2_{[H^1(\Omega)]^3} \le c \int_\Omega |\boldsymbol{\epsilon}(\boldsymbol{u})|^2 dx \quad \forall\, \boldsymbol{u} \in [H^1_{\Gamma_0}(\Omega)]^3, \tag{5.22}$$

where Γ_0 is a measurable subset of $\partial\Omega$ with meas $(\Gamma_0) > 0$, and

$$[H^1_{\Gamma_0}(\Omega)]^3 = \{\boldsymbol{v} \in [H^1(\Omega)]^3 : \ \boldsymbol{v} = \boldsymbol{0} \text{ a.e. on } \Gamma_0\}.$$

The space $W^{-m,p}(\Omega)$. Let m be a positive integer, p a real number satisfying $1 \le p < \infty$, and q the conjugate number of p, i.e., $q = (1-1/p)^{-1}$ if $1 < p < \infty$, and $q = \infty$ if $p = 1$. Then $W^{-m,q}(\Omega)$ is defined to be the dual space of $W_0^{m,p}(\Omega)$. Since $\mathcal{D}(\Omega)$ is dense in $W_0^{m,p}(\Omega)$, it follows that $W^{-m,q}(\Omega) \subset \mathcal{D}'(\Omega)$; in other words, $W^{-m,q}(\Omega)$ is a space of distributions. This property is made more concrete in the following result.

THEOREM 5.18. *A distribution ℓ belongs to $W^{-m,q}(\Omega)$ if and only if it can be expressed in the form*

$$\ell = \sum_{|\alpha| \le m} D^\alpha u_\alpha,$$

where u_α are functions in $L^q(\Omega)$.

A space with negative index that will be used often is $H^{-1}(\Omega)$, the dual space of $H_0^1(\Omega)$. Given any function $f \in L^2(\Omega)$, we can naturally define an $H^{-1}(\Omega)$ function ℓ through the relation

$$\langle \ell, v\rangle = \int_\Omega f\, v\, dx \quad \forall\, v \in H_0^1(\Omega),$$

and we identify ℓ with f. For this reason, for $f \in H^{-1}(\Omega)$ and $v \in H_0^1(\Omega)$, sometimes we use the notation $\int_\Omega f\, v\, dx$ to represent the duality pairing $\langle f, v\rangle$ on $H^{-1}(\Omega) \times H_0^1(\Omega)$.

5.2.3 Spaces of Vector-Valued Functions

When dealing with initial–boundary value problems, it makes a great deal of sense to treat functions of space and time as maps from a time interval into a Banach space such as those that have been discussed earlier in this section. To begin, let X be a Banach space and T a positive number; then the space $C^m([0,T];X)$ $(m = 0, 1, \dots)$ consists of all continuous functions v from $[0,T]$ to X that have continuous derivatives of order less than or equal to m. This is a Banach space when endowed with the norm

$$\|v\|_{C^m([0,T],X)} = \sum_{k=0}^m \max_{0\le t\le T} \|v^{(k)}(t)\|_X,$$

where $v^{(k)}(t)$ denotes the kth time derivative of v. We write $C([0,T],X)$ for the case $m = 0$.

Turning next to Lebesgue spaces, for $1 \leq p < \infty$ the space $L^p(0,T;X)$ consists of all measurable functions v from $[0,T]$ to X for which

$$\|v\|_{L^p(0,T,X)} \equiv \left(\int_0^T \|v(t)\|_X^p \, dt \right)^{1/p} < \infty.$$

This is a Banach space with the norm $\|v\|_{L^p(0,T,X)}$, provided that the members are understood to represent equivalence classes of functions that are equal a.e. on $(0,T)$.

The extension of this definition to include the case $p = \infty$ is carried out in the usual way: The space $L^\infty(0,T;X)$ consists of all measurable functions v from $[0,T]$ to X that are essentially bounded. This is a Banach space with the norm

$$\|v\|_{L^\infty(0,T,X)} \equiv \text{ess sup}_{0 \leq t \leq T} \|v(t)\|_X.$$

If X is a *Hilbert space* with inner product $(\cdot,\cdot)_X$, then $L^2(0,T;X)$ is a Hilbert space with the inner product

$$(u,v)_{L^2(0,T;X)} = \int_0^T (u(t),v(t))_X \, dt.$$

The following theorem summarizes some properties of these spaces.

THEOREM 5.19. *Let* $m = 0, 1, \ldots,$ *and* $1 \leq p \leq \infty$. *Then*

(a) $C([0,T];X)$ *is dense in* $L^p(0,T;X)$, *and the embedding is continuous.*

(b) *If* $X \hookrightarrow Y$, *then* $L^p(0,T;X) \hookrightarrow L^q(0,T;Y)$ *for* $1 \leq q \leq p \leq \infty$.

Let X' be the topological dual of a separable normed space X. Then for $1 < p < \infty$ the dual space of $L^p(0,T;X)$ is given by

$$[L^p(0,T;X)]' = L^q(0,T;X') \quad \text{with } \frac{1}{p} + \frac{1}{q} = 1. \tag{5.23}$$

Furthermore, if X is *reflexive*, then so is $L^p(0,T;X)$.

It is necessary to define in an appropriate way derivatives with respect to the time variable for functions that lie in the spaces $L^p(0,T;X)$. The approach is similar to that taken in the case of generalized derivatives of distributions; that is, we take as a starting point the elementary integration by parts formula

$$\int_0^T \phi^{(m)}(t)\, v(t)\, dt = (-1)^m \int_0^T \phi(t)\, v^{(m)}(t)\, dt,$$

which holds for all functions $\phi \in C_0^\infty(0,T)$ and $v \in C^m([0,T];X)$; here $(\cdot)^{(m)} \equiv d^m(\cdot)/dt^m$. A function $v \in L^1_{\text{loc}}(0,T;X)$ is then said to possess an

mth *generalized derivative* if there exists a function $w \in L^1_{\text{loc}}(0,T;Y)$ such that

$$\int_0^T \phi^{(m)}(t)\, v(t)\, dt = (-1)^m \int_0^T \phi(t)\, w(t)\, dt \quad \forall\, \phi \in C_0^\infty(0,T), \tag{5.24}$$

where X and Y are appropriate Banach spaces. When (5.24) holds, we write simply $w = v^{(m)}$. For the case in which $Y = X = \mathbb{R}$ and $v \in C^m(0,T)$, (5.24) reduces to the classical integration by parts formula.

The following lemma gives an important property of generalized derivatives.

LEMMA 5.20. *Let V be a reflexive Banach space and H a Hilbert space with the property that $V \hookrightarrow H \hookrightarrow V'$, the continuous embedding $V \hookrightarrow H$ being dense. Let $1 \le p, q \le \infty$, with $1/p + 1/q = 1$. Then any function $u \in L^p(0,T;V)$ possesses a unique generalized derivative $u^{(m)} \in L^q(0,T;V')$ if and only if there is a function $w \in L^q(0,T;V')$ such that*

$$\int_0^T (u(t), v)_H \phi^{(m)}(t)\, dt = (-1)^m \int_0^T \phi(t) \langle w(t), v\rangle_{V' \times V}\, dt$$

for all $v \in V$, $\phi \in C_0^\infty(0,T)$. Then $u^{(m)} = w$, and for almost all $t \in (0,T)$,

$$\frac{d^m}{dt^m}(u(t), v)_H = \langle w(t), v\rangle_{V' \times V} \quad \forall\, v \in V.$$

For an integer $m \ge 0$ and a real $p \ge 1$, we define by $W^{m,p}(0,T;X)$ the space of functions $f \in L^p(0,T;X)$ such that $f^{(i)} \in L^p(0,T;X)$, $i \le m$. This is a Banach space with the norm

$$\|f\|_{W^{m,p}(0,T;X)} = \left\{ \sum_{i=0}^m \|f^{(i)}\|^p_{L^p(0,T;X)} \right\}^{1/p}.$$

We use the shorthand notation $H^m(0,T;X)$ for $W^{m,p}(0,T;X)$ when $p = 2$. If X is a Hilbert space, $H^m(0,T;X)$ is also a Hilbert space with the inner product

$$(f, g)_{H^m(0,T;X)} = \int_0^T \sum_{i=0}^m \left(f^{(i)}(t), g^{(i)}(t) \right)_X dt.$$

We record the fundamental inequality

$$\|f(t) - f(s)\|_X \le \int_s^t \|\dot f(\tau)\|_X\, d\tau, \tag{5.25}$$

which holds for $s < t$ and $f \in W^{1,p}(0,T;X)$, $p \ge 1$. Here, $\dot f = df/dt$. On several occasions we will also need the continuous embedding property

$$H^1(0,T;X) \hookrightarrow C([0,T], X); \tag{5.26}$$

in particular, there exists a constant $c > 0$ such that

$$\|v\|_{C([0,T],X)} \le c\,\|v\|_{H^1(0,T;X)} \quad \forall\, v \in H^1(0,T;X).$$

We will also need the property

$$C^\infty([0,T];X) \text{ is dense in } H^1(0,T;X). \tag{5.27}$$

A theorem of Lebesgue. The following result is useful when we localize a global relation; in particular, it will be used in proving Theorem 7.3 in Section 7.2 on the existence of a solution for an abstract variational inequality.

THEOREM 5.21. *Assume that X is a normed space, $f \in L^1(a,b;X)$. Then*

$$\lim_{h\to 0} \frac{1}{h} \int_{t_0}^{t_0+h} \|f(t) - f(t_0)\|_X \, dt = 0 \quad \text{for almost all } t_0 \in (a,b).$$

We see that the theorem implies that

$$\lim_{h\to 0} \frac{1}{h} \int_{t_0}^{t_0+h} f(t)\, dt = f(t_0) \quad \text{for almost all } t_0 \in (a,b),$$

where the limit is understood in the sense of the norm of X: that is,

$$\lim_{h\to 0} \left\| \frac{1}{h} \int_{t_0}^{t_0+h} f(t)\, dt - f(t_0) \right\|_X = 0 \quad \text{for almost all } t_0 \in (a,b).$$

6
Variational Equations and Inequalities

In this chapter we review some standard results for boundary value and initial–boundary value problems, paying particular attention to weak or variational formulations. The first section will be concerned with elliptic variational equations, and this will be followed by a review of some material on elliptic variational inequalities. Because the variational form taken by elastoplastic problems resembles that of parabolic variational inequalities, we also include some material on this class of problems.

The numerical approximation of variational problems by the finite element method will be considered later, in Chapter 10.

6.1 Variational Formulation of Elliptic Boundary Value Problems

We begin with some model elliptic boundary value problems. Let Ω be a bounded domain in $\mathbb{R}^d$ with a Lipschitz continuous boundary Γ. The unit outward normal vector $\boldsymbol{n} = (n_1, \ldots, n_d)^T$ exists a.e. on Γ, and we will use $\partial u/\partial n$ to denote the normal derivative of u on Γ.

Consider the boundary value problem corresponding to the Poisson equation with homogeneous Dirichlet boundary condition

$$\begin{aligned} -\Delta u &= f \quad \text{in } \Omega, \\ u &= 0 \quad \text{on } \Gamma. \end{aligned} \tag{6.1}$$

Here Δ denotes the Laplacian operator, defined by

$$\Delta u = \frac{\partial^2 u}{\partial x_i \partial x_i}.$$

A classical solution of the problem (6.1) is a smooth function $u \in C^2(\Omega) \cap C(\overline{\Omega})$ that satisfies the differential equation $(6.1)_1$ and the boundary condition $(6.1)_2$ pointwise. Necessarily we have to assume $f \in C(\Omega)$, but this condition does not guarantee the existence of a classical solution of the problem. One purpose of the introduction of the weak formulation is the removal of the high smoothness requirement on the solution; once this (possibly unrealistic) restriction is removed, it is easier to obtain results on the existence of a (weak) solution.

To derive the weak formulation corresponding to (6.1), we temporarily assume that it has a classical solution $u \in C^2(\Omega) \cap C(\overline{\Omega})$. We multiply the differential equation $(6.1)_1$ by an arbitrary function $v \in C_0^\infty(\Omega)$ (the space of so-called smooth test functions), and integrate the relation over Ω, to obtain

$$-\int_\Omega \Delta u \, v \, dx = \int_\Omega f \, v \, dx.$$

Next, we integrate by parts, and recalling that $v = 0$ on Γ, we have

$$\int_\Omega \nabla u \cdot \nabla v \, dx = \int_\Omega f \, v \, dx. \tag{6.2}$$

This relation has been derived under the assumptions $u \in C^2(\Omega) \cap C(\overline{\Omega})$ and $v \in C_0^\infty(\Omega)$. However, for the relation (6.2) to make sense, we require only that $u, v \in H^1(\Omega)$, assuming that $f \in L^2(\Omega)$. Then since $H_0^1(\Omega)$ is the closure of $C_0^\infty(\Omega)$ in $H^1(\Omega)$, (6.2) is valid for any $v \in H_0^1(\Omega)$. Meanwhile, the solution u is sought in the space $H_0^1(\Omega)$. Therefore, the weak formulation of the boundary value problem (6.1) is

$$u \in H_0^1(\Omega), \quad \int_\Omega \nabla u \cdot \nabla v \, dx = \int_\Omega f \, v \, dx \quad \forall \, v \in H_0^1(\Omega). \tag{6.3}$$

Actually, we do not even need the assumption $f \in L^2(\Omega)$. It suffices to assume that $f \in H^{-1}(\Omega)$, as long as we interpret the integral $\int_\Omega f \, v \, dx$ as the duality pairing $\langle f, v \rangle$ between $H^{-1}(\Omega)$ and $H_0^1(\Omega)$.

We have shown that if u is a classical solution of (6.1), then it is also a solution of the weak formulation (6.3). Conversely, suppose that u is a weak solution with the additional regularity $u \in C^2(\Omega) \cap C(\overline{\Omega})$. Then for any $v \in C_0^\infty(\Omega) \subset H_0^1(\Omega)$, from (6.3) we obtain

$$\int_\Omega (-\Delta u - f) \, v \, dx = 0.$$

So we must have $-\Delta u = f$ in Ω; that is, the differential equation $(6.1)_1$ is satisfied. Also, u satisfies the homogeneous Dirichlet boundary condition pointwise. Thus a weak solution of (6.3) with the additional regularity condition is also a classical solution of the boundary value problem (6.1). In the event that the weak solution u does not have the regularity $u \in C^2(\Omega) \cap C(\overline{\Omega})$, we will say that u formally solves the boundary value problem (6.1).

Now we set $V = H_0^1(\Omega)$ and let $a(\cdot,\cdot) : V \times V \to \mathbb{R}$ be the bilinear form defined by

$$a(u,v) = \int_\Omega \nabla u \cdot \nabla v \, dx \quad \text{for } u, v \in V,$$

and $\ell : V \to \mathbb{R}$ the linear functional defined by

$$\langle \ell, v \rangle = \int_\Omega f \, v \, dx \quad \text{for } v \in V.$$

Then the weak formulation of the problem is

$$u \in V, \quad a(u,v) = \langle \ell, v \rangle \quad \forall \, v \in V. \tag{6.4}$$

The bilinear form $a(\cdot,\cdot)$ is V-elliptic, thanks to the Poincaré–Friedrichs inequality (5.19); and it is also continuous, as is readily verified. Finally, the functional ℓ is bounded and linear. By the Lax–Milgram lemma (Theorem 5.8), therefore, the problem (6.4) has a unique solution $u \in V$.

A formulation of the kind (6.1), that is, in the form of a partial differential equation and a set of boundary conditions, will be referred to henceforth as the *classical formulation* of a boundary value problem, while a formulation of the kind (6.4) will be known as a *weak* or *variational* formulation. The term "weak" derives from the fact that less regularity is sought in solutions to (6.4), since u need belong only to $H_0^1(\Omega)$. On the other hand, for a solution u of (6.1) to make sense, it is required that $u \in C^2(\Omega) \cap C(\overline{\Omega})$.

We established above a formal equivalence between the classical formulation (6.1) and the weak formulation (6.4). Such a formal equivalence result is not completely satisfactory, since the classical pointwise formulation is actually not the physically natural form. Indeed, for physical processes in general, the classical pointwise formulation is usually derived from a fundamental physical law that is posed as an *integral balance law*, that is, as an equation or inequality involving integrals over the domain (or arbitrary subdomains) and its boundary. By assuming appropriate smoothness of the relevant quantities in the integral balance law, one can then obtain the pointwise statement of the balance law. Now, both the *integral balance law* and the *weak formulation* make sense as long as each expression in the formulations is well-defined; also, the required regularity assumptions are far weaker than those used in the derivation of the weak formulation from the boundary value problem, which in turn is obtained from the integral balance law. Naturally, one may ask the question, are the integral balance law and the weak formulation equivalent under the weakest possible regularity assumption, namely, that the integrals appearing in these formulations make sense as Lebesgue integrals? A satisfactory answer can be found in the work of Antman and Osborn [3] (see also the monograph [2]), where it is shown that precisely formulated versions of the integral balance laws of motion and of the principle of virtual work (that is, the weak formulation) are equivalent. Thus we see that between a classical formulation and a weak formulation for a physical process, the weak formulation is the form that is physically more natural.

The approach taken in the remainder of this work will always be to regard variational or weak formulations as fundamental. In particular, the elastoplastic problems formulated in classical form in Chapter 4 will be recast in variational form later in the following chapters, and it is the variational forms that will be studied in detail.

The case of nonhomogeneous Dirichlet boundary conditions may be dealt with as follows. Suppose that instead of $(6.1)_2$ the boundary condition is

$$u = g \quad \text{on } \Gamma, \tag{6.5}$$

where $g \in H^{1/2}(\Gamma)$ is given. Since there exists a surjection from $H^1(\Omega)$ onto $H^{1/2}(\Gamma)$ (see Theorem 5.15), it follows that there is a function $G \in H^1(\Omega)$ such that $\gamma G = g$. Thus, setting

$$u = w + G,$$

the problem may be transformed into one of seeking w that satisfies

$$\begin{aligned} -\Delta w &= f + \Delta G \quad \text{in } \Omega, \\ w &= 0 \quad \text{on } \Gamma. \end{aligned}$$

There is no problem in posing this problem in a weak form, since $f + \Delta G$ belongs to $H^{-1}(\Omega)$. Indeed, the variational formulation for the transformed problem takes the following form: Find $w \in H_0^1(\Omega)$ such that

$$\int_\Omega \nabla w \cdot \nabla v \, dx = \int_\Omega (f\, v - \nabla G \cdot \nabla v) \, dx \quad \forall\, v \in H_0^1(\Omega).$$

Since nonhomogeneous Dirichlet boundary conditions can be rendered homogeneous in this way, for convenience we consider henceforth only problems with homogeneous Dirichlet conditions.

Consider next the *Neumann* problem of determining u that satisfies

$$\begin{aligned} -\Delta u + u &= f \quad \text{in } \Omega, \\ \partial u/\partial n &= g \quad \text{on } \Gamma. \end{aligned} \tag{6.6}$$

For simplicity, assume that $f \in L^2(\Omega)$, $g \in L^2(\Gamma)$. The appropriate space in which to formulate this problem in weak form is $H^1(\Omega)$. Multiplying $(6.6)_1$ by an arbitrary smooth test function $v \in C^\infty(\overline{\Omega})$, integrating over Ω, and performing an integration by parts, we obtain

$$\int_\Omega (\nabla u \cdot \nabla v + u\, v)\, dx = \int_\Omega f\, v\, dx + \int_\Gamma \frac{\partial u}{\partial n} v\, ds.$$

Then, use of the Neumann boundary condition $(6.6)_2$ in the boundary term leads to the relation

$$\int_\Omega (\nabla u \cdot \nabla v + u\, v)\, dx = \int_\Omega f\, v\, dx + \int_\Gamma g\, v\, ds \quad \forall\, v \in C^\infty(\overline{\Omega}).$$

Since $C^\infty(\overline{\Omega})$ is dense in $H^1(\Omega)$ and the trace operator γ is continuous from $H^1(\Omega)$ to $L^2(\Gamma)$, we see that the above relation holds for any $v \in H^1(\Omega)$. Thus the weak formulation of the boundary value problem (6.6) is to find $u \in H^1(\Omega)$ such that

$$\int_\Omega (\nabla u \cdot \nabla v + u\, v)\, dx = \int_\Omega f\, v\, dx + \int_\Gamma g\, v\, ds \quad \forall\, v \in H^1(\Omega). \tag{6.7}$$

This problem has the form

$$u \in V, \quad a(u, v) = \langle \ell, v \rangle \quad \forall\, v \in V, \tag{6.8}$$

where $V = H^1(\Omega)$ and $a(\cdot,\cdot)$ and $\langle \ell, \cdot \rangle$ are defined by

$$\begin{aligned} a(u, v) &= \int_\Omega (\nabla u \cdot \nabla v + u\, v)\, dx, \\ \langle \ell, v \rangle &= \int_\Omega f\, v\, dx + \int_\Gamma g\, v\, ds. \end{aligned}$$

Again, applying the Lax–Milgram lemma, it is not difficult to show that the weak formulation (6.8) has a unique solution. Thus, a classical solution $u \in C^2(\Omega) \cap C^1(\overline{\Omega})$ of the boundary value problem (6.6) is also the solution of the weak formulation (6.8). Conversely, it can be shown that a solution to (6.8) with the additional regularity $u \in C^2(\Omega) \cap C^1(\overline{\Omega})$ is a classical solution of the boundary value problem (6.6).

The Neumann problem for the Poisson equation is given by

$$\begin{aligned} -\Delta u &= f \quad \text{in } \Omega, \\ \partial u / \partial n &= g \quad \text{on } \Gamma, \end{aligned} \tag{6.9}$$

where $f \in L^2(\Omega)$ and $g \in L^2(\Gamma)$ are given. The study of this problem is more delicate than that of (6.6). We will see that in general, (6.9) does not have a solution, and when the problem does have a solution u, this solution is not unique, since any function of the form $u+c$, $c \in \mathbb{R}$, also satisfies (6.9).

Formally, the corresponding weak formulation is

$$u \in H^1(\Omega), \quad \int_\Omega \nabla u \cdot \nabla v \, dx = \int_\Omega f \, v \, dx + \int_\Gamma g \, v \, ds \quad \forall \, v \in H^1(\Omega). \tag{6.10}$$

A necessary condition for (6.10) to have a solution is that the given data satisfy

$$\int_\Omega f \, dx + \int_\Gamma g \, ds = 0; \tag{6.11}$$

this is easily seen by taking $v = 1$ in (6.10). The condition (6.11) is also a sufficient condition for the problem (6.10) to have a solution. Indeed, the problem (6.10) is most conveniently studied in the quotient space $V = H^1(\Omega)/\mathbb{R}$, where each element $\dot{v} \in V$ is an equivalence class $\dot{v} = \{v + t : t \in \mathbb{R}\}$, and any $v \in \dot{v}$ is called a representative element. Applying Theorem 5.17, it is not difficult to show that over the space V, the quotient norm $\|\dot{v}\|_V \equiv \inf_t \|v + t\|_1$ is equivalent to the seminorm $|v|_1$ for any $v \in \dot{v}$. It is then easy to see that

$$\overline{a}(\dot{u}, \dot{v}) = \int_\Omega \nabla u \cdot \nabla v \, dx, \quad u \in \dot{u}, \ v \in \dot{v},$$

defines a bilinear form on V, which is continuous and V-elliptic. Because of the condition (6.11),

$$\langle \overline{\ell}, \dot{v} \rangle = \int_\Omega f \, v \, dx + \int_\Gamma g \, v \, ds$$

is a well-defined linear form on V. To see that ℓ is continuous, we have

$$\langle \overline{\ell}, \dot{v} \rangle = \int_\Omega f \, (v + t) \, dx + \int_\Gamma g \, (v + t) \, ds \quad \forall \, t \in \mathbb{R}$$

and hence

$$|\langle \overline{\ell}, \dot{v} \rangle| \le \inf_t \left\{ \|f\|_{0,\Omega} \|v + t\|_{0,\Omega} + \|g\|_{0,\Gamma} \|v + t\|_{0,\Gamma} \right\}.$$

Using the definition of the quotient norm, we have

$$\inf_t \left\{ \|v + t\|_{0,\Omega} + \|v + t\|_{0,\Gamma} \right\} \le c \, \inf_t \|v + t\|_{1,\Omega} = c \, \|\dot{v}\|_V.$$

Thus,

$$|\langle \overline{\ell}, \dot{v} \rangle| \le c \, \|\dot{v}\|_V,$$

i.e., $\overline{\ell}$ is continuous on V.

Hence, we can apply the Lax–Milgram lemma to conclude that the problem

$$\dot{u} \in V, \quad \overline{a}(\dot{u}, \dot{v}) = \langle \overline{\ell}, \dot{v} \rangle \quad \forall \dot{v} \in V$$

has a unique solution $\dot{u}$. It is easy to see that any $u \in \dot{u}$ is a solution of (6.10).

Mixed boundary conditions are also possible, such as in the problem

$$\begin{aligned} -\Delta u + u &= f \quad \text{in } \Omega, \\ u &= 0 \quad \text{on } \Gamma_D, \\ \partial u / \partial n &= g \quad \text{on } \Gamma_N. \end{aligned} \tag{6.12}$$

Here, Γ_D and Γ_N form a nonoverlapping decomposition of the boundary $\partial\Omega$. That is, Γ_D and Γ_N are relatively open, $\partial\Omega = \overline{\Gamma}_D \cup \overline{\Gamma}_N$, and $\Gamma_D \cap \Gamma_N = \varnothing$. The appropriate space in which to pose this problem in weak form is now

$$V \equiv H^1_{\Gamma_D}(\Omega) = \{v \in H^1(\Omega) : v = 0 \text{ on } \Gamma_D\}. \tag{6.13}$$

Then the weak problem becomes one of finding $u \in V$ such that (6.8) holds, with

$$a(u, v) = \int_\Omega (\nabla u \cdot \nabla v + u\, v)\, dx$$

and

$$\langle \ell, v \rangle = \int_\Omega f\, v\, dx + \int_{\Gamma_N} g\, v\, ds.$$

Under suitable assumptions, say $f \in L^2(\Omega)$ and $g \in L^2(\Gamma_N)$, we can again apply the Lax-Milgram lemma to conclude that the weak problem has a unique solution.

The issue of existence and uniqueness of solutions to the problems just discussed may be treated in the more general framework of arbitrary linear elliptic PDEs of second order. Suppose that the boundary Γ is partitioned according to $\Gamma = \overline{\Gamma}_D \cup \overline{\Gamma}_N$ with $\Gamma_D \cap \Gamma_N = \varnothing$, Γ_D and Γ_N being open subsets of Γ. Consider the boundary value problem

$$\begin{aligned} -D_j(a_{ij} D_i u) + b_i D_i u + cu &= f \quad \text{in } \Omega, \\ u &= 0 \quad \text{on } \Gamma_D, \\ a_{ij} D_i u\, n_j &= g \quad \text{on } \Gamma_N. \end{aligned} \tag{6.14}$$

Here $\boldsymbol{n} = (n_1, \ldots, n_d)^T$ is the unit outward normal on Γ_N, D_i denotes the derivative $\partial/\partial x_i$, D_{ij} stands for the second derivative $\partial^2/\partial x_i \partial x_j$, and the indices i and j range between 1 and d.

The given functions a_{ij}, b_i, c, f, and g are assumed to satisfy the following conditions:

$$\begin{aligned}
&a_{ij}, b_i, c \in L^\infty(\Omega);\\
&\text{the partial differential operator is uniformly elliptic in}\\
&\quad\text{the sense that there exists a constant } \theta > 0 \text{ such that}\\
&\quad a_{ij}\xi_i\xi_j \ge \theta\,|\boldsymbol{\xi}|^2 \quad \forall\,\boldsymbol{\xi} = (\xi_i) \in \mathbb{R}^d, \text{ a.e. in } \Omega;\\
&f \in L^2(\Omega);\\
&g \in L^2(\Gamma_N).
\end{aligned} \tag{6.15}$$

The weak formulation of the problem (6.14) is obtained again in the usual way by multiplying the differential equation in (6.14) by an arbitrary test function $v \in H^1_{\Gamma_D}(\Omega)$, integrating over Ω, performing an integration by parts, and applying the specified boundary conditions. As a result, we get the following weak formulation: Find $u \in H^1_{\Gamma_D}(\Omega)$ such that

$$\begin{aligned}
\int_\Omega (a_{ij} D_i u D_j v + b_i (D_i u)\, v + c\, u\, v)\, dx = \int_\Omega f\, v\, dx + \int_{\Gamma_N} g\, v\, ds \\
\forall\, v \in H^1_{\Gamma_D}(\Omega).
\end{aligned} \tag{6.16}$$

The issue of well-posedness of this problem is settled by appealing to the Lax–Milgram lemma. The space $V = H^1_{\Gamma_D}(\Omega)$ is a Hilbert space, with the standard H^1-norm. The assumptions (6.15) ensure that the lefthand side of (6.16) defines a bounded bilinear form on V, and the right-hand side a bounded linear form on V. What remains to be established is the V-ellipticity of the bilinear form.

By elementary manipulations, it can be shown that the bilinear form is V-elliptic if additionally, one of the following three conditions is satisfied, with $\boldsymbol{b} = (b_1, \ldots, b_d)^T$ and θ the ellipticity constant in (6.15):

$$c \ge c_0 > 0, \ |\boldsymbol{b}| \le B \text{ a.e. in } \Omega, \text{ and } B^2 < 4\,\theta\, c_0$$

or

$$\boldsymbol{b} \cdot \boldsymbol{n} \ge 0 \text{ a.e. on } \Gamma_N, \text{ and } c - \tfrac{1}{2}\operatorname{div} \boldsymbol{b} \ge c_0 > 0 \text{ a.e. in } \Omega$$

or

$$\operatorname{meas}(\Gamma_D) > 0, \ \boldsymbol{b} = \boldsymbol{0}, \text{ and } \inf_\Omega c > -\theta/\bar{c},$$

where $\bar{c}$ is the best constant in the Poincaré inequality

$$\int_\Omega v^2\, dx \le \bar{c} \int_\Omega |\nabla v|^2\, dx \quad \forall\, v \in H^1_{\Gamma_D}(\Omega).$$

This best constant can be computed by solving a linear elliptic eigenvalue problem: $\bar{c} = 1/\lambda_1$, with $\lambda_1 > 0$ the smallest eigenvalue of the eigenvalue problem

$$\begin{aligned} -\Delta u &= \lambda u \quad \text{in } \Omega, \\ u &= 0 \quad \text{on } \Gamma_D, \\ \frac{\partial u}{\partial n} &= 0 \quad \text{on } \Gamma_N. \end{aligned}$$

A special and important case is that corresponding to $b_i = 0$; in this case the bilinear form is symmetric, and V-ellipticity is assured if $c \geq c_0 > 0$, or $c \geq 0$ and meas $(\Gamma_D) > 0$.

Linear elasticity. Since many of the developments later on can be seen in some ways as an extension of the basic theory for the boundary value problem of linear elasticity, this problem and its well-posedness are discussed here.

The equations that govern the behavior of elastic bodies have been presented earlier in Chapter 2 and are repeated here for convenience. Let Ω be a bounded domain in $\mathbb{R}^d$ with a Lipschitz continuous boundary Γ. The governing equations for static behavior are

$$\left.\begin{array}{ll} \text{the equation of equilibrium} & -\operatorname{div} \boldsymbol{\sigma} = \boldsymbol{f} \\ \text{the elastic constitutive law} & \boldsymbol{\sigma} = \boldsymbol{C}\boldsymbol{\epsilon}(\boldsymbol{u}) \\ \text{the strain–displacement equation} & \boldsymbol{\epsilon}(\boldsymbol{u}) = \frac{1}{2}\left(\nabla \boldsymbol{u} + (\nabla \boldsymbol{u})^T\right) \end{array}\right\} \text{ in } \Omega. \tag{6.17}$$

Suppose that the boundary Γ is divided into two complementary parts $\overline{\Gamma}_u$ and $\overline{\Gamma}_g$, where Γ_u and Γ_g are open, $\Gamma_u \cap \Gamma_g = \varnothing$, and $\Gamma_u \neq \varnothing$. The boundary conditions are assumed to be

$$\begin{aligned} \boldsymbol{u} &= \boldsymbol{0} \quad \text{on } \Gamma_u, \\ \boldsymbol{\sigma n} &= \boldsymbol{g} \quad \text{on } \Gamma_g. \end{aligned} \tag{6.18}$$

In order to formulate this problem in a weak form we introduce the space of admissible displacements V defined by

$$V = [H^1_{\Gamma_u}(\Omega)]^d \equiv \{\boldsymbol{v} = (v_i) : \ v_i \in H^1(\Omega), \ v_i = 0 \text{ on } \Gamma_u, \ 1 \leq i \leq d\}.$$

We first eliminate the variable $\boldsymbol{\sigma}$ from the first two equations of (6.17) to obtain

$$-\operatorname{div}\left(\boldsymbol{C}\,\boldsymbol{\epsilon}(\boldsymbol{u})\right) = \boldsymbol{f} \quad \text{in } \Omega.$$

Then multiplication of the above equation by an arbitrary member $\boldsymbol{v}$ of V, integration over Ω, use of the integration by parts formula, and imposition of the boundary condition $(6.18)_2$ lead to the problem of finding $\boldsymbol{u} \in V$ such that

$$a(\boldsymbol{u}, \boldsymbol{v}) = \langle \boldsymbol{\ell}, \boldsymbol{v} \rangle \quad \forall \boldsymbol{v} \in V, \tag{6.19}$$

where

$$a(\boldsymbol{u}, \boldsymbol{v}) = \int_\Omega \boldsymbol{C}\boldsymbol{\epsilon}(\boldsymbol{u}) : \boldsymbol{\epsilon}(\boldsymbol{v})\, dx, \tag{6.20}$$

$$\langle \boldsymbol{\ell}, \boldsymbol{v} \rangle = \int_\Omega \boldsymbol{f} \cdot \boldsymbol{v}\, dx + \int_{\Gamma_g} \boldsymbol{g} \cdot \boldsymbol{v}\, ds. \tag{6.21}$$

The question of well-posedness of the problem (6.19) is once again settled by appealing to the Lax–Milgram lemma. The continuity of the bilinear form and the linear functional are fairly straightforward to verify, while the V-ellipticity of $a(\cdot,\cdot)$ follows from the assumption that $\boldsymbol{C}$ is pointwise stable (see (2.29)) and the use of Korn's inequality (5.22). We thus have the following result.

THEOREM 6.1. *The problem defined by* (6.19)–(6.21) *has a unique solution* $\boldsymbol{u} \in V$ *under the stated hypotheses. Furthermore, there is a constant* $c > 0$ *such that*

$$\|\boldsymbol{u}\|_V \le c\,(\|\boldsymbol{f}\|_{L^2(\Omega)} + \|\boldsymbol{g}\|_{L^2(\Gamma_g)}). \tag{6.22}$$

Minimization problems. The term "variational" in the description of problems of the form (6.4) derives from the association with minimization problems, for the case in which the bilinear form $a(\cdot,\cdot)$ is symmetric. Indeed, assuming the symmetry of $a(\cdot,\cdot)$, we can consider the problem of minimizing the functional $J : V \to \mathbb{R}$, defined by

$$J(v) = \tfrac{1}{2}a(v,v) - \langle \ell, v \rangle,$$

among all functions in V. It is not difficult to show that the condition satisfied by a minimizer of $J(\cdot)$ is precisely (6.4). Conversely, under the additional assumption that $a(\cdot,\cdot)$ is V-elliptic, a solution of (6.4) is also a minimizer of the functional $J(\cdot)$. Thus, under the stated assumptions, the weak formulation and the minimization problem are equivalent, and the existence of a unique minimizer of J may be inferred from the unique solvability of the weak formulation concluded by the Lax–Milgram lemma. We remark that the framework of weak formulations is more general than that of minimization problems in that the bilinear forms are not assumed to be symmetric.

For functionals of a more general nature, the following proposition gives conditions under which a unique minimizer exists (cf. [99], Proposition 2.2.1).

PROPOSITION 6.2. *Let X be a reflexive Banach space, K a nonempty,* *closed, convex subset of X, and f a proper, convex, l.s.c. functional on K. Assume that*

$$f(x) \to \infty \quad as \ \|x\|_X \to \infty, \ x \in K. \tag{6.23}$$

Then there exists $x_0 \in K$ such that

$$f(x_0) = \min_{x \in K} f(x).$$

If f is strictly convex on K, then the solution x_0 is unique.

PROOF. *Existence.* Define $\alpha = \inf_{x \in K} f(x)$. Since f is proper, it follows that $-\infty \le \alpha < \infty$. From the definition of infimum, there exists a sequence $\{x_n\}$ in K such that

$$f(x_n) \to \alpha \quad \text{as } n \to \infty.$$

By the coerciveness assumption (6.23), we see that the sequence is bounded, and so by Theorem 5.1 there exists a subsequence, again denoted by $\{x_n\}$, such that $x_n \rightharpoonup x_0$ for some $x_0 \in X$. Since K is closed and convex, it is sequentially weakly closed (cf. [34]). Hence $x_n \rightharpoonup x_0$ and $x_0 \in K$.

Since f is convex and l.s.c., it is weakly l.s.c. (see Section 4.1), and thus

$$f(x_0) \le \liminf_{n \to \infty} f(x_n) = \alpha,$$

which implies that $f(x_0) = \alpha$, i.e., $x_0 \in K$ is a minimizer of f over K.

Uniqueness. Suppose that the problem has two distinct solutions, x_1 and x_2. Then $\frac{1}{2}(x_1 + x_2) \in K$ and, since f is strictly convex,

$$f\left(\tfrac{1}{2}(x_1 + x_2)\right) < \tfrac{1}{2} f(x_1) + \tfrac{1}{2} f(x_2) = f(x_1).$$

This contradicts the assumption that x_1 minimizes f over K. □

Mixed variational problems. This class of variational problems may be regarded as an extension of the standard elliptic problem (6.4). It arises generally in one of two ways: as a result of either the introduction of additional dependent variables as unknowns in a problem or the introduction of extra unknown variables to eliminate constraints of a problem. The mixed problem is then posed on the Cartesian product $V \times Q$ of two spaces V and Q, Q being the space for the additional unknowns, and the problem takes the following form. Let $a(\cdot, \cdot)$ be a bilinear form defined on V as before, and let $b(\cdot, \cdot) : V \times Q \to \mathbb{R}$ be another bilinear form, and ℓ and m linear functionals defined respectively on V and on Q. Then the problem is one of finding $u \in V$ and $p \in Q$ that satisfy

$$\begin{aligned} a(u, v) + b(v, p) &= \langle \ell, v \rangle \quad \forall\, v \in V, \\ b(u, q) &= \langle m, q \rangle \quad \forall\, q \in Q. \end{aligned} \tag{6.24}$$

are sometimes referred to as *saddle-point problems* because in the event that $a(\cdot,\cdot)$ is symmetric, the problem (6.24) can be shown to be equivalent to the saddle-point, or minimax, problem of finding $(u,p) \in V \times Q$ such that

$$L(u,q) \le L(u,p) \le L(v,p) \quad \forall\, v \in V,\ q \in Q,$$

where $L(v,q) = \frac{1}{2}a(v,v) + b(v,q) - \langle \ell, v\rangle - \langle m, q\rangle$. However, it should be stressed that not every mixed problem may be posed as a saddle-point problem, so that the formulation (6.24) is more general, and will in fact be the formulation favored in these developments.

As an example of a mixed variational problem arising in linear elasticity, we return to (6.17), but this time agree to use as variables both the displacement $\boldsymbol{u}$ and the stress $\boldsymbol{\sigma}$. Multiplying the equilibrium equation by an arbitrary function $\boldsymbol{v} \in V = [H^1_{\Gamma_u}(\Omega)]^d$, integrating over Ω, performing an integration by parts, and using the boundary condition $(6.18)_2$, we obtain the equation

$$\int_\Omega \boldsymbol{\sigma} : \boldsymbol{\epsilon}(\boldsymbol{v})\, dx = \int_\Omega \boldsymbol{f} \cdot \boldsymbol{v}\, dx + \int_{\Gamma_g} \boldsymbol{g} \cdot \boldsymbol{v}\, ds \quad \forall\, \boldsymbol{v} \in V. \tag{6.25}$$

Next, set

$$Q = [L^2(\Omega)]^{d\times d}_{\mathrm{sym}} \equiv \left\{ \boldsymbol{\tau} \in [L^2(\Omega)]^{d\times d} :\ \tau_{ij} = \tau_{ji},\ 1 \le i,j \le d \right\};$$

this is the space of admissible stresses. The second equation from the constitutive law $(6.17)_2$ is employed in the form

$$\boldsymbol{\epsilon}(\boldsymbol{u}) = \boldsymbol{A}\boldsymbol{\sigma},$$

where $\boldsymbol{A} = \boldsymbol{C}^{-1}$ is the elastic compliance tensor. Now take the scalar product of this constitutive equation with an arbitrary member $\boldsymbol{\tau} \in Q$ and integrate to obtain

$$\int_\Omega \boldsymbol{A}\boldsymbol{\sigma} : \boldsymbol{\tau}\, dx - \int_\Omega \boldsymbol{\epsilon}(\boldsymbol{u}) : \boldsymbol{\tau}\, dx = 0 \quad \forall\, \boldsymbol{\tau} \in Q. \tag{6.26}$$

Equations (6.25) and (6.26) make up the mixed variational problem of determining $(\boldsymbol{u}, \boldsymbol{\sigma}) \in V \times Q$ that satisfy

$$\begin{aligned} a(\boldsymbol{\sigma},\boldsymbol{\tau}) + b(\boldsymbol{\tau},\boldsymbol{u}) &= 0 \quad \forall\, \boldsymbol{\tau} \in Q,\\ b(\boldsymbol{\sigma},\boldsymbol{v}) &= \langle \boldsymbol{\ell}, \boldsymbol{v}\rangle \quad \forall\, \boldsymbol{v} \in V, \end{aligned} \tag{6.27}$$

in which

$$\begin{aligned} a(\boldsymbol{\sigma},\boldsymbol{\tau}) &= \int_\Omega \boldsymbol{A}\boldsymbol{\sigma} : \boldsymbol{\tau}\, dx,\\ b(\boldsymbol{\tau},\boldsymbol{v}) &= -\int_\Omega \boldsymbol{\epsilon}(\boldsymbol{v}) : \boldsymbol{\tau}\, dx,\\ \langle \boldsymbol{\ell}, \boldsymbol{\tau}\rangle &= -\int_\Omega \boldsymbol{f}\cdot\boldsymbol{v}\, dx - \int_{\Gamma_g} \boldsymbol{g}\cdot\boldsymbol{v}\, ds. \end{aligned} \tag{6.28}$$

Returning to the general case, clearly the question of well-posedness of mixed problems will be closely tied to the choices of V and Q, and to the properties of the two bilinear forms that appear in the problem. The key to the question of existence is Theorem 6.3 below. First, assume that both bilinear forms are continuous, so that

$$\begin{aligned} |a(u,v)| &\leq \|a\|\,\|u\|\,\|v\| \quad \forall\, u,v \in V, \\ |b(v,q)| &\leq \|b\|\,\|v\|\,\|q\| \quad \forall\, v \in V, q \in Q; \end{aligned} \tag{6.29}$$

then it is possible to define bounded linear operators $A : V \to V'$, $B : V \to Q'$, and $B' : Q \to V'$ that satisfy

$$\begin{aligned} \langle Au, v\rangle &= a(u,v) \quad \forall\, u,v \in V, \\ \langle Bv, q\rangle &= \langle B'q, v\rangle = b(v,q) \quad \forall\, v \in V,\ q \in Q. \end{aligned} \tag{6.30}$$

Furthermore, define the kernels $\operatorname{Ker} B$ and $\operatorname{Ker} B'$ of B and B' by

$$\begin{aligned} \operatorname{Ker} B &= \{v \in V:\ b(v,q) = 0 \quad \forall\, q \in Q\}, \\ \operatorname{Ker} B' &= \{q \in Q:\ b(v,q) = 0 \quad \forall\, v \in V\}. \end{aligned} \tag{6.31}$$

Then the following result holds.

THEOREM 6.3 (BABUŠKA [4], BREZZI [17]). *Let V and Q be Banach spaces. Suppose that the bilinear form $a(\cdot,\cdot)$ is symmetric, continuous and* $\operatorname{Ker} B$*-elliptic. Suppose furthermore that $b(\cdot,\cdot)$ is continuous, and that there exists a constant $\beta > 0$ such that*

$$\sup_{v\in V} \frac{b(v,q)}{\|v\|} \geq \beta\, \|q\|_{Q\setminus \operatorname{Ker} B'} \quad \forall\, q \in Q. \tag{6.32}$$

Then there exists a solution (u,p) to (6.24) for any $\ell \in V'$ and $m \in Q'$, with u being unique and p being uniquely determined up to a member of $\operatorname{Ker} B'$.

In (6.32), the quotient norm $\|q\|_{Q\setminus \operatorname{Ker} B'}$ is defined by

$$\|q\|_{Q\setminus \operatorname{Ker} B'} = \inf_{q_0 \in \operatorname{Ker} B'} \|q + q_0\|.$$

It is quite straightforward to show that if $\operatorname{meas}(\Gamma_u) \neq 0$, then the mixed variational problem of linear elasticity (6.27)–(6.28) satisfies all the conditions of Theorem 6.3, with $\operatorname{Ker} B' = \{\mathbf{0}\}$, and therefore has a unique solution.

6.2 Elliptic Variational Inequalities

The analysis of variational inequalities has its origins in the work of Fichera [40], who studied inequalities arising in unilateral problems of elasticity.

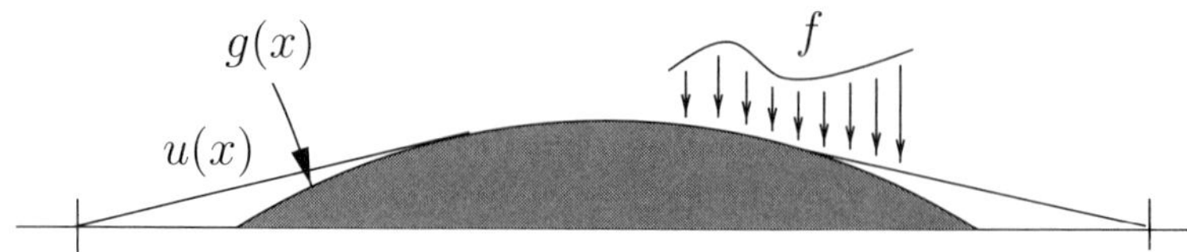

Figure 6.1: A membrane supported by an obstacle; one-dimensional view

Significant contributions were also made by Lions and Stampacchia [79]. In the literature there are several monographs devoted the theory and numerical solution of variational inequalities; see, for example, Duvaut and Lions [33], Friedman [42], Glowinski [44], Glowinski, Lions, and Trémolières [45], Kikuchi and Oden [70], Kinderlehrer and Stampacchia [71], and Panagiotopoulos [99]. In this section we give a brief introduction to some well-known results on the existence and uniqueness of solutions to standard elliptic variational inequalities (EVI). The presentation on well-posedness given here follows that of [44].

As an example of a problem that leads to an elliptic variational inequality, let us consider the obstacle problem. We need to determine the equilibrium position of an elastic membrane that (1) passes through a curve Γ, the boundary of a planar domain Ω; (2) lies above an obstacle of height g; and (3) is subject to the action of a vertical force of density f, where f is a given function.

The unknown variable of the problem is the vertical displacement u of the membrane. Since the membrane is fixed along the boundary Γ, we see that $u = 0$ on Γ. To make the problem meaningful, we assume that the obstacle function satisfies the condition $g \le 0$ on Γ. We will assume that $g \in H^1(\Omega)$ and $f \in H^{-1}(\Omega)$. Thus the set of admissible displacements is

$$K = \{v \in H_0^1(\Omega) : \ v \ge g \text{ a.e. in } \Omega\}.$$

The principle of minimum potential energy asserts that the displacement u is a minimizer of the total energy; that is,

$$u \in K : \quad J(u) = \inf\{J(v) : v \in K\}, \tag{6.33}$$

where the energy functional is given by

$$J(v) = \int_\Omega \left(\frac{1}{2}\,|\nabla v|^2 - f\,v\right) dx.$$

The set K is nonempty, because the function $\max\{0, g\}$ belongs to K. Also, it is easy to see that K is closed and convex. Now, the energy functional J is strictly convex, coercive, and continuous on K. Hence the minimization problem (6.33) has a unique solution $u \in K$, which is also characterized by the variational inequality

$$u \in K, \quad \int_\Omega \nabla u \cdot \nabla(v-u)\,dx \ge \int_\Omega f\,(v-u)\,dx \quad \forall\, v \in K. \tag{6.34}$$

Now we derive the corresponding boundary value problem for the weak formulation (6.34). Assume that $f \in C(\Omega)$ and $g \in C(\Omega)$. Suppose that the solution u of (6.34) is sufficiently smooth; more precisely, $u \in C^2(\Omega) \cap C(\overline{\Omega})$. We then perform an integration by parts in (6.34) to obtain

$$\int_\Omega (-\Delta u - f)\,(v-u)\,dx \ge 0 \quad \forall\, v \in K. \tag{6.35}$$

We let $v = u + \phi$, $\phi \in C_0^\infty(\Omega)$, and $\phi \ge 0$, in (6.35),

$$\int_\Omega (-\Delta u - f)\,\phi\,dx \ge 0 \quad \forall\, \phi \in C_0^\infty(\Omega),\ \phi \ge 0.$$

We see then that

$$-\Delta u - f \ge 0 \quad \text{in } \Omega.$$

If $u(\boldsymbol{x}_0) > g(\boldsymbol{x}_0)$ at $\boldsymbol{x}_0 \in \Omega$, then we can find a neighborhood $U(\boldsymbol{x}_0) \subset \Omega$ of $\boldsymbol{x}_0$ and a number $\delta > 0$ such that $u(\boldsymbol{x}) > g(\boldsymbol{x}) + \delta$ for $\boldsymbol{x} \in U(\boldsymbol{x}_0)$. Then in (6.35) we take $v = u \pm \delta\,\phi$ with $\phi \in C_0^\infty(U(\boldsymbol{x}_0))$ and $\|\phi\|_\infty \le 1$,

$$\pm \int_\Omega (-\Delta u - f)\,\phi\,dx \ge 0 \quad \forall\, \phi \in C_0^\infty(U(\boldsymbol{x}_0)),\ \|\phi\|_\infty \le 1.$$

Therefore,

$$\int_\Omega (-\Delta u - f)\,\phi\,dx = 0 \quad \forall\, \phi \in C_0^\infty(U(\boldsymbol{x}_0)),$$

from which we get

$$(-\Delta u - f)(\boldsymbol{x}_0) = 0.$$

In conclusion, we have shown that if the weak solution of the problem (6.34) is sufficiently smooth, then it satisfies the relations

$$u - g \ge 0, \quad -\Delta u - f \ge 0, \quad (u-g)\,(-\Delta u - f) = 0 \quad \text{in } \Omega. \tag{6.36}$$

Thus the domain Ω can be decomposed into two parts. On one part, Ω_1, the membrane has no contact with the obstacle, and so

$$u > g \quad \text{and} \quad -\Delta u - f = 0 \quad \text{in } \Omega_1.$$

The second of these equations expresses the condition of equilibrium, or balance of forces. On the other part, Ω_2, there is contact between the membrane and the obstacle, so that

$$u = g \quad \text{and} \quad -\Delta u - f > 0 \quad \text{in } \Omega_2.$$

The second relation expresses the fact that the net force exerted by the obstacle on the membrane is positive. We notice that the region of contact, $\{\boldsymbol{x} \in \Omega : u(\boldsymbol{x}) = g(\boldsymbol{x})\}$, is an unknown a priori.

Conversely, from (6.36) we can derive the variational inequality (6.34). To see this, we first have

$$\int_\Omega (-\Delta u - f)\,(v - g)\,dx \geq 0 \quad \forall\, v \in K.$$

Since

$$\int_\Omega (-\Delta u - f)\,(u - g)\,dx = 0,$$

we have

$$\int_\Omega (-\Delta u - f)\,(v - u)\,dx \geq 0 \quad \forall\, v \in K,$$

and after an integration by parts,

$$\int_\Omega \nabla u \cdot \nabla (v - u)\,dx \geq \int_\Omega f\,(v - u)\,dx \quad \forall\, v \in K.$$

The obstacle problem is a canonical example of a class of inequality problems known as *elliptic variational inequalities (EVIs) of the first kind*. We remark that not every variational inequality is derived from a minimization principle. The feature of a variational inequality arising from a quadratic minimization problem is that the bilinear form of the inequality is symmetric. In our general framework, we do not need the symmetry assumption on the bilinear form.

The abstract form of the EVI of the first kind is the following. Let V be a real Hilbert space with inner product $(\cdot,\cdot)$ and associated norm $\|\cdot\|$, K a subset of V. Let $a : V \times V \to \mathbb{R}$ be a continuous, V-elliptic bilinear form. Given a linear functional $\ell : V \to \mathbb{R}$, it is required to find $u \in K$ satisfying

$$a(u, v - u) \geq \langle \ell, v - u \rangle \quad \forall\, v \in K. \tag{6.37}$$

Variational inequalities of the first kind may be characterized by the fact that they are posed on convex subsets; indeed, if the set K is in fact a subspace of V, then the problem becomes a variational equation.

THEOREM 6.4. *Let V be a real Hilbert space, $a : V \times V \to \mathbb{R}$ a continuous, V-elliptic bilinear form, $\ell : V \to \mathbb{R}$ a bounded linear functional, and $K \subset V$ a nonempty, closed and convex set. Then the EVI (6.37) has a unique solution $u \in K$.*

In the proof of Theorem 6.4 we will use the following well-known result.

THEOREM 6.5 (BANACH FIXED-POINT THEOREM). *Let X be a Banach* *space. Assume that $f : X \to X$ is a contractive mapping, that is, for some $\kappa \in [0, 1)$,*

$$\|f(x) - f(y)\| \le \kappa \|x - y\| \quad \forall\, x, y \in X.$$

Then f has a unique fixed point $x \in X$, $f(x) = x$.

PROOF OF THEOREM 6.4. We rewrite the inequality (6.37) in the form of an equivalent fixed-point problem. To do this, we first apply the Riesz representation theorem (Theorem 5.7) to claim that there exists a unique member $L \in V$ such that $\|L\| = \|\ell\|$ and

$$\langle \ell, v \rangle = (L, v) \quad \forall\, v \in V.$$

For any fixed $u \in V$, the mapping $v \mapsto a(u, v)$ defines a linear, continuous form on V. Thus applying the Riesz representation theorem again, we have a mapping $A : V \to V$ such that

$$a(u, v) = (Au, v) \quad \forall\, v \in V.$$

Since $a(\cdot, \cdot)$ is bilinear and continuous, it is easy to verify that A is linear and bounded, with

$$\|A\| \le M.$$

For any $\theta > 0$, the problem (6.37) is therefore equivalent to one of finding $u \in K$ such that

$$((u - \theta\,(Au - L)) - u, v - u) \le 0 \quad \forall\, v \in K. \tag{6.38}$$

If P_K denotes the orthogonal projection onto K, then (6.38) may be written in the form

$$u = P_K\,(u - \theta\,(Au - L))\,. \tag{6.39}$$

Recall that the projection operator P_K is nonexpansive. We show that by choosing $\theta > 0$ sufficiently small, the operator defined by the right-hand side of (6.39) is a contraction. Indeed, for any $v_1, v_2 \in V$, we have

$$\begin{aligned}
&\|P_K\,(v_1 - \theta\,(Av_1 - L)) - P_K\,(v_2 - \theta\,(Av_2 - L))\,\|^2 \\
&\quad \le \|\,(v_1 - \theta\,(Av_1 - L)) - (v_2 - \theta\,(Av_2 - L))\,\|^2 \\
&\quad = \|(v_1 - v_2) - \theta\,A(v_1 - v_2)\|^2 \\
&\quad = \|v_1 - v_2\|^2 - 2\,\theta\,a(v_1 - v_2, v_1 - v_2) + \theta^2\|A(v_1 - v_2)\|^2 \\
&\quad \le (1 - 2\,\theta\,\alpha + \theta^2 M^2)\,\|v_1 - v_2\|^2.
\end{aligned}$$

Here $\alpha > 0$ is the V-ellipticity constant for the bilinear form $a(\cdot, \cdot)$.

Thus if we choose $\theta \in (0, 2\,\alpha/M^2)$, then the operator defined by the right-hand side of (6.39) is a contraction. Then by Theorem 6.4, the problem (6.39), and equivalently the problem (6.37), has a unique solution. □

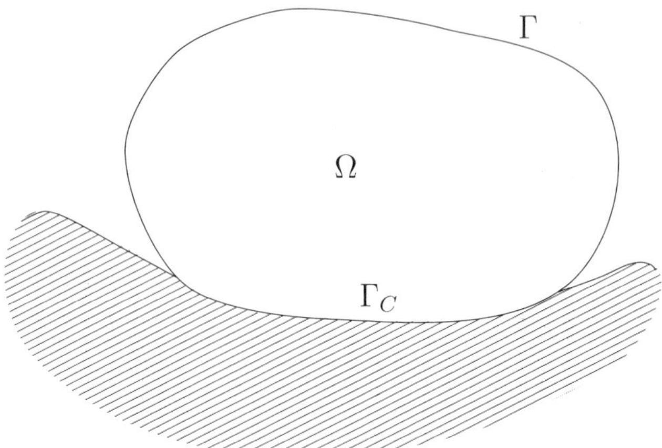

Figure 6.2: An elastic body in frictional contact with a rigid obstacle

We note that Theorem 6.4 is a generalization of the Lax–Milgram lemma. The obstacle problem (6.34) is clearly an elliptic variational inequality of the first kind. All the conditions stated in Theorem 6.4 are satisfied, and hence the obstacle problem (6.34) has a unique solution.

A second class of variational inequalities arises as a result of the presence of nondifferentiable functions. As an example, we consider a reduced problem arising in frictional contact between an elastic body and a rigid foundation (cf. [70]). The elastic body occupies a bounded domain Ω with a Lipschitz boundary Γ. A part Γ_C of the boundary is in contact with a rigid obstacle (Figure 6.2); contact between the body and the obstacle is assumed to frictional, with friction governed by a reduced Coulomb law. For this problem Γ_C is assumed to be known in advance, as is the normal surface traction on Γ_C. The differential equations of the problem are given by (6.17) on the domain Ω, but the boundary conditions now differ: The boundary is assumed to be partitioned into three nonoverlapping regions Γ_u, Γ_g, and Γ_C. The boundary conditions on Γ_u and Γ_g are given in (6.18). To describe the boundary condition on Γ_C, we first introduce some notation. On the boundary Γ we define the normal displacement to be $u_n = \boldsymbol{u} \cdot \boldsymbol{n}$ and the tangential displacement $\boldsymbol{u}_t = \boldsymbol{u} - u_n\boldsymbol{n}$. Then we have the decomposition

$$\boldsymbol{u} = u_n\boldsymbol{n} + \boldsymbol{u}_t$$

for the displacement. Similarly, $\boldsymbol{\sigma n}$ is the stress vector, and we define its normal component $\sigma_n = (\boldsymbol{\sigma n}) \cdot \boldsymbol{n}$ and the tangential stress vector $\boldsymbol{\sigma}_t =$

$\boldsymbol{\sigma n} - \sigma_n \boldsymbol{n}$. In this way, we have the decomposition

$$\boldsymbol{\sigma n} = \sigma_n \boldsymbol{n} + \boldsymbol{\sigma}_t$$

for the stress vector. On Γ_C, we impose the condition

$$\begin{aligned} &\sigma_n = -G, \\ &|\boldsymbol{\sigma}_t| \le G, \\ &|\boldsymbol{\sigma}_t| < \nu_F G \Longrightarrow \boldsymbol{u}_t = \boldsymbol{0}, \\ &|\boldsymbol{\sigma}_t| = \nu_F G \Longrightarrow \boldsymbol{u}_t = -\lambda\, \boldsymbol{\sigma}_t \text{ for some } \lambda \ge 0. \end{aligned} \tag{6.40}$$

Here, $G > 0$ and the friction coefficient $\nu_F > 0$ are prescribed functions, $G, \nu_F \in L^\infty(\Gamma_C)$. From (6.40), we see that

$$\boldsymbol{\sigma}_t \cdot \boldsymbol{u}_t = -\nu_F G\, |\boldsymbol{u}_t| \quad \text{on } \Gamma_C.$$

Then the variational problem corresponding to (6.17), (6.18), and (6.40) becomes one of finding the displacement field $\boldsymbol{u} \in V \equiv [H^1_{\Gamma_u}(\Omega)]^3$ (see (6.13) for the definition of the space) that satisfies

$$\begin{aligned} &\int_\Omega \boldsymbol{C\epsilon}(\boldsymbol{u}) : \boldsymbol{\epsilon}(\boldsymbol{v})\, dx + \int_{\Gamma_C} \nu_F G\, |\boldsymbol{v}_t|\, ds - \int_{\Gamma_C} \nu_F G\, |\boldsymbol{u}_t|\, ds \\ &\qquad \ge \int_\Omega \boldsymbol{f} \cdot (\boldsymbol{v} - \boldsymbol{u})\, dx + \int_{\Gamma_g} \boldsymbol{g} \cdot (\boldsymbol{v} - \boldsymbol{u})\, ds \\ &\qquad - \int_{\Gamma_C} G\, (v_n - u_n)\, ds \qquad \forall\, \boldsymbol{v} \in V. \end{aligned} \tag{6.41}$$

For simplicity, we assume as before that $\boldsymbol{f} \in [L^2(\Omega)]^3$ and $\boldsymbol{g} \in [L^2(\Gamma_g)]^3$.

The problem (6.41) is an example of an EVI of the *second* kind. To give the general framework for this class of problems, in addition to the bilinear form $a(\cdot,\cdot)$ and the linear functional ℓ, we introduce a proper, convex, and lower semicontinuous (l.s.c.) functional $j : V \to \overline{\mathbb{R}}$ (see Chapter 4 for the definitions). The functional j is not assumed to be differentiable. Then the problem of finding $u \in V$ that satisfies

$$a(u, v - u) + j(v) - j(u) \ge \langle \ell, v - u \rangle \quad \forall\, v \in V \tag{6.42}$$

is referred to as an EVI of the second kind.

The key difference between an EVI of the first kind and one of the second kind is that the former is an inequality as a result of the problem being formulated on a convex subset rather than on the whole space; the latter is an inequality due to the presence of the nondifferentiable term $j(\cdot)$.

THEOREM 6.6. *Let V be a real Hilbert space, $a : V \times V \to \mathbb{R}$ a continuous, V-elliptic bilinear form, $\ell : V \to \mathbb{R}$ a bounded linear functional, and $j : V \to \overline{\mathbb{R}}$ a proper, convex, and l.s.c. functional on V. Then the EVI of the second kind* (6.42) *has a unique solution.*

PROOF. Showing uniqueness is straightforward. Since j is proper, $j(v_0) < \infty$ for some $v_0 \in V$. Thus a solution u of (6.42) satisfies

$$j(u) \le a(u, v_0 - u) + j(v_0) - \langle \ell, v_0 - u \rangle < \infty,$$

that is, $j(u)$ is a real number. Now let u_1 and u_2 denote two solutions of the problem (6.42); then

$$\begin{aligned} a(u_1, u_2 - u_1) + j(u_2) - j(u_1) &\ge \langle \ell, u_2 - u_1 \rangle, \\ a(u_2, u_1 - u_2) + j(u_1) - j(u_2) &\ge \langle \ell, u_1 - u_2 \rangle. \end{aligned}$$

Adding the two inequalities, we get

$$-a(u_1 - u_2, u_1 - u_2) \ge 0,$$

which implies, by the V-ellipticity of $a(\cdot,\cdot)$, that $u_1 = u_2$.

The proof of existence is more involved. First consider the case in which $a(\cdot,\cdot)$ is symmetric; under this additional assumption, the inequality (6.42) is equivalent to the minimization problem

$$u \in V, \quad J(u) = \inf\{J(v) : v \in V\}, \tag{6.43}$$

where

$$J(v) = \tfrac{1}{2}\, a(v, v) + j(v) - \langle \ell, v \rangle.$$

Since j is proper, convex, and l.s.c., from a result in convex analysis (cf. [34]) it is bounded below by a bounded affine functional, that is,

$$j(v) \ge \langle \ell_j, v \rangle + c_0 \quad \forall\, v \in V,$$

where ℓ_j is a continuous linear form on V and $c_0 \in \mathbb{R}$. Thus by the stated assumptions on a, j, and ℓ, we see that J is proper, convex, and l.s.c., and has the property that

$$J(v) \to \infty \quad \text{as } \|v\| \to \infty.$$

Applying Proposition 6.2, we see that the problem (6.43), and hence the problem (6.42), has a solution.

Consider next the general case without the symmetry assumption. Again we will convert the problem into an equivalent fixed-point problem. For any $\theta > 0$, the problem (6.42) is equivalent to

$$\begin{aligned} &(u, v - u) + \theta\, j(v) - \theta\, j(u) \\ &\qquad \ge (u, v - u) - \theta\, a(u, v - u) + \theta\, \langle \ell, v - u \rangle \quad \forall\, v \in V. \end{aligned}$$

Now for any $u \in V$, consider the problem of finding $w \in V$ such that

$$\begin{aligned} &(w, v - w) + \theta\, j(v) - \theta\, j(w) \\ &\qquad \ge (u, v - w) - \theta\, a(u, v - w) + \theta\, \langle \ell, v - w \rangle \quad \forall\, v \in V. \end{aligned} \tag{6.44}$$

From the previous discussion we see that this problem has a unique solution, which is denoted by $w = P_\theta u$. Obviously, a fixed point of the mapping P_θ is a solution of the problem (6.42). We will show that for sufficiently small $\theta > 0$, P_θ is a contraction and hence has a unique fixed point by Theorem 6.4.

For any $u_1, u_2 \in V$, let $w_1 = P_\theta u_1$ and $w_2 = P_\theta u_2$. Then we have

$$\begin{aligned}
&(w_1, w_2 - w_1) + \theta\, j(w_2) - \theta\, j(w_1) \\
&\qquad \ge (u_1, w_2 - w_1) - \theta\, a(u_1, w_2 - w_1) + \theta\, \langle \ell, w_2 - w_1 \rangle, \\
&(w_2, w_1 - w_2) + \theta\, j(w_1) - \theta\, j(w_2) \\
&\qquad \ge (u_2, w_1 - w_2) - \theta\, a(u_2, w_1 - w_2) + \theta\, \langle \ell, w_1 - w_2 \rangle.
\end{aligned}$$

Adding the two inequalities and simplifying, we get

$$\begin{aligned}
\|w_1 - w_2\|^2 &\le (u_1 - u_2, w_1 - w_2) - \theta\, a(u_1 - u_2, w_1 - w_2) \\
&= ((I - \theta\, A)(u_1 - u_2), w_1 - w_2),
\end{aligned}$$

where the operator A is defined by the relation $a(u, v) = (Au, v)$ for any $u, v \in V$, as in the proof of Theorem 6.4. Hence

$$\|w_1 - w_2\| \le \|(I - \theta\, A)(u_1 - u_2)\|.$$

Now for any $u \in V$,

$$\begin{aligned}
\|(I - \theta\, A)u\|^2 &= \|u - \theta\, Au\|^2 \\
&= \|u\|^2 - 2\,\theta\, a(u, u) + \theta^2 \|Au\|^2 \\
&\le (1 - 2\,\theta\,\alpha + \theta^2 M^2)\, \|u\|^2.
\end{aligned}$$

Here M and α are the continuity and V-ellipticity constants of the bilinear form $a(\cdot,\cdot)$. Therefore, again, for $\theta \in (0, 2\,\alpha/M^2)$ the mapping P_θ is a contraction on the Hilbert space V. $\square$

With the identification

$$\begin{aligned}
a(\boldsymbol{u}, \boldsymbol{v}) &= \int_\Omega C_{ijkl} u_{i,j} v_{k,l}\, dx, \\
j(\boldsymbol{v}) &= \int_{\Gamma_C} \nu_F G\, |\boldsymbol{v}_t|\, ds, \\
\langle \boldsymbol{\ell}, \boldsymbol{v} \rangle &= \int_\Omega \boldsymbol{f} \cdot \boldsymbol{v}\, dx + \int_{\Gamma_g} \boldsymbol{g} \cdot \boldsymbol{v}\, ds - \int_{\Gamma_C} G\, (v_n - u_n)\, ds,
\end{aligned}$$

we see that the contact problem (6.41) is an elliptic variational inequality of the second kind. It is easy to verify that the conditions stated in Theorem 6.6 are satisfied, and hence the problem (6.41) has a unique solution.

It happens in some applications that the bilinear form a satisfies the V-ellipticity condition only on the subset K; that is, there is a constant $c_0 > 0$ such that

$$a(v,v) \geq c_0 \|v\|^2 \quad \forall\, v \in K.$$

In such a situation we cannot apply Theorems 6.4 and 6.6 to draw conclusions about the solvability of the variational inequality, and instead Proposition 6.2 must be invoked.

Results on the regularity of solutions to EVIs are important in deriving optimal order error estimates of numerical solutions. Some regularity results can be found in the references introduced at the beginning of the section.

6.3 Parabolic Variational Inequalities

Parabolic variational inequalities arise in a way very similar to that of the elliptic variational inequalities that have been discussed, when time is also present as an independent variable. We briefly review some abstract results for parabolic variational inequalities.

Let V and H be two real Hilbert spaces such that $V \subset H$ and V is dense in H. We identify H with its dual space H'. Let K be a nonempty, closed and convex subset of V. Let A be a linear continuous functional from V to V' with $\langle Av, v\rangle \geq \alpha \|v\|_V^2$. The definition $a(u,v) = \langle Au, v\rangle$ induces a bilinear form $a : V \times V \to \mathbb{R}$ that is continuous and V-elliptic. Let $f \in L^2(0,T;V')$ for some time interval $[0,T]$, and suppose that the time derivative $\dot{f} \in L^2(0,T;V')$. Finally, let $u_0 \in K$ be a given initial value. Then a parabolic variational inequality of the first kind is a problem of the following form: Find a function $u \in L^2(0,T;V)$ with $\dot{u} \in L^2(0,T;V')$ and $u(0) = u_0$, such that for almost all (a.a.) $t \in [0,T]$, $u(t) \in K$ and

$$(\dot{u}(t), v - u(t)) + a(u(t), v - u(t)) \geq \langle f(t), v - u(t)\rangle \quad \forall\, v \in K. \tag{6.45}$$

The following result is found in Glowinski, Lions, and Trémolières [45] (Chapter 6, Section 2).

THEOREM 6.7. *In addition to the above assumptions, assume further that* $f(0) - Au_0 \in H$. *Then the parabolic variational inequality of the first kind* (6.45) *has a unique solution. Furthermore,*

$$u, \dot{u} \in L^2(0,T;V) \cap L^\infty(0,T;H).$$

To describe the problem corresponding to a parabolic variational inequality of the second kind, we introduce a functional $j : V \to \overline{\mathbb{R}}$. Following Duvaut and Lions [33] (Chapter 1, Section 5), we assume that $j : V \to \overline{\mathbb{R}}$ is proper, convex, and l.s.c. and that there exists a family of differentiable functions j_k on V such that the following three conditions are satisfied:

- $\int_0^T j_k(v(t))\,dt \to \int_0^T j(v(t))\,dt$ for any $v \in L^2(0,T;V)$;
- there is a sequence u_k bounded in V such that $j_k'(u_k) = 0$ for any k;
- if $v_k \rightharpoonup v$, $\dot{v}_k \rightharpoonup \dot{v}$ in $L^2(0,T;V)$ and $\int_0^T j_k(v_k)\,dt$ is bounded from above, then

$$\liminf_{k\to\infty} \int_0^T j_k(v_k)\,dt \ge \int_0^T j(v)\,dt.$$

Ellipticity of the bilinear form a can be weakened to the coerciveness condition

$$a(v,v) + \lambda \|v\|_H^2 \ge \alpha \|v\|_V^2 \quad \forall\, v \in V$$

for some constants $\lambda \ge 0$ and $\alpha > 0$. Finally, with regard to the initial value function u_0, assume that $j(u_0) \in \mathbb{R}$ and that there exists a sequence $\{u_{0k}\}$ such that $u_{0k} \to u_0$ in V, and $\|Au_{0k} + j_k'(u_{0k})\|_H$ is bounded.

THEOREM 6.8. *Under the above assumptions, there is a unique solution to the problem of finding a function* $u \in L^2(0,T;V)$ *with* $\partial u/\partial t \in L^2(0,T;V')$ *and* $u(0) = u_0$ *such that for a.a.* $t \in [0,T]$,

$$\begin{aligned} &(\dot{u}(t), v - u(t)) + a(u(t), v - u(t)) + j(v) - j(u(t)) \\ &\quad \ge \langle f(t), v - u(t)\rangle \quad \forall\, v \in V. \end{aligned} \tag{6.46}$$

Furthermore, $\dot{u} \in L^2(0,T;V) \cap L^\infty(0,T;H)$.

It is possible to formulate the problems (6.45) and (6.46) in a unified manner. Indeed, set

$$\begin{aligned} \mathcal{K} &= \Big\{ v : v \in L^2(0,T;V),\ \dot{v} \in L^2(0,T;V'),\ v(t) \in K \text{ a.e. } t \in [0,T] \Big\}, \\ \mathcal{K}_{u_0} &= \{ v : v \in \mathcal{K},\ v(0) - u_0 \}, \end{aligned}$$

where $u_0 \in H$ is given. A general form of the parabolic variational inequality is then

$$\begin{aligned} u \in \mathcal{K}_{u_0}, \quad & \int_0^T (\dot{u}, v - u)\,dt + \int_0^T [a(u, v - u) + j(v) - j(u)]\,dt \\ & \ge \int_0^T \langle f, v - u\rangle\,dt \quad \forall\, v \in \mathcal{K}, \end{aligned} \tag{6.47}$$

and its equivalent local form is

$$\begin{aligned} u \in \mathcal{K}_{u_0}, \quad & (\dot{u}(t), v - u(t)) + a(u(t), v - u(t)) + j(v) - j(u(t)) \\ & \ge \langle f(t), v - u(t)\rangle \quad \forall\, v \in K, \text{ for a.a. } t \in [0,T]. \end{aligned} \tag{6.48}$$

To avoid the unnatural assumptions on $f(0)$ and u_0 in Theorem 6.7 and on the nondifferentiable functional j in Theorem 6.8, one can resort to an analysis of the *weak* form of the problem of finding $u \in L^2(0,T;V)$ with $u(t) \in K$ for a.a. $t \in [0,T]$ such that

$$\int_0^T (\dot{v}, v-u)\,dt + \int_0^T [a(u, v-u) + j(v) - j(u)]\,dt \\ \geq \int_0^T \langle f, v-u\rangle\,dt \quad \forall\, v \in \mathcal{K}_{u_0}. \tag{6.49}$$

It is easy to show that a solution of the problem (6.47) satisfies the relation (6.49), but not conversely. The following result on the weak formulation (6.49) is found in [45].

THEOREM 6.9. *Assume that K is a nonempty closed convex subset of V, $u_0 \in K$. Let $a : V \times V \to \mathbb{R}$ be a bilinear elliptic form on V, and $j : K \to \mathbb{R}$ a convex l.s.c. functional with the property that $|\int_0^T j(v)\,dt| < \infty$ for any $v \in L^2(0,T;K)$. Then for any $f \in L^2(0,T;V')$, there exists a unique function $u \in L^2(0,T;V)$ with $u(t) \in K$ for a.a. $t \in [0,T]$ such that* (6.49) *is satisfied.*

Results on the regularity of solutions to parabolic variational inequalities exist and are very useful in accurately predicting convergence orders of numerical approximations. The following is one such example ([16]).

THEOREM 6.10. *For the problem* (6.45) *in which $H = L^2(\Omega)$, $V = H_0^1(\Omega)$,*

$$a(u,v) = \int_\Omega \nabla u \cdot \nabla v\,dx, \quad \langle f, v\rangle = \int_\Omega f\,v\,dx,$$

and $K = \{v \in H_0^1(\Omega) : v \geq 0 \text{ a.e. on } \Omega\}$, assume that

$$f \in C([0,T];L^\infty(\Omega)), \quad \dot{f} \in L^2([0,T];L^\infty(\Omega)),$$

and

$$u_0 \in W^{2,\infty}(\Omega) \cap K.$$

Then there is a unique solution of (6.45) *satisfying*

$$u \in L^2([0,T];H^2(\Omega)), \quad \dot{u} \in L^2([0,T];H_0^1(\Omega)) \cap L^\infty([0,T];L^\infty(\Omega)),$$

and

$$\left(\frac{\partial u^+(t)}{\partial t}, v - u(t)\right) + a(u(t), v-u(t)) \geq \langle f(t), v - u(t)\rangle$$

for all $v \in K$, $t \in [0,T]$, where $\partial u^+(t)/\partial t$ denotes the right-hand derivative of u with respect to t.

The elastoplasticity problem studied in this work will be formulated in two alternative forms as time-dependent variational inequalities involving the first time derivatives of certain quantities. However, our variational inequalities differ in several aspects from the standard parabolic variational inequalities presented above. First, the plasticity problems to be considered are quasistatic, so that we do not have the first term on the lefthand side of either (6.45) or (6.46), while the time derivative appears in other terms of the formulation. Secondly, one of our variational inequality formulations (the primal variational formulation) is of mixed kind, that is, it is a variational inequality both by virtue of the presence of a nondifferentiable term and by the fact that the problem is posed over a convex subset of the whole space.

7

The Primal Variational Problem of Elastoplasticity

The initial–boundary value problem of elastoplasticity may be formulated in two alternative ways, depending on which of the two forms of the plastic flow law (see Section 4.2) is adopted. We describe as the *primal problem* the version that takes as its point of departure the flow law in the form (4.38), while the *dual problem* is formulated using the form (4.35) of the flow law.

This chapter is devoted to the formulation and analysis of the primal variational problem of elastoplasticity. In Section 7.1 we introduce the variational formulation of the primal problem. This is followed, in Section 7.2, by the formulation and analysis of an abstract variational inequality of which the primal variational problem is a special case. In Section 7.3 the results on the abstract problem obtained in Section 7.2 are applied to the primal problem. Finally, in Section 7.4 we consider the continuous dependence of the solution of the primal problem on the input data; an estimate is derived for the continuous dependence of the solution of a particular primal problem PRIM1, defined in Section 7.1, on various input data.

7.1 The Primal Variational Problem

Rather than work at the greatest possible level of generality, we focus on the problem corresponding to linear elastic behavior and linear hardening laws. We will pay particular attention to the special case of an elastoplastic material with either or both of linear kinematic hardening and linear isotropic hardening.

Basic relations. The flow law of the problem takes the form (4.38), which involves the dissipation function D. For convenience we reproduce here the full set of governing equations, which are assumed to be posed on a bounded Lipschitz domain Ω with boundary Γ.

The unknown variables are the displacement $\boldsymbol{u}$, the plastic strain $\boldsymbol{p}$, and the internal hardening variables $\boldsymbol{\xi}$, which are required to satisfy, in Ω, the equilibrium equation

$$\operatorname{div} \boldsymbol{\sigma} + \boldsymbol{f} = \boldsymbol{0}, \tag{7.1}$$

the strain–displacement relation

$$\boldsymbol{\epsilon}(\boldsymbol{u}) = \tfrac{1}{2}(\nabla \boldsymbol{u} + (\nabla \boldsymbol{u})^T), \tag{7.2}$$

the constitutive relations

$$\boldsymbol{\sigma} = \boldsymbol{C}\,(\boldsymbol{\epsilon}(\boldsymbol{u}) - \boldsymbol{p}), \tag{7.3}$$

$$\boldsymbol{\chi} = -\boldsymbol{H}\,\boldsymbol{\xi}, \tag{7.4}$$

and the flow law

$$\begin{aligned} &(\dot{\boldsymbol{p}}, \dot{\boldsymbol{\xi}}) \in K_p, \\ &D(\boldsymbol{q}, \boldsymbol{\eta}) \geq D(\dot{\boldsymbol{p}}, \dot{\boldsymbol{\xi}}) + \boldsymbol{\sigma} : (\boldsymbol{q} - \dot{\boldsymbol{p}}) + \boldsymbol{\chi} : (\boldsymbol{\eta} - \dot{\boldsymbol{\xi}}) \quad \forall\, (\boldsymbol{q}, \boldsymbol{\eta}) \in K_p, \end{aligned} \tag{7.5}$$

where $K_p = \operatorname{dom} D$. The dissipation function D is a gauge, that is, it is nonnegative, convex, positively homogeneous, and l.s.c., with $D(\boldsymbol{0}) = 0$.

Properties of material parameters. When analyzing the elastoplasticity problem, we need to make some assumptions about the material parameters. These assumptions encapsulate realistic properties of elastoplastic materials.

The elasticity tensor $\boldsymbol{C}$ has the symmetry properties

$$C_{ijkl} = C_{jikl} = C_{klij}; \tag{7.6}$$

it is assumed, furthermore, that $\boldsymbol{C}$ has bounded and measurable components, that is,

$$C_{ijkl} \in L^\infty(\Omega), \tag{7.7}$$

and that it is pointwise stable: There exists a constant $C_0 > 0$ such that

$$C_{ijkl}(\boldsymbol{x})\zeta_{ij}\zeta_{kl} \geq C_0|\boldsymbol{\zeta}|^2 \quad \forall\, \boldsymbol{\zeta} = (\zeta_{ij}) \in M^3, \quad \text{a.e. in } \Omega. \tag{7.8}$$

The hardening modulus $\boldsymbol{H}$, viewed as a linear operator from $\mathbb{R}^m$ into itself, is assumed to be symmetric in the sense that

$$\boldsymbol{\xi} : \boldsymbol{H}\boldsymbol{\lambda} = \boldsymbol{\lambda} : \boldsymbol{H}\boldsymbol{\xi} \quad \forall\, \boldsymbol{\xi}, \boldsymbol{\lambda} \in \mathbb{R}^m. \tag{7.9}$$

It is further assumed that $\boldsymbol{H}$ has bounded and measurable components, that is,

$$H_{ij} \in L^\infty(\Omega), \tag{7.10}$$

and that it is positive definite in the sense that a constant $H_0 > 0$ exists such that

$$\boldsymbol{\xi} : \boldsymbol{H}\boldsymbol{\xi} \geq H_0|\boldsymbol{\xi}|^2 \quad \forall \boldsymbol{\xi} \in \mathbb{R}^m, \text{ a.e. in } \Omega. \tag{7.11}$$

We see that the compliance tensor $\boldsymbol{C}^{-1}$ has the same symmetry properties as $\boldsymbol{C}$ and is also pointwise stable in the sense that a constant $C_0' > 0$ exists such that

$$C_{ijkl}^{-1}(\boldsymbol{x})\zeta_{ij}\zeta_{kl} \geq C_0'|\boldsymbol{\zeta}|^2 \quad \forall \boldsymbol{\zeta} \in M^3, \text{ a.e. in } \Omega.$$

The inverse $\boldsymbol{H}^{-1}$ of the hardening modulus possesses the same properties as $\boldsymbol{H}$: It is a symmetric operator whose matrix representation has uniformly bounded components. Furthermore, there exists a constant $H_0' > 0$ such that

$$\boldsymbol{\chi} : \boldsymbol{H}^{-1}\boldsymbol{\chi} \geq H_0'|\boldsymbol{\chi}|^2 \quad \forall \boldsymbol{\chi} \in \mathbb{R}^m, \text{ a.e. in } \Omega.$$

Initial and boundary value conditions. For convenience we confine our attention to the homogeneous Dirichlet (displacement) boundary condition

$$\boldsymbol{u} = \boldsymbol{0} \quad \text{on } \Gamma. \tag{7.12}$$

The treatment of other boundary conditions does not present any essential difficulty. The initial condition is

$$\boldsymbol{u}(\boldsymbol{x}, 0) = \boldsymbol{0}. \tag{7.13}$$

Function spaces. We introduce here the function spaces corresponding to the variables of interest.

The space V of displacements is defined by

$$V = [H_0^1(\Omega)]^3.$$

To define the space of plastic strains we first introduce the space

$$Q = \{\boldsymbol{q} = (q_{ij})_{3\times 3} : \ q_{ji} = q_{ij}, \ q_{ij} \in L^2(\Omega)\}$$

with the usual inner product and norm of the space $[L^2(\Omega)]^{3\times 3}$. Then the space Q_0 of plastic strains is the closed subspace of Q defined by

$$Q_0 = \{\boldsymbol{q} \in Q : \ \operatorname{tr} \boldsymbol{q} = 0 \text{ a.e. in } \Omega\}.$$

The space M of internal variables is defined by

$$M = [L^2(\Omega)]^m$$

with the usual $L^2(\Omega)$-based inner product and norm. We will also need the product space $Z = V \times Q_0 \times M$, which is a Hilbert space with the inner product

$$(\boldsymbol{w}, \boldsymbol{z})_Z = (\boldsymbol{u}, \boldsymbol{v})_V + (\boldsymbol{p}, \boldsymbol{q})_Q + (\boldsymbol{\xi}, \boldsymbol{\eta})_M$$

and the norm $\|\boldsymbol{z}\|_Z = (\boldsymbol{z}, \boldsymbol{z})_Z^{1/2}$, where $\boldsymbol{w} = (\boldsymbol{u}, \boldsymbol{p}, \boldsymbol{\xi})$ and $\boldsymbol{z} = (\boldsymbol{v}, \boldsymbol{q}, \boldsymbol{\eta})$. Corresponding to the set $K_p = \text{dom}\,(D)$, we define

$$Z_p = \{\boldsymbol{z} = (\boldsymbol{v}, \boldsymbol{q}, \boldsymbol{\eta}) \in Z : \ (\boldsymbol{q}, \boldsymbol{\eta}) \in K_p \text{ a.e. in } \Omega\}, \tag{7.14}$$

which is a nonempty, closed, convex cone in Z.

Functionals and the bilinear form. We introduce the bilinear form $a : Z \times Z \to \mathbb{R}$ defined by

$$a(\boldsymbol{w}, \boldsymbol{z}) = \int_\Omega [\boldsymbol{C}(\boldsymbol{\epsilon}(\boldsymbol{u}) - \boldsymbol{p}) : (\boldsymbol{\epsilon}(\boldsymbol{v}) - \boldsymbol{q}) + \boldsymbol{\xi} : \boldsymbol{H}\boldsymbol{\eta}]\, dx, \tag{7.15}$$

the linear functional

$$\boldsymbol{\ell}(t) : Z \to \mathbb{R}, \quad \langle \boldsymbol{\ell}(t), \boldsymbol{z} \rangle = \int_\Omega \boldsymbol{f}(t) \cdot \boldsymbol{v}\, dx, \tag{7.16}$$

and the functional

$$j : Z \to \mathbb{R}, \quad j(\boldsymbol{z}) = \int_\Omega D(\boldsymbol{q}, \boldsymbol{\eta})\, dx, \tag{7.17}$$

where as before, $\boldsymbol{w} = (\boldsymbol{u}, \boldsymbol{p}, \boldsymbol{\xi})$ and $\boldsymbol{z} = (\boldsymbol{v}, \boldsymbol{q}, \boldsymbol{\eta})$.

The bilinear form $a(\cdot, \cdot)$ is symmetric as a result of the symmetry properties of $\boldsymbol{C}$ and $\boldsymbol{H}$. From the properties of D, $j(\cdot)$ is a convex, positively homogeneous, nonnegative, and l.s.c. functional. Note, however, that in general, j *is not differentiable* (cf. (4.51) for the case of combined linear kinematic and isotropic hardening with the von Mises yield function).

The primal variational formulation. To arrive at a primal variational formulation of the problem, we begin by integrating the relation (7.5) and use the expressions (7.3) and (7.4) to obtain $(\dot{\boldsymbol{p}}, \dot{\boldsymbol{\xi}}) \in Z_p$ and

$$\begin{aligned} \int_\Omega D(\boldsymbol{q}, \boldsymbol{\eta})\, dx \geq & \int_\Omega D(\dot{\boldsymbol{p}}, \dot{\boldsymbol{\xi}})\, dx \\ & + \int_\Omega \left[\boldsymbol{C}\,(\boldsymbol{\epsilon}(\boldsymbol{u}) - \boldsymbol{p}) : (\boldsymbol{q} - \dot{\boldsymbol{p}}) - \boldsymbol{H}\,\boldsymbol{\xi} : (\boldsymbol{\eta} - \dot{\boldsymbol{\xi}})\right] dx \\ & \forall\, (\boldsymbol{q}, \boldsymbol{\eta}) \in Z_p. \end{aligned} \tag{7.18}$$

We next take the scalar product of (7.1) with $\boldsymbol{v} - \dot{\boldsymbol{u}}$ for arbitrary $\boldsymbol{v} \in V$, integrate over Ω, and perform an integration by parts with the use of the expression (7.3) for $\boldsymbol{\sigma}$ to obtain

$$\int_\Omega \boldsymbol{C}\,(\boldsymbol{\epsilon}(\boldsymbol{u}) - \boldsymbol{p}) : (\boldsymbol{\epsilon}(\boldsymbol{v}) - \boldsymbol{\epsilon}(\dot{\boldsymbol{u}}))\,dx = \int_\Omega \boldsymbol{f} \cdot (\boldsymbol{v} - \dot{\boldsymbol{u}})\,dx \quad \forall\, \boldsymbol{v} \in V. \tag{7.19}$$

We now add (7.18) and (7.19) to obtain the variational inequality

$$a(\boldsymbol{w}(t), \boldsymbol{z} - \dot{\boldsymbol{w}}(t)) + j(\boldsymbol{z}) - j(\dot{\boldsymbol{w}}(t)) \geq \langle \boldsymbol{\ell}(t), \boldsymbol{z} - \dot{\boldsymbol{w}}(t)\rangle,$$

which is posed on the space Z_p.

The primal variational problem of elastoplasticity thus takes the following form.

Problem Prim. Given $\boldsymbol{\ell} \in H^1(0,T;Z')$, $\boldsymbol{\ell}(0) = \boldsymbol{0}$, find $\boldsymbol{w} = (\boldsymbol{u}, \boldsymbol{p}, \boldsymbol{\xi}) : [0,T] \to Z$, $\boldsymbol{w}(0) = \boldsymbol{0}$, such that for almost all $t \in (0,T)$, $\dot{\boldsymbol{w}}(t) \in Z_p$ and

$$a(\boldsymbol{w}(t), \boldsymbol{z} - \dot{\boldsymbol{w}}(t)) + j(\boldsymbol{z}) - j(\dot{\boldsymbol{w}}(t)) \geq \langle \boldsymbol{\ell}(t), \boldsymbol{z} - \dot{\boldsymbol{w}}(t)\rangle \quad \forall\, \boldsymbol{z} \in Z_p. \tag{7.20}$$

We have seen that if $\boldsymbol{w}$ is a classical solution of the problem defined by (7.1)–(7.5) and (7.12)–(7.13), then it is a solution of the problem Prim. Conversely, reversing the argument leading to the inequality (7.20), we see that if $\boldsymbol{w}$ is a smooth solution of Problem Prim, then $\boldsymbol{w}$ is also a classical solution of the problem defined by (7.1)–(7.5) and (7.12)–(7.13). Thus the two problems are formally equivalent.

From the point of view of a theoretical analysis it is more convenient to view the inequality (7.20) as one posed on the whole space Z, rather than on Z_p. Observing that $j(\boldsymbol{z}) = \infty$ for $\boldsymbol{z} \notin Z_p$ and bearing in mind the requirement $\dot{\boldsymbol{w}}(t) \in Z_p$, we can express the relation (7.20) in the following equivalent form:

$$a(\boldsymbol{w}(t), \boldsymbol{z} - \dot{\boldsymbol{w}}(t)) + j(\boldsymbol{z}) - j(\dot{\boldsymbol{w}}(t)) \geq \langle \boldsymbol{\ell}(t), \boldsymbol{z} - \dot{\boldsymbol{w}}(t)\rangle \quad \forall\, \boldsymbol{z} \in Z. \tag{7.21}$$

The definition

$$\partial j(\dot{\boldsymbol{w}}) = \{\boldsymbol{w}^* \in Z' : \ j(\boldsymbol{z}) \geq j(\dot{\boldsymbol{w}}) + \langle \boldsymbol{w}^*, \boldsymbol{z} - \dot{\boldsymbol{w}}\rangle \ \ \forall\, \boldsymbol{z} \in Z\} \tag{7.22}$$

of the subdifferential of $j(\cdot)$ (cf. (4.12)) permits us to rewrite (7.21) in the form of an equation and an inclusion. Indeed, by using (7.22) and introducing the variable $\boldsymbol{w}^*$, we see that the problem Prim is equivalent to the problem of finding functions $\boldsymbol{w}$: $[0,T] \to Z$ and $\boldsymbol{w}^*$: $[0,T] \to Z'$ such that for almost all $t \in (0,T)$,

$$a(\boldsymbol{w}(t), \boldsymbol{z}) + \langle \boldsymbol{w}^*(t), \boldsymbol{z}\rangle = \langle \boldsymbol{\ell}(t), \boldsymbol{z}\rangle \quad \forall\, \boldsymbol{z} \in Z, \tag{7.23}$$

$$\boldsymbol{w}^*(t) \in \partial j(\dot{\boldsymbol{w}}(t)). \tag{7.24}$$

From the definition of the subdifferential and the positive homogeneity of j, we observe that the relation (7.24) is equivalent to the pair of conditions

$$\langle \boldsymbol{w}^*(t), \boldsymbol{z}\rangle \leq j(\boldsymbol{z}) \quad \forall\, \boldsymbol{z} \in Z \tag{7.25}$$

and

$$\langle \boldsymbol{w}^*(t), \dot{\boldsymbol{w}}(t)\rangle = j(\dot{\boldsymbol{w}}(t)). \tag{7.26}$$

We notice that in the case where there is no plastic deformation, the problem PRIM reduces to the boundary value problem of linear elasticity. To see this, setting $\boldsymbol{p} = \boldsymbol{0}$ and $\boldsymbol{\xi} = \boldsymbol{0}$, we obtain from (7.20) that

$$\int_\Omega \boldsymbol{C}\boldsymbol{\epsilon}(\boldsymbol{u}) : \boldsymbol{\epsilon}(\boldsymbol{v} - \dot{\boldsymbol{u}})\, dx = \int_\Omega \boldsymbol{f} \cdot (\boldsymbol{v} - \dot{\boldsymbol{u}})\, dx \quad \forall \boldsymbol{v} \in [H_0^1(\Omega)]^3,$$

i.e.,

$$\int_\Omega \boldsymbol{C}\boldsymbol{\epsilon}(\boldsymbol{u}) : \boldsymbol{\epsilon}(\boldsymbol{v})\, dx = \int_\Omega \boldsymbol{f} \cdot \boldsymbol{v}\, dx \quad \forall \boldsymbol{v} \in [H_0^1(\Omega)]^3,$$

which is the linear elasticity problem (6.19) when the homogeneous displacement condition is specified on the whole boundary.

Combined linear kinematic and isotropic hardening with von Mises yield condition. Later on, in order to make the results more accessible, they will be presented in the context of the special cases of combined linear kinematic and isotropic hardening, or linear kinematic hardening only, together with the von Mises yield condition (see Examples 4.8 and 4.9). Besides their simplicity, these special cases, owing to their popular usage, are also important particular applications of the general theory developed in this work. Another important special case, namely, that corresponding to linear isotropic hardening, can be analyzed in a way very similar to that for the combined linear kinematic and isotropic hardening material.

We now turn to the formulation of the problem for the case of combined linear kinematic and isotropic hardening with the von Mises yield function. From Example 4.8 it is seen that the unknown variables for this special case are the displacement $\boldsymbol{u}$, the plastic strain $\boldsymbol{p}$, and the isotropic hardening variable γ. The spaces V and Q_0 of displacements and plastic strains are unchanged, and the space M of internal variables is simply $M = L^2(\Omega)$.

In this special context the constraint set Z_p of Z is given by

$$Z_p = \{\boldsymbol{z} = (\boldsymbol{v}, \boldsymbol{q}, \mu) \in Z : \ |\boldsymbol{q}| \le \mu \text{ a.e. in } \Omega\}.$$

The bilinear form $a : Z \times Z \to \mathbb{R}$ becomes

$$\begin{aligned} a(\boldsymbol{w}, \boldsymbol{z}) &= \int_\Omega [\boldsymbol{C}(\boldsymbol{\epsilon}(\boldsymbol{u}) - \boldsymbol{p}) : (\boldsymbol{\epsilon}(\boldsymbol{v}) - \boldsymbol{q}) + k_1 \boldsymbol{p} : \boldsymbol{q} + k_2 \gamma\mu]\, dx \\ &= \int_\Omega [C_{ijkl}(\epsilon_{ij}(\boldsymbol{u}) - p_{ij})(\epsilon_{kl}(\boldsymbol{v}) - q_{kl}) + k_1 p_{ij} q_{ij} + k_2 \gamma\mu]\, dx, \end{aligned} \tag{7.27}$$

where $\boldsymbol{w} = (\boldsymbol{u}, \boldsymbol{p}, \gamma)$ and $\boldsymbol{z} = (\boldsymbol{v}, \boldsymbol{q}, \mu)$. The functional j defined in (7.17) is

$$j(\boldsymbol{z}) = \int_\Omega D(\boldsymbol{q}, \mu)\, dx$$

with the dissipation function D corresponding to the von Mises yield function being given by

$$D(\boldsymbol{q}, \mu) = \begin{cases} c_0|\boldsymbol{q}| & \text{if } |\boldsymbol{q}| \le \mu, \\ +\infty & \text{if } |\boldsymbol{q}| > \mu. \end{cases} \tag{7.28}$$

The linear functional $\boldsymbol{\ell}(t)$ is as in (7.16). With these identifications, we arrive at the following primal variational problem of elastoplasticity for the combined linear kinematic–isotropic hardening material with the von Mises yield function.

Problem Prim1. Given $\boldsymbol{\ell} \in H^1(0,T;Z')$, $\boldsymbol{\ell}(0) = \boldsymbol{0}$, find $\boldsymbol{w} = (\boldsymbol{u}, \boldsymbol{p}, \gamma) : [0,T] \to Z$ with $\boldsymbol{w}(0) = \boldsymbol{0}$ such that for almost all $t \in (0,T)$, $\dot{\boldsymbol{w}}(t) \in Z_p$ and

$$a(\boldsymbol{w}(t), \boldsymbol{z} - \dot{\boldsymbol{w}}(t)) + j(\boldsymbol{z}) - j(\dot{\boldsymbol{w}}(t)) \ge \langle \boldsymbol{\ell}(t), \boldsymbol{z} - \dot{\boldsymbol{w}}(t)\rangle \quad \forall\, \boldsymbol{z} \in Z_p. \tag{7.29}$$

Again, the formal equivalence of Problem Prim1 to the classical form of the problem can be established by a standard procedure. We take the variational problem Prim1 as fundamental.

Linear kinematic hardening with von Mises yield condition. The problem for a material undergoing linear kinematic hardening only, together with the von Mises yield function, can be formally viewed as a degenerate case of Problem Prim1 with $k_2 = 0$. The unknown variables in this case are the displacement $\boldsymbol{u}$ and the plastic strain $\boldsymbol{p}$, and the spaces V and Q_0 are as previously defined. The solution space is now $Z = V \times Q_0$, with the inner product

$$(\boldsymbol{w}, \boldsymbol{z})_Z = (\boldsymbol{u}, \boldsymbol{v})_V + (\boldsymbol{p}, \boldsymbol{q})_Q$$

and the norm $\|\boldsymbol{z}\|_Z = (\boldsymbol{z}, \boldsymbol{z})_Z^{1/2}$, where $\boldsymbol{w} = (\boldsymbol{u}, \boldsymbol{p})$ and $\boldsymbol{z} = (\boldsymbol{v}, \boldsymbol{q})$. This time, instead of (7.28), the dissipation function takes the simple form

$$D(\boldsymbol{q}) = c_0|\boldsymbol{q}| \quad \forall\, \boldsymbol{q} \in Q_0, \tag{7.30}$$

so that the function

$$j(\boldsymbol{z}) = \int_\Omega D(\boldsymbol{q})\, dx \quad \text{for } \boldsymbol{z} = (\boldsymbol{v}, \boldsymbol{q}) \in Z$$

is finite on the whole space Z. The bilinear form $a : Z \times Z \to \mathbb{R}$ is

$$\begin{aligned} a(\boldsymbol{w}, \boldsymbol{z}) &= \int_\Omega [\boldsymbol{C}(\boldsymbol{\epsilon}(\boldsymbol{u}) - \boldsymbol{p}) \cdot (\boldsymbol{\epsilon}(\boldsymbol{v}) - \boldsymbol{q}) + k_1 \boldsymbol{p} \cdot \boldsymbol{q}]\, dx \\ &= \int_\Omega [C_{ijkl}(\epsilon_{ij}(\boldsymbol{u}) - p_{ij})(\epsilon_{kl}(\boldsymbol{v}) - q_{kl}) + k_1 p_{ij} q_{ij}]\, dx, \end{aligned} \tag{7.31}$$

while the linear functional $\boldsymbol{\ell}(t)$ is unchanged from (7.16). We can now define the primal variational problem corresponding to linear kinematic hardening material with the von Mises yield function.

PROBLEM PRIM2. Given $\boldsymbol{\ell} \in H^1(0,T;Z')$, $\boldsymbol{\ell}(0) = \mathbf{0}$, find $\boldsymbol{w} = (\boldsymbol{u}, \boldsymbol{p}) : [0,T] \to Z$ with $\boldsymbol{w}(0) = \mathbf{0}$ such that for almost all $t \in (0,T)$,

$$a(\boldsymbol{w}(t), \boldsymbol{z} - \dot{\boldsymbol{w}}(t)) + j(\boldsymbol{z}) - j(\dot{\boldsymbol{w}}(t)) \geq \langle \boldsymbol{\ell}(t), \boldsymbol{z} - \dot{\boldsymbol{w}}(t) \rangle \quad \forall\, \boldsymbol{z} \in Z. \quad (7.32)$$

7.2 Qualitative Analysis of an Abstract Problem

We find it convenient to study the primal variational problem in the framework of an abstract variational inequality. Apart from elastoplasticity, another application in which this variational inequality may be found is some contact problems with frictions. This section is devoted to the study of the well-posedness of the abstract problem. Once the issues of existence and uniqueness of a solution to the abstract problem have been settled, we will, in the next section, return to the primal problem of elastoplasticity and will apply the abstract results to that problem.

The abstract problem takes the following form.

PROBLEM ABS. *Find* $w : [0,T] \to H$, $w(0) = 0$, *such that for almost all* $t \in (0,T)$, $\dot{w}(t) \in K$ *and*

$$a(w(t), z - \dot{w}(t)) + j(z) - j(\dot{w}(t)) \geq \langle \ell(t), z - \dot{w}(t) \rangle \quad \forall\, z \in K. \quad (7.33)$$

Here H denotes a Hilbert space and K a nonempty, closed, convex cone in H. The bilinear form $a : H \times H \to \mathbb{R}$ is symmetric, bounded, and H-elliptic, that is,

$$a(w,z) = a(z,w) \quad \forall\, w, z \in H,$$

and there exist constants $c_0, c_1 > 0$ such that

$$|a(w,z)| \leq c_1 \|w\|_H \|z\|_H, \quad a(z,z) \geq c_0 \|z\|_H^2 \quad \forall\, w, z \in H.$$

We assume that $\ell \in H^1(0,T;H')$, $\ell(0) = 0$, and that $j : K \to \mathbb{R}$ is nonnegative, convex, positively homogeneous, and Lipschitz continuous, but *not* necessarily differentiable.

We remark that a solution $w(t)$ of the problem ABS actually lies in the set K for almost all $t \in [0,T]$. This follows from the elementary formula

$$w(t) = \int_0^t \dot{w}(t)\, dt,$$

the condition $\dot{w}(t) \in K$ for a.a. $t \in [0,T]$, and the property that K is a closed, convex cone in H.

We call the problem ABS a variational inequality of the mixed kind because it has features of variational inequalities of both the first kind (the presence of the convex set K) and the second kind (the presence of the nondifferentiable functional j).

Questions of existence and uniqueness of solutions to this problem were first investigated in the context of elastoplasticity with linear kinematic hardening by Reddy [104]. The results were extended in Han, Reddy, and Schroeder [56] to cover the elastoplasticity problem with combined linear kinematic and isotropic hardening. The present treatment follows closely that of Han and Reddy [55].

The functional j may be extended from K to the whole space H by introducing the functional $J : H \to \mathbb{R} \cup \{+\infty\}$ through the formula

$$J(z) = \begin{cases} j(z) & \text{if } z \in K, \\ +\infty & \text{if } z \notin K. \end{cases}$$

Since K is a nonempty, closed, and convex cone, and since j is convex, positively homogeneous, and Lipschitz continuous on K, the extended functional J is proper, positively homogeneous, convex, and l.s.c. From now on, we will identify j with J; that is, we will use the same notation $j(z)$ to denote the extension of $j(z)$ from K to H by ∞ for $z \notin K$. With this identification, (7.33) is equivalent to

$$a(w(t), z - \dot{w}(t)) + j(z) - j(\dot{w}(t)) \geq \langle \ell(t), z - \dot{w}(t) \rangle \quad \forall\, z \in H;$$

in other words, the form of the problem is not affected by whether the test functions z are taken in H or only in K. There is an advantage in posing the variational inequality on the whole space, though, in that the standard solvability result, Theorem 6.6, can be applied directly to a sequence of approximation problems; see the proof of Lemma 7.1 below. We also observe that Problem ABS is equivalent to the problem of finding functions w: $[0, T] \to H$, $w(0) = 0$, and $w^*(t)$: $[0, T] \to H'$ such that for almost all $t \in (0, T)$,

$$a(w(t), z) + \langle w^*(t), z \rangle = \langle \ell(t), z \rangle \quad \forall\, z \in H, \tag{7.34}$$

$$w^*(t) \in \partial j(\dot{w}(t)), \tag{7.35}$$

where $\partial j(\dot{w}(t))$ denotes the subdifferential of $j(\cdot)$ at $\dot{w}(t)$.

As has been observed in the last section, because of the positive homogeneity of j, the relation $w^*(t) \in \partial j(\dot{w}(t))$ is equivalent to

$$\langle w^*(t), z \rangle \leq j(z) \ \forall\, z \in H \quad \text{and} \quad \langle w^*(t), \dot{w}(t) \rangle = j(\dot{w}(t)). \tag{7.36}$$

A feature of the proof of the existence result presented below is that it employs a discretization method closely related to one that is used in practice for computational purposes (see, for example, Reddy and Martin [107], [108]). The method of proof has interesting parallels with the time-discrete approximations of Problem ABS, for which an estimate of the rate of convergence of the approximations is derived in Chapter 11.

We now prove that the problem ABS is uniquely solvable in an appropriate space setting to be made precise below, and that the solution is stable with respect to perturbations in the data ℓ.

Existence. The proof of existence involves two stages: the first entails discretizing in time and establishing the existence of a family of solutions $\{w_n\}_{n=1}^N$ to the discrete problems. The second stage involves constructing piecewise linear interpolants w^k of the discrete solutions $\{w_n\}_{n=1}^N$ and showing that as the time step-size k approaches zero, the limit of these interpolants is in fact a solution of Problem ABS.

Time-discretization involves a uniform partitioning of the time interval $[0, T]$ according to

$$0 = t_0 < t_1 < \cdots < t_N = T, \quad \text{where} \quad t_n - t_{n-1} = k, \quad k = T/N.$$

We write $\ell_n = \ell(t_n)$, which is well-defined, since $\ell \in H^1(0, T; H')$ implies $\ell \in C([0, T]; H')$ by the embedding theorem

$$H^1(0, T; X) \hookrightarrow C([0, T]; X)$$

for any Banach space X. Corresponding to a sequence $\{w_n\}_{n=0}^N$, we define Δw_n to be the backward difference $w_n - w_{n-1}$, and $\delta w_n = \Delta w_n / k$ to be the backward divided difference, $n = 1, 2, \ldots, N$.

We first study a problem that is a semidiscrete counterpart of the continuous problem ABS. Notice that no summation is implied over the repeated index n.

LEMMA 7.1. *For any given $\{\ell_n\}_{n=0}^N \subset H'$, $\ell_0 = 0$, there exists a unique sequence $\{w_n\}_{n=0}^N \subset H$ with $w_0 = 0$ such that for $n = 1, 2, \ldots, N$, $\Delta w_n \in K$ and*

$$a(w_n, z - \Delta w_n) + j(z) - j(\Delta w_n) \geq \langle \ell_n, z - \Delta w_n \rangle \quad \forall\, z \in H. \tag{7.37}$$

Furthermore, there exists a constant c, independent of k, such that

$$\|\Delta w_n\|_H \leq c \, \|\Delta \ell_n\|_{H'}, \quad n = 1, \ldots, N. \tag{7.38}$$

PROOF. The inequality (7.37) may be rewritten as

$$\begin{aligned} a(\Delta w_n, z - \Delta w_n) &+ j(z) - j(\Delta w_n) \\ &\geq \langle \ell_n, z - \Delta w_n \rangle - a(w_{n-1}, z - \Delta w_n). \end{aligned} \tag{7.39}$$

We proceed inductively. For $n = 1$, since the bilinear form $a(\cdot, \cdot)$ is continuous and H-elliptic, the functional $j(\cdot)$ is proper, convex, and l.s.c., and the functional defined by the right-hand side of (7.39) is bounded and linear, the problem (7.39) has a unique solution $\Delta w_1 = w_1$ by Theorem 6.6. Obviously, $j(\Delta w_1) < \infty$. Hence, $\Delta w_1 \in K$. Assuming now that the solution

w_{n-1} is known, we can similarly show the existence and uniqueness of the solution $w_n = \Delta w_n + w_{n-1}$.

To derive the estimate (7.38), set $z = 0$ in (7.39) to get

$$a(\Delta w_n, \Delta w_n) \le \langle \Delta \ell_n, \Delta w_n \rangle - a(w_{n-1}, \Delta w_n) \\ - j(\Delta w_n) + \langle \ell_{n-1}, \Delta w_n \rangle. \tag{7.40}$$

We now show that $-a(w_{n-1}, \Delta w_n) - j(\Delta w_n) + \langle \ell_{n-1}, \Delta w_n \rangle \le 0$. By replacing n by $(n-1)$ and setting $z = \Delta w_{n-1} + \Delta w_n \in K$ in (7.37) we obtain

$$\begin{aligned} 0 &\le a(w_{n-1}, \Delta w_n) - \langle \ell_{n-1}, \Delta w_n \rangle + j(\Delta w_{n-1} + \Delta w_n) - j(\Delta w_{n-1}) \\ &\le a(w_{n-1}, \Delta w_n) - \langle \ell_{n-1}, \Delta w_n \rangle + j(\Delta w_n), \end{aligned}$$

where we used the convexity and positive homogeneity of $j(\cdot)$. Hence from (7.40) we obtain the inequality

$$a(\Delta w_n, \Delta w_n) \le \langle \Delta \ell_n, \Delta w_n \rangle,$$

from which the estimate (7.38) follows by the H-ellipticity of $a(\cdot, \cdot)$. □

LEMMA 7.2. *Assume that $\ell \in H^1(0, T; H')$ with $\ell(0) = 0$. Then the solution $\{w_n\}_{n=0}^N$ defined in Lemma 7.1 satisfies*

$$\max_{1 \le n \le N} \|w_n\|_H \le c \|\dot{\ell}\|_{L^1(0,T;H')}, \tag{7.41}$$

$$\sum_{n=1}^N \|\delta w_n\|_H^2 k \le c \|\dot{\ell}\|_{L^2(0,T;H')}^2. \tag{7.42}$$

PROOF. We write

$$w_n = \sum_{k=1}^n \Delta w_k.$$

Using (7.38) and (5.25) we have

$$\|w_n\|_H \le \sum_{k=1}^n \|\Delta w_k\|_H \le c \sum_{k=1}^n \|\Delta \ell_k\|_{H'} \le c \int_0^T \|\dot{\ell}(\tau)\|_{H'} \, d\tau.$$

The inequality (7.41) now follows by taking the maximum over all n.

To derive (7.42) we again begin by using (5.25) to get

$$\|\Delta w_n\|_H \le c \|\Delta \ell_n\|_{H'} \le c \int_{t_{n-1}}^{t_n} \|\dot{\ell}(\tau)\|_{H'} \, d\tau;$$

thus

$$\|\delta w_n\|_H^2 \, k \le c \int_{t_{n-1}}^{t_n} \|\dot{\ell}(\tau)\|_{H'}^2 \, d\tau$$

using the Cauchy–Schwarz inequality. We now sum over n to obtain

$$\sum_{n=1}^{N} \|\delta w_n\|_H^2 \, k \le c \int_0^T \|\dot{\ell}(\tau)\|_{H'}^2 \, d\tau$$

as desired. □

We now construct a piecewise linear interpolant w^k of $\{w_n\}_{n=0}^N$ by setting

$$w^k(t) = w_{n-1} + \delta w_n \, (t - t_{n-1})$$

for $t_{n-1} \le t \le t_n$, $1 \le n \le N$. Clearly, $w^k \in L^\infty(0,T;H)$, while $\dot{w}^k \in L^2(0,T;H)$. For any sequence $\{z_n\}_{n=1}^N \subset H$, we define a step function $z(t)$ by

$$z(t) = z_n \quad \text{for} \;\; t_{n-1} \le t < t_n, \; n = 1, \dots, N-1,$$
$$z(t) = z_N \quad \text{for} \;\; t_{N-1} \le t \le t_N.$$

Let $z_{N+1} = 0$. We divide both sides of (7.37) by k and use the positive homogeneity of j to obtain

$$a(w_n, z - \delta w_n) + j(z) - j(\delta w_n) - \langle \ell_n, z - \delta w_n \rangle \ge 0 \quad \forall \, z \in H.$$

Taking $z = (z_n + z_{n+1})/2$ in the above inequality, multiplying by k, and summing over n from 1 to N, we find that

$$\sum_{n=1}^{N} k \, a(w_n, (z_n + z_{n+1})/2 - \delta w_n) + \sum_{n=1}^{N} k \, j((z_n + z_{n+1})/2)$$
$$- \sum_{n=1}^{N} k \, j(\delta w_n) - \sum_{n=1}^{N} k \, \langle \ell_n, (z_n + z_{n+1})/2 - \delta w_n \rangle \ge 0. \quad (7.43)$$

Let us manipulate each of the sums in (7.43). For the first sum, we have

$$\sum_{n=1}^{N} k \, a(w_n, (z_n + z_{n+1})/2) = \int_0^T a(w^k(t), z(t)) \, dt,$$

$$\sum_{n=1}^{N} k \, a(w_n, \delta w_n) \ge \int_0^T a(w^k(t), \dot{w}^k(t)) \, dt.$$

Using the convexity of j, we can bound the second sum,

$$\sum_{n=1}^{N} k \, j(\tfrac{1}{2}(z_n + z_{n+1})) \le \sum_{n=1}^{N} \frac{k}{2} \, (j(z_n) + j(z_{n+1})) = \int_0^T j(z(t)) \, dt - \frac{k}{2} \, j(z_1).$$

Easily, the third sum can be rewritten as

$$\sum_{n=1}^{N} k\, j(\delta w_n) = \int_0^T j(\dot{w}^k(t))\, dt.$$

To deal with the last sum, we use the following relations:

$$\sum_{n=1}^{N} k\, \langle \ell_n, \tfrac{1}{2}(z_n + z_{n+1})\rangle = \int_0^T \langle \ell^k(t), z(t)\rangle\, dt,$$

and

$$\begin{aligned}\sum_{n=1}^{N} k\, \langle \ell_n, \delta w_n\rangle &= \int_0^T \langle \ell^k(t), \dot{w}^k(t)\rangle\, dt + \sum_{n=1}^{N} \langle \Delta\ell_n, \Delta w_n\rangle \\ &\le \int_0^T \langle \ell^k(t), \dot{w}^k(t)\rangle\, dt + c\, k \int_0^T \|\dot{\ell}(t)\|_{H'}^2\, dt,\end{aligned}$$

where $\ell^k(t)$ represents the piecewise linear interpolant of $\{\ell_n\}_{n=0}^N$ and c is the constant appearing in (7.38).

Thus, from (7.43) we see that w^k satisfies the variational inequality

$$\begin{aligned}0 &\le J^k \\ &\equiv \int_0^T \Big[a(w^k(t), z(t) - \dot{w}^k(t)) + j(z(t)) - j(\dot{w}^k(t)) \qquad (7.44) \\ &\qquad - \langle \ell^k(t), z(t) - \dot{w}^k(t)\rangle\Big]\, dt - \tfrac{1}{2}\, k\, j(z_1) + \tfrac{1}{2} c\, k \int_0^T \|\dot{\ell}(t)\|_{H'}^2\, dt.\end{aligned}$$

From (7.41), (7.42), and the definition of w^k, we see by direct evaluation that for some constant c independent of k,

$$\|w^k\|_{L^\infty(0,T;H)} \le c \quad \text{and} \quad \|\dot{w}^k\|_{L^2(0,T;H)} \le c.$$

Now we fix a step-size $k_0 > 0$ and consider the sequence of step-sizes $k_l = 2^{-l} k_0$, $l = 0, 1, \ldots$. It follows that there exists a subsequence $\{w^{k_{l_i}}\}$ of the sequence $\{w^{k_l}\}$ and a function $w \in H^1(0,T;H)$ such that

$$w^{k_{l_i}} \overset{*}{\rightharpoonup} w \text{ in } L^\infty(0,T;H) \quad \text{and} \quad \dot{w}^{k_{l_i}} \rightharpoonup \dot{w} \text{ in } L^2(0,T;H) \quad \text{as } i \to \infty.$$

It remains to show that w satisfies the variational inequality (7.33). We return to (7.44) and consider each of the terms appearing there.

First, using the fact that $w^{k_{l_i}}(0) = 0$ we obtain

$$\begin{aligned}\limsup_{i\to\infty} - \int_0^T a(w^{k_{l_i}}(t), \dot{w}^{k_{l_i}}(t))\, dt &= -\liminf_{i\to\infty} \tfrac{1}{2}\, a(w^{k_{l_i}}(T), w^{k_{l_i}}(T)) \\ &\le -\tfrac{1}{2}\, a(w(T), w(T)) \\ &= -\int_0^T a(w(t), \dot{w}(t))\, dt.\end{aligned}$$

Next,

$$\limsup_{i\to\infty}\int_0^T a(w^{k_{l_i}}(t),z(t))\,dt = \lim_{i\to\infty}\int_0^T a(w^{k_{l_i}}(t),z(t))\,dt$$
$$= \int_0^T a(w(t),z(t))\,dt.$$

From the properties of j, it is easy to verify that the functional $\int_0^T j(z(t))\,dt$ is convex and l.s.c. on $L^1(0,T;K)$, and so is weakly l.s.c. on $L^1(0,T;H)$. Since we also have

$$\dot{w}^{k_{l_i}} \rightharpoonup \dot{w} \quad \text{in} \quad L^1(0,T;H) \quad \text{as} \quad i\to\infty,$$

it follows that

$$\int_0^T j(\dot{w}(t))\,dt \le \liminf_{i\to\infty}\int_0^T j(\dot{w}^{k_{l_i}}(t))\,dt.$$

This inequality in turn implies that $\dot{w}(t)\in K$ for almost all $t\in[0,T]$.

This leaves the terms involving the approximation $\ell^{k_{l_i}}(t)$ to the linear functional $\ell(t)$. By the assumption and the construction we have that $\ell, \ell^{k_{l_i}} \in L^2(0,T;H')$; furthermore, since for $t_{n-1}\le t\le t_n$ we have

$$\|\ell(t)-\ell^{k_{l_i}}(t)\|_{H'} \le \|\ell(t)-\ell(t_{n-1})\|_{H'} + \frac{|t-t_{n-1}|}{k_{l_i}}\|\triangle\ell_n\|_{H'}$$
$$\le 2\int_{t_{n-1}}^{t_n}\|\dot{\ell}(\tau)\|_{H'}\,d\tau,$$

and thus

$$\|\ell(t)-\ell^{k_{l_i}}(t)\|_{H'}^2 \le 4\,k_{l_i}\int_{t_{n-1}}^{t_n}\|\dot{\ell}(\tau)\|_{H'}^2\,d\tau,$$
$$\int_0^T\|\ell(t)-\ell^{k_{l_i}}(t)\|_{H'}^2\,dt \le c\,{k_{l_i}}^2\int_0^T\|\dot{\ell}(\tau)\|_{H'}^2\,d\tau.$$

It follows that $\ell^{k_{l_i}}\to\ell$ in $L^2(0,T;H')$ as $i\to\infty$.

Hence, as $i\to\infty$,

$$\int_0^T\langle\ell^{k_{l_i}}(t),z(t)-\dot{w}^{k_{l_i}}(t)\rangle\,dt \to \int_0^T\langle\ell(t),z(t)-\dot{w}(t)\rangle\,dt.$$

The groundwork is now complete. Using the above results we have

$$0 \le \limsup_{i\to\infty} J^{k_{l_i}}$$
$$\le \int_0^T\big[a(w(t),z(t)-\dot{w}(t)) + j(z(t))$$
$$- j(\dot{w}(t)) - \langle\ell(t),z(t)-\dot{w}(t)\rangle\big]\,dt$$

for any step function z corresponding to a step-size k_{l_i}, $i = 1, 2, \ldots$. Approximating any $z \in L^2(0,T;K)$ by its piecewise averaging step functions $z^{k_{l_i}}$, it then follows that

$$\int_0^T \Big[a(w(t), z(t) - \dot{w}(t)) + j(z(t)) - j(\dot{w}(t)) - \langle \ell(t), z(t) - \dot{w}(t) \rangle \Big]\, dt \geq 0 \tag{7.45}$$

for all $z \in L^2(0,T;K)$. Here we have used the Lipschitz continuity of j on K and the fact that

$$z \in L^2(0,T;K) \Longrightarrow z^{k_{l_i}}(t) \in K \text{ for a.a. } t \in [0,T].$$

Now, for any $t_0 \in (0,T)$, let $h > 0$ be such that $t_0 + h < T$. For an arbitrary $z \in K$, we define

$$z(t) = \begin{cases} z & t_0 \leq t \leq t_0 + h, \\ \dot{w}(t) & \text{otherwise.} \end{cases}$$

Obviously, $z(t) \in L^2(0,T;K)$. We take this $z(t)$ in (7.45) to obtain

$$\frac{1}{h} \int_{t_0}^{t_0+h} [a(w(t), z - \dot{w}(t)) + j(z) - j(\dot{w}(t)) - \langle \ell(t), z - \dot{w}(t) \rangle]\, dt \geq 0.$$

Then we take the limit $h \to 0$. Applying the Lebesgue theorem (Theorem 5.21), we find that w satisfies the variational inequality (7.33) a.e. on $[0,T]$. By the Sobolev embedding theorem, $H^1(0,T;H) \hookrightarrow C([0,T];H)$, and we observe that $w \in L^\infty(0,T;H)$ and $\dot{w} \in L^2(0,T;H)$ is equivalent to $w \in H^1(0,T;H)$.

Uniqueness. The technique for the proof of uniqueness is standard. Suppose that Problem ABS has two solutions, w_1 and w_2. Denote by Δw the difference $w_1 - w_2$. From (7.33), on setting $w = w_1$, $z = \dot{w}_2 \in K$ and then $w = w_2$, $z = \dot{w}_1 \in K$, respectively, we have

$$\begin{aligned} a(w_1, \Delta \dot{w}) + j(\dot{w}_1) - j(\dot{w}_2) &\leq \langle \ell, \Delta \dot{w} \rangle, \\ -a(w_2, \Delta \dot{w}) + j(\dot{w}_2) - j(\dot{w}_1) &\leq -\langle \ell, \Delta \dot{w} \rangle. \end{aligned}$$

Adding the two inequalities, we get

$$a(\Delta w, \Delta \dot{w}) = \frac{1}{2} \frac{d}{dt} a(\Delta w, \Delta w) \leq 0.$$

We integrate the above inequality and use the initial conditions $w_1(0) = w_2(0) = 0$ to find that

$$a(\Delta w(t), \Delta w(t)) \leq 0, \quad t \in [0,T].$$

Then the H-ellipticity of $a(\cdot,\cdot)$ yields $\Delta w(t) = 0$ for $t \in [0,T]$, as required.

The above results are summarized in the following theorem.

THEOREM 7.3. (EXISTENCE AND UNIQUENESS) *Let H be a Hilbert space; $K \subset H$ a nonempty, closed, convex cone; $a\colon H \times H \to \mathbb{R}$ a bilinear form that is symmetric, bounded, and H-elliptic; $\ell \in H^1(0,T;H')$ with $\ell(0) = 0$; and $j : K \to \mathbb{R}$ nonnegative, convex, positively homogeneous, and Lipschitz continuous. Then there exists a unique solution w of Problem* ABS *satisfying $w \in H^1(0,T;H)$. Furthermore, $w : [0,T] \to H$ is the solution to Problem* ABS *if and only if there is a function $w^*(t)\colon [0,T] \to H'$ such that for almost all $t \in (0,T)$,*

$$a(w(t), z) + \langle w^*(t), z\rangle = \langle \ell(t), z\rangle \quad \forall\, z \in H, \tag{7.46}$$

$$w^*(t) \in \partial j(\dot{w}(t)). \tag{7.47}$$

We observe that from (7.46), w^* has the regularity property

$$w^* \in H^1(0,T;H'). \tag{7.48}$$

We remark that the existence proof above can be trivially modified for Problem ABS under the more general assumption $\ell \in W^{1,p}(0,T;H')$, $1 \le p \le \infty$. We notice that for $1 \le p \le \infty$, $W^{1,p}(0,T;H') \hookrightarrow C([0,T];H')$. When $1 \le p < \infty$, (7.42) can be replaced by

$$\sum_{n=1}^{N} \|\Delta w_n\|_H^p \le c\, k^{p-1} \|\dot{\ell}\|^p_{L^p(0,T;H')}.$$

As a result, $\{w^k\}$ is uniformly bounded in $W^{1,p}(0,T;H)$. Then the solution w belongs to $W^{1,p}(0,T;H)$. Similarly, if $\ell \in W^{1,\infty}(0,T;H')$, then $w \in W^{1,\infty}(0,T;H)$. We confine ourselves, however, to the Hilbert space case $p = 2$.

Stability. Let $\ell^{(1)}$, $\ell^{(2)} \in H^1(0,T;H')$ be given, with $\ell^{(1)}(0) = \ell^{(2)}(0) = 0$, and let $w^{(1)}$ and $w^{(2)}$ be the corresponding solutions whose existence is assured by Theorem 7.3. Thus we have, for almost all $t \in (0,T)$, $\dot{w}^{(1)}(t) \in K$, $\dot{w}^{(2)}(t) \in K$, and for all $z \in K$,

$$\begin{aligned} &a(w^{(1)}(t), z - \dot{w}^{(1)}(t)) + j(z) - j(\dot{w}^{(1)}(t)) \\ &\qquad \ge \langle \ell^{(1)}(t), z - \dot{w}^{(1)}(t)\rangle, \end{aligned} \tag{7.49}$$

$$\begin{aligned} &a(w^{(2)}(t), z - \dot{w}^{(2)}(t)) + j(z) - j(\dot{w}^{(2)}(t)) \\ &\qquad \ge \langle \ell^{(2)}(t), z - \dot{w}^{(2)}(t)\rangle. \end{aligned} \tag{7.50}$$

Set $e = w^{(1)} - w^{(2)}$. Taking $z = \dot{w}^{(2)}(t) \in K$ in (7.49) and $z = \dot{w}^{(1)}(t) \in K$ in (7.50), and adding the two resultant inequalities, we obtain

$$\frac{1}{2}\frac{d}{dt} a\,(e(t), e(t)) \le \langle \ell^{(1)}(t) - \ell^{(2)}(t), \dot{w}^{(1)}(t) - \dot{w}^{(2)}(t)\rangle.$$

Observing that $e(0) = 0$, we have

$$\frac{1}{2} a(e(t), e(t)) \leq \int_0^t \langle \ell^{(1)}(t) - \ell^{(2)}(t), \dot{e}(t) \rangle \, dt$$
$$= \langle \ell^{(1)}(t) - \ell^{(2)}(t), e(t) \rangle - \int_0^t \langle \dot{\ell}^{(1)}(t) - \dot{\ell}^{(2)}(t), e(t) \rangle \, dt.$$

Since $a(\cdot, \cdot)$ is H-elliptic, we have

$$\|e(t)\|_H^2$$
$$\leq c \|\ell^{(1)}(t) - \ell^{(2)}(t)\|_{H'} \|e(t)\|_H + c \int_0^t \|\dot{\ell}^{(1)}(t) - \dot{\ell}^{(2)}(t)\|_{H'} \|e(t)\|_H dt.$$

Set $M = \sup_{0 \leq t \leq T} \|e(t)\|_H$; then

$$\|e(t)\|_H^2 \leq c \|\ell^{(1)}(t) - \ell^{(2)}(t)\|_{H'} M + c \int_0^t \|\dot{\ell}^{(1)}(t) - \dot{\ell}^{(2)}(t)\|_{H'} M \, dt.$$

Hence

$$M^2 \leq c M \|\ell^{(1)} - \ell^{(2)}\|_{L^\infty(0,T;H')} + c M \|\dot{\ell}^{(1)} - \dot{\ell}^{(2)}\|_{L^1(0,T;H')},$$

and so

$$M \leq c \left(\|\ell^{(1)} - \ell^{(2)}\|_{L^\infty(0,T;H')} + \|\dot{\ell}^{(1)} - \dot{\ell}^{(2)}\|_{L^1(0,T;H')} \right).$$

Now, $\ell^{(1)}(0) = \ell^{(2)}(0) = 0$ by assumption; we have

$$\ell^{(1)}(t) - \ell^{(2)}(t) = \int_0^t \left(\dot{\ell}^{(1)}(t) - \dot{\ell}^{(2)}(t) \right) dt,$$

and hence

$$\|\ell^{(1)} - \ell^{(2)}\|_{L^\infty(0,T;H')} \leq \|\dot{\ell}^{(1)} - \dot{\ell}^{(2)}\|_{L^1(0,T;H')}.$$

In conclusion, we have proved the following stability result.

THEOREM 7.4. (STABILITY) *Under the assumptions of Theorem 7.3, the solution of Problem* ABS *depends continuously on* ℓ; *more precisely, for* $\ell^{(1)}, \ell^{(2)} \in H^1(0,T;H')$ *with* $\ell^{(1)}(0) = \ell^{(2)}(0) = 0$, *the corresponding solutions* $w^{(1)}$ *and* $w^{(2)}$ *satisfy*

$$\|w^{(1)} - w^{(2)}\|_{L^\infty(0,T;H)} \leq c \|\dot{\ell}^{(1)} - \dot{\ell}^{(2)}\|_{L^1(0,T;H')}.$$

7.3 Analysis of the Primal Problem

The groundwork has now been laid for a proper analysis of the primal problem of elastoplasticity. Indeed, since this problem is a special case of the abstract problem analyzed in Section 7.2, all that is required is to verify the validity of various assumptions for the primal problem. This will then lead to a result on the well-posedness of the primal variational problem.

Consider then the primal variational problem PRIM (cf. Section 7.1). The associated bilinear form $a(\cdot,\cdot)$ is given in (7.15). In general, we cannot expect $a(\cdot,\cdot)$ to be Z-elliptic, since

$$a(\boldsymbol{z},\boldsymbol{z}) = \int_\Omega [\boldsymbol{C}(\boldsymbol{\epsilon}(\boldsymbol{v}) - \boldsymbol{q}) : (\boldsymbol{\epsilon}(\boldsymbol{v}) - \boldsymbol{q}) + \boldsymbol{\eta} : \boldsymbol{H}\boldsymbol{\eta}]\, dx$$

for $\boldsymbol{z} = (\boldsymbol{v},\boldsymbol{q},\boldsymbol{\eta}) \in Z$, and we have at best

$$a(\boldsymbol{z},\boldsymbol{z}) \geq c\,(\|\boldsymbol{\epsilon}(\boldsymbol{v}) - \boldsymbol{q}\|^2_{[L^2(\Omega)]^{3\times 3}} + \|\boldsymbol{\eta}\|^2_{[L^2(\Omega)]^m}).$$

On the other hand, it is possible to exploit the fact that the problem is actually posed on the set Z_p defined in (7.14), and $\boldsymbol{z} \in Z_p$ imposes a constraint on the relation between the components $\boldsymbol{q}$ and $\boldsymbol{\eta}$. We introduce the assumption

$$\boldsymbol{z} = (\boldsymbol{v},\boldsymbol{q},\boldsymbol{\eta}) \in K_p \Longrightarrow \beta\,|\boldsymbol{q}|^2 \leq \boldsymbol{\eta} : \boldsymbol{H}\boldsymbol{\eta} \quad \text{for some constant } \beta > 0. \tag{7.51}$$

The assumption is satisfied in important special cases that are of frequent use in practice. As an example, for the problem with linear kinematic hardening, that is, the problem PRIM2, we have $K_p = Q_0$, $\boldsymbol{\eta} = \boldsymbol{q}$, and $\boldsymbol{H} = k_1 \boldsymbol{I}$. Since $k_1 > 0$, the condition (7.51) is satisfied with $\beta = k_1$. As another example, consider the problem with linear isotropic hardening. In this case, we have $K_p = \{(\boldsymbol{q},\eta) \in Q_0 \times M : |\boldsymbol{q}| \leq \eta\}$ and $\boldsymbol{H} = H = k_2$. Hence, $\eta\, H \eta = k_2 |\eta|^2 \geq k_2 |\boldsymbol{q}|^2$ for any $(\boldsymbol{q},\eta) \in K_p$, i.e., the condition (7.51) is satisfied with $\beta = k_2$.

With the assumption (7.51), we can show that the bilinear form $a(\cdot,\cdot)$ defined in (7.15) is Z-elliptic on Z_p, in that for some constant $c_0 > 0$,

$$\begin{aligned} a(\boldsymbol{z},\boldsymbol{z}) &= \int_\Omega [\boldsymbol{C}(\boldsymbol{\epsilon}(\boldsymbol{v}) - \boldsymbol{q}) : (\boldsymbol{\epsilon}(\boldsymbol{v}) - \boldsymbol{q}) + \boldsymbol{\eta} \cdot \boldsymbol{H}\boldsymbol{\eta}]\, dx \\ &\geq c_0\,(\|\boldsymbol{v}\|^2_V + \|\boldsymbol{q}\|^2_Q + \|\boldsymbol{\eta}\|^2_M) \quad \forall\, \boldsymbol{z} = (\boldsymbol{v},\boldsymbol{q},\boldsymbol{\eta}) \in Z_p. \end{aligned} \tag{7.52}$$

Indeed, using (7.8), (7.11), and (7.51), we have, for $\boldsymbol{z} = (\boldsymbol{v},\boldsymbol{q},\boldsymbol{\eta}) \in Z_p$,

$$\begin{aligned} &\boldsymbol{C}(\boldsymbol{\epsilon}(\boldsymbol{v}) - \boldsymbol{q}) : (\boldsymbol{\epsilon}(\boldsymbol{v}) - \boldsymbol{q}) + \boldsymbol{\eta} : \boldsymbol{H}\boldsymbol{\eta} \\ &\quad = \boldsymbol{C}(\boldsymbol{\epsilon}(\boldsymbol{v}) - \boldsymbol{q}) : (\boldsymbol{\epsilon}(\boldsymbol{v}) - \boldsymbol{q}) + \tfrac{1}{2}\,\boldsymbol{\eta} : \boldsymbol{H}\boldsymbol{\eta} + \tfrac{1}{2}\,\boldsymbol{\eta} : \boldsymbol{H}\boldsymbol{\eta} \\ &\quad \geq C_0 |\boldsymbol{\epsilon}(\boldsymbol{v}) - \boldsymbol{q}|^2 + \tfrac{1}{2}\,\beta\,|\boldsymbol{q}|^2 + \tfrac{1}{2}\,H\,|\boldsymbol{\eta}|^2 \\ &\quad = C_0(|\boldsymbol{\epsilon}(\boldsymbol{v})|^2 + |\boldsymbol{q}|^2 - 2\,\boldsymbol{\epsilon}(\boldsymbol{v}) : \boldsymbol{q}) + \tfrac{1}{2}\,\beta\,|\boldsymbol{q}|^2 + \tfrac{1}{2}\,H\,|\boldsymbol{\eta}|^2 \\ &\quad \geq C_0(|\boldsymbol{\epsilon}(\boldsymbol{v})|^2 + |\boldsymbol{q}|^2 - d\,|\boldsymbol{\epsilon}(\boldsymbol{v})|^2 - d^{-1}|\boldsymbol{q}|^2) + \tfrac{1}{2}\,\beta\,|\boldsymbol{q}|^2 + \tfrac{1}{2}\,H\,|\boldsymbol{\eta}|^2 \\ &\qquad (0 < d < 1) \\ &\quad = C_0(1-d)\,|\boldsymbol{\epsilon}(\boldsymbol{v})|^2 + [C_0(1-d^{-1}) + \tfrac{1}{2}\,\beta]|\boldsymbol{q}|^2 + \tfrac{1}{2}\,H\,|\boldsymbol{\eta}|^2. \end{aligned}$$

Choosing $d \in (0, 2C_0/(2C_0 + \beta))$, we obtain

$$\boldsymbol{C}(\boldsymbol{\epsilon}(\boldsymbol{v}) - \boldsymbol{q}) : (\boldsymbol{\epsilon}(\boldsymbol{v}) - \boldsymbol{q}) + \boldsymbol{\eta} : \boldsymbol{H}\boldsymbol{\eta} \geq c\left(|\boldsymbol{\epsilon}(\boldsymbol{v})|^2 + |\boldsymbol{q}|^2 + |\boldsymbol{\eta}|^2\right)$$

for all $\boldsymbol{z} = (\boldsymbol{v}, \boldsymbol{q}, \boldsymbol{\eta}) \in Z_p$, from which an application of Korn's inequality (5.21) yields (7.52).

THEOREM 7.5. *Under the assumption* (7.51) *and the assumptions made in Section 7.1, the problem* PRIM *has a solution.*

We will only give a sketch of the proof, since it can be carried out in a manner that parallels that of the existence proof in Section 7.2. In Section 7.2, the bilinear form is assumed to be elliptic on the whole space; this condition is now replaced by a weaker condition, (7.52). Thus, we cannot apply Theorem 6.6 to claim the existence of a solution to a semidiscrete approximation. Instead, we will use Proposition 6.2.

As in Section 7.2, we divide the time interval $[0, T]$ into N equal parts with step-size $k = T/N$. Since we do not have the ellipticity of the bilinear form on the whole space, Lemma 7.1 is replaced by the following result.

LEMMA 7.6. *For any given* $\{\boldsymbol{\ell}_n\}_{n=0}^N \subset Z'$, $\boldsymbol{\ell}_0 = \boldsymbol{0}$, *there exists a sequence* $\{\boldsymbol{w}_n\}_{n=0}^N \subset Z$ *with* $\boldsymbol{w}_0 = \boldsymbol{0}$ *such that for* $n = 1, \ldots, N$, $\Delta \boldsymbol{w}_n \in Z_p$ *and*

$$a(\boldsymbol{w}_n, \boldsymbol{z} - \Delta\boldsymbol{w}_n) + j(\boldsymbol{z}) - j(\Delta\boldsymbol{w}_n) \geq \langle \boldsymbol{\ell}_n, \boldsymbol{z} - \Delta\boldsymbol{w}_n \rangle \quad \forall\, \boldsymbol{z} \in Z_p, \tag{7.53}$$

and there is a constant $c > 0$ *independent of* k *such that*

$$\|\Delta\boldsymbol{w}_n\|_Z \leq c\,\|\Delta\boldsymbol{\ell}_n\|_{Z'}, \quad n = 1, \ldots, N. \tag{7.54}$$

PROOF. We rewrite (7.53) as

$$\begin{aligned} a(\Delta\boldsymbol{w}_n, \boldsymbol{z} - \Delta\boldsymbol{w}_n) &+ j(\boldsymbol{z}) - j(\Delta\boldsymbol{w}_n) \\ &\geq \langle \boldsymbol{\ell}_n, \boldsymbol{z} - \Delta\boldsymbol{w}_n \rangle - a(\boldsymbol{w}_{n-1}, \boldsymbol{z} - \Delta\boldsymbol{w}_n) \end{aligned} \tag{7.55}$$

for all $\boldsymbol{z} \in Z_p$. Now define the functional

$$f(\boldsymbol{z}) = \tfrac{1}{2}\, a(\boldsymbol{z}, \boldsymbol{z}) + j(\boldsymbol{z}) - \langle \boldsymbol{\ell}_n, \boldsymbol{z} \rangle - a(\boldsymbol{w}_{n-1}, \boldsymbol{z}).$$

Then f is proper, convex, and l.s.c. Furthermore, from the inequality (7.52) we have

$$f(\boldsymbol{z}) \to \infty \quad \text{as } \|\boldsymbol{z}\|_Z \to \infty \quad \text{for } \boldsymbol{z} \in Z_p.$$

An application of Proposition 6.2 yields the existence of an element in Z_p, denoted by $\Delta\boldsymbol{w}_n = \boldsymbol{w}_n - \boldsymbol{w}_{n-1}$, such that

$$f(\Delta\boldsymbol{w}_n) = \inf_{\boldsymbol{z} \in Z_p} f(\boldsymbol{z}).$$

Equivalently, $\Delta\boldsymbol{w}_n \in Z_p$ is a solution of (7.55), and hence $\boldsymbol{w}_n$ a solution of (7.53). The estimate (7.54) follows from the same argument given in the proof of Lemma 7.1, and from the properties that $\Delta \boldsymbol{w}_n \in Z_p$ and $a(\cdot,\cdot)$ is Z-elliptic on Z_p. □

The estimate (7.54) is crucial for the existence proof. Once we have this estimate, the rest of the proof follows in a manner similar to that given in Section 7.2.

In general, if we do not have ellipticity of $a(\cdot,\cdot)$ on the whole space Z, we cannot prove uniqueness and stability of a solution to the problem. For the rest of the section we will focus on the analysis of two important special cases for which we *do* have ellipticity on the whole space: combined linear kinematic and isotropic hardening, and linear kinematic hardening only. Thus the results of Section 7.2 may be applied directly to these two cases.

The problem with combined linear kinematic and isotropic hardening. We are concerned here with the variational problem PRIM1 stated in Section 7.1. We identify H in Theorem 7.3 with $Z = (H_0^1(\Omega))^3 \times Q_0 \times L^2(\Omega)$, and define

$$K = \{\boldsymbol{z} = (\boldsymbol{v}, \boldsymbol{q}, \mu) \in Z : |\boldsymbol{q}| \le \mu \text{ a.e. in } \Omega\}.$$

In addition to the assumptions on the material made in Section 7.1, we assume for the hardening coefficients k_1 and k_2 that there exist positive constants $\overline{k}_1$ and $\overline{k}_2$ such that

$$k_1 \ge \overline{k}_1 > 0, \quad k_2 \ge \overline{k}_2 > 0 \quad \text{a.e. on } \Omega.$$

We will show that the bilinear form $a(\cdot,\cdot)$ defined in (7.27) is Z-elliptic. The remaining assumptions of Theorem 7.3 are obviously true; in particular, the functional $j(\cdot)$ inherits the properties that Theorem 7.3 requires of it from the corresponding properties of the dissipation function D.

LEMMA 7.7. *The bilinear form $a : Z \times Z \to \mathbb{R}$ defined in (7.27) is Z-elliptic, that is, there exists $\alpha > 0$ such that*

$$a(\boldsymbol{z}, \boldsymbol{z}) \ge \alpha \, \|\boldsymbol{z}\|_Z^2 \quad \forall\, \boldsymbol{z} \in Z.$$

PROOF. For any $\boldsymbol{z} = (\boldsymbol{v}, \boldsymbol{q}, \mu) \in Z$ we have, using the pointwise stability assumption on $\boldsymbol{C}$ (cf. (7.8)),

$$\begin{aligned} a(\boldsymbol{z}, \boldsymbol{z}) &\ge C_0 \int_\Omega |\boldsymbol{\epsilon}(\boldsymbol{v}) - \boldsymbol{q}|^2 dx + \overline{k}_1 \int_\Omega |\boldsymbol{q}|^2 dx + \overline{k}_2 \int_\Omega |\mu|^2 dx \\ &\ge C_0 \theta \int_\Omega |\boldsymbol{\epsilon}(\boldsymbol{v})|^2 dx + \left(\overline{k}_1 - \frac{1}{1-\theta}\right) \int_\Omega |\boldsymbol{q}|^2 dx + \overline{k}_2 \int_\Omega |\mu|^2 dx, \end{aligned}$$

for every $\theta \in (0,1)$. The result then follows by choosing $\theta = \overline{k}_1/(2C_0 + \overline{k}_1)$ and using Korn's inequality (5.21). □

Applying Theorems 7.3 and 7.4 to Problem PRIM1, we thus have the following result.

THEOREM 7.8. *Under the assumptions made on the data in Section 7.1, the quasistatic elastoplasticity problem* PRIM1 *has a unique solution* $\boldsymbol{w} = (\boldsymbol{u}, \boldsymbol{p}, \gamma) \in H^1(0,T;Z)$. *Furthermore, if* $\boldsymbol{w}^{(1)}$ *and* $\boldsymbol{w}^{(2)}$ *are the solutions corresponding to* $\boldsymbol{\ell}^{(1)}, \boldsymbol{\ell}^{(2)} \in H^1(0,T;Z')$ *with* $\boldsymbol{\ell}^{(1)}(0) = \boldsymbol{\ell}^{(2)}(0) = \mathbf{0}$, *then*

$$\|\boldsymbol{w}^{(1)} - \boldsymbol{w}^{(2)}\|_{L^\infty(0,T;Z)} \le c\,\|\dot{\boldsymbol{\ell}}^{(1)} - \dot{\boldsymbol{\ell}}^{(2)}\|_{L^1(0,T;Z')}.$$

The problem with linear kinematic hardening. The quasistatic problem of elastoplasticity with linear kinematic hardening, Problem PRIM2 in Section 7.1, is a special case of the more general problem with combined kinematic and isotropic hardening treated above. Besides its importance in certain applications, the problem with linear kinematic hardening allows a simpler treatment. We still assume, for some positive constant $\overline{k}_1$,

$$k_1 \ge \overline{k}_1 > 0 \quad \text{a.e. on } \Omega.$$

The details of this problem are summarized in (7.30)–(7.32) and (7.16). From Lemma 7.7 (with $k_2 = 0$) it is seen that $a(\cdot,\cdot)$ is Z-elliptic, and we have the following theorem.

THEOREM 7.9. *Under the assumptions made in Section 7.1, the quasistatic elastoplasticity problem with linear kinematic hardening, that is, the problem* PRIM2, *has a unique solution* $\boldsymbol{w} = (\boldsymbol{u}, \boldsymbol{p}) \in H^1(0,T;Z)$. *Furthermore, if* $\boldsymbol{w}^{(1)}$ *and* $\boldsymbol{w}^{(2)}$ *are the solutions corresponding to* $\boldsymbol{\ell}^{(1)}, \boldsymbol{\ell}^{(2)} \in H^1(0,T;Z')$ *with* $\boldsymbol{\ell}^{(1)}(0) = \boldsymbol{\ell}^{(2)}(0) = \mathbf{0}$, *then*

$$\|\boldsymbol{w}^{(1)} - \boldsymbol{w}^{(2)}\|_{L^\infty(0,T;Z)} \le c\,\|\dot{\boldsymbol{\ell}}^{(1)} - \dot{\boldsymbol{\ell}}^{(2)}\|_{L^1(0,T;Z')}.$$

REMARK. It is important to observe that the presence of some kind of hardening is essential in order that the problem be well-posed in Sobolev spaces. For example, if we return to the proof of Lemma 7.7, it is seen there that the scalar θ is zero for the case of perfect plasticity ($k_1 = k_2 = 0$). This is a clear warning that an alternative approach is required for perfect plasticity. The physical situation also alerts one to the need for a different approach: In materials that are realistically idealized as perfectly plastic, it is found that singularities such as shear bands and slip lines occur; these amount to discontinuities in components of displacement, so that one no longer can expect to have $\boldsymbol{u}(t) \in V$. Rather, it is necessary to seek solutions in the space $BD(\Omega)$ of functions of bounded deformation; these are vector-valued functions that are integrable, and the corresponding strains of which are bounded measures. A treatment of this class of problems may be found in [88, 122, 123].

7.4 Stability Analysis

The application of plasticity theory in the solution of engineering problems involves the selection of the input data such as the material constants. These data are mostly determined through physical experiments, and are subject to various errors. In reality, it is impossible to specify the data associated with a plasticity problem exactly. As a result, the problem in hand is only an approximation of the real problem. Evidently, it is important to know whether small changes in the input data cause only small changes in the solution of the plasticity problem.

Since the plasticity problem describes complicated deformation processes, it is reasonable to expect that a quantitative analysis of the continuous dependence of the solution on the input data is a difficult task. Nevertheless, it is still possible to provide some analysis on the stability of the problem in the context of particular situations. For the abstract variational inequality studied in Section 7.2, one of the results proved is a stability estimate of the effect of perturbations in the linear functional associated with applied loads. In the context of a concrete plasticity problem, usually some more results on stability can be proved. Here we take the problem PRIM1 as an example and derive a stability estimate for the solution with respect to perturbations in the material properties $\boldsymbol{C}$, k_1, k_2, c_0, and the load functional $\boldsymbol{\ell}$.

Thus for the deformation of an elastoplastic material with combined linear kinematic–isotropic hardening and subject to the von Mises yield criterion, suppose that we are given two sets of data, viz. $\boldsymbol{C}^{(1)}$, $k_1^{(1)}$, $k_2^{(1)}$, $c_0^{(1)}$, $\boldsymbol{\ell}^{(1)}$ and $\boldsymbol{C}^{(2)}$, $k_1^{(2)}$, $k_2^{(2)}$, $c_0^{(2)}$, $\boldsymbol{\ell}^{(2)}$. These data satisfy the conditions stated in Section 7.1. We denote the corresponding solutions by $\boldsymbol{w}^{(1)} = (\boldsymbol{u}^{(1)}, \boldsymbol{p}^{(1)}, \gamma^{(1)})$ and $\boldsymbol{w}^{(2)} = (\boldsymbol{u}^{(2)}, \boldsymbol{p}^{(2)}, \gamma^{(2)})$. By Theorem 7.8, $\boldsymbol{w}^{(1)} \in H^1(0,T;Z)$ with $\boldsymbol{w}^{(1)}(0) = \boldsymbol{0}$ is the unique function with the property that for almost all $t \in (0,T)$, $\dot{\boldsymbol{w}}^{(1)}(t) \in Z_p$ and

$$\begin{aligned} a^{(1)}(\boldsymbol{w}^{(1)}(t), \boldsymbol{z} - \dot{\boldsymbol{w}}^{(1)}(t)) + j^{(1)}(\boldsymbol{z}) - j^{(1)}(\dot{\boldsymbol{w}}^{(1)}(t)) \\ \geq \langle \boldsymbol{\ell}^{(1)}(t), \boldsymbol{z} - \dot{\boldsymbol{w}}^{(1)}(t) \rangle \end{aligned} \tag{7.56}$$

for all $\boldsymbol{z} = (\boldsymbol{v}, \boldsymbol{q}, \mu) \in Z_p$. Here the space setting is as before, so that

$$\begin{aligned} Z &= (H_0^1(\Omega))^3 \times Q_0 \times L^2(\Omega), \\ Z_p &= \{\boldsymbol{z} = (\boldsymbol{v}, \boldsymbol{q}, \mu) \in Z : |\boldsymbol{q}| \leq \mu \text{ a.e. in } \Omega\}. \end{aligned}$$

The bilinear form $a^{(1)}(\cdot,\cdot)$ and functional $j^{(1)}(\cdot)$ are defined by

$$a^{(1)}(\boldsymbol{w}, \boldsymbol{z}) = \int_\Omega \left[\boldsymbol{C}^{(1)}(\boldsymbol{\epsilon}(\boldsymbol{u}) - \boldsymbol{p}) : (\boldsymbol{\epsilon}(\boldsymbol{v}) - \boldsymbol{q}) + k_1^{(1)} \boldsymbol{p} : \boldsymbol{q} + k_2^{(1)} \gamma \mu \right] dx$$

and

$$j^{(1)}(\boldsymbol{z}) = c_0^{(1)} j_0(\boldsymbol{z})$$

with

$$j_0(\boldsymbol{z}) = \int_\Omega D_0(\boldsymbol{q}, \mu)\, dx$$

and

$$D_0(\boldsymbol{q}, \mu) = \begin{cases} |\boldsymbol{q}| & \text{if } |\boldsymbol{q}| \le \mu, \\ +\infty & \text{if } |\boldsymbol{q}| > \mu. \end{cases}$$

Similarly, $\boldsymbol{w}^{(2)} \in H^1(0,T;Z)$, $\boldsymbol{w}^{(2)}(0) = \boldsymbol{0}$ is the unique function with the properties that for almost all $t \in (0,T)$, $\dot{\boldsymbol{w}}^{(2)}(t) \in Z_p$ and

$$\begin{aligned} a^{(2)}(\boldsymbol{w}^{(2)}(t), \boldsymbol{z} - \dot{\boldsymbol{w}}^{(2)}(t)) + j^{(2)}(\boldsymbol{z}) - j^{(2)}(\dot{\boldsymbol{w}}^{(2)}(t)) \\ \ge \langle \boldsymbol{\ell}^{(2)}(t), \boldsymbol{z} - \dot{\boldsymbol{w}}^{(2)}(t) \rangle \end{aligned} \tag{7.57}$$

for all $\boldsymbol{z} = (\boldsymbol{v}, \boldsymbol{q}, \mu) \in Z_p$, where

$$a^{(2)}(\boldsymbol{w}, \boldsymbol{z}) = \int_\Omega \left[\boldsymbol{C}^{(2)}(\boldsymbol{\epsilon}(\boldsymbol{u}) - \boldsymbol{p}) : (\boldsymbol{\epsilon}(\boldsymbol{v}) - \boldsymbol{q}) + k_1^{(2)} \boldsymbol{p} : \boldsymbol{q} + k_2^{(2)} \gamma \mu \right] dx$$

and

$$j^{(2)}(\boldsymbol{z}) = c_0^{(2)} j_0(\boldsymbol{z}).$$

We are interested in estimating the difference

$$\boldsymbol{e} = \boldsymbol{w}^{(1)} - \boldsymbol{w}^{(2)} \equiv (\boldsymbol{u}_e, \boldsymbol{p}_e, \gamma_e).$$

We take $\boldsymbol{z} = \dot{\boldsymbol{w}}^{(2)}(t) \in Z_p$ in (7.56) and divide the inequality by $c_0^{(1)}$ to obtain

$$\begin{aligned} -\frac{1}{c_0^{(1)}}\, a^{(1)}(\boldsymbol{w}^{(1)}(t), \dot{\boldsymbol{e}}(t)) + j_0(\dot{\boldsymbol{w}}^{(2)}(t)) - j_0(\dot{\boldsymbol{w}}^{(1)}(t)) \\ \ge -\frac{1}{c_0^{(1)}} \langle \boldsymbol{\ell}^{(1)}(t), \dot{\boldsymbol{e}}(t) \rangle. \end{aligned} \tag{7.58}$$

Similarly, from (7.57) we obtain

$$\begin{aligned} \frac{1}{c_0^{(2)}}\, a^{(2)}(\boldsymbol{w}^{(2)}(t), \dot{\boldsymbol{e}}(t)) + j_0(\dot{\boldsymbol{w}}^{(1)}(t)) - j_0(\dot{\boldsymbol{w}}^{(2)}(t)) \\ \ge \frac{1}{c_0^{(2)}} \langle \boldsymbol{\ell}^{(2)}(t), \dot{\boldsymbol{e}}(t) \rangle. \end{aligned} \tag{7.59}$$

We now add (7.58) and (7.59) to find that

$$\begin{aligned} -\frac{1}{c_0^{(1)}}\, a^{(1)}(\boldsymbol{w}^{(1)}(t), \dot{\boldsymbol{e}}(t)) + \frac{1}{c_0^{(2)}}\, a^{(2)}(\boldsymbol{w}^{(2)}(t), \dot{\boldsymbol{e}}(t)) \\ \ge -\frac{1}{c_0^{(1)}} \langle \boldsymbol{\ell}^{(1)}(t), \dot{\boldsymbol{e}}(t) \rangle + \frac{1}{c_0^{(2)}} \langle \boldsymbol{\ell}^{(2)}(t), \dot{\boldsymbol{e}}(t) \rangle. \end{aligned}$$

Then

$$a^{(2)}(\boldsymbol{e}(t), \dot{\boldsymbol{e}}(t)) \leq a^{(2)}(\boldsymbol{w}^{(1)}(t), \dot{\boldsymbol{e}}(t)) - \frac{c_0^{(2)}}{c_0^{(1)}} a^{(1)}(\boldsymbol{w}^{(1)}(t), \dot{\boldsymbol{e}}(t))$$
$$+ \left\langle \frac{c_0^{(2)}}{c_0^{(1)}} \boldsymbol{\ell}^{(1)}(t) - \boldsymbol{\ell}^{(2)}(t), \dot{\boldsymbol{e}}(t) \right\rangle.$$

Using the fact that $a^{(2)}(\cdot,\cdot)$ is symmetric, the above inequality can be rewritten as

$$\frac{1}{2}\frac{d}{dt}a^{(2)}(\boldsymbol{e}(t), \boldsymbol{e}(t))$$
$$\leq \int_\Omega \Big[\Big(\boldsymbol{C}^{(1)} - \frac{c_0^{(2)}}{c_0^{(1)}} \boldsymbol{C}^{(2)}\Big)\Big(\boldsymbol{\epsilon}(\boldsymbol{u}^{(1)}) - \boldsymbol{p}^{(1)}\Big) : (\boldsymbol{\epsilon}(\dot{\boldsymbol{u}}_e) - \dot{\boldsymbol{p}}_e)$$
$$+ \Big(k_1^{(1)} - \frac{c_0^{(2)}}{c_0^{(1)}} k_1^{(2)}\Big) \boldsymbol{p}^{(1)} : \dot{\boldsymbol{p}}_e + \Big(k_2^{(1)} - \frac{c_0^{(2)}}{c_0^{(1)}} k_2^{(2)}\Big) \gamma^{(1)} \dot{\gamma}_e\Big]\, dx$$
$$+ \left\langle \frac{c_0^{(2)}}{c_0^{(1)}} \boldsymbol{\ell}^{(1)}(t) - \boldsymbol{\ell}^{(2)}(t), \dot{\boldsymbol{e}}(t) \right\rangle.$$

We now integrate the relation from 0 to t and use the initial condition $\boldsymbol{e}(0) = \boldsymbol{0}$ to obtain

$$\frac{1}{2}a^{(2)}(\boldsymbol{e}(t), \boldsymbol{e}(t))$$
$$\leq \int_\Omega \Big[\Big(\boldsymbol{C}^{(1)} - \frac{c_0^{(2)}}{c_0^{(1)}} \boldsymbol{C}^{(2)}\Big)\Big(\boldsymbol{\epsilon}(\boldsymbol{u}^{(1)}) - \boldsymbol{p}^{(1)}\Big) : (\boldsymbol{\epsilon}(\boldsymbol{u}_e) - \boldsymbol{p}_e)$$
$$+ \Big(k_1^{(1)} - \frac{c_0^{(2)}}{c_0^{(1)}} k_1^{(2)}\Big) \boldsymbol{p}^{(1)} : \boldsymbol{p}_e + \Big(k_2^{(1)} - \frac{c_0^{(2)}}{c_0^{(1)}} k_2^{(2)}\Big) \gamma^{(1)} \gamma_e\Big]\, dx$$
$$- \int_0^t \int_\Omega \Big[\Big(\boldsymbol{C}^{(1)} - \frac{c_0^{(2)}}{c_0^{(1)}} \boldsymbol{C}^{(2)}\Big)\Big(\boldsymbol{\epsilon}(\dot{\boldsymbol{u}}^{(1)}) - \dot{\boldsymbol{p}}^{(1)}\Big) : (\boldsymbol{\epsilon}(\boldsymbol{u}_e) - \boldsymbol{p}_e)$$
$$+ \Big(k_1^{(1)} - \frac{c_0^{(2)}}{c_0^{(1)}} k_1^{(2)}\Big) \dot{\boldsymbol{p}}^{(1)} : \boldsymbol{p}_e + \Big(k_2^{(1)} - \frac{c_0^{(2)}}{c_0^{(1)}} k_2^{(2)}\Big) \dot{\gamma}^{(1)} \gamma_e\Big]\, dx\, dt$$
$$+ \left\langle \frac{c_0^{(2)}}{c_0^{(1)}} \boldsymbol{\ell}^{(1)}(t) - \boldsymbol{\ell}^{(2)}(t), \boldsymbol{e}(t) \right\rangle$$
$$- \int_0^t \int_\Omega \left\langle \frac{c_0^{(2)}}{c_0^{(1)}} \dot{\boldsymbol{\ell}}^{(1)}(t) - \dot{\boldsymbol{\ell}}^{(2)}(t), \boldsymbol{e}(t) \right\rangle dx\, dt.$$

Set $M = \|\boldsymbol{e}\|_{L^\infty(0,T;Z)}$ and

$$C(\boldsymbol{C}, c_0, k_1, k_2)$$
$$= \max\Big\{ \Big\| \boldsymbol{C}^{(1)} - \frac{c_0^{(2)}}{c_0^{(1)}} \boldsymbol{C}^{(2)} \Big\|_{L^\infty(\Omega)}, \Big| k_1^{(1)} - \frac{c_0^{(2)}}{c_0^{(1)}} k_1^{(2)} \Big|, \Big| k_2^{(1)} - \frac{c_0^{(2)}}{c_0^{(1)}} k_2^{(2)} \Big| \Big\}.$$

Using the Z-ellipticity of the bilinear form $a^{(2)}(\cdot,\cdot)$, we find that for some constants $c_1, c_2 > 0$,

$$\|\boldsymbol{e}(t)\|_Z^2$$
$$\leq c_1\, C(\boldsymbol{C}, c_0, k_1, k_2) \left(\|\boldsymbol{w}^{(1)}(t)\|_Z + \int_0^t \|\dot{\boldsymbol{w}}^{(1)}(t)\|_Z dt \right) M$$
$$+ c_2 \left(\Big\| \frac{c_0^{(2)}}{c_0^{(1)}} \boldsymbol{\ell}^{(1)}(t) - \boldsymbol{\ell}^{(2)}(t) \Big\|_{Z'} + \int_0^t \Big\| \frac{c_0^{(2)}}{c_0^{(1)}} \dot{\boldsymbol{\ell}}^{(1)}(t) - \dot{\boldsymbol{\ell}}^{(2)}(t) \Big\|_{Z'} dt \right) M.$$

Since $\boldsymbol{w}^{(1)}(0) = \boldsymbol{0}$ and $\boldsymbol{\ell}^{(1)}(0) = \boldsymbol{\ell}^{(2)}(0) = \boldsymbol{0}$, we have, for some constants $c_1, c_2 > 0$,

$$\|\boldsymbol{e}(t)\|_Z^2 \leq c_1\, C(\boldsymbol{C}, c_0, k_1, k_2) \int_0^t \|\dot{\boldsymbol{w}}^{(1)}(t)\|_Z dt\, M$$
$$+ c_2 \int_0^t \Big\| \frac{c_0^{(2)}}{c_0^{(1)}} \dot{\boldsymbol{\ell}}^{(1)}(t) - \dot{\boldsymbol{\ell}}^{(2)}(t) \Big\|_{Z'} dt\, M.$$

It is then easy to see that

$$\|\boldsymbol{w}^{(1)} - \boldsymbol{w}^{(2)}\|_{L^\infty(0,T;Z)}$$
$$\leq c_1\, C(\boldsymbol{C}, c_0, k_1, k_2)\, \|\dot{\boldsymbol{w}}^{(1)}\|_{L^1(0,T;Z)} + c_2 \Big\| \frac{c_0^{(2)}}{c_0^{(1)}} \dot{\boldsymbol{\ell}}^{(1)} - \dot{\boldsymbol{\ell}}^{(2)} \Big\|_{L^1(0,T;Z')}. \tag{7.60}$$

The estimate (7.60) clearly shows that the solution of the problem PRIM1 depends Lipschitz continuously on the material properties and the applied forces. The constants c_1 and c_2 in the estimate depend only on the continuity constant and Z-ellipticity constant of the bilinear form $a^{(2)}(\cdot,\cdot)$. A more careful derivation of the estimate (7.60) will reveal concrete expressions of these constants, so (7.60) can be used both as an a priori error estimate (by taking $\boldsymbol{w}^{(1)}$ as the unknown solution and $\boldsymbol{w}^{(2)}$ as its approximation) and an a posteriori error estimate (by taking $\boldsymbol{w}^{(2)}$ as the unknown solution and $\boldsymbol{w}^{(1)}$ as its approximation).

The estimate (7.60) can also be used in a stability analysis of the stress $\boldsymbol{\sigma}$ with respect to perturbations in the input data. For the two sets of data, the corresponding stresses are $\boldsymbol{\sigma}^{(1)} = \boldsymbol{C}^{(1)}(\boldsymbol{\epsilon}(\boldsymbol{u}^{(1)}) - \boldsymbol{p}^{(1)})$ and $\boldsymbol{\sigma}^{(2)} = \boldsymbol{C}^{(2)}(\boldsymbol{\epsilon}(\boldsymbol{u}^{(2)}) - \boldsymbol{p}^{(2)})$. Evidently, the stress difference $\boldsymbol{\sigma}^{(1)} - \boldsymbol{\sigma}^{(2)}$ depends on the perturbations in the data also in a Lipschitz manner.

8

The Dual Variational Problem of Elastoplasticity

This chapter has a purpose parallel to that of Chapter 7, in that the dual variational problem of elastoplasticity will be studied in detail. This problem takes as its point of departure the flow law in the form (4.35), that is, the statement of the flow law that makes use of the yield surface and the normality law.

Since the dual and primal forms are two different formulations of the same problem, the two formulations are equivalent, in a sense to be made precise in Theorem 8.3 below. Once such a correspondence is established, it follows that well-posedness of the dual variational problem may be inferred from the results of Chapter 7. Nevertheless, a direct analysis of the dual variational problem is of interest in its own right, and the main purpose of the chapter is to give such a qualitative analysis *de novo*.

The dual problem has been the most popular framework for both mathematical analyses of, and computational approaches to, the problem of elastoplasticity. The unknown variables in the dual variational problem are the generalized stress and the displacement. Our analysis starts with a consideration of the well-posedness of the so-called stress problem, in Section 8.2, in which the explicit appearance of the displacement is eliminated from the formulation. We study the full dual variational problem, where the displacement variable is present, in Section 8.3. Error analyses of various numerical approximation schemes will be taken up later, in Chapter 13.

The analysis presented here draws on those of Johnson [66] and Matthies [88], but there are some significant differences. First, the safe load condition, an essential ingredient of an analysis of well-posedness, is phrased in a different way here. Second, the problem is firmly embedded in the mixed variational framework of Babuška and Brezzi [4, 17].

8.1 The Dual Variational Problem

We begin by formulating the dual problem that makes use of the flow law in the form (4.35). This will lead to a problem in which the unknown variables are the generalized stress $\boldsymbol{\Sigma} = (\boldsymbol{\sigma}, \boldsymbol{\chi})$ and the displacement $\boldsymbol{u}$.

The space V of displacements is, as before,

$$V = [H_0^1(\Omega)]^3,$$

and the space of stresses is defined by

$$S = \{\boldsymbol{\tau} = (\tau_{ij})_{3\times 3} : \ \tau_{ji} = \tau_{ij}, \ \tau_{ij} \in L^2(\Omega)\}.$$

We continue to regard internal variables at a point as being members of a finite-dimensional space X isomorphic to $\mathbb{R}^m$. The conjugate forces likewise are regarded as members of X (strictly speaking, they are members of X', the topological dual of X, but we may ignore this distinction in the finite-dimensional case). Thus we introduce the space M of conjugate forces

$$M = \{\boldsymbol{\mu} = (\mu_j) : \ \mu_j \in L^2(\Omega), \ j = 1, \ldots, m\}.$$

Further, let

$$\mathcal{T} = S \times M.$$

This space is endowed with the inner products induced by the natural inner products on S and M. Admissible generalized stresses are those that belong to the set K pointwise. We accordingly define the convex subset

$$\mathcal{P} = \{\boldsymbol{T} = (\boldsymbol{\tau}, \boldsymbol{\mu}) \in \mathcal{T} : \ (\boldsymbol{\tau}, \boldsymbol{\mu}) \in K \text{ a.e. in } \Omega\}. \tag{8.1}$$

We now introduce the bilinear forms associated with the dual problem: These are

$$\bar{a} : S \times S \to \mathbb{R}, \quad \bar{a}(\boldsymbol{\sigma}, \boldsymbol{\tau}) = \int_\Omega \boldsymbol{\sigma} : \boldsymbol{C}^{-1}\boldsymbol{\tau}\, dx, \tag{8.2}$$

$$b : V \times S \to \mathbb{R}, \quad b(\boldsymbol{v}, \boldsymbol{\tau}) = -\int_\Omega \boldsymbol{\epsilon}(\boldsymbol{v}) : \boldsymbol{\tau}\, dx, \tag{8.3}$$

$$c : M \times M \to \mathbb{R}, \quad c(\boldsymbol{\chi}, \boldsymbol{\mu}) = \int_\Omega \boldsymbol{\chi} : \boldsymbol{H}^{-1}\boldsymbol{\mu}\, dx, \tag{8.4}$$

and the bilinear form

$$A : \mathcal{T} \times \mathcal{T} \to \mathbb{R}, \quad A(\boldsymbol{\Sigma}, \boldsymbol{T}) = \bar{a}(\boldsymbol{\sigma}, \boldsymbol{\tau}) + c(\boldsymbol{\chi}, \boldsymbol{\mu}) \tag{8.5}$$

for $\boldsymbol{\Sigma} = (\boldsymbol{\sigma}, \boldsymbol{\chi})$ and $\boldsymbol{T} = (\boldsymbol{\tau}, \boldsymbol{\mu})$. Here $\boldsymbol{C}^{-1}$ is the *compliance tensor*, which is inverse to the elasticity tensor $\boldsymbol{C}$ in the sense that

$$\boldsymbol{C}^{-1}(\boldsymbol{C}\boldsymbol{\epsilon}) = \boldsymbol{\epsilon} \quad \text{and} \quad \boldsymbol{C}(\boldsymbol{C}^{-1}\boldsymbol{\sigma}) = \boldsymbol{\sigma}$$

for all matrices or second-order tensors $\boldsymbol{\epsilon}$ and $\boldsymbol{\sigma}$. Likewise, $\boldsymbol{H}^{-1}$ is the inverse to the hardening modulus $\boldsymbol{H}$. Recalling the material properties stated in Section 7.1, we see that the compliance tensor $\boldsymbol{C}^{-1}$ has the same symmetry properties as $\boldsymbol{C}$ and is pointwise stable in the sense that a constant $C_0' > 0$ exists such that

$$C_{ijkl}^{-1}(\boldsymbol{x})\zeta_{ij}\zeta_{kl} \geq C_0'|\boldsymbol{\zeta}|^2 \quad \forall \boldsymbol{\zeta} \in M^3, \text{ a.e. in } \Omega. \tag{8.6}$$

Also, the inverse $\boldsymbol{H}^{-1}$ of the hardening modulus possesses the same properties as $\boldsymbol{H}$: It is a symmetric operator whose matrix representation has uniformly bounded components. Furthermore, there exists a constant $H_0' > 0$ such that

$$\boldsymbol{\chi} : \boldsymbol{H}^{-1}\boldsymbol{\chi} \geq H_0'|\boldsymbol{\chi}|^2 \;\; \forall \boldsymbol{\chi} \in \mathbb{R}^m, \text{ a.e. in } \Omega. \tag{8.7}$$

The bilinear form $A(\cdot,\cdot)$ is symmetric, continuous, and $\mathcal{T}$-elliptic; that is, there exist constants α_A, $\beta_A > 0$ such that

$$|A(\boldsymbol{\Sigma}, \boldsymbol{T})| \leq \alpha_A \|\boldsymbol{\Sigma}\|_{\mathcal{T}} \|\boldsymbol{T}\|_{\mathcal{T}} \quad \forall \boldsymbol{\Sigma}, \boldsymbol{T} \in \mathcal{T}, \tag{8.8}$$

$$A(\boldsymbol{T}, \boldsymbol{T}) \geq \beta_A \|\boldsymbol{T}\|_{\mathcal{T}}^2 \quad \forall \boldsymbol{T} \in \mathcal{T}. \tag{8.9}$$

The ellipticity property (8.9) follows easily from (8.6) and (8.7) of the moduli $\boldsymbol{C}^{-1}$ and $\boldsymbol{H}^{-1}$; indeed, we may take $\beta_A = \min\{C_0', H'\}$.

The bilinear form $b(\cdot,\cdot)$ is continuous; that is, for some constant $\alpha_b > 0$,

$$|b(\boldsymbol{v}, \boldsymbol{\tau})| \leq \alpha_b \|\boldsymbol{v}\|_V \|\boldsymbol{\tau}\|_S \quad \forall \boldsymbol{v} \in V, \, \boldsymbol{\tau} \in S. \tag{8.10}$$

Furthermore, for some constant $\beta_b > 0$,

$$\sup_{0 \neq \boldsymbol{\tau} \in S} \frac{|b(\boldsymbol{v}, \boldsymbol{\tau})|}{\|\boldsymbol{\tau}\|_S} \geq \beta_b \|\boldsymbol{v}\|_V \quad \forall \boldsymbol{v} \in V. \tag{8.11}$$

This property is readily derived by setting $\boldsymbol{\tau} = \boldsymbol{\epsilon}(\boldsymbol{v})$; then $|b(\boldsymbol{v}, \boldsymbol{\tau})|/\|\boldsymbol{\tau}\|_S = \|\boldsymbol{\epsilon}(\boldsymbol{v})\|_S \geq c\,\|\boldsymbol{v}\|_V$, using Korn's inequality (5.21).

We will also need the linear functional

$$\boldsymbol{\ell}(t) : V \to \mathbb{R}, \qquad \langle \boldsymbol{\ell}(t), \boldsymbol{v} \rangle = -\int_\Omega \boldsymbol{f}(t) \cdot \boldsymbol{v}\, dx. \tag{8.12}$$

The dual variational problem is obtained through the standard procedure from the equilibrium equation

$$\operatorname{div} \boldsymbol{\sigma} + \boldsymbol{f} = \boldsymbol{0} \quad \text{in } \Omega$$

and the flow law

$$(\dot{\boldsymbol{\sigma}}^E - \dot{\boldsymbol{\sigma}}) : \boldsymbol{C}^{-1}(\boldsymbol{\tau} - \boldsymbol{\sigma}) - \dot{\boldsymbol{\chi}} : \boldsymbol{H}^{-1}(\boldsymbol{\mu} - \boldsymbol{\chi}) \leq 0 \quad \forall\, (\boldsymbol{\tau}, \boldsymbol{\mu}) \in K,$$

in which $\dot{\boldsymbol{\sigma}}^E = \boldsymbol{C}\boldsymbol{\epsilon}(\dot{\boldsymbol{u}})$ is the elastic stress rate.

Problem Dual. Given $\boldsymbol{\ell} \in H^1(0, T; V')$ with $\boldsymbol{\ell}(0) = \boldsymbol{0}$, find $(\boldsymbol{u}, \boldsymbol{\Sigma}) = (\boldsymbol{u}, \boldsymbol{\sigma}, \boldsymbol{\chi}) : [0, T] \to V \times \mathcal{P}$ with $(\boldsymbol{u}(0), \boldsymbol{\Sigma}(0)) = (\boldsymbol{0}, \boldsymbol{0})$ such that for almost all $t \in (0, T)$,

$$b(\boldsymbol{v}, \boldsymbol{\sigma}(t)) = \langle \boldsymbol{\ell}(t), \boldsymbol{v} \rangle \quad \forall\, \boldsymbol{v} \in V, \tag{8.13}$$

$$A(\dot{\boldsymbol{\Sigma}}(t), \boldsymbol{T} - \boldsymbol{\Sigma}(t)) + b(\dot{\boldsymbol{u}}(t), \boldsymbol{\tau} - \boldsymbol{\sigma}(t)) \geq 0 \quad \forall\, \boldsymbol{T} = (\boldsymbol{\tau}, \boldsymbol{\mu}) \in \mathcal{P}. \tag{8.14}$$

The formal equivalence of Problem Dual to the classical problem can be readily established.

We notice that when there is no plastic deformation, the problem Dual is reduced to a linear elasticity problem. To see this, we set $\boldsymbol{\chi} = \boldsymbol{0}$ and allow $\boldsymbol{\sigma}, \boldsymbol{\tau} \in S$. Then from (8.14), we obtain

$$\int_\Omega \dot{\boldsymbol{\sigma}} : \boldsymbol{C}^{-1}(\boldsymbol{\tau} - \boldsymbol{\sigma})\, dx - \int_\Omega \boldsymbol{\epsilon}(\dot{\boldsymbol{u}}) : (\boldsymbol{\tau} - \boldsymbol{\sigma})\, dx \geq 0 \quad \forall\, \boldsymbol{\tau} \in S.$$

Since S is a linear space, we get

$$\int_\Omega \dot{\boldsymbol{\sigma}} : \boldsymbol{C}^{-1}\boldsymbol{\tau}\, dx - \int_\Omega \boldsymbol{\epsilon}(\dot{\boldsymbol{u}}) : \boldsymbol{\tau}\, dx = 0 \quad \forall\, \boldsymbol{\tau} \in S,$$

and upon integrating with respect to time,

$$\int_\Omega \boldsymbol{\sigma} : \boldsymbol{C}^{-1}\boldsymbol{\tau}\, dx - \int_\Omega \boldsymbol{\epsilon}(\boldsymbol{u}) : \boldsymbol{\tau}\, dx = 0 \quad \forall\, \boldsymbol{\tau} \in S.$$

Similarly, from (8.13), we get

$$-\int_\Omega \boldsymbol{\epsilon}(\boldsymbol{v}) : \boldsymbol{\sigma}\, dx = -\int_\Omega \boldsymbol{f} \cdot \boldsymbol{v}\, dx \quad \forall\, \boldsymbol{v} \in [H_0^1(\Omega)]^3.$$

Thus we have recovered the mixed formulation (6.25)–(6.26) of the linear elasticity problem with the homogeneous displacement boundary condition.

In dealing with the constraint (8.13) of the dual problem, Theorem 5.5 will play a crucial role. For convenience, we restate the result here.

Proposition 8.1. *Let V and S be two Hilbert spaces. Let $b : V \times S \to \mathbb{R}$ be a continuous bilinear form. Define two bounded linear operators $B : S \to V'$ and $B' : V \to S'$ by*

$$b(v, s) = \langle Bs, v \rangle = \langle B'v, s \rangle \quad \text{for } v \in V,\ s \in S.$$

Then the following statements are equivalent:

(a) *the bilinear form* $b(\cdot,\cdot)$ *satisfies the Babuška–Brezzi condition*

$$\sup_{0\neq s\in S}\frac{|b(v,s)|}{\|s\|_S}\geq c_0\|v\|_V\quad\forall\, v\in V;$$

(b) *the operator* B *is an isomorphism from* $(\operatorname{Ker} B)^\perp$ *onto* V', *where*

$$\operatorname{Ker} B=\{s\in S: b(v,s)=0\ \forall\, v\in V\};$$

(c) *the operator* B' *is an isomorphism from* V *onto* $(\operatorname{Ker} B)^\circ$, *where*

$$(\operatorname{Ker} B)^\circ=\{f\in S':\ \langle f,s\rangle=0\ \ \forall\, s\in \operatorname{Ker} B\}.$$

For the bilinear form $b(\cdot,\cdot)$ defined in (8.3), the Babuška–Brezzi condition is satisfied (cf. (8.11)). Thus, if we define the operators $B: S\to V'$ and $B': V\to S'$ by

$$b(\boldsymbol{v},\boldsymbol{\tau})=\langle B\boldsymbol{\tau},\boldsymbol{v}\rangle=\langle B'\boldsymbol{v},\boldsymbol{\tau}\rangle\quad\text{for } \boldsymbol{v}\in V,\ \boldsymbol{\tau}\in S,$$

then we have the following result.

LEMMA 8.2. *For the bilinear form* b *defined in* (8.3), *the operator* B *is an isomorphism from* $(\operatorname{Ker} B)^\perp$ *onto* V', *and the operator* B' *is an isomorphism from* V *onto* $(\operatorname{Ker} B)^\circ$. *Here*

$$\begin{aligned}\operatorname{Ker} B&=\{\boldsymbol{\tau}\in S: b(\boldsymbol{v},\boldsymbol{\tau})=0\quad \forall\, \boldsymbol{v}\in V\},\\ (\operatorname{Ker} B)^\circ&=\{f\in S':\ \langle f,\boldsymbol{\tau}\rangle=0\quad\forall\,\boldsymbol{\tau}\in\operatorname{Ker} B\}.\end{aligned}$$

We now address the issue of the equivalence of the dual variational problem to the primal problem, in a precise sense. The primal problem is formulated in Section 7.1. The dual form and the primal form are two different formulations of the same elastoplasticity problem, with the only distinction that two different, *yet equivalent*, forms of the flow law are employed. Hence, we have the following equivalence theorem.

THEOREM 8.3. *Assume* $\boldsymbol{f}\in H^1(0,T;V')$. *Then* $(\boldsymbol{u},\boldsymbol{p},\boldsymbol{\xi})\in H^1(0,T;Z)$ *is a solution of the problem* PRIM *if and only if* $(\boldsymbol{u},\boldsymbol{\sigma},\boldsymbol{\chi})\in H^1(0,T;V\times\mathcal{T})$ *is a solution of the problem* DUAL, *where* $(\boldsymbol{u},\boldsymbol{p},\boldsymbol{\xi})$ *and* $(\boldsymbol{u},\boldsymbol{\sigma},\boldsymbol{\chi})$ *are related by*

$$\begin{aligned}\boldsymbol{\sigma}&=\boldsymbol{C}\,(\boldsymbol{\epsilon}(\boldsymbol{u})-\boldsymbol{p}),\\ \boldsymbol{\chi}&=-\boldsymbol{H}\,\boldsymbol{\xi},\end{aligned}$$

or equivalently,

$$\begin{aligned}\boldsymbol{p}&=\boldsymbol{\epsilon}(\boldsymbol{u})-\boldsymbol{C}^{-1}\boldsymbol{\sigma},\\ \boldsymbol{\xi}&=-\boldsymbol{H}^{-1}\boldsymbol{\chi}.\end{aligned}$$

Theorem 8.3 allows us to deduce the well-posedness of the dual form of an elastoplasticity problem from that of its primal form, as long as the well-posedness of the primal form of the problem has been established. In this chapter, however, our main purpose is to give a qualitative analysis of the dual variational problem by considering this problem directly.

8.2 Analysis of the Stress Problem

In this section we analyze the stress problem that is a reduced form of the problem DUAL. By a stress problem we mean a problem in which the sole unknown variable is the (generalized) stress; that is, the displacement variable is formally eliminated. To derive the stress problem corresponding to the constraint set (8.1) we introduce the time-dependent constraint set

$$\mathcal{P}(t) = \{\boldsymbol{T} = (\boldsymbol{\tau}, \boldsymbol{\mu}) \in \mathcal{P} : \ b(\boldsymbol{v}, \boldsymbol{\tau}) = \langle \boldsymbol{\ell}(t), \boldsymbol{v} \rangle \quad \forall\, \boldsymbol{v} \in V\}. \tag{8.15}$$

We then view (8.13) as a constraint on the variable $\boldsymbol{\sigma}(t)$, and the variable $\boldsymbol{w}(t) \equiv \dot{\boldsymbol{u}}(t)$ as a Lagrangian multiplier for the constraint. Eliminating the variable $\boldsymbol{w}(t)$ from Problem DUAL, we then have the following stress problem.

PROBLEM DUAL1. Given $\boldsymbol{\ell} \in H^1(0,T;V')$, $\boldsymbol{\ell}(0) = 0$, find $\boldsymbol{\Sigma} = (\boldsymbol{\sigma}, \boldsymbol{\chi}) : [0,T] \to \mathcal{P}$ with $\boldsymbol{\Sigma}(0) = \boldsymbol{0}$ such that for almost all $t \in (0,T)$, $\boldsymbol{\Sigma}(t) \in \mathcal{P}(t)$ and

$$A(\dot{\boldsymbol{\Sigma}}(t), \boldsymbol{T} - \boldsymbol{\Sigma}(t)) \geq 0 \quad \forall\, \boldsymbol{T} \in \mathcal{P}(t). \tag{8.16}$$

In order to show the well-posedness of the stress problem DUAL1, we impose the following assumption on the structure of the set $\mathcal{P}$.

ASSUMPTION 8.4. *There is a constant $c > 0$ with the property that for any $\boldsymbol{\Sigma}_1 = (\boldsymbol{\sigma}_1, \boldsymbol{\chi}_1) \in \mathcal{P}$ and any $\boldsymbol{\sigma}_2 \in S$, there exists $\boldsymbol{\chi}_2 \in M$ such that $|\boldsymbol{\chi}_2| \leq c\,|\boldsymbol{\sigma}_2|$ and $\boldsymbol{\Sigma}_1 + \boldsymbol{\Sigma}_2 \in \mathcal{P}$, where $\boldsymbol{\Sigma}_2 = (\boldsymbol{\sigma}_2, \boldsymbol{\chi}_2)$.*

Assumption 8.4 is an alternative, and more transparent, way of stating the safe load condition used, for example, in [66]. For materials undergoing combined linear kinematic and isotropic hardening, the set K in (8.1) is defined by the relation (cf. Example 4.8)

$$\phi(\boldsymbol{\Sigma}) = \Phi(\boldsymbol{\sigma} + \boldsymbol{a}) + g - c_0 \leq 0 \quad \text{for } \boldsymbol{\Sigma} = (\boldsymbol{\sigma}, \boldsymbol{a}, g).$$

Here we have identified $\boldsymbol{\chi}$ with the pair $(\boldsymbol{a}, g)$. Now suppose that $\boldsymbol{\Sigma}_1 = (\boldsymbol{\sigma}_1, \boldsymbol{a}_1, g_1) \in \mathcal{P}$, $\boldsymbol{\sigma}_2 \in S$. Then at any point $\boldsymbol{x} \in \Omega$,

$$\phi(\boldsymbol{\Sigma}_1) = \Phi(\boldsymbol{\sigma}_1 + \boldsymbol{a}_1) + g_1 - c_0 \leq 0.$$

We need to find a pair $(\boldsymbol{a}_2, g_2)$ such that

$$\phi(\boldsymbol{\Sigma}_1 + \boldsymbol{\Sigma}_2) = \Phi(\boldsymbol{\sigma}_1 + \boldsymbol{\sigma}_2 + \boldsymbol{a}_1 + \boldsymbol{a}_2) + g_1 + g_2 - c_0 \leq 0.$$

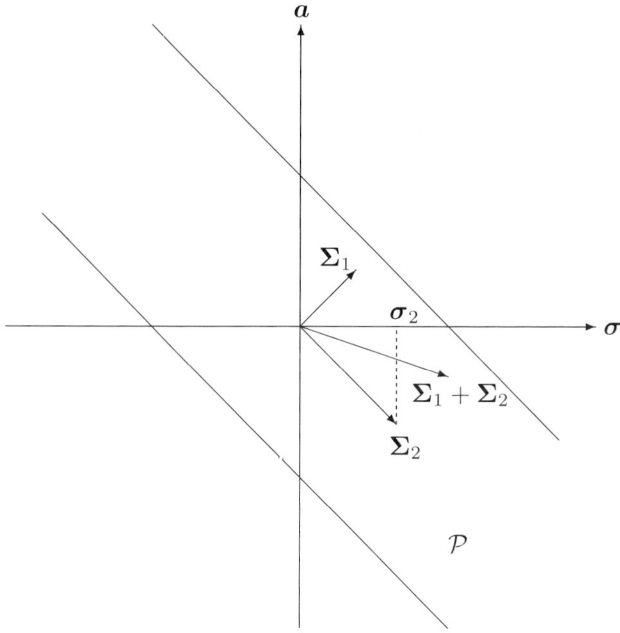

Figure 8.1: Illustration of the safe load condition

Obviously, it suffices to take

$$\boldsymbol{a}_2 = -\boldsymbol{\sigma}_2, \quad g_2 = 0.$$

Then we have $\boldsymbol{\Sigma}_1 + \boldsymbol{\Sigma}_2 \in \mathcal{P}$, and $|(\boldsymbol{a}_2, g_2)| = |\boldsymbol{\sigma}_2|$, that is, the constant c in Assumption 8.4 for this case can be taken to be equal to 1 (see Figure 8.1). In a similar way, Assumption 8.4 can be shown to hold for the special cases of linear kinematic or isotropic hardening only, but it is of course degenerate for perfect plasticity.

We note that in Assumption 8.4, the element $\boldsymbol{\Sigma}_1$ is arbitrary in $\mathcal{P}$. In particular, we can take $\boldsymbol{\Sigma}_1 = \mathbf{0}$, and then Assumption 8.4 states that for any $\boldsymbol{\sigma} \in S$, there exists $\boldsymbol{\chi} \in M$ such that $|\boldsymbol{\chi}| \le c\,|\boldsymbol{\sigma}|$ and $\boldsymbol{\Sigma} = (\boldsymbol{\sigma}, \boldsymbol{\chi}) \in \mathcal{P}$.

The following result is a simple consequence of Lemma 8.2 and Assumption 8.4.

Lemma 8.5. *The set $\mathcal{P}(t)$ defined in* (8.15) *is nonempty, closed, and convex.*

Proof. The closedness and convexity of the set $\mathcal{P}(t)$ follow from the corresponding properties of the set $\mathcal{P}$ defined in (8.1). So we need to prove only that the set $\mathcal{P}(t)$ is nonempty. Lemma 8.2 assures the existence of $\boldsymbol{\sigma} \in S$

such that

$$b(\boldsymbol{v}, \boldsymbol{\sigma}) = \langle \boldsymbol{\ell}(t), \boldsymbol{v} \rangle \quad \forall \boldsymbol{v} \in V.$$

Using Assumption 8.4 (with $\boldsymbol{\Sigma}_1 = \boldsymbol{0}$ there), we can find $\boldsymbol{\chi} \in M$ such that $(\boldsymbol{\sigma}, \boldsymbol{\chi}) \in \mathcal{P}$. Hence, $(\boldsymbol{\sigma}, \boldsymbol{\chi}) \in \mathcal{P}(t)$. □

Later on we will employ a regularization method as in [64, 88] by making use of the Yosida regularization J_ϵ defined by

$$J_\epsilon(\boldsymbol{T}) = \frac{1}{2\,\epsilon} \|\boldsymbol{T} - \Pi \boldsymbol{T}\|_{\mathcal{T}}^2,$$

where Π is the projection operator onto $\mathcal{P}$ and $\epsilon > 0$ is a small regularization parameter. Some basic properties of the functional J_ϵ are summarized in the following lemma.

LEMMA 8.6. *The functional J_ϵ is convex and is Gâteaux differentiable with the derivative*

$$J_\epsilon'(\boldsymbol{T}) = \frac{1}{\epsilon}(\boldsymbol{T} - \Pi \boldsymbol{T}). \tag{8.17}$$

Here a member of $\mathcal{T}'$ is identified with its image in $\mathcal{T}$ under the Riesz isomorphism. The Gâteaux derivative is monotone.

PROOF. By the definition of the projection,

$$\|\boldsymbol{T} - \Pi \boldsymbol{T}\| = \inf\{\|\boldsymbol{T} - \boldsymbol{T}_1\| : \boldsymbol{T}_1 \in \mathcal{P}\}.$$

The projection is characterized by the inequality

$$(\boldsymbol{T} - \Pi \boldsymbol{T}, \boldsymbol{T}_1 - \Pi \boldsymbol{T}) \le 0 \quad \forall \boldsymbol{T}_1 \in \mathcal{P}.$$

By definition of the Gâteaux derivative, for any $\boldsymbol{T}_1 \in \mathcal{T}$,

$$(J_\epsilon'(\boldsymbol{T}), \boldsymbol{T}_1) = \lim_{t \to 0} \frac{1}{t} \left[J_\epsilon(\boldsymbol{T} + t\,\boldsymbol{T}_1) - J_\epsilon(\boldsymbol{T})\right].$$

We have

$$\begin{aligned}
&\|\boldsymbol{T} + t\,\boldsymbol{T}_1 - \Pi(\boldsymbol{T} + t\,\boldsymbol{T}_1)\|^2 \\
&\quad = \|(\boldsymbol{T} - \Pi \boldsymbol{T}) + t\,\boldsymbol{T}_1 + \Pi \boldsymbol{T} - \Pi(\boldsymbol{T} + t\,\boldsymbol{T}_1)\|^2 \\
&\quad = \|\boldsymbol{T} - \Pi \boldsymbol{T}\|^2 + 2\,t\,(\boldsymbol{T} - \Pi \boldsymbol{T}, \boldsymbol{T}_1) \\
&\qquad + (\boldsymbol{T} - \Pi \boldsymbol{T}, \Pi \boldsymbol{T} - \Pi(\boldsymbol{T} + t\,\boldsymbol{T}_1)) + \|t\,\boldsymbol{T}_1 + \Pi \boldsymbol{T} - \Pi(\boldsymbol{T} + t\,\boldsymbol{T}_1)\|^2.
\end{aligned}$$

The last term is of order $O(t^2)$, since Π is nonexpansive (cf. Section 5.1), and so

$$\|t\,\boldsymbol{T}_1 + \Pi \boldsymbol{T} - \Pi(\boldsymbol{T} + t\,\boldsymbol{T}_1)\| \le |t|\,\|\boldsymbol{T}_1\| + \|\Pi \boldsymbol{T} - \Pi(\boldsymbol{T} + t\,\boldsymbol{T}_1)\| \le O(|t|).$$

Now on one hand, by the characterizing inequality of the projection,

$$(\boldsymbol{T} - \Pi\boldsymbol{T}, \Pi\boldsymbol{T} - \Pi(\boldsymbol{T} + t\,\boldsymbol{T}_1)) \geq 0.$$

On the other hand,

$$\begin{aligned}(\boldsymbol{T} &- \Pi\boldsymbol{T}, \Pi\boldsymbol{T} - \Pi(\boldsymbol{T} + t\,\boldsymbol{T}_1)) \\ &= (\boldsymbol{T} - \Pi(\boldsymbol{T} + t\,\boldsymbol{T}_1), \Pi\boldsymbol{T} - \Pi(\boldsymbol{T} + t\,\boldsymbol{T}_1)) - \|\Pi\boldsymbol{T} - \Pi(\boldsymbol{T} + t\,\boldsymbol{T}_1)\|^2.\end{aligned}$$

Again, by the characterizing inequality and the nonexpansiveness of the projection,

$$(\boldsymbol{T} - \Pi\boldsymbol{T}, \Pi\boldsymbol{T} - \Pi(\boldsymbol{T} + t\,\boldsymbol{T}_1)) \leq O(t^2).$$

Thus,

$$\left|\frac{1}{t}\left[J_\epsilon(\boldsymbol{T} + t\,\boldsymbol{T}_1) - J_\epsilon(\boldsymbol{T})\right] - \frac{1}{\epsilon}\left(\boldsymbol{T} - \Pi\boldsymbol{T}, \boldsymbol{T}_1\right)\right| - O(|t|).$$

So the Gâteaux derivative $J_\epsilon'(\boldsymbol{T})$ is given by the formula (8.17).

It is easy to see that $J_\epsilon'(\boldsymbol{T})$ is monotone; hence the functional $J_\epsilon(\boldsymbol{T})$ is convex. □

As in the proof of the existence result for the primal variational problem, existence of a solution to Problem DUAL1 is approached by first considering a time-discrete approximation to the problem. Again, we use a uniform partition of the time interval $[0, T]$: $0 = t_0 < t_1 < \cdots < t_N = T$, with $t_n = n\,k$, $n = 0, \ldots, N$, and $k = T/N$.

PROBLEM DUAL1^k. Given $\{\boldsymbol{\ell}_n\}_{n=0}^N \subset V'$, $\boldsymbol{\ell}_0 = \boldsymbol{0}$, find $\boldsymbol{\Sigma}_n^k = (\boldsymbol{\sigma}_n^k, \boldsymbol{\chi}_n^k) \in \mathcal{P}_n$, $\boldsymbol{\Sigma}_0^k = \boldsymbol{0}$ such that for $n = 1, 2, \ldots, N$,

$$A(\Delta\boldsymbol{\Sigma}_n^k, \boldsymbol{T} - \boldsymbol{\Sigma}_n^k) \geq 0 \quad \forall\, \boldsymbol{T} \in \mathcal{P}_n. \tag{8.18}$$

The constraint set $\mathcal{P}_n$ is defined by

$$\mathcal{P}_n \equiv \mathcal{P}(t_n) = \{\boldsymbol{T} = (\boldsymbol{\tau}, \boldsymbol{\mu}) \in \mathcal{P} :\ b(\boldsymbol{v}, \boldsymbol{\tau}) = \langle \boldsymbol{\ell}_n, \boldsymbol{v}\rangle \quad \forall\, \boldsymbol{v} \in V\}.$$

LEMMA 8.7. *There exists a unique solution to Problem* DUAL1^k.

PROOF. We notice that (8.18) is equivalent to

$$\boldsymbol{\Sigma}_n^k \in \mathcal{P}_n, \quad A(\boldsymbol{\Sigma}_n^k, \boldsymbol{T} - \boldsymbol{\Sigma}_n^k) \geq A(\boldsymbol{\Sigma}_{n-1}^k, \boldsymbol{T} - \boldsymbol{\Sigma}_n^k) \quad \forall\, \boldsymbol{T} \in \mathcal{P}_n.$$

This is an elliptic variational inequality of the first kind, and we can apply Theorem 6.4 to get the existence of a unique solution for this inequality. □

Let us give another proof of Lemma 8.7 without making use of Theorem 6.4. This second proof is interesting in that the same idea will be used to prove the existence result for the problem DUAL in the next section. For this reason, we provide the second proof next.

SECOND PROOF OF LEMMA 8.7. We first prove that for any n, the variational inequality (8.18) has a solution $\boldsymbol{\Sigma}_n^k \in \mathcal{P}_n$. By Lemma 8.2, there exists a unique element $\boldsymbol{\sigma}_0 \in (\operatorname{Ker} B)^\perp$ such that

$$b(\boldsymbol{v}, \boldsymbol{\sigma}_0) = \langle \boldsymbol{\ell}_n, \boldsymbol{v} \rangle \quad \forall \boldsymbol{v} \in V$$

and

$$\|\boldsymbol{\sigma}_0\|_S \le c\,\|\boldsymbol{\ell}_n\|$$

for some constant c independent of $\boldsymbol{\ell}$. Here and below, the notation $\boldsymbol{\ell}_n = \boldsymbol{\ell}(t_n)$ is used. Using Assumption 8.4, it is then possible to find a $\boldsymbol{\chi}_0 \in M$ such that $\boldsymbol{\Sigma}_0 = (\boldsymbol{\sigma}_0, \boldsymbol{\chi}_0) \in \mathcal{P}$ and

$$\|\boldsymbol{\chi}_0\| \le c\,\|\boldsymbol{\sigma}_0\| \le c\,\|\boldsymbol{\ell}_n\|.$$

Now we rewrite (8.18) as

$$A(\boldsymbol{\Sigma}_n^k, \boldsymbol{T} - \boldsymbol{\Sigma}_n^k) \ge A(\boldsymbol{\Sigma}_{n-1}^k, \boldsymbol{T} - \boldsymbol{\Sigma}_n^k) \quad \forall \boldsymbol{T} \in \mathcal{P}_n \tag{8.19}$$

and consider its regularization

$$A(\boldsymbol{\Sigma}_{n,\epsilon}^k, \boldsymbol{T}) + \Big(J_\epsilon'(\boldsymbol{\Sigma}_{n,\epsilon}^k), \boldsymbol{T}\Big) = A(\boldsymbol{\Sigma}_{n-1}^k, \boldsymbol{T}) \quad \forall \boldsymbol{T} \in \operatorname{Ker} B \times M, \tag{8.20}$$

using the functional J_ϵ. We write $\boldsymbol{\Sigma}_{n,\epsilon}^k = \boldsymbol{\Sigma}_0 + \boldsymbol{\Sigma}_{1,\epsilon}$, with $\boldsymbol{\Sigma}_{1,\epsilon} = (\boldsymbol{\sigma}_{1,\epsilon}, \boldsymbol{\chi}_{1,\epsilon})$ and $\boldsymbol{\sigma}_{1,\epsilon} \in \operatorname{Ker} B$. The variational equation

$$\begin{gathered} A(\boldsymbol{\Sigma}_{1,\epsilon}, \boldsymbol{T}) + (J_\epsilon'(\boldsymbol{\Sigma}_0 + \boldsymbol{\Sigma}_{1,\epsilon}), \boldsymbol{T}) = A(\boldsymbol{\Sigma}_{n-1}^k - \boldsymbol{\Sigma}_0, \boldsymbol{T}) \\ \forall \boldsymbol{T} \in \operatorname{Ker} B \times M \end{gathered} \tag{8.21}$$

follows from (8.20). Now define the operator $L : \mathcal{T} \to \mathcal{T}'$ by

$$\langle L\boldsymbol{\Sigma}_{1,\epsilon}, \boldsymbol{T} \rangle = A(\boldsymbol{\Sigma}_{1,\epsilon}, \boldsymbol{T}) + (J_\epsilon'(\boldsymbol{\Sigma}_0 + \boldsymbol{\Sigma}_{1,\epsilon}), \boldsymbol{T}), \quad \boldsymbol{T} \in \mathcal{T}.$$

Since $A(\cdot,\cdot)$ is $\mathcal{T}$-elliptic and J_ϵ' is monotone (see Lemma 8.6), the operator thus defined is strongly monotone on $\operatorname{Ker} K \times M$. Furthermore, it is easy to show that L is Lipschitz continuous. Therefore, by Theorem 5.10, the problem (8.21) has a unique solution $\boldsymbol{\Sigma}_{1,\epsilon} \in \operatorname{Ker} K \times M$.

Next, we derive a uniform bound for the sequence $\{\boldsymbol{\Sigma}_{n,\epsilon}^k\}_\epsilon$. To do this, set $\boldsymbol{T} = \boldsymbol{\Sigma}_{1,\epsilon}$ in (8.21) to obtain

$$A(\boldsymbol{\Sigma}_{1,\epsilon}, \boldsymbol{\Sigma}_{1,\epsilon}) + (J_\epsilon'(\boldsymbol{\Sigma}_0 + \boldsymbol{\Sigma}_{1,\epsilon}), \boldsymbol{\Sigma}_{1,\epsilon}) = A(\boldsymbol{\Sigma}_{n-1}^k - \boldsymbol{\Sigma}_0, \boldsymbol{\Sigma}_{1,\epsilon}). \tag{8.22}$$

Since J_ϵ is convex, we have

$$J_\epsilon(\boldsymbol{\Sigma}_0) \ge J_\epsilon(\boldsymbol{\Sigma}_0 + \boldsymbol{\Sigma}_{1,\epsilon}) + (J_\epsilon'(\boldsymbol{\Sigma}_0 + \boldsymbol{\Sigma}_{1,\epsilon}), -\boldsymbol{\Sigma}_{1,\epsilon}).$$

Noting that $\mathbf{\Sigma}_0 \in \mathcal{P}$, we have $J_\epsilon(\mathbf{\Sigma}_0) = 0$. Therefore,

$$(J'_\epsilon(\mathbf{\Sigma}_0 + \mathbf{\Sigma}_{1,\epsilon}), \mathbf{\Sigma}_{1,\epsilon}) \geq J_\epsilon(\mathbf{\Sigma}_0 + \mathbf{\Sigma}_{1,\epsilon}) \geq 0. \tag{8.23}$$

Then from (8.22), we find that

$$A(\mathbf{\Sigma}_{1,\epsilon}, \mathbf{\Sigma}_{1,\epsilon}) \leq A(\mathbf{\Sigma}^k_{n-1} - \mathbf{\Sigma}_0, \mathbf{\Sigma}_{1,\epsilon}).$$

The $\mathcal{T}$-ellipticity of the bilinear form A and the uniform boundedness of $\mathbf{\Sigma}^k_{n-1}$ and $\mathbf{\Sigma}_0$ independent of ϵ lead to the bound

$$\|\mathbf{\Sigma}_{1,\epsilon}\| \leq c\left(\|\mathbf{\Sigma}^k_{n-1}\| + \|\mathbf{\Sigma}_0\|\right) \leq c,$$

so that

$$\|\mathbf{\Sigma}^k_{n,\epsilon}\| \leq \|\mathbf{\Sigma}_0\| + \|\mathbf{\Sigma}_{1,\epsilon}\| \leq c;$$

in other words, the sequence $\{\mathbf{\Sigma}^k_{n,\epsilon}\}_\epsilon$ is uniformly bounded. Hence, there is a subsequence of $\{\mathbf{\Sigma}^k_{n,\epsilon}\}_\epsilon$, still denoted by $\{\mathbf{\Sigma}^k_{n,\epsilon}\}_\epsilon$, and an element $\mathbf{\Sigma}^k_n \in \mathcal{T}$ such that

$$\mathbf{\Sigma}^k_{n,\epsilon} \rightharpoonup \mathbf{\Sigma}^k_n \quad \text{in } \mathcal{T} \text{ as } \epsilon \to 0.$$

Since $\mathbf{\Sigma}^k_{n,\epsilon} = \mathbf{\Sigma}_0 + \mathbf{\Sigma}_{1,\epsilon}$ satisfies the constraint

$$b(\boldsymbol{v}, \mathbf{\Sigma}^k_{n,\epsilon}) = b(\boldsymbol{v}, \mathbf{\Sigma}_0) = \langle \boldsymbol{\ell}_n, \boldsymbol{v}\rangle \quad \forall \boldsymbol{v} \in V,$$

we see immediately that

$$b(\boldsymbol{v}, \mathbf{\Sigma}^k_n) = \langle \boldsymbol{\ell}_n, \boldsymbol{v}\rangle \quad \forall \boldsymbol{v} \in V.$$

Let us prove next that the limit $\mathbf{\Sigma}^k_n$ satisfies the inequality (8.19). Again, because of the convexity of the functional J_ϵ, we have

$$0 = J_\epsilon(\boldsymbol{T}) \geq J_\epsilon(\mathbf{\Sigma}^k_{n,\epsilon}) + \left(J'_\epsilon(\mathbf{\Sigma}^k_{n,\epsilon}), \boldsymbol{T} - \mathbf{\Sigma}^k_{n,\epsilon}\right) \quad \forall \boldsymbol{T} \in \mathcal{P}_n,$$

that is,

$$\left(J'_\epsilon(\mathbf{\Sigma}^k_{n,\epsilon}), \boldsymbol{T} - \mathbf{\Sigma}^k_{n,\epsilon}\right) \leq -J_\epsilon(\mathbf{\Sigma}^k_{n,\epsilon}) \leq 0 \quad \forall \boldsymbol{T} \in \mathcal{P}_n.$$

Thus, if in (8.21) we replace $\boldsymbol{T} \in \operatorname{Ker} B \times M$ by $\boldsymbol{T} - \mathbf{\Sigma}^k_{n,\epsilon}$ with $\boldsymbol{T} \in \mathcal{P}_n$, then

$$A(\mathbf{\Sigma}^k_{n,\epsilon}, \boldsymbol{T} - \mathbf{\Sigma}^k_{n,\epsilon}) \geq A(\mathbf{\Sigma}^k_{n-1}, \boldsymbol{T} - \mathbf{\Sigma}^k_{n,\epsilon}) \quad \forall \boldsymbol{T} \in \mathcal{P}_n. \tag{8.24}$$

Since $\mathbf{\Sigma}^k_{n,\epsilon} \rightharpoonup \mathbf{\Sigma}^k_n$ as $\epsilon \to 0$, we have, as $\epsilon \to 0$,

$$A(\mathbf{\Sigma}^k_{n,\epsilon}, \boldsymbol{T}) \to A(\mathbf{\Sigma}^k_n, \boldsymbol{T})$$

and

$$A(\boldsymbol{\Sigma}_{n-1}^k, \boldsymbol{T} - \boldsymbol{\Sigma}_{n,\epsilon}^k) \to A(\boldsymbol{\Sigma}_{n-1}^k, \boldsymbol{T} - \boldsymbol{\Sigma}_n^k)$$

for arbitrary $\boldsymbol{T}$. From the identity

$$2\,A(\boldsymbol{\Sigma}_{n,\epsilon}^k, \boldsymbol{\Sigma}_n^k) - A(\boldsymbol{\Sigma}_n^k, \boldsymbol{\Sigma}_n^k) = A(\boldsymbol{\Sigma}_{n,\epsilon}^k, \boldsymbol{\Sigma}_{n,\epsilon}^k) - A(\boldsymbol{\Sigma}_{n,\epsilon}^k - \boldsymbol{\Sigma}_n^k, \boldsymbol{\Sigma}_{n,\epsilon}^k - \boldsymbol{\Sigma}_n^k),$$

we find that

$$2\,A(\boldsymbol{\Sigma}_{n,\epsilon}^k, \boldsymbol{\Sigma}_n^k) - A(\boldsymbol{\Sigma}_n^k, \boldsymbol{\Sigma}_n^k) \le A(\boldsymbol{\Sigma}_{n,\epsilon}^k, \boldsymbol{\Sigma}_{n,\epsilon}^k),$$

and by taking the limit $\epsilon \to 0$,

$$A(\boldsymbol{\Sigma}_n^k, \boldsymbol{\Sigma}_n^k) \le \liminf_{\epsilon \to 0} A(\boldsymbol{\Sigma}_{n,\epsilon}^k, \boldsymbol{\Sigma}_{n,\epsilon}^k).$$

Thus, from (8.24) with $\epsilon \to 0$, we see that

$$A(\boldsymbol{\Sigma}_n^k, \boldsymbol{T} - \boldsymbol{\Sigma}_n^k) \ge A(\boldsymbol{\Sigma}_{n-1}^k, \boldsymbol{T} - \boldsymbol{\Sigma}_n^k) \quad \forall \boldsymbol{T} \in \mathcal{P}_n,$$

that is, $\boldsymbol{\Sigma}_n^k$ satisfies (8.19).

Finally, it is required to show that $\boldsymbol{\Sigma}_n^k \in \mathcal{P}_n$. As we have seen above, the constraint related to the bilinear form $b(\cdot,\cdot)$ is satisfied; so we only need to show that $\boldsymbol{\Sigma}_n^k \in \mathcal{P}$. From (8.23),

$$J_\epsilon(\boldsymbol{\Sigma}_{n,\epsilon}^k) \le \left(J_\epsilon'(\boldsymbol{\Sigma}_{n,\epsilon}^k), \boldsymbol{\Sigma}_{1,\epsilon}\right).$$

Using (8.22), we then have

$$J_\epsilon(\boldsymbol{\Sigma}_{n,\epsilon}^k) \le A(\boldsymbol{\Sigma}_{n-1}^k - \boldsymbol{\Sigma}_{n,\epsilon}^k, \boldsymbol{\Sigma}_{1,\epsilon}).$$

Now the uniform boundedness of $\boldsymbol{\Sigma}_{n,\epsilon}^k$ and $\boldsymbol{\Sigma}_{1,\epsilon}$ yields

$$J_\epsilon(\boldsymbol{\Sigma}_{n,\epsilon}^k) \le c,$$

that is,

$$\|\boldsymbol{\Sigma}_{n,\epsilon}^k - \Pi \boldsymbol{\Sigma}_{n,\epsilon}^k\|^2 \le c\,\epsilon. \tag{8.25}$$

Since the functional $f(\boldsymbol{\Sigma}) = \|\boldsymbol{\Sigma} - \Pi\boldsymbol{\Sigma}\|^2$ is convex and l.s.c., it is also weakly l.s.c. Thus, as $\epsilon \to 0$ we find from (8.25) that

$$\|\boldsymbol{\Sigma}_n^k - \Pi \boldsymbol{\Sigma}_n^k\| = 0;$$

in other words, $\boldsymbol{\Sigma}_n^k \in \mathcal{P}$.

UNIQUENESS. Assume that there are two elements $\boldsymbol{\Sigma}_n^{k,1}, \boldsymbol{\Sigma}_n^{k,2} \in \mathcal{P}_n$ such that

$$A(\boldsymbol{\Sigma}_n^{k,1}, \boldsymbol{T} - \boldsymbol{\Sigma}_n^{k,1}) \ge A(\boldsymbol{\Sigma}_{n-1}^k, \boldsymbol{T} - \boldsymbol{\Sigma}_n^{k,1}) \quad \forall \boldsymbol{T} \in \mathcal{P}_n, \tag{8.26}$$

$$A(\boldsymbol{\Sigma}_n^{k,2}, \boldsymbol{T} - \boldsymbol{\Sigma}_n^{k,2}) \ge A(\boldsymbol{\Sigma}_{n-1}^k, \boldsymbol{T} - \boldsymbol{\Sigma}_n^{k,2}) \quad \forall \boldsymbol{T} \in \mathcal{P}_n. \tag{8.27}$$

We set $\boldsymbol{T} = \boldsymbol{\Sigma}_n^{k,2}$ in (8.26), $\boldsymbol{T} = \boldsymbol{\Sigma}_n^{k,1}$ in (8.27), and add the two resulting inequalities to obtain

$$A(\boldsymbol{\Sigma}_n^{k,1} - \boldsymbol{\Sigma}_n^{k,2}, \boldsymbol{\Sigma}_n^{k,1} - \boldsymbol{\Sigma}_n^{k,2}) \leq 0.$$

Then the $\mathcal{T}$-ellipticity of $A(\cdot,\cdot)$ immediately yields $\boldsymbol{\Sigma}_n^{k,1} - \boldsymbol{\Sigma}_n^{k,2} = \mathbf{0}$. □

The next step is to derive some useful bounds for the time-discrete solutions.

LEMMA 8.8. *For the solution* $\{\boldsymbol{\Sigma}_n^k = (\boldsymbol{\sigma}_n^k, \boldsymbol{\chi}_n^k)\}_{n=0}^N$ *of the problem* DUAL1^k, *the estimate*

$$\|\Delta \boldsymbol{\Sigma}_n^k\| \leq c\,\|\Delta \boldsymbol{\ell}_n\|, \quad n = 1, \ldots, N \tag{8.28}$$

holds for some constant c. *Consequently, assuming* $\boldsymbol{\ell} \in H^1(0,T;V')$, *we have*

$$\max_{0 \leq n \leq N} \|\boldsymbol{\Sigma}_n^k\| \leq c\,\|\dot{\boldsymbol{\ell}}\|_{L^1(0,T;V')} \tag{8.29}$$

and

$$\sum_{n=1}^N \|\Delta \boldsymbol{\Sigma}_n^k\|^2 \leq c\,k\,\|\dot{\boldsymbol{\ell}}\|_{L^2(0,T;V')}^2. \tag{8.30}$$

PROOF. The estimates (8.29) and (8.30) are simple consequences of (8.28) and $\boldsymbol{\Sigma}_0^k = \mathbf{0}$. So here we give a proof only for the estimate (8.28).

From the fact that $\boldsymbol{\Sigma}_{n-1}^k = (\boldsymbol{\sigma}_{n-1}^k, \boldsymbol{\chi}_{n-1}^k) \in \mathcal{P}_{n-1}$, we know that $\boldsymbol{\sigma}_{n-1}^k$ satisfies

$$b(\boldsymbol{v}, \boldsymbol{\sigma}_{n-1}^k) = \langle \boldsymbol{\ell}_{n-1}, \boldsymbol{v} \rangle \quad \forall\, \boldsymbol{v} \in V. \tag{8.31}$$

By Lemma 8.2, there is a unique element $\boldsymbol{\sigma}_2 \in (\operatorname{Ker} B)^\perp$ such that

$$b(\boldsymbol{v}, \boldsymbol{\sigma}_2) = \langle \Delta \boldsymbol{\ell}_n, \boldsymbol{v} \rangle \quad \forall\, \boldsymbol{v} \in V \tag{8.32}$$

and

$$\|\boldsymbol{\sigma}_2\| \leq c\,\|\Delta \boldsymbol{\ell}_n\|. \tag{8.33}$$

Using Assumption 8.4, we can choose $\boldsymbol{\chi}_2 \in M$ such that

$$\|\boldsymbol{\chi}_2\| \leq c\,\|\boldsymbol{\sigma}_2\| \tag{8.34}$$

and with the notation $\boldsymbol{\Sigma}_2 = (\boldsymbol{\sigma}_2, \boldsymbol{\chi}_2)$,

$$\boldsymbol{\Sigma}_{n-1}^k + \boldsymbol{\Sigma}_2 \in \mathcal{P}. \tag{8.35}$$

By (8.31), (8.32), and (8.35), we see that $\mathbf{\Sigma}^k_{n-1} + \mathbf{\Sigma}_2 \in \mathcal{P}_n$. From (8.33) and (8.34) we have

$$\|\mathbf{\Sigma}_2\| \le c\,\|\Delta \boldsymbol{\ell}_n\|. \tag{8.36}$$

Thus, taking $\boldsymbol{T} = \mathbf{\Sigma}^k_{n-1} + \mathbf{\Sigma}_2$ in (8.18) we get

$$A(\Delta\mathbf{\Sigma}^k_n, \Delta\mathbf{\Sigma}^k_n) \le A(\Delta\mathbf{\Sigma}^k_n, \mathbf{\Sigma}_2).$$

The continuity and $\mathcal{T}$-ellipticity of $A(\cdot,\cdot)$ then yield

$$\|\Delta\mathbf{\Sigma}^k_n\| \le c\,\|\mathbf{\Sigma}_2\|,$$

and the estimate (8.28) follows from (8.36). □

Corresponding to the partition of the time interval $[0,T]$ with the step-size k, we define piecewise linear functions $\boldsymbol{\ell}^k(t)$ and $\mathbf{\Sigma}^k(t)$ by the formulae

$$\begin{aligned} \boldsymbol{\ell}^k(t) &= \frac{t-t_{n-1}}{k}\,\boldsymbol{\ell}(t_n) + \frac{t_n - t}{k}\,\boldsymbol{\ell}(t_{n-1}), \\ \mathbf{\Sigma}^k(t) &= \frac{t-t_{n-1}}{k}\,\mathbf{\Sigma}^k_n + \frac{t_n - t}{k}\,\mathbf{\Sigma}^k_{n-1} \end{aligned}$$

for $t \in [t_{n-1}, t_n]$, $n = 1, \dots, N$. For the derivative of $\mathbf{\Sigma}^k(t)$, we have the formula

$$\dot{\mathbf{\Sigma}}^k(t) = \frac{1}{k}\,\Delta\mathbf{\Sigma}^k_n,\ t \in (t_{n-1}, t_n), \quad n = 1, \dots, N.$$

So from Lemma 8.8 we see that the sequence $\{\mathbf{\Sigma}^k\}_k$ is uniformly bounded in $H^1(0,T;\mathcal{T})$:

$$\|\mathbf{\Sigma}^k\|_{H^1(0,T;\mathcal{T})} \le c\,\|\dot{\boldsymbol{\ell}}\|_{L^2(0,T;V')}. \tag{8.37}$$

Therefore, there exists a subsequence of $\{\mathbf{\Sigma}^k\}_k$, still denoted by $\{\mathbf{\Sigma}^k\}_k$, and an element $\mathbf{\Sigma} \in H^1(0,T;\mathcal{T})$ such that

$$\mathbf{\Sigma}^k \rightharpoonup \mathbf{\Sigma} \quad \text{in } H^1(0,T;\mathcal{T}) \text{ as } k \to 0. \tag{8.38}$$

In particular, four simple consequences of (8.37) and (8.38) (resorting to a subsequence if necessary) are that

$$\begin{aligned} \mathbf{\Sigma}^k &\to \mathbf{\Sigma} \quad \text{in } L^2(0,T;\mathcal{T}) \text{ as } k \to 0, && (8.39) \\ \mathbf{\Sigma}^k &\to \mathbf{\Sigma} \quad \text{in } \mathcal{T} \text{ for a.a. } t \in [0,T] \text{ as } k \to 0, && (8.40) \\ \dot{\mathbf{\Sigma}}^k &\rightharpoonup \dot{\mathbf{\Sigma}} \quad \text{in } L^2(0,T;\mathcal{T}) \text{ as } k \to 0, && (8.41) \\ \|\mathbf{\Sigma}\|_{H^1(0,T;\mathcal{T})} &\le \liminf_{k\to 0} \|\mathbf{\Sigma}^k\|_{H^1(0,T;\mathcal{T})} \le c\,\|\dot{\boldsymbol{\ell}}\|_{L^2(0,T;V')}. && (8.42) \end{aligned}$$

The aim now is to prove that the limit $\boldsymbol{\Sigma}$ is a solution of the problem DUAL1. First, since $\{\boldsymbol{\Sigma}_n^k\}_{n=0}^N \subset \mathcal{P}$, by the convexity of $\mathcal{P}$ it follows that $\boldsymbol{\Sigma}^k(t) \in \mathcal{P}$ for $t \in [0,T]$. Then the closedness of the set $\mathcal{P}$ and the limit relation (8.40) imply that $\boldsymbol{\Sigma}(t) \in \mathcal{P}$ for a.a. $t \in [0,T]$.

Secondly, we show that

$$b(\boldsymbol{v}, \boldsymbol{\sigma}(t)) = \langle \boldsymbol{\ell}(t), \boldsymbol{v} \rangle \quad \forall\, \boldsymbol{v} \in V, \text{ a.a. } t \in [0,T]. \tag{8.43}$$

To do this, let $\boldsymbol{v} \in L^2(0,T;V)$ be an arbitrary function. Corresponding to the time-interval partition for $\boldsymbol{\Sigma}^k(t)$, we can construct a piecewise constant function $\boldsymbol{v}^k(t)$ (for example, by piecewise averaging) such that

$$\boldsymbol{v}^k \to \boldsymbol{v} \quad \text{in } L^2(0,T;V) \quad \text{as } k \to 0.$$

From the defining relation

$$b(\boldsymbol{v}, \boldsymbol{\sigma}_n^k) - \langle \boldsymbol{\ell}_n, \boldsymbol{v} \rangle \quad \forall\, \boldsymbol{v} \subset V$$

we can easily show that

$$\int_0^T b(\boldsymbol{v}^k(t), \boldsymbol{\sigma}^k(t))\, dt = \int_0^T \langle \boldsymbol{\ell}^k(t), \boldsymbol{v}^k(t) \rangle\, dt.$$

Taking the limit $k \to 0$ in the above relation, we get

$$\int_0^T b(\boldsymbol{v}(t), \boldsymbol{\sigma}(t))\, dt = \int_0^T \langle \boldsymbol{\ell}(t), \boldsymbol{v}(t) \rangle\, dt.$$

Since this relation holds for any $\boldsymbol{v} \in L^2(0,T;V)$, we can localize the relation and, using arguments similar to those in the proof of Theorem 7.3, conclude that (8.43) holds. Thus, $\boldsymbol{\Sigma}(t) \in \mathcal{P}(t)$ for a.a. $t \in [0,T]$.

Finally, we need to show that (8.16) holds. Let $\boldsymbol{T} = (\boldsymbol{\tau}, \boldsymbol{\mu}) \in H^1(0,T;\mathcal{T})$ with the property that $\boldsymbol{T}(t) \in \mathcal{P}(t)$ for $t \in [0,T]$. Let $\boldsymbol{T}^k(t)$ be the piecewise linear interpolant of $\boldsymbol{T}(t)$ based on the time-interval partition for $\boldsymbol{\Sigma}^k(t)$. Then we have

$$\boldsymbol{T}^k(t_n) = \boldsymbol{T}(t_n) \in \mathcal{P}_n, \quad n = 1, \ldots, N.$$

It is well known that for the piecewise linear interpolation,

$$\boldsymbol{T}^k \to \boldsymbol{T} \quad \text{in } L^2(0,T;\mathcal{T}) \quad \text{as } k \to 0. \tag{8.44}$$

On the other hand, applying (5.25) and Cauchy–Schwarz inequality, we have

$$\sum_{n=1}^N \|\Delta \boldsymbol{T}(t_n)\|^2 \le c\, k \int_0^T \|\dot{\boldsymbol{T}}(t)\|^2\, dt. \tag{8.45}$$

Then we consider the expression

$$\begin{aligned}
\int_0^T & A(\dot{\mathbf{\Sigma}}^k(t), \boldsymbol{T}^k(t) - \mathbf{\Sigma}^k(t))\, dt \\
&= \sum_{n=1}^N \int_{t_{n-1}}^{t_n} A((1/k)\, \Delta\mathbf{\Sigma}_n^k, \boldsymbol{T}^k(t) - \mathbf{\Sigma}^k(t))\, dt \\
&= \sum_{n=1}^N A(\Delta\mathbf{\Sigma}_n^k, \tfrac{1}{2}\,(\boldsymbol{T}(t_n) + \boldsymbol{T}(t_{n-1})) - \tfrac{1}{2}\,(\mathbf{\Sigma}_n^k + \mathbf{\Sigma}_{n-1}^k)) \\
&= \sum_{n=1}^N A(\Delta\mathbf{\Sigma}_n^k, \boldsymbol{T}(t_n) - \mathbf{\Sigma}_n^k) \\
&\quad + \tfrac{1}{2} \sum_{n=1}^N A(\Delta\mathbf{\Sigma}_n^k, \Delta\mathbf{\Sigma}_n^k - \Delta\boldsymbol{T}(t_n)).
\end{aligned} \tag{8.46}$$

By (8.18),

$$A(\Delta\mathbf{\Sigma}_n^k, \boldsymbol{T}(t_n) - \mathbf{\Sigma}_n^k) \geq 0, \quad n = 1, \ldots, N.$$

Using (8.30) and (8.45), we have

$$\begin{aligned}
&\left| \sum_{n=1}^N A(\Delta\mathbf{\Sigma}_n^k, \Delta\mathbf{\Sigma}_n^k - \Delta\boldsymbol{T}(t_n)) \right| \\
&\qquad \leq c \sum_{n=1}^N \|\Delta\mathbf{\Sigma}_n^k\| \left(\|\Delta\mathbf{\Sigma}_n^k\| + \|\Delta\boldsymbol{T}(t_n)\| \right) \\
&\qquad \leq c \left(\sum_{n=1}^N \|\Delta\mathbf{\Sigma}_n^k\|^2 + \sum_{n=1}^N \|\Delta\boldsymbol{T}(t_n)\|^2 \right) \\
&\qquad \leq c\, k \\
&\qquad \to 0 \quad \text{as } k \to 0.
\end{aligned}$$

Thus,

$$\int_0^T A(\dot{\mathbf{\Sigma}}(t), \boldsymbol{T}(t) - \mathbf{\Sigma}(t))\, dt = \lim_{k\to 0} \int_0^T A(\dot{\mathbf{\Sigma}}^k(t), \boldsymbol{T}^k(t) - \mathbf{\Sigma}^k(t))\, dt \geq 0. \tag{8.47}$$

This relation holds for any $\boldsymbol{T} = (\boldsymbol{\tau}, \boldsymbol{\mu}) \in H^1(0, T; \mathcal{T})$ with the property that $\boldsymbol{T}(t) \in \mathcal{P}(t)$ for $t \in [0, T]$. By a standard density argument, we can claim that (8.47) holds for any $\boldsymbol{T} \in L^2(0, T; \mathcal{T})$ with the property that $\boldsymbol{T}(t) \in \mathcal{P}(t)$ for almost all $t \in [0, T]$. Then, again by the standard procedure of passing to the pointwise inequality, we obtain (8.16), and so $\mathbf{\Sigma}$ is a solution of the problem DUAL1.

Uniqueness of the solution to DUAL1 can be proved in much the same way as in the case of the abstract problem in Section 7.2, and the proof is therefore omitted here.

We summarize the main results proved so far in the following theorem.

THEOREM 8.9. *The problem* DUAL1 *has a unique solution* $\mathbf{\Sigma}$, *and* $\mathbf{\Sigma} \in H^1(0,T;\mathcal{T})$ *with*

$$\|\mathbf{\Sigma}\|_{H^1(0,T;\mathcal{T})} \le c\,\|\dot{\boldsymbol{\ell}}\|_{L^2(0,T;V')}.$$

REMARKS.

1. As in the case for the abstract problem studied in Section 7.2, if we assume that $\boldsymbol{\ell} \in W^{1,p}(0,T;V')$ for some $p \in (1,\infty]$, then the problem DUAL1 has a unique solution $\mathbf{\Sigma}$, and $\mathbf{\Sigma} \in W^{1,p}(0,T;\mathcal{T})$. This is due to the fact that the regularity of the solution $\mathbf{\Sigma}$ is obtained from the estimate (8.28) and the regularity assumption on $\boldsymbol{\ell}$.

2. Since the solution to the problem DUAL1 is unique, any sequence of the time-discrete approximations, and not merely a subsequence, converges to the solution.

3. A curious feature of the proof of well-posedness of the dual problem is the fact that Assumption 8.4 plays an essential role, yet the equivalent primal problem does not require an equivalent or analogous assumption.

The last result of the section is a stability estimate for the solution of the problem DUAL1. Thus, suppose that the functions $\boldsymbol{\ell}^{(1)}, \boldsymbol{\ell}^{(2)} \in H^1(0,T;V')$ are given, and define

$$\mathcal{P}^{(1)}(t) = \{\boldsymbol{T} = (\boldsymbol{\tau},\boldsymbol{\mu}) \in \mathcal{P} :\ b(\boldsymbol{v},\boldsymbol{\tau}) = \langle \boldsymbol{\ell}^{(1)}(t),\boldsymbol{v}\rangle \quad \forall \boldsymbol{v} \in V\},$$
$$\mathcal{P}^{(2)}(t) = \{\boldsymbol{T} = (\boldsymbol{\tau},\boldsymbol{\mu}) \in \mathcal{P} :\ b(\boldsymbol{v},\boldsymbol{\tau}) = \langle \boldsymbol{\ell}^{(2)}(t),\boldsymbol{v}\rangle \quad \forall \boldsymbol{v} \in V\}.$$

By Theorem 8.9, there exist unique functions $\mathbf{\Sigma}^{(1)}, \mathbf{\Sigma}^{(2)} \in H^1(0,T;\mathcal{T})$, $\mathbf{\Sigma}^{(1)}(0) = \mathbf{\Sigma}^{(2)}(0) = \mathbf{0}$, such that for almost all $t \in (0,T)$, $\mathbf{\Sigma}^{(1)}(t) \in \mathcal{P}^{(1)}(t)$, $\mathbf{\Sigma}^{(2)}(t) \in \mathcal{P}^{(2)}(t)$, and

$$A(\dot{\mathbf{\Sigma}}^{(1)}(t), \boldsymbol{T} - \mathbf{\Sigma}^{(1)}(t)) \ge 0 \quad \forall \boldsymbol{T} \in \mathcal{P}^{(1)}(t), \tag{8.48}$$
$$A(\dot{\mathbf{\Sigma}}^{(2)}(t), \boldsymbol{T} - \mathbf{\Sigma}^{(2)}(t)) \ge 0 \quad \forall \boldsymbol{T} \in \mathcal{P}^{(2)}(t). \tag{8.49}$$

By Lemma 8.2, there exists a unique element $\boldsymbol{\sigma}_1(t) \in (\operatorname{Ker} B)^\perp$ such that

$$b(\boldsymbol{v}, \boldsymbol{\sigma}_1(t)) = \langle \boldsymbol{\ell}^{(1)}(t) - \boldsymbol{\ell}^{(2)}(t), \boldsymbol{v}\rangle \quad \forall \boldsymbol{v} \in V$$

and

$$\|\boldsymbol{\sigma}_1(t)\|_S \le c\,\|\boldsymbol{\ell}^{(1)}(t) - \boldsymbol{\ell}^{(2)}(t)\|_{V'}.$$

By Assumption 8.4, we can find a $\boldsymbol{\chi}_1(t) \in M$ such that

$$\|\boldsymbol{\chi}_1(t)\|_M \le c\,\|\boldsymbol{\sigma}_1(t)\|_S$$

and $\boldsymbol{\Sigma}^{(2)}(t) + \boldsymbol{\Sigma}_1(t) \in \mathcal{P}$, where $\boldsymbol{\Sigma}_1(t) = (\boldsymbol{\sigma}_1(t), \boldsymbol{\chi}_1(t))$. Thus

$$\|\boldsymbol{\Sigma}_1(t)\|_{\mathcal{T}} \le c\,\|\boldsymbol{\ell}^{(1)}(t) - \boldsymbol{\ell}^{(2)}(t)\|_{V'} \tag{8.50}$$

and $\boldsymbol{\Sigma}^{(2)}(t) + \boldsymbol{\Sigma}_1(t) \in \mathcal{P}^{(1)}(t)$. Now we set $\boldsymbol{T} = \boldsymbol{\Sigma}^{(2)}(t) + \boldsymbol{\Sigma}_1(t)$ in (8.48) to obtain

$$A(\dot{\boldsymbol{\Sigma}}^{(1)}(t), \boldsymbol{\Sigma}^{(2)}(t) + \boldsymbol{\Sigma}_1(t) - \boldsymbol{\Sigma}^{(1)}(t)) \ge 0,$$

which can be rewritten as

$$A(\dot{\boldsymbol{\Sigma}}^{(1)}(t), \boldsymbol{\Sigma}^{(1)}(t) - \boldsymbol{\Sigma}^{(2)}(t)) \le A(\dot{\boldsymbol{\Sigma}}^{(1)}(t), \boldsymbol{\Sigma}_1(t)). \tag{8.51}$$

Similarly, we have the inequality

$$A(\dot{\boldsymbol{\Sigma}}^{(2)}(t), \boldsymbol{\Sigma}^{(2)}(t) - \boldsymbol{\Sigma}^{(1)}(t)) \le A(\dot{\boldsymbol{\Sigma}}^{(2)}(t), \boldsymbol{\Sigma}_2(t)) \tag{8.52}$$

for some $\boldsymbol{\Sigma}_2(t)$ satisfying

$$\|\boldsymbol{\Sigma}_2(t)\|_{\mathcal{T}} \le c\,\|\boldsymbol{\ell}^{(1)}(t) - \boldsymbol{\ell}^{(2)}(t)\|_{V'}. \tag{8.53}$$

We now add (8.51) and (8.52) to obtain

$$\begin{aligned} A(\dot{\boldsymbol{\Sigma}}^{(1)}(t) - \dot{\boldsymbol{\Sigma}}^{(2)}(t), \boldsymbol{\Sigma}^{(1)}(t) - \boldsymbol{\Sigma}^{(2)}(t)) \\ \le A(\dot{\boldsymbol{\Sigma}}^{(1)}(t), \boldsymbol{\Sigma}_1(t)) + A(\dot{\boldsymbol{\Sigma}}^{(2)}(t), \boldsymbol{\Sigma}_2(t)), \end{aligned}$$

i.e.,

$$\begin{aligned} \frac{1}{2}\frac{d}{dt} A(\boldsymbol{\Sigma}^{(1)}(t) - \boldsymbol{\Sigma}^{(2)}(t), \boldsymbol{\Sigma}^{(1)}(t) - \boldsymbol{\Sigma}^{(2)}(t)) \\ \le A(\dot{\boldsymbol{\Sigma}}^{(1)}(t), \boldsymbol{\Sigma}_1(t)) + A(\dot{\boldsymbol{\Sigma}}^{(2)}(t), \boldsymbol{\Sigma}_2(t)). \end{aligned}$$

We then integrate the above inequality and use the condition $\boldsymbol{\Sigma}^{(1)}(0) - \boldsymbol{\Sigma}^{(2)}(0) = \mathbf{0}$. After some manipulations, we obtain

$$\begin{aligned} &\frac{1}{2} A(\boldsymbol{\Sigma}^{(1)}(t) - \boldsymbol{\Sigma}^{(2)}(t), \boldsymbol{\Sigma}^{(1)}(t) - \boldsymbol{\Sigma}^{(2)}(t)) \\ &\quad \le \int_0^t c\,\left(\|\dot{\boldsymbol{\Sigma}}^{(1)}(t)\|_{\mathcal{T}} + \|\dot{\boldsymbol{\Sigma}}^{(2)}(t)\|_{\mathcal{T}}\right) \|\boldsymbol{\ell}^{(1)}(t) - \boldsymbol{\ell}^{(2)}(t)\|_{V'} dt \\ &\quad \le c \left\{ \int_0^t \left(\|\dot{\boldsymbol{\Sigma}}^{(1)}(t)\|_{\mathcal{T}}^2 + \|\dot{\boldsymbol{\Sigma}}^{(2)}(t)\|_{\mathcal{T}}^2\right) dt \right\}^{1/2} \\ &\qquad \times \left\{ \int_0^t \|\boldsymbol{\ell}^{(1)}(t) - \boldsymbol{\ell}^{(2)}(t)\|_{V'}^2 dt \right\}^{1/2}, \end{aligned}$$

which implies

$$\begin{aligned}&\|\boldsymbol{\Sigma}^{(1)}-\boldsymbol{\Sigma}^{(2)}\|^2_{L^\infty(0,T;\mathcal{T})}\\&\quad\le c\left(\|\dot{\boldsymbol{\ell}}^{(1)}\|_{L^2(0,T;V')}+\|\dot{\boldsymbol{\ell}}^{(2)}\|_{L^2(0,T;V')}\right)\|\boldsymbol{\ell}^{(1)}-\boldsymbol{\ell}^{(2)}\|_{L^2(0,T;V')}.\end{aligned}$$

We have thus proved the following stability result.

THEOREM 8.10. *For given* $\boldsymbol{\ell}^{(1)},\boldsymbol{\ell}^{(2)}\in H^1(0,T;V')$, $\boldsymbol{\ell}^{(1)}(0)=\boldsymbol{\ell}^{(2)}(0)=\mathbf{0}$, *the corresponding solutions* $\boldsymbol{\Sigma}^{(1)}$ *and* $\boldsymbol{\Sigma}^{(2)}$ *of the problem* DUAL1 *satisfy the inequality*

$$\begin{aligned}&\|\boldsymbol{\Sigma}^{(1)}-\boldsymbol{\Sigma}^{(2)}\|^2_{L^\infty(0,T;\mathcal{T})}\\&\quad\le c\left(\|\dot{\boldsymbol{\ell}}^{(1)}\|_{L^2(0,T;V')}+\|\dot{\boldsymbol{\ell}}^{(2)}\|_{L^2(0,T;V')}\right)\|\boldsymbol{\ell}^{(1)}-\boldsymbol{\ell}^{(2)}\|_{L^2(0,T;V')}.\end{aligned}$$

Thus, the solution of the problem DUAL1 *is locally stable with respect to the data* $\boldsymbol{\ell}$.

8.3 Analysis of the Dual Problem

In this section we extend the existence and uniqueness result for the stress problem to that of the dual variational problem DUAL defined in Section 8.1. In addition to Assumption 8.4, we need another assumption on the structure of the constraint set K. Recall that K is a subset of a finite-dimensional space, defined by

$$K=\{\boldsymbol{\Sigma}\in M^3\times X:\phi(\boldsymbol{\Sigma})\le 0\}.$$

Here, the dimension of the space X is $m<\infty$.

ASSUMPTION 8.11. *For any* $\boldsymbol{\Sigma}\in K$, *and any* $\kappa\in[0,1)$, *we have* $\kappa\boldsymbol{\Sigma}\in K$ *and*

$$\inf_{\boldsymbol{x}\in\Omega}\operatorname{dist}(\kappa\boldsymbol{\Sigma}(\boldsymbol{x}),\partial K)>0.$$

This assumption is easy to verify for practically important situations. For example, for combined linear kinematic–isotropic hardening materials, $\boldsymbol{\Sigma}=(\boldsymbol{\sigma},\boldsymbol{a},g)$, and with a positive constant c_0,

$$\phi(\boldsymbol{\Sigma})=\Phi(\boldsymbol{\sigma}+\boldsymbol{a})+g-c_0.$$

If the function Φ is positively homogeneous and Lipschitz continuous, Assumption 8.11 is satisfied. Indeed, let $\lambda>0$ be the Lipschitz constant for the function Φ. Let $\boldsymbol{\Sigma}=(\boldsymbol{\sigma},\boldsymbol{a},g)\in K$, $\kappa\in[0,1)$, and define $\boldsymbol{\Sigma}_1=(\boldsymbol{\sigma}_1,\boldsymbol{a}_1,g_1)$.

We have

$$\begin{aligned}
&\phi(\kappa\,\boldsymbol{\Sigma}+\boldsymbol{\Sigma}_1)\\
&\quad=\Phi(\kappa\,(\boldsymbol{\sigma}+\boldsymbol{a})+\boldsymbol{\sigma}_1+\boldsymbol{a}_1)+\kappa\,g+g_1-c_0\\
&\quad=\kappa\,[\Phi(\boldsymbol{\sigma}+\boldsymbol{a})+g-c_0]+\Phi(\kappa\,(\boldsymbol{\sigma}+\boldsymbol{a})+\boldsymbol{\sigma}_1+\boldsymbol{a}_1)-\Phi(\kappa\,(\boldsymbol{\sigma}+\boldsymbol{a}))\\
&\qquad+g_1-(1-\kappa)\,c_0\\
&\quad\le\lambda\,|\boldsymbol{\sigma}_1+\boldsymbol{a}_1|+g_1-(1-\kappa)\,c_0.
\end{aligned}$$

Obviously, for $\boldsymbol{\Sigma}_1$ in a neighborhood of the origin, defined by the relation

$$|\boldsymbol{\Sigma}_1|\equiv|\boldsymbol{\sigma}_1|+|\boldsymbol{a}_1|+|g_1|\le\frac{1-\kappa}{\max\{\lambda,1\}}\,c_0,$$

we have $\phi(\kappa\,\boldsymbol{\Sigma}+\boldsymbol{\Sigma}_1)\le 0$, that is, $\kappa\,\boldsymbol{\Sigma}+\boldsymbol{\Sigma}_1\in K$. Hence Assumption 8.11 is satisfied.

The main purpose of the section is to prove the following result.

THEOREM 8.12. *Under Assumptions 8.4 and 8.11, the dual variational problem* DUAL *has a solution* $(\boldsymbol{u},\boldsymbol{\Sigma})$, $\boldsymbol{u}\in H^1(0,T;V)$, $\boldsymbol{\Sigma}\in H^1(0,T;\mathcal{T})$, *and* $\boldsymbol{\Sigma}$ *is unique.*

PROOF. Let $\boldsymbol{w}$ denote the velocity variable $\dot{\boldsymbol{u}}$. As in the last section, we will prove the result by first analyzing a time-discrete approximation of the problem DUAL. We use the uniform partition of the time interval $[0,T]$ with the step-size $k=T/N$, where N is the number of subintervals. Then the time-discrete approximation problem is the following.

PROBLEM DUAL^k. Let $\boldsymbol{w}_0^k=\boldsymbol{0}$ and $\boldsymbol{\Sigma}_0^k=\boldsymbol{0}$. For $n=1,2,\ldots,N$, find $(\boldsymbol{w}_n^k,\boldsymbol{\Sigma}_n^k)=(\boldsymbol{w}_n^k,\boldsymbol{\sigma}_n^k,\boldsymbol{\chi}_n^k)\in V\times\mathcal{P}$ satisfying

$$b(\boldsymbol{v},\boldsymbol{\sigma}_n^k)=\langle\boldsymbol{\ell}_n,\boldsymbol{v}\rangle\quad\forall\,\boldsymbol{v}\in V,\tag{8.54}$$

$$A(\delta\boldsymbol{\Sigma}_n^k,\boldsymbol{T}-\boldsymbol{\Sigma}_n^k)+b(\boldsymbol{w}_n^k,\boldsymbol{\tau}-\boldsymbol{\sigma}_n^k)\ge 0\quad\forall\,\boldsymbol{T}=(\boldsymbol{\tau},\boldsymbol{\mu})\in\mathcal{P}.\tag{8.55}$$

We first prove that the problem DUAL^k has a solution. This is done by introducing a regularization of the problem (8.54)–(8.55): Find $(\boldsymbol{w}_{n,\epsilon}^k,\boldsymbol{\Sigma}_{n,\epsilon}^k)\in V\times\mathcal{T}$ such that

$$b(\boldsymbol{v},\boldsymbol{\sigma}_{n,\epsilon}^k)=\langle\boldsymbol{\ell}_n,\boldsymbol{v}\rangle\quad\forall\,\boldsymbol{v}\in V,\tag{8.56}$$

$$A(\delta\boldsymbol{\Sigma}_{n,\epsilon}^k,\boldsymbol{T})+b(\boldsymbol{w}_{n,\epsilon}^k,\boldsymbol{\tau})+(J'_\epsilon(\boldsymbol{\Sigma}_{n,\epsilon}^k),\boldsymbol{T})=0\;\forall\,\boldsymbol{T}=(\boldsymbol{\tau},\boldsymbol{\mu})\in\mathcal{T}.\tag{8.57}$$

Again by Lemma 8.2 we have a unique element $\boldsymbol{\sigma}_0\in(\operatorname{Ker}B)^\perp$ such that

$$b(\boldsymbol{v},\boldsymbol{\sigma}_0)=\langle\boldsymbol{\ell}_n,\boldsymbol{v}\rangle\quad\forall\,\boldsymbol{v}\in V,$$

and for some constant c,

$$\|\boldsymbol{\sigma}_0\|_S\le c\,\|\boldsymbol{\ell}_n\|.$$

We now apply Assumption 8.4 to $2\,\boldsymbol{\sigma}_0$, rather than to $\boldsymbol{\sigma}_0$ as in the analysis of the stress problem. Then we have an element in M, denoted by $2\,\boldsymbol{\chi}_0$, such that $(2\,\boldsymbol{\sigma}_0, 2\,\boldsymbol{\chi}_0) \in \mathcal{P}$ and

$$\|\boldsymbol{\chi}_0\| \le c\,\|\boldsymbol{\sigma}_0\| \le c\,\|\boldsymbol{\ell}_n\|.$$

Set $\boldsymbol{\Sigma}_0 = (\boldsymbol{\sigma}_0, \boldsymbol{\chi}_0)$. By Assumption 8.11 (with $\kappa = \frac{1}{2}$), we claim that

$$\boldsymbol{\Sigma}_0 \in K \text{ a.e. in } \Omega \text{ and } \; d_0 \equiv \inf_{\boldsymbol{x}\in\Omega} \operatorname{dist}(\boldsymbol{\Sigma}_0(\boldsymbol{x}), \partial K) > 0. \tag{8.58}$$

As in the proof of Lemma 8.7, we write $\boldsymbol{\Sigma}^k_{n,\epsilon} = \boldsymbol{\Sigma}_0 + \boldsymbol{\Sigma}_{1,\epsilon}$. The element $\boldsymbol{\Sigma}_{1,\epsilon} = (\boldsymbol{\sigma}_{1,\epsilon}, \boldsymbol{\chi}_{1,\epsilon}) \in \operatorname{Ker} B \times M$ is determined from

$$\begin{gathered}\frac{1}{k}\,A(\boldsymbol{\Sigma}_{1,\epsilon}, \boldsymbol{T}) + (J'_\epsilon(\boldsymbol{\Sigma}_0 + \boldsymbol{\Sigma}_{1,\epsilon}), \boldsymbol{T}) = \frac{1}{k}\,A(\boldsymbol{\Sigma}^k_{n-1} - \boldsymbol{\Sigma}_0, \boldsymbol{T}) \\ \forall\, \boldsymbol{T} = (\boldsymbol{\tau}, \boldsymbol{\mu}) \in \operatorname{Ker} B \times M,\end{gathered} \tag{8.59}$$

which is the restriction of equation (8.57) to $\operatorname{Ker} B \times M$. Furthermore, the problem (8.59) has a unique solution $\boldsymbol{\Sigma}_{1,\epsilon} \in \operatorname{Ker} B \times M$. In other words, we have proved the existence of $\boldsymbol{\Sigma}^k_{n,\epsilon} \in \mathcal{T}$ such that

$$A(\delta\boldsymbol{\Sigma}^k_{n,\epsilon}, \boldsymbol{T}) + (J'_\epsilon(\boldsymbol{\Sigma}^k_{n,\epsilon}), \boldsymbol{T}) = 0 \quad \forall\, \boldsymbol{T} = (\boldsymbol{\tau}, \boldsymbol{\mu}) \in \operatorname{Ker} B \times M. \tag{8.60}$$

The lefthand side of (8.60) defines a linear continuous form on $\mathcal{T}$. Hence, by Lemma 8.2 there exists an element $\boldsymbol{w}^k_{n,\epsilon} \in V$ such that (8.57) is satisfied. Therefore, the regularized problem (8.56)–(8.57) has a solution.

We will take the limit $\epsilon \to 0$ on the pair $(\boldsymbol{w}^k_{n,\epsilon}, \boldsymbol{\Sigma}^k_{n,\epsilon})$ to obtain a solution to the problem (8.54)–(8.55). To do this, we need to bound $(\boldsymbol{w}^k_{n,\epsilon}, \boldsymbol{\Sigma}^k_{n,\epsilon})$ uniformly with respect to ϵ. Such a uniform bound on $\boldsymbol{\Sigma}^k_{n,\epsilon}$ is given in the proof of Lemma 8.7. So here we need only derive a uniform bound for $\boldsymbol{w}^k_{n,\epsilon}$.

We prove first that $\|J'_\epsilon(\boldsymbol{\Sigma}^k_{n,\epsilon})\|$ is uniformly bounded. For $\boldsymbol{x} \in \Omega$, if $\boldsymbol{\Sigma}^k_{n,\epsilon}(\boldsymbol{x}) \in K$, then by (8.17), $J'_\epsilon(\boldsymbol{\Sigma}^k_{n,\epsilon}(\boldsymbol{x})) = \boldsymbol{0}$. Now assume that $\boldsymbol{\Sigma}^k_{n,\epsilon}(\boldsymbol{x}) \notin K$. Define

$$\boldsymbol{j}(\boldsymbol{x}) = \frac{\boldsymbol{\Sigma}^k_{n,\epsilon}(\boldsymbol{x}) - \Pi\boldsymbol{\Sigma}^k_{n,\epsilon}(\boldsymbol{x})}{|\boldsymbol{\Sigma}^k_{n,\epsilon}(\boldsymbol{x}) - \Pi\boldsymbol{\Sigma}^k_{n,\epsilon}(\boldsymbol{x})|}.$$

Since K is a closed convex set in $M^3 \times X$, $\boldsymbol{j}(\boldsymbol{x})$ is normal to a hyperplane L separating K and $\boldsymbol{\Sigma}^k_{n,\epsilon}(\boldsymbol{x})$, and for some constant $\delta_0 > 0$,

$$L = \{\boldsymbol{T} \in M^3 \times X : \; \boldsymbol{j}(\boldsymbol{x}) : \boldsymbol{T} = \delta_0\}.$$

Now noting that $\boldsymbol{\Sigma}_0(\boldsymbol{x}) + d_0\boldsymbol{j}(\boldsymbol{x}) \in K$ and $\boldsymbol{0} \in K$, we have

$$\boldsymbol{j}(\boldsymbol{x}) : \boldsymbol{\Sigma}^k_{n,\epsilon}(\boldsymbol{x}) \ge \delta_0$$

and

$$\boldsymbol{j}(\boldsymbol{x}) : (\boldsymbol{\Sigma}_0(\boldsymbol{x}) + d_0\boldsymbol{j}(\boldsymbol{x})) \leq \delta_0.$$

Hence

$$\boldsymbol{j}(\boldsymbol{x}) : (\boldsymbol{\Sigma}_{n,\epsilon}^k(\boldsymbol{x}) - \boldsymbol{\Sigma}_0(\boldsymbol{x})) \geq d_0,$$

that is,

$$|\boldsymbol{\Sigma}_{n,\epsilon}^k(\boldsymbol{x}) - \Pi\boldsymbol{\Sigma}_{n,\epsilon}^k(\boldsymbol{x})| \leq \frac{1}{d_0}\,(\boldsymbol{\Sigma}_{n,\epsilon}^k(\boldsymbol{x}) - \Pi\boldsymbol{\Sigma}_{n,\epsilon}^k(\boldsymbol{x})) : (\boldsymbol{\Sigma}_{n,\epsilon}^k(\boldsymbol{x}) - \boldsymbol{\Sigma}_0(\boldsymbol{x})).$$

This inequality holds for any $\boldsymbol{x} \in \Omega$, whether $\boldsymbol{\Sigma}_{n,\epsilon}^k(\boldsymbol{x}) \in K$ or not. Therefore,

$$\|J_\epsilon'(\boldsymbol{\Sigma}_{n,\epsilon}^k)\| \leq \frac{1}{d_0}\,(J_\epsilon'(\boldsymbol{\Sigma}_{n,\epsilon}^k), \boldsymbol{\Sigma}_{n,\epsilon}^k - \boldsymbol{\Sigma}_0). \tag{8.61}$$

We next let $\boldsymbol{T} = \boldsymbol{\Sigma}_{n,\epsilon}^k - \boldsymbol{\Sigma}_0 \in \operatorname{Ker} B \times M$ in (8.60). From (8.61) and the uniform boundedness (with respect to ϵ) of $\boldsymbol{\Sigma}_{n,\epsilon}^k$, we obtain

$$\|J_\epsilon'(\boldsymbol{\Sigma}_{n,\epsilon}^k)\|_{\mathcal{T}} \leq -\frac{1}{d_0}\,A(\delta\boldsymbol{\Sigma}_{n,\epsilon}^k, \boldsymbol{\Sigma}_{n,\epsilon}^k - \boldsymbol{\Sigma}_0) \leq c \tag{8.62}$$

for some constant c independent of ϵ. Returning to (8.57) with $\boldsymbol{\mu} = \boldsymbol{0}$, by the Babuška–Brezzi condition (8.11), the bound (8.62), and the uniform boundedness of $\boldsymbol{\Sigma}_{n,\epsilon}^k$, we find that

$$\begin{aligned}
\beta_b\|\boldsymbol{w}_{n,\epsilon}^k\|_V &\leq \sup_{\boldsymbol{\tau}\in S} \frac{|b(\boldsymbol{w}_{n,\epsilon}^k, \boldsymbol{\tau})|}{\|\boldsymbol{\tau}\|_S} \\
&\leq c\left(\|\delta\boldsymbol{\Sigma}_{n,\epsilon}^k\|_{\mathcal{T}} + \|J_\epsilon'(\boldsymbol{\Sigma}_{n,\epsilon}^k)\|_{\mathcal{T}}\right) \\
&\leq c,
\end{aligned}$$

where c is independent of ϵ.

Now that $\boldsymbol{\Sigma}_{n,\epsilon}^k$ and $\boldsymbol{w}_{n,\epsilon}^k$ have been shown to be uniformly bounded independent of ϵ, we can extract a subsequence of $\{(\boldsymbol{w}_{n,\epsilon}^k, \boldsymbol{\Sigma}_{n,\epsilon}^k)\}_\epsilon$, still denoted by $\{(\boldsymbol{w}_{n,\epsilon}^k, \boldsymbol{\Sigma}_{n,\epsilon}^k)\}_\epsilon$, and find an element $(\boldsymbol{w}_n^k, \boldsymbol{\Sigma}_n^k) \in V \times \mathcal{T}$ such that

$$(\boldsymbol{w}_{n,\epsilon}^k, \boldsymbol{\Sigma}_{n,\epsilon}^k) \rightharpoonup (\boldsymbol{w}_n^k, \boldsymbol{\Sigma}_n^k) \quad \text{as } \epsilon \to 0.$$

Then proceeding as in the proof of Lemma 8.7, we can show that $\boldsymbol{\Sigma}_n^k \in \mathcal{P}$ and that $(\boldsymbol{w}_n^k, \boldsymbol{\Sigma}_n^k)$ is a solution of the problem (8.54)–(8.55). Here, the limit of the term $b(\boldsymbol{w}_{n,\epsilon}^k, \boldsymbol{\sigma}_{n,\epsilon}^k)$ is computed as follows (recalling the decomposition $\boldsymbol{\sigma}_{n,\epsilon}^k = \boldsymbol{\sigma}_0 + \boldsymbol{\sigma}_{1,\epsilon}$ with $\boldsymbol{\sigma}_{1,\epsilon} \in \operatorname{Ker} B$):

$$b(\boldsymbol{w}_{n,\epsilon}^k, \boldsymbol{\sigma}_{n,\epsilon}^k) = b(\boldsymbol{w}_{n,\epsilon}^k, \boldsymbol{\sigma}_0) \to b(\boldsymbol{w}_n^k, \boldsymbol{\sigma}_0) = b(\boldsymbol{w}_n^k, \boldsymbol{\sigma}_n^k) \quad \text{as } \epsilon \to 0.$$

So far, we have proved that the time-discrete approximation problem DUAL^k has a solution. The next step is to take the limit as $k \to 0$ of the sequence $\{(\boldsymbol{w}_n^k, \boldsymbol{\Sigma}_n^k)\}_n$ to obtain a solution of the dual variational problem DUAL. To do this, we need some uniform bounds (with respect to the step-size k) on the sequence $\{(\boldsymbol{w}_n^k, \boldsymbol{\Sigma}_n^k)\}_n$. From Lemma 8.8, we have the estimate (8.28) for the quantity $\|\Delta\boldsymbol{\Sigma}_n^k\|_{\mathcal{T}}$. It remains for us to derive a uniform bound for $\|\boldsymbol{w}_n^k\|_V$. We apply the Babuška–Brezzi condition (8.11) for this purpose. For any $\boldsymbol{\tau} \in S$, Assumption 8.4 guarantees the existence of a $\boldsymbol{\mu} \in M$ such that $(-\boldsymbol{\tau}, \boldsymbol{\mu}) + \boldsymbol{\Sigma}_n^k \in \mathcal{P}$ and $\|\boldsymbol{\mu}\| \le c\,\|\boldsymbol{\tau}\|$, for some constant $c > 0$. We take $\boldsymbol{T} = (-\boldsymbol{\tau}, \boldsymbol{\mu}) + \boldsymbol{\Sigma}_n^k$ in (8.55) to find that

$$\begin{aligned} b(\boldsymbol{w}_n^k, \boldsymbol{\tau}) &\le A(\delta\boldsymbol{\Sigma}_n^k, (-\boldsymbol{\tau}, \boldsymbol{\mu})) \\ &\le c\,\|\delta\boldsymbol{\Sigma}_n^k\|_{\mathcal{T}}\,(\|\boldsymbol{\tau}\|_S + \|\boldsymbol{\mu}\|_M) \\ &\le c\,\|\delta\boldsymbol{\Sigma}_n^k\|_{\mathcal{T}}\|\boldsymbol{\tau}\|_S. \end{aligned}$$

Thus

$$\|\boldsymbol{w}_n^k\|_V \le \frac{1}{\beta_b} \sup_{\boldsymbol{\tau} \in S} \frac{b(\boldsymbol{v}, \boldsymbol{\tau})}{\|\boldsymbol{\tau}\|_S} \le c\,\|\delta\boldsymbol{\Sigma}_n^k\|_{\mathcal{T}}. \tag{8.63}$$

Then as in the last section, corresponding to the partition of the time interval $[0, T]$ with the step-size k, we define piecewise linear functions $\boldsymbol{\ell}^k(t)$, $\boldsymbol{\Sigma}^k(t)$, and $\boldsymbol{w}^k(t)$. We have shown the existence of a subsequence of $\{\boldsymbol{\Sigma}^k\}_k$, still denoted by $\{\boldsymbol{\Sigma}^k\}_k$, and an element $\boldsymbol{\Sigma} \in H^1(0, T; \mathcal{T})$ such that the relation (8.38) holds. Using the estimates (8.63), (8.28) and the inequality (5.25), we have

$$\begin{aligned} \|\boldsymbol{w}^k\|_{L^2(0,T;V)}^2 &\le c \sum_{n=1}^N k\,\|\boldsymbol{w}_n^k\|_V^2 \\ &\le c \sum_{n=1}^N k\,\|\delta\boldsymbol{\Sigma}_n^k\|_{\mathcal{T}}^2 \\ &\le c \sum_{n=1}^N k\,\|\delta\boldsymbol{\ell}_n\|_{V'}^2 \\ &\le c\,\|\dot{\boldsymbol{\ell}}\|_{L^2(0,T;V')}^2. \end{aligned}$$

Thus the sequence $\{\boldsymbol{w}^k\}_k$ is bounded in $L^2(0, T; V)$, and we can extract a subsequence, still denoted by $\{\boldsymbol{w}^k\}_k$, such that

$$\boldsymbol{w}^k \rightharpoonup \boldsymbol{w} \quad \text{in } L^2(0, T; V) \text{ as } k \to 0, \tag{8.64}$$

for some $\boldsymbol{w} \in L^2(0, T; V)$. An argument similar to that used in the proof of Theorem 8.9 yields the result that the limit $(\boldsymbol{w}, \boldsymbol{\Sigma})$ is a solution of the problem DUAL; here we identify $\boldsymbol{w}$ with $\dot{\boldsymbol{u}}$. The limit

$$\int_0^T b(\boldsymbol{w}^k(t), \boldsymbol{\sigma}^k(t))\,dt \to \int_0^T b(\boldsymbol{w}(t), \boldsymbol{\sigma}(t))\,dt \quad \text{as } k \to 0$$

follows from (8.64) and (8.39). From the relation

$$\boldsymbol{u}(t) = \int_0^t \boldsymbol{w}(t)\, dt$$

and $\boldsymbol{w} \in L^2(0,T;V)$, we see that $\boldsymbol{u} \in H^1(0,T;V)$.

In conclusion, we have shown the existence of a solution $(\boldsymbol{u}, \boldsymbol{\Sigma})$ for the dual variational problem DUAL. The uniqueness of $\boldsymbol{\Sigma}$ has been proved in the last section. □

8.4 Rate Form of Stress–Strain Relation

In the literature, a popular approach in studying the dual variational formulation is through the use of the rate form of the stress–strain relation, which is obtained by eliminating the plastic multiplier λ (see Section 3.2). For this approach, we have to assume that the yield surface is smooth. More precisely, we assume in this section that the yield surface ϕ is continuously differentiable. Then from the discussion in Section 3.2, the flow law is of the form

$$\dot{\boldsymbol{P}} = \lambda\, \nabla\phi(\boldsymbol{\Sigma}), \tag{8.65}$$

where the plastic multiplier λ and the yield function ϕ satisfy the complementarity condition

$$\lambda \geq 0, \quad \phi \leq 0, \quad \lambda\, \phi = 0 \tag{8.66}$$

and the consistency condition that when $\phi = 0$,

$$\lambda \geq 0, \quad \dot{\phi} \leq 0, \quad \lambda\, \dot{\phi} = 0. \tag{8.67}$$

Also we recall the relations

$$\dot{\boldsymbol{p}} = \boldsymbol{\epsilon}(\dot{\boldsymbol{u}}) - \boldsymbol{C}^{-1}\dot{\boldsymbol{\sigma}} \tag{8.68}$$

and

$$\dot{\boldsymbol{\xi}} = -\boldsymbol{H}^{-1}\dot{\boldsymbol{\chi}}. \tag{8.69}$$

From (8.65), (8.68), and (8.69), we find that

$$\boldsymbol{\epsilon}(\boldsymbol{w}) - \boldsymbol{C}^{-1}\dot{\boldsymbol{\sigma}} = \lambda\, \frac{\partial\phi}{\partial\boldsymbol{\sigma}} \tag{8.70}$$

and

$$-\boldsymbol{H}^{-1}\dot{\boldsymbol{\chi}} = \lambda\, \frac{\partial\phi}{\partial\boldsymbol{\chi}}. \tag{8.71}$$

Evidently, from (8.66) and (8.67) we see that $\lambda = 0$ if $\phi < 0$ or, if $\phi = 0$ and $\dot{\phi} < 0$. Let us try to find a formula for λ when $\phi = \dot{\phi} = 0$. We have

$$\dot{\phi}(\boldsymbol{\Sigma}) = \frac{\partial \phi}{\partial \boldsymbol{\sigma}} : \dot{\boldsymbol{\sigma}} + \frac{\partial \phi}{\partial \boldsymbol{\chi}} : \dot{\boldsymbol{\chi}} = 0,$$

which, together with the relations (8.68) and (8.69), implies

$$\lambda = \frac{\partial \phi}{\partial \boldsymbol{\sigma}} : \boldsymbol{C}\, \boldsymbol{\epsilon}(\boldsymbol{w}) \Big/ \left(\frac{\partial \phi}{\partial \boldsymbol{\sigma}} : \boldsymbol{C} \frac{\partial \phi}{\partial \boldsymbol{\sigma}} + \frac{\partial \phi}{\partial \boldsymbol{\chi}} : \boldsymbol{H} \frac{\partial \phi}{\partial \boldsymbol{\chi}} \right). \tag{8.72}$$

This formula is derived under the assumption that $\lambda \geq 0$ exists. By the positive definiteness of the tensors $\boldsymbol{C}$ and $\boldsymbol{H}$, the denominator in the formula (8.72) is always positive. Thus the formula makes sense only if the numerator is nonnegative:

$$\frac{\partial \phi}{\partial \boldsymbol{\sigma}} : \boldsymbol{C}\, \boldsymbol{\epsilon}(\boldsymbol{w}) \geq 0.$$

We distinguish three cases according to the sign of the numerator.

Case 1. Assume that the numerator is negative. Let us show that $\lambda = 0$.
We have, no matter what is the value of $\lambda \geq 0$,

$$\begin{aligned}\dot{\phi}(\boldsymbol{\Sigma}) &= \frac{\partial \phi}{\partial \boldsymbol{\sigma}} : \dot{\boldsymbol{\sigma}} + \frac{\partial \phi}{\partial \boldsymbol{\chi}} : \dot{\boldsymbol{\chi}} \\ &= \frac{\partial \phi}{\partial \boldsymbol{\sigma}} : \boldsymbol{C}\, \boldsymbol{\epsilon}(\boldsymbol{w}) - \lambda \left(\frac{\partial \phi}{\partial \boldsymbol{\sigma}} : \boldsymbol{C} \frac{\partial \phi}{\partial \boldsymbol{\sigma}} + \frac{\partial \phi}{\partial \boldsymbol{\chi}} : \boldsymbol{H} \frac{\partial \phi}{\partial \boldsymbol{\chi}} \right) < 0.\end{aligned}$$

Thus, by the consistency condition (8.67), $\lambda = 0$.

Case 2. Assume that the numerator is positive. Let us show that $\lambda > 0$, and consequently, λ is indeed given by the formula (8.72).
We argue by contradiction. Suppose $\lambda = 0$. Then

$$\begin{aligned}\dot{\phi}(\boldsymbol{\Sigma}) &= \frac{\partial \phi}{\partial \boldsymbol{\sigma}} : \dot{\boldsymbol{\sigma}} + \frac{\partial \phi}{\partial \boldsymbol{\chi}} : \dot{\boldsymbol{\chi}} \\ &= \frac{\partial \phi}{\partial \boldsymbol{\sigma}} : \boldsymbol{C}\, \boldsymbol{\epsilon}(\boldsymbol{w}) \\ &> 0,\end{aligned}$$

which is not allowed given that $\phi = 0$.

Case 3. Assume that the numerator is zero. Let us show that $\lambda = 0$.
We use the consistency condition (8.67) to find that

$$-\lambda^2 \left(\frac{\partial \phi}{\partial \boldsymbol{\sigma}} : \boldsymbol{C} \frac{\partial \phi}{\partial \boldsymbol{\sigma}} + \frac{\partial \phi}{\partial \boldsymbol{\chi}} : \boldsymbol{H} \frac{\partial \phi}{\partial \boldsymbol{\chi}} \right) = 0.$$

Therefore, $\lambda = 0$.

Summarizing the above discussion, we conclude the following formula

$$\lambda = \begin{cases} \left(\dfrac{\partial \phi}{\partial \boldsymbol{\sigma}} : \boldsymbol{C}\, \boldsymbol{\epsilon}(\boldsymbol{w}) \right)_+ \Big/ \left(\dfrac{\partial \phi}{\partial \boldsymbol{\sigma}} : \boldsymbol{C} \dfrac{\partial \phi}{\partial \boldsymbol{\sigma}} + \dfrac{\partial \phi}{\partial \boldsymbol{\chi}} : \boldsymbol{H} \dfrac{\partial \phi}{\partial \boldsymbol{\chi}} \right) \\ \qquad \text{if } \phi(\boldsymbol{\sigma}, \boldsymbol{\chi}) = 0, \\ 0 \quad \text{if } \phi(\boldsymbol{\sigma}, \boldsymbol{\chi}) < 0, \end{cases} \tag{8.73}$$

where $(x)_+ = \max\{x, 0\}$. This formula was mentioned, but not proved, in [115].

Based on the formulae (8.73) and (8.70), we find that if $\phi(\boldsymbol{\sigma}, \boldsymbol{\chi}) < 0$ or if $\phi(\boldsymbol{\sigma}, \boldsymbol{\chi}) = 0$ and $\dfrac{\partial \phi}{\partial \boldsymbol{\sigma}} : \dot{\boldsymbol{\sigma}} \le 0$, then

$$\boldsymbol{\epsilon}(\boldsymbol{w}) = \boldsymbol{C}^{-1} \dot{\boldsymbol{\sigma}},$$

whereas if $\phi(\boldsymbol{\sigma}, \boldsymbol{\chi}) = 0$ and $\dfrac{\partial \phi}{\partial \boldsymbol{\sigma}} : \dot{\boldsymbol{\sigma}} > 0$, then

$$\boldsymbol{\epsilon}(\boldsymbol{w}) = \tilde{\boldsymbol{C}}^{-1} \dot{\boldsymbol{\sigma}},$$

where

$$\tilde{\boldsymbol{C}} = \boldsymbol{C} - \frac{\boldsymbol{C} \dfrac{\partial \phi}{\partial \boldsymbol{\sigma}} \otimes \boldsymbol{C} \dfrac{\partial \phi}{\partial \boldsymbol{\sigma}}}{\dfrac{\partial \phi}{\partial \boldsymbol{\sigma}} : \boldsymbol{C} \dfrac{\partial \phi}{\partial \boldsymbol{\sigma}} + \dfrac{\partial \phi}{\partial \boldsymbol{\chi}} : \boldsymbol{H} \dfrac{\partial \phi}{\partial \boldsymbol{\chi}}},$$

which is invertible by the assumptions on $\boldsymbol{C}$ and $\boldsymbol{H}$. From these formulae and the uniqueness of $\boldsymbol{\sigma}$ (Theorem 8.12) we see that $\boldsymbol{\epsilon}(\boldsymbol{w})$ is uniquely determined. Together with the condition $\boldsymbol{w} \in V$, this implies that $\boldsymbol{w}$, and therefore also $\boldsymbol{u}$, is uniquely determined. Therefore, under the additional assumption that the yield function is continuously differentiable, the result of Theorem 8.12 can be strengthened to include the uniqueness for the displacement variable also.

The relation

$$\dot{\boldsymbol{\sigma}} = \begin{cases} \boldsymbol{C}\, \boldsymbol{\epsilon}(\dot{\boldsymbol{u}}) & \text{if } \phi(\boldsymbol{\sigma}, \boldsymbol{\chi}) < 0, \text{ or } \phi(\boldsymbol{\sigma}, \boldsymbol{\chi}) = 0 \text{ and } \dfrac{\partial \phi}{\partial \boldsymbol{\sigma}} : \dot{\boldsymbol{\sigma}} \le 0, \\ \tilde{\boldsymbol{C}}\, \boldsymbol{\epsilon}(\dot{\boldsymbol{u}}) & \text{if } \phi(\boldsymbol{\sigma}, \boldsymbol{\chi}) = 0 \text{ and } \dfrac{\partial \phi}{\partial \boldsymbol{\sigma}} : \dot{\boldsymbol{\sigma}} > 0 \end{cases} \tag{8.74}$$

is called the rate form of the stress–strain relation.

Part III

Numerical Analysis of the Variational Problems

9

Introduction to Finite Element Analysis

In the previous two chapters we have formulated and analyzed the primal and dual variational formulations of the elastoplasticity problem. Later on, we will study various numerical methods to solve the variational problems. In all the numerical methods to be considered, we will use finite differences to approximate the time derivative and use the finite element method to discretize the spatial variables. The finite element method is widely used for solving boundary value problems of partial differential equations arising in physics and engineering, especially solid mechanics. The method is derived from discretizing the weak formulation of a boundary value problem. The analysis of the finite element method is closely related to that of the boundary value problem.

The development of a finite element algorithm for solving a boundary value problem includes four main steps. First, the boundary value problem is reformulated into an equivalent variational problem. Second, the domain of the independent variables (or usually the domain of the spatial variables, for a time-dependent problem) is partitioned into subdomains called finite elements, and then a finite-dimensional space, called the finite element space, is constructed as a collection of piecewise smooth functions with a certain degree of global smoothness. Third, the variational problem is projected to the finite element space, and in this way, a finite element system is formed. Finally, the finite element system is solved, say by some iterative method, and various conclusions are drawn from the solution of the finite element system. The mathematical theory of the finite element method also addresses issues such as a priori and a posteriori error estimates, and superconvergence.

Compared to the classical finite difference method, the development of the finite element method is a relatively recent event. It is generally agreed that the real engineering practice of the finite element method started in the mid-1950s, and the mathematical analysis of the method began in the mid-1960s. Interesting historical accounts of the method can be found in Oden [97] and in Zienkiewicz [133]. The former gives a balanced presentation on the development of the theory and practice of the method, while the latter is written from the viewpoint of an engineer. For readers interested in the implementation of the finite element method, several engineering books on the method can be consulted, e.g., Bathe [7], Hughes [63], Szabó and Babuška [121], Zienkiewicz and Taylor [134, 135]. In particular, [121] includes discussions on the so-called p-version of the finite element method, where convergence of the method is achieved by increasing polynomial degrees, and the h-p-version of the method, where convergence is obtained by increasing polynomial degrees and refining the mesh simultaneously. Mathematical foundations of the finite element method can be found in Babuška and Aziz [5], Strang and Fix [119], Oden and Reddy [98], Ciarlet [23], and more recently, Ciarlet [25], which is an updated edition of most parts of [23]. For the particular application to solving Navier–Stokes equations, see Girault and Raviart [43], and for a comprehensive treatment of mixed and hybrid finite element methods, see Brezzi and Fortin [18]. The texts by Johnson [69], Brenner and Scott [15], and Braess [14], offer easily accessible expository accounts of basic theoretical aspects of finite element methods.

The main purpose of this chapter is to introduce basic aspects of the finite element method and sample results on finite element interpolation theory. A reader familiar with the basic theoretical results of the finite element method may skip this chapter.

The focus of this chapter is to provide a mathematical assessment for the method; more precisely, we will discuss issues related to the convergence and error estimations. As will be seen in later chapters, the problem of the finite element solution error estimation can be reduced to one of estimating finite element interpolation errors. Therefore, we will discuss how to obtain such interpolation estimates. The theory is developed in the context of elements that are obtained by affine maps from a reference element, so that the domain Ω will be assumed to have a boundary that is polygonal in $\mathbb{R}^2$ and polyhedral in $\mathbb{R}^3$. Otherwise, the theory presented here will be quite general in nature. For the case of a nonpolygonal boundary, curved elements can usually be used to increase the accuracy of the solution; the reader can consult the references mentioned above for detailed discussion. In forming finite element systems, numerical integrations are usually performed to compute integrals appearing in the formulations. Again, the reader can consult the references mentioned above for discussions of effects of numerical integrations on the accuracy of approximate solutions.

This chapter draws heavily on the work by Ciarlet [23]. In the first section we give a brief introduction to the basic ideas underlying the finite element method, as well as a number of issues that arise in practice. Section 9.2 is devoted to a discussion of affine families of elements and of interpolation operators. The aim of Section 9.3 is to derive estimates of the interpolation error on a single element. This estimate will take the form of a bound on the H^m-seminorm of the error in terms of geometrical properties of the element. Finally, in Section 9.4 global interpolation error estimates are derived in appropriate Sobolev norms. Applications of the theory of finite element interpolation error estimations will be given in the next chapter in deriving order error estimates for finite element solutions of various variational problems, and in later chapters for error estimates of numerical solutions of the elastoplasticity problem.

9.1 Basics of the Finite Element Method

The finite element method is based on a discretization of a weak formulation associated with a boundary value problem. For a linear elliptic boundary value problem defined on a Lipschitz domain Ω, the general form of the weak formulation is

$$u \in V, \quad a(u, v) = \langle \ell, v \rangle \quad \forall v \in V. \tag{9.1}$$

Here V is a Sobolev space on Ω. For a second-order differential equation problem, $V = H^1(\Omega)$ if the given boundary condition is natural (i.e., if the condition involves first-order derivatives), and $V = H_0^1(\Omega)$ if the homogeneous Dirichlet boundary condition is specified over the whole boundary. As is discussed in Chapter 6, a problem with a nonhomogeneous Dirichlet boundary condition on a part of the boundary $\Gamma_D \subset \partial\Omega$ can be converted to one with the homogeneous Dirichlet boundary condition on Γ_D after a change of the dependent variable. In this case, then, the space $V = H^1_{\Gamma_D}(\Omega)$. The form $a(\cdot, \cdot)$ is assumed to be bilinear, continuous, and V elliptic, while ℓ is a given linear continuous form on V.

Since V is infinite-dimensional, it is usually impossible to find the solution of the problem (9.1) exactly. The idea of the Galerkin method is to approximate (9.1) by its discrete analogue:

$$u_N \in V_N, \quad a(u_N, v) = \langle \ell, v \rangle \quad \forall v \in V_N, \tag{9.2}$$

where V_N is a finite-dimensional space and is used to approximate the space V. When V_N consists of piecewise polynomials (or more precisely, piecewise images of polynomials) associated with a partition of the domain Ω, the Galerkin method (9.2) becomes the celebrated finite element method. Convergence of the finite element method may be achieved by progressively refining the mesh or by increasing polynomial degrees or by doing both simultaneously; then we get the h-version, p-version, or h-p-version of the finite element method, respectively. It is customary to use h as the parameter for the mesh-size and p the parameter for the polynomial degree(s). Efficient selection among the three versions of the method depends on the a priori knowledge about the regularity of the exact solution of the problem. Roughly speaking, over a region where the solution is smooth, high-degree polynomials with large elements can be used, while in a region where the solution has singularities, low-order elements together with a refined mesh should be used. In this work we will only consider the application of the h-version method, mainly for the reason that the solution of the elastoplasticity problem does not enjoy high regularity. Conventionally, for the h-version finite element method, we use V^h instead of V_N to denote the finite element space. Thus, with a finite element space V^h chosen, the finite element method is

$$u^h \in V^h, \quad a(u^h, v^h) = \langle \ell, v^h \rangle \quad \forall v \in V^h. \tag{9.3}$$

Expressing the trial function u^h in terms of a basis of the space V^h and taking each of the basis functions for the test function v^h, we obtain an equivalent linear system, called the finite element system, from (9.3) for the coefficients in the expansion of u^h with respect to the basis. Once the finite element system is solved, we obtain the finite element solution u^h.

The quality of the finite element solution u^h, i.e., whether u^h is a good approximation of u, is determined by the regularity of the exact solution u, the construction of the finite element space V^h, and the way we solve the finite element system resulting from (9.3). We will discuss in greater detail the construction of V^h.

To begin with, we need to define a partition $\mathcal{T}_h = \{\Omega_e\}_{e=1}^E$ of the domain $\overline{\Omega}$ into a finite number of *closed* subsets Ω_e, $e = 1, \ldots, E$, called *elements*. By this we mean that the following properties are satisfied.

(1) Each Ω_e is a closed nonempty set, with a Lipschitz continuous boundary.

(2) $\overline{\Omega} = \cup_e \Omega_e$.

(3) For $e_1 \neq e_2$, $\mathring{\Omega}_{e_1} \cap \mathring{\Omega}_{e_2} = \varnothing$.

Over each element Ω_e we associate a finite-dimensional function space X_e. We will only consider the case where each function $v \in X_e$ is uniquely determined by its values at a finite number of points in Ω_e: $\boldsymbol{x}_i^{(e)}$, $1 \le i \le I$, called the (local) nodal points. For example, if Ω_e is a triangle and X_e is the space of linear functions on Ω_e, then we can choose the three vertices of the triangle as the nodal points; if X_e consists of all the quadratic functions on Ω_e, then we can use the three vertices and the three side midpoints as the nodal points. We require that any two neighboring elements Ω_{e_1} and Ω_{e_2} have the same nodal points on $\Omega_{e_1} \cap \Omega_{e_2}$. This requirement usually leads to a regularity condition on the finite element partition that the intersection of any two elements must be empty, a vertex, or a common side (or face).

For convenience in practical implementation as well as in theoretical analysis, we will also assume that there exists a fixed number of closed Lipschitz domains, ambiguously represented by one symbol $\hat{\Omega}$, such that for each element Ω_e, there is a smooth mapping function F_e with $\Omega_e = F_e(\hat{\Omega})$. A finite-dimensional function space $\hat{X}$ will be introduced on $\hat{\Omega}$, together with the nodal points on $\hat{\Omega}$ used to uniquely determine functions in $\hat{X}$. Then the local function space X_e on Ω_e will be obtained through the mapping function F_e from $\hat{X}$ by $X_e \circ F_e = \hat{X}$. In most applications of the finite element method, $\hat{X}$ is a space of polynomials of certain degrees. In this work, *we will always assume that $\hat{X}$ is a polynomial space*. In the case of a polygonal domain Ω, we usually partition it into triangles and quadrilaterals. Then we may choose a right isosceles triangle or a square for $\hat{\Omega}$, and correspondingly, the mapping function F_e is linear if Ω_e is a triangle, and bilinear if Ω_e is a quadrilateral. Then the finite element space can be defined to be

$$V^h = \{v^h \in V : \ v^h \circ F_e \in \hat{X} \ \ \forall\, e\}. \tag{9.4}$$

We see that if $\hat{X}$ consists of polynomials, then a function from the space V^h is a piecewise image of polynomials. When F_e is linear, $v^h|_{\Omega_e}$, the restriction of a function $v^h \in V^h$ on Ω_e is a polynomial, while if F_e is nonlinear (e.g., bilinear for quadrilateral elements), $v^h|_{\Omega_e}$ is in general not a polynomial.

A function from the space V^h is called a *finite element*. Sometimes a function in the local function space X_e is also called a finite element. When the associated function space X_e is understood from the context, we even call the element Ω_e a finite element.

A few sentences are in order on the requirement $v^h \in V$. For a second-order boundary value problem, V is a subspace of $H^1(\Omega)$. Since the restriction of v^h on each element Ω_e is a smooth function, a necessary and sufficient condition that $v^h \in V$ is $v^h \in C(\overline{\Omega})$ and v^h satisfies any possible Dirichlet boundary condition specified in V (cf. [23]).

As a consequence of the defining relations (9.1) and (9.3) for u and u^h, together with the continuity and V-ellipticity of the bilinear form $a(\cdot,\cdot)$, Ceá's lemma holds: (Theorem 10.1 in Chapter 10) this result estimates the error of the finite element solution according

$$\|u - u^h\|_V \le c \inf_{v^h \in V^h} \|u - v^h\|_V. \tag{9.5}$$

That is, up to a multiplicative constant, the finite element solution u^h is an optimal approximation to u among the functions from the finite element space V^h. Thus the problem of estimating the finite element solution error

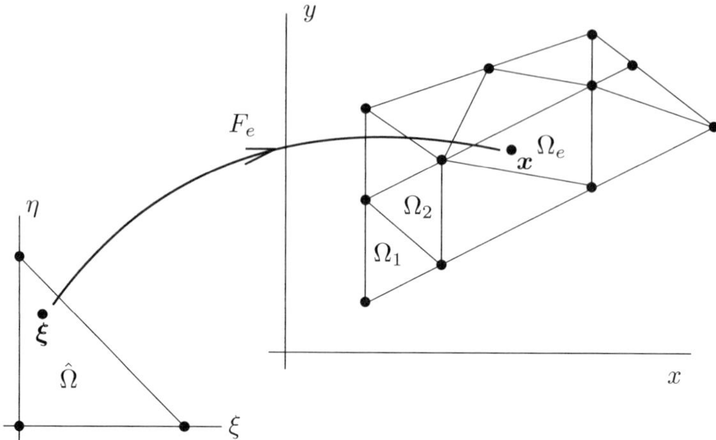

Figure 9.1: Generation of a finite element mesh by a family of affine maps

can be reduced to one of estimating the approximation error

$$\|u - u^h\|_V \leq c\,\|u - \Pi^h u\|_V, \tag{9.6}$$

where $\Pi^h u$ is a finite element interpolant of u.

In the next several sections we will consider in some detail affine families of finite elements and derive order error estimates for the finite element interpolants.

9.2 Affine Families of Finite Elements

In this section we set up the machinery that is vital to a proper development of error estimates for finite element approximations.

Affine-equivalent elements. We consider a situation in which a domain Ω is partitioned into E finite elements, all elements being of the same geometrical type (for example, all triangles) and having the same degree of approximation. Such a finite element mesh may be generated simply by setting up a single *reference element* $\hat{\Omega}$, say, and by mapping or transforming $\hat{\Omega}$ into each one of the elements Ω_e in turn (Figure 9.1).

The basic idea is this: First, we define the reference element $\hat{\Omega}$, this element being of the same geometrical type as the elements that make up Ω. Next, we define an *affine transformation*, that is, a transformation that maps straight lines into straight lines, by

$$F_e : \hat{\Omega} \rightarrow \Omega_e \subset \mathbb{R}^d, \quad \boldsymbol{\xi} \mapsto \boldsymbol{x} = F_e(\boldsymbol{\xi}) \equiv \boldsymbol{T}_e\boldsymbol{\xi} + \boldsymbol{b}_e \tag{9.7}$$

and such that F_e is a bijection between $\hat{\Omega}$ and Ω_e. Here $\boldsymbol{b}_e$ is a translation vector, $\boldsymbol{T}_e$ is an invertible $d \times d$ matrix, and furthermore, $\det(\boldsymbol{T}_e) > 0$. We

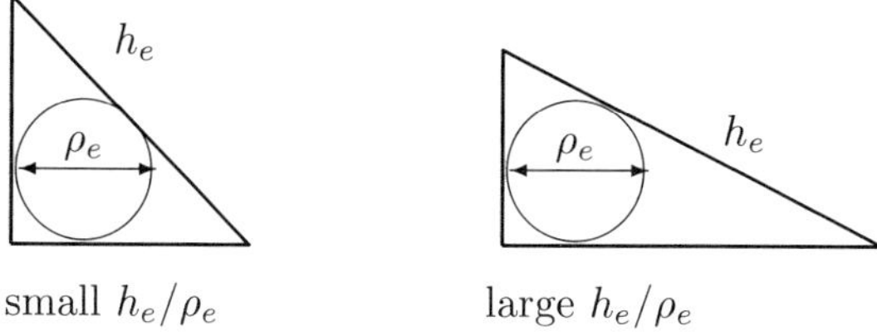

Figure 9.2: The constants h_e and ρ_e associated with an element

also require of F_e that it map the nodal point $\boldsymbol{\xi}_i$, $1 \leq i \leq I$, of $\hat{\Omega}$ to the (locally numbered) nodal point $\boldsymbol{x}_i^{(e)}$, $1 \leq i \leq I$, of Ω_e:

$$F_e(\boldsymbol{\xi}_i) = \boldsymbol{x}_i^{(e)}, \quad i = 1, \ldots, I. \tag{9.8}$$

Once a set of affine transformations has been constructed in this way for each element, we need to focus attention only on the reference element $\hat{\Omega}$ and the family of transformations F_1, F_2, ..., F_E, since these provide a complete description of the mesh. Since $\hat{X}$ consists of polynomials and F_e is an affine mapping, we see that X_e also consists of polynomials.

When two elements $\hat{\Omega}$ and Ω_e are related to each other by a transformation of the type (9.7) with the property (9.8), they are said to be *affine-equivalent*. A set of finite elements Ω_1, ..., Ω_E is called an *affine family* if all elements are affine-equivalent to a single reference element $\hat{\Omega}$.

The relative size and shape of an arbitrary element Ω_e are quantified in a natural way by defining the quantities

$$h_e = \text{diam}\,(\Omega_e) = \max\{\|\boldsymbol{x} - \boldsymbol{y}\| : \boldsymbol{x}, \boldsymbol{y} \in \Omega_e\} \tag{9.9}$$

and

$$\rho_e = \text{diameter of the largest sphere } S_e \text{ inscribed in } \Omega_e. \tag{9.10}$$

Here and below, the vector norm $\|\boldsymbol{x}\|$ is the Euclidean norm.

When dealing with the reference element $\hat{\Omega}$, we denote the corresponding quantities by $\hat{h}$ and $\hat{\rho}$. These quantities are illustrated in Figure 9.2: Whereas h_e gives some idea of the "size" of Ω_e, the ratio h_e/ρ_e gives an indication of how "thin" the element is.

We now summarize some useful properties of the affine transformation (9.7).

LEMMA 9.1. *Let $\hat{\Omega}$, $\Omega_e \subset \mathbb{R}^d$, and $F_e : \hat{\Omega} \to \Omega_e$ be the affine map from $\hat{\Omega}$ to Ω_e defined by (9.7). If the matrix norm $\|\boldsymbol{T}_e\|$ is defined by*

$$\|\boldsymbol{T}_e\| = \sup\left\{\frac{\|\boldsymbol{T}_e\boldsymbol{\xi}\|}{\|\boldsymbol{\xi}\|} : \boldsymbol{\xi} \neq \boldsymbol{0}\right\},$$

then

$$\|\boldsymbol{T}_e\| \le \frac{h_e}{\hat{\rho}} \quad \text{and} \quad \|\boldsymbol{T}_e^{-1}\| \le \frac{\hat{h}}{\rho_e}.$$

PROOF. For $\boldsymbol{\xi} \neq \boldsymbol{0}$, let $\boldsymbol{z} = \hat{\rho}\boldsymbol{\xi}/\|\boldsymbol{\xi}\|$; then $\|\boldsymbol{z}\| = \hat{\rho}$ and

$$\|\boldsymbol{T}_e\| = \sup\left\{\frac{\|\boldsymbol{T}_e\boldsymbol{\xi}\|}{\|\boldsymbol{\xi}\|} : \boldsymbol{\xi} \neq \boldsymbol{0}\right\} = \hat{\rho}^{-1}\sup\{\|\boldsymbol{T}_e\boldsymbol{z}\| : \|\boldsymbol{z}\| = \hat{\rho}\}.$$

Now for an arbitrary $\boldsymbol{z}$ with $\|\boldsymbol{z}\| = \hat{\rho}$, pick up any two points $\boldsymbol{\xi}$ and $\boldsymbol{\eta}$ that lie on the largest sphere $\hat{S}$ of diameter $\hat{\rho}$, which is inscribed in $\hat{\Omega}$, such that $\boldsymbol{z} = \boldsymbol{\xi} - \boldsymbol{\eta}$. Then

$$\begin{aligned}
\|\boldsymbol{T}_e\| &= \hat{\rho}^{-1}\sup\left\{\|\boldsymbol{T}_e(\boldsymbol{\xi} - \boldsymbol{\eta})\| : \boldsymbol{\xi}, \boldsymbol{\eta} \in \hat{S}\right\} \\
&= \hat{\rho}^{-1}\sup\left\{\|(\boldsymbol{T}_e\boldsymbol{\xi} + \boldsymbol{b}_e) - (\boldsymbol{T}_e\boldsymbol{\eta} + \boldsymbol{b}_e)\| : \boldsymbol{\xi}, \boldsymbol{\eta} \in \hat{S}\right\} \\
&\le \hat{\rho}^{-1}\sup\{\|\boldsymbol{x} - \boldsymbol{y}\| : \boldsymbol{x}, \boldsymbol{y} \in \Omega_e\} \\
&\le h_e/\hat{\rho}.
\end{aligned}$$

The second inequality is proved similarly. □

Mappings of functions. By making use of the affine map (9.7), we can set up an operator K_e that maps a function v defined on Ω_e to a function $\hat{v}$ on $\hat{\Omega}$, the function $\hat{v}$ being defined by

$$\hat{v}(\boldsymbol{\xi}) = v(\boldsymbol{x}), \quad \boldsymbol{x} = F_e(\boldsymbol{\xi}). \tag{9.11}$$

Since F_e is a bijective mapping from $\hat{\Omega}$ to Ω_e, the operator K_e is invertible with inverse K_e^{-1} mapping functions on $\hat{\Omega}$ to functions on Ω_e, so that

$$K_e^{-1}\hat{v} = v. \tag{9.12}$$

Now let $\{\boldsymbol{\xi}_i\}_{i=1}^I$ be the nodal points on $\hat{\Omega}$, and $\{\hat{N}_i\}_{i=1}^I$ be a set of *local basis functions* defined on $\hat{\Omega}$ with the property that

$$\hat{N}_i(\boldsymbol{\xi}_j) = \begin{cases} 1 & \text{if } j = i, \\ 0 & \text{otherwise.} \end{cases}$$

Usually, the function $\hat{N}_i$ is chosen to be a polynomial, say of degree k. By using (9.12), we can define

$$N_i^{(e)} = K_e^{-1}\hat{N}_i, \quad i = 1, \ldots, I.$$

Here $\{N_i^{(e)}\}_{i=1}^I$ is the corresponding set of polynomial *local basis functions* defined on Ω_e; these functions also have the property that $N_i^{(e)}(\boldsymbol{x}_i^{(e)}) = 1$ and $N_i^{(e)}(\boldsymbol{x}_j^{(e)}) = 0$ for $j \neq i$, since (9.11) implies that $\hat{N}_i(\boldsymbol{\xi}_j) = N_i^{(e)}(\boldsymbol{x}_j^{(e)})$.

The local basis functions $\{\hat{N}_i\}_{i=1}^{I}$ span a space $\hat{X}$ (of polynomials, in our case) on $\hat{\Omega}$. We can construct a *projection operator* $\hat{\Pi}$ that maps any $\hat{v} \in C(\hat{\Omega})$ to its *interpolant* $\hat{\Pi}\hat{v}$ in $\hat{X}$, according to

$$\hat{\Pi} : C(\hat{\Omega}) \to \hat{X}, \quad \hat{\Pi}\hat{v} = \sum_{i=1}^{I} \hat{v}(\boldsymbol{\xi}_i)\hat{N}_i. \tag{9.13}$$

Similarly, we define the projection operator Π_e by

$$\Pi_e : C(\Omega_e) \to X_e, \quad \Pi_e v = \sum_{i=1}^{I} v(\boldsymbol{x}_i^{(e)})N_i^{(e)}, \tag{9.14}$$

where $X_e = \text{span}\,\{N_i^{(e)}\}_{i=1}^{I}$ and $\Pi_e v$ is the interpolant of v in X_e. We come now to a crucial question about such interpolations: Given a function v in $C(\Omega_e)$ and its image $\hat{v} = K_e v$ in $C(\hat{\Omega})$, are $\hat{\Pi}(K_e v)$ and $K_e(\Pi_e v)$ the same functions? That is, if we map v to $\hat{v}$ and then interpolate in $\hat{\Omega}$, is this the same as first interpolating v and then mapping it? The next result answers this question.

THEOREM 9.2. *Let $\hat{\Omega}$ and Ω_e be affine-equivalent subsets in $\mathbb{R}^d$. Then the interpolation operators $\hat{\Pi}$ and Π_e satisfy the relation*

$$\hat{\Pi}(K_e v) = K_e(\Pi_e v),$$

i.e.,

$$\hat{\Pi}\hat{v} = \widehat{\Pi_e v}.$$

PROOF. We have from the definition (9.14),

$$\Pi_e v = \sum_{i=1}^{I} v(\boldsymbol{x}_i^{(e)})N_i^{(e)} = \sum_{i=1}^{I} \hat{v}(\boldsymbol{\xi}_i)N_i^{(e)}.$$

Hence

$$\begin{aligned} K_e(\Pi_e v) &= K_e\left(\sum_{i=1}^{I} \hat{v}(\boldsymbol{\xi}_i)N_i^{(e)}\right) \\ &= \sum_{i=1}^{I} \hat{v}(\boldsymbol{\xi}_i)K_e N_i^{(e)} \quad (K_e \text{ is a linear operator}) \\ &= \sum_{i=1}^{I} \hat{v}(\boldsymbol{\xi}_i)\hat{N}_i, \end{aligned}$$

which is precisely $\hat{\Pi}\hat{v}$. $\square$

9.3 Local Interpolation Error Estimates

Recall from the estimate (9.6) that the error $\|u - u^h\|_V$, measured in the norm of the space V, can be bounded above by a constant multiple of the *interpolation error* $\|u - \Pi^h u\|_V$, where $\Pi^h u$ is the interpolant of u in V^h, defined piecewise by the formula $(\Pi^h u)|_{\Omega_e} = \Pi_e u$. The task of estimating the finite element solution error consequently reduces to one of estimating the interpolation error. We go one step further towards obtaining such an estimate by deriving in this section an estimate of the interpolation error $\|v - \Pi_e v\|$ for functions defined on a *single* finite element Ω_e. Once this estimate is found, it can be used to derive an estimate for the error of the global interpolation of a function, defined over the entire domain Ω.

We assume that the finite-dimensional space X_e spanned by the local basis functions $\{N_i^{(e)}\}_{i=1}^I$ contains polynomials of degree less than or equal to k, for some $k \geq 1$. In other words,

$$X_e \supset P_k(\Omega_e).$$

Here, $P_k(\Omega_e)$ denotes the space of the polynomials on Ω_e of degree less than or equal to k. We will show that for $m \leq k+1$, an interpolation error estimate in the $H^m(\Omega_e)$-norm can be derived for a function v that is in $H^{k+1}(\Omega_e)$. Here and below, we use the notation $H^m(\Omega_e)$ to stand for $H^m(\mathring{\Omega}_e)$. So we consider the situation in which there are two spaces $H^{k+1}(\Omega_e)$ and $H^m(\Omega_e)$ with $k+1 \geq m$, and a projection operator Π_e that maps members of $H^{k+1}(\Omega_e)$ to $H^m(\Omega_e)$, the images $\Pi_e v$ all lying in X_e,

$$\Pi_e : H^{k+1}(\Omega_e) \to X_e \subset H^m(\Omega_e). \tag{9.15}$$

In the following, we assume $k+1 > d/2$; then from the Sobolev embedding theorem (Theorem 5.14), $H^{k+1}(\Omega_e) \hookrightarrow C(\Omega_e)$. Thus for $v \in H^{k+1}(\Omega_e)$, pointwise values $v(\boldsymbol{x})$ are well-defined. Let the projection operator Π_e be defined by (9.14). Since $P_k(\Omega_e) \subset X_e$ by assumption, the operator Π_e has the property that

$$\Pi_e v = v \quad \forall\, v \in P_k(\Omega_e). \tag{9.16}$$

Similarly,

$$\hat{\Pi}\hat{v} = \hat{v} \quad \forall\, \hat{v} \in P_k(\hat{\Omega}). \tag{9.17}$$

A property of the form (9.16) or (9.17) is called a *polynomial invariance property* of the finite element interpolation operator. We remark that only (9.17), the polynomial invariance property on the reference element $\hat{\Omega}$, is essential in deriving finite element interpolation error estimates. The polynomial invariance property (9.16) on the real element is a consequence of (9.17) and the fact that F_e is an affine mapping. In the more general case when F_e is not affine, functions from X_e may not be polynomials, so (9.16) cannot hold; nevertheless, (9.16) does not play any role in deriving error estimates for finite element interpolations.

The main result in this section will be the following: For $v \in H^{k+1}(\Omega_e)$ and Π_e satisfying the above properties, the interpolation error in the H^m-norm, $0 \le m \le k+1$, can be estimated by

$$\|v - \Pi_e v\|_{m,\Omega_e} \le c\, h_e^{k+1-m} |v|_{k+1,\Omega_e},$$

where h_e is defined in (9.9) and for an integer l, $|\cdot|_{l,\Omega_e}$ denotes the Sobolev *seminorm*

$$|v|_{l,\Omega_e}^2 = \sum_{|\alpha|=l} \int_{\Omega_e} [D^\alpha v(\boldsymbol{x})]^2 \, dx.$$

Recall also that the Sobolev norm $\|\cdot\|_{l,\Omega_e}$ is given by

$$\|v\|_{l,\Omega_e}^2 = \sum_{j=0}^{l} |v|_{j,\Omega_e}^2.$$

We start the development by presenting an important result that will be required later.

THEOREM 9.3. *For any bounded set $\Omega_0 \subset \mathbb{R}^d$ with a Lipschitz continuous boundary, there is a constant c, depending only on the geometry of Ω_0, such that*

$$\inf_{p \in P_k(\Omega_0)} \|v + p\|_{k+1,\Omega_0} \le c\,|v|_{k+1,\Omega_0} \quad \forall\, v \in H^{k+1}(\Omega_0). \tag{9.18}$$

PROOF. First, we apply Theorem 5.17 to get the inequality

$$\|u\|_{k+1,\Omega_0} \le c \left(|u|_{k+1,\Omega_0} + \sum_{|\alpha| \le k} \left| \int_{\Omega_0} D^\alpha u(\boldsymbol{x}) \, dx \right| \right) \quad \forall\, u \in H^{k+1}(\Omega_0).$$

Now, replacing u by $v + p$ with $v \in H^{k+1}(\Omega_0)$ and $p \in P_k(\Omega_0)$, and noting that $D^\alpha p = 0$ for $|\alpha| = k+1$, we have

$$\begin{aligned} \|v + p\|_{k+1,\Omega_0} &\le c \left(|v|_{k+1,\Omega_0} + \sum_{|\alpha| \le k} \left| \int_{\Omega_0} D^\alpha (v + p) \, dx \right| \right) \\ &\forall\, v \in H^{k+1}(\Omega_0),\ p \in P_k(\Omega_0). \end{aligned} \tag{9.19}$$

Now construct a polynomial $\bar{p}$ in $P_k(\Omega_0)$ that has the property that

$$\int_{\Omega_0} D^\alpha (v + \bar{p}) \, dx = 0 \ \text{ for } \ |\alpha| \le k. \tag{9.20}$$

This can always be done: Set $|\alpha| = k$; then $D^\alpha \bar{p}$ equals $\alpha_1! \cdots \alpha_d!$ times the coefficient of $\boldsymbol{x}^\alpha$. The coefficient can be computed by using (9.20). Having found all the coefficients of the terms of degree k, we set $|\alpha| = k - 1$, and use (9.20) to compute all the coefficients of the terms of degree $k - 1$. Proceeding in this way, we find the polynomial $\bar{p}$ for the given v.

With $p = \bar{p}$ in (9.19), we have

$$\inf_{p \in P_k(\Omega_0)} \|v + p\|_{k+1,\Omega_0} \le \|v + \bar{p}\|_{k+1,\Omega_0} \le c\,|v|_{k+1,\Omega_0},$$

from which (9.18) follows. □

Later on, the result of Theorem 9.3 will be applied for $\Omega_0 = \hat{\Omega}$, the reference element.

Next we need to know how the seminorms of the functions v and of $\hat{v}$ are related.

THEOREM 9.4. *Let Ω_e and $\hat{\Omega}$ be two affine-equivalent subsets of $\mathbb{R}^d$, and l a nonnegative integer. Then $v \in H^l(\Omega_e)$ if and only if $\hat{v} = K_e v \in H^l(\hat{\Omega})$ and for some constant c independent of Ω_e and $\hat{\Omega}$, the following estimates hold:*

$$|\hat{v}|_{l,\hat{\Omega}} \le c\,\|\boldsymbol{T}_e\|^l\,|\det \boldsymbol{T}_e|^{-1/2}|v|_{l,\Omega_e}, \tag{9.21}$$

$$|v|_{l,\Omega_e} \le c\,\|\boldsymbol{T}_e^{-1}\|^l\,|\det \boldsymbol{T}_e|^{1/2}|\hat{v}|_{l,\hat{\Omega}}, \tag{9.22}$$

where $\boldsymbol{T}_e$ is the matrix occurring in the affine map (9.7).

PROOF. We prove (9.21); (9.22) is proved in a similar fashion. Recall the result from multivariable calculus that if $\xi_i = f_i(x_1, \ldots, x_d)$, $1 \le i \le d$, then

$$d\xi \equiv d\xi_1 d\xi_2 \cdots d\xi_d = |\det(\partial f_i/\partial x_j)|\,dx_1 dx_2 \cdots dx_d.$$

We have

$$\begin{aligned} |\hat{v}|^2_{l,\hat{\Omega}} &= \sum_{|\alpha|=l} \int_{\hat{\Omega}} \left(D^\alpha_{\boldsymbol{\xi}} \hat{v}(\boldsymbol{\xi})\right)^2 d\xi \\ &= \sum_{|\alpha|=l} \int_{\Omega_e} \left(D^\alpha_{\boldsymbol{\xi}} \hat{v}(\boldsymbol{\xi}(\boldsymbol{x}))\right)^2 |\det \boldsymbol{T}_e|^{-1}\,dx, \end{aligned} \tag{9.23}$$

where

$$D^\alpha_{\boldsymbol{\xi}} = \frac{\partial^{|\alpha|}}{\partial \xi_1^{\alpha_1} \cdots \partial \xi_d^{\alpha_d}} \quad \text{for } \alpha = (\alpha_1, \ldots, \alpha_d).$$

By an application of the chain rule we have

$$\sum_{|\alpha|=l} |D^\alpha_{\boldsymbol{\xi}} \hat{v}(\boldsymbol{\xi}(\boldsymbol{x}))|^2 \le c\,\|\boldsymbol{T}_e\|^{2l} \sum_{|\alpha|=l} |D^\alpha_{\boldsymbol{x}} v(\boldsymbol{x})|^2, \tag{9.24}$$

where

$$D_{\boldsymbol{x}}^{\alpha} = \frac{\partial^{|\alpha|}}{\partial x_1^{\alpha_1} \cdots \partial x_d^{\alpha_d}} \quad \text{for } \alpha = (\alpha_1, \ldots, \alpha_d).$$

Hence from (9.23),

$$\begin{aligned} |\hat{v}|_{l,\hat{\Omega}}^2 &\le c \sum_{|\alpha|=l} \int_{\Omega_e} (D_{\boldsymbol{x}}^{\alpha} v(\boldsymbol{x}))^2 \, \|\boldsymbol{T}_e\|^{2\,l} (\det \boldsymbol{T}_e)^{-1} \, dx \\ &= c \, \|\boldsymbol{T}_e\|^{2\,l} (\det \boldsymbol{T}_e)^{-1} \, |v|_{l,\Omega_e}^2, \end{aligned}$$

from which (9.21) follows. □

We now estimate the interpolation error in the seminorm $|v - \Pi_e v|_{m,\Omega_e}$.

THEOREM 9.5. *Let k and m be nonnegative integers such that $k+1 > d/2$, $k+1 \ge m$, and*

$$P_k(\hat{\Omega}) \subset \hat{X} \subset H^m(\hat{\Omega}).$$

Let Π_e and $\hat{\Pi}$ be the operators defined in (9.13) *and* (9.14). *Then for any affine equivalent element Ω_e,*

$$|v - \Pi_e v|_{m,\Omega_e} \le c \, \frac{h_e^{k+1}}{\rho_e^m} |v|_{k+1,\Omega_e} \quad \forall \, v \in H^{k+1}(\Omega_e), \tag{9.25}$$

where h_e and ρ_e are defined in (9.9) *and* (9.10), *and c is a constant depending only on $\hat{\Omega}$ and $\hat{\Pi}$.*

PROOF. Notice that $k+1 > d/2$ implies $H^{k+1}(\hat{\Omega}) \subset C(\hat{\Omega})$, so $\hat{v} = K_e v \in H^{k+1}(\hat{\Omega})$ is continuous and the point values of $\hat{\Pi}\hat{v}$ are well-defined. Using (9.17), we have, for all $\hat{v} \in H^{k+1}(\hat{\Omega})$ and all $\hat{p} \in P_k(\hat{\Omega})$,

$$\begin{aligned} |\hat{v} - \hat{\Pi}\hat{v}|_{m,\hat{\Omega}} &\le \|\hat{v} - \hat{\Pi}\hat{v}\|_{m,\hat{\Omega}} = \|\hat{v} - \hat{\Pi}\hat{v} + \hat{p} - \hat{\Pi}\hat{p}\|_{m,\hat{\Omega}} \\ &\le \|(\hat{v} + \hat{p}) - \hat{\Pi}(\hat{v} + \hat{p})\|_{m,\hat{\Omega}} \\ &\le \|\hat{v} + \hat{p}\|_{m,\hat{\Omega}} + \|\hat{\Pi}(\hat{v} + \hat{p})\|_{m,\hat{\Omega}} \\ &\le (1 + \|\hat{\Pi}\|) \, \|\hat{v} + \hat{p}\|_{k+1,\hat{\Omega}}. \end{aligned}$$

Notice that $\|\hat{\Pi}\| < \infty$, i.e., $\hat{\Pi}$ is a bounded operator from $H^{k+1}(\hat{\Omega})$ to $H^m(\hat{\Omega})$:

$$\|\hat{\Pi}\hat{v}\|_{m,\hat{\Omega}} \le \sum_{i=1}^{I} |\hat{v}(\boldsymbol{\xi}_i)| \, \|\hat{N}_i\|_{m,\hat{\Omega}} \le c \, \|\hat{v}\|_{C(\hat{\Omega})} \le c \, \|\hat{v}\|_{k+1,\hat{\Omega}}.$$

The use of Theorem 9.3 now yields

$$|\hat{v} - \hat{\Pi}\hat{v}|_{m,\hat{\Omega}} \le c \inf_{\hat{p} \in P_k(\hat{\Omega})} \|\hat{v} + \hat{p}\|_{k+1,\hat{\Omega}} \le c \, |\hat{v}|_{k+1,\hat{\Omega}}. \tag{9.26}$$

From Theorem 9.2 we have $\hat{\Pi}(K_e v) = K_e(\Pi_e v)$, so that

$$\hat{v} - \hat{\Pi}\hat{v} = K_e v - \hat{\Pi}(K_e v) = K_e(v - \Pi_e v). \tag{9.27}$$

Consequently, using (9.22) (replacing v by $v - \Pi_e v$ and setting $l = m$) and (9.27), we obtain

$$\begin{aligned} |v - \Pi_e v|_{m,\Omega_e} &\le c\,\|\boldsymbol{T}_e^{-1}\|^m |\det \boldsymbol{T}_e|^{1/2} |K_e(v - \Pi_e v)|_{m,\hat{\Omega}} \\ &= c\,\|\boldsymbol{T}_e^{-1}\|^m |\det \boldsymbol{T}_e|^{1/2} |\hat{v} - \hat{\Pi}\hat{v}|_{m,\hat{\Omega}}. \end{aligned} \tag{9.28}$$

Furthermore, from (9.21) with $l = k+1$,

$$|\hat{v}|_{k+1,\hat{\Omega}} \le c\,\|\boldsymbol{T}_e\|^{k+1} |\det \boldsymbol{T}_e|^{-1/2} |v|_{k+1,\Omega_e}. \tag{9.29}$$

Finally, substituting (9.26) in (9.28), then using (9.29) in that result, we obtain

$$|v - \Pi_e v|_{m,\Omega_e} \le c\,\|\boldsymbol{T}_e^{-1}\|^m \|\boldsymbol{T}_e\|^{k+1} |v|_{k+1,\Omega_e},$$

which, together with the result of Lemma 9.1, leads to (9.25). □

REMARKS.

1. Since we wish to evaluate $|v - \Pi_e v|_{m,\Omega_e}$, it follows that both v and $\Pi_e v$ must be in $H^m(\Omega_e)$ for this term to make sense. Equivalently, $\hat{v}$ and $\hat{\Pi}\hat{v}$ must be in $H^m(\hat{\Omega})$. This accounts for the assumptions $H^{k+1}(\hat{\Omega}) \subset H^m(\hat{\Omega})$ and $\hat{X} \subset H^m(\hat{\Omega})$. Note that $v \in H^{k+1}(\Omega_e)$ implies $\hat{v} \in H^{k+1}(\hat{\Omega})$. The inclusion $H^{k+1}(\hat{\Omega}) \subset H^m(\hat{\Omega})$ of course holds if $m \le k+1$.

2. In evaluating the interpolant $\Pi_e v$ of v, it is necessary to use the nodal values of v. This in turn requires that v be continuous, so that we must assume $v \in H^{k+1}(\Omega_e) \hookrightarrow C(\Omega_e)$, or equivalently, $\hat{v} \in H^{k+1}(\hat{\Omega}) \hookrightarrow C(\hat{\Omega})$. By the Sobolev embedding theorem, this inclusion holds if $k + 1 > d/2$ for a problem in $\mathbb{R}^d$.

3. The error estimate (9.25) is proved through the use of the reference element $\hat{\Omega}$. This method of proof can be termed the *reference element technique*. We notice that in the proof we use only the polynomial invariance property (9.17) of the finite element interpolation on the reference element, and we do not need to use the polynomial invariance property on the real finite element. This feature is important when we analyze finite element spaces that are not based on affine-equivalent elements. For example, suppose the domain is partitioned into quadrilateral elements $\{\Omega_1, \ldots, \Omega_E\}$. Then a reference element can be taken to be the unit square $\hat{\Omega} = [0,1]^2$. For each element Ω_e,

the mapping function F_e is bilinear and maps each vertex of the reference element $\hat{\Omega}$ to a corresponding vertex of Ω_e. The first-degree finite element space for approximating $V = H^1(\Omega)$ is

$$V^h = \{v^h \in C(\overline{\Omega}) : v^h \circ F_e \in Q_1(\hat{\Omega}),\ e = 1, \ldots, E\},$$

where

$$Q_1(\hat{\Omega}) = \{v : v(\boldsymbol{\xi}) = a + b\,\xi_1 + c\,\xi_2 + d\,\xi_1\xi_2,\ \boldsymbol{\xi} \in \hat{\Omega},\ a, b, c, d \in \mathbb{R}\}$$

is the space of bilinear functions. We see that for $v^h \in V^h$, on each element Ω_e, $v^h|_{\Omega_e}$ is not a polynomial (as a function of the variable $\boldsymbol{x}$), but rather the image of a polynomial on the reference element. Obviously, (9.16) does not hold, but (9.17) is still valid. For such a finite element space, the proof of Theorem 9.5 still goes through.

4. Continuing the last remark, we comment that the reference element technique is useful not only for theoretical error analysis, but also for actual implementation of the finite element method. By employing the technique, all the calculations in forming the finite element system are done over the reference element. This simplifies the whole implementation process, e.g., numerical integrations need to be performed only on the reference element.

The two parameters h_e and ρ_e appearing in (9.25) may be reduced to one if attention is restricted to a family of finite elements for which the ratio h_e/ρ_e is bounded above, so that elements are not allowed to become too "flat." For this purpose we introduce the notion of a *regular* family of finite elements: A family of partitions $\{\Omega_e\}_{e=1}^E$ is said to be regular if

(i) there exists a constant σ such that $h_e/\rho_e \le \sigma$ for all elements Ω_e;

(ii) the mesh-size $h = \max_{1 \le e \le E} h_e$ approaches zero.

In the case of a regular family of finite elements, the error estimate of Theorem 9.5 can be stated in terms of a *norm*; this is recorded in the following result.

COROLLARY 9.6. *Let the conditions of Theorem 9.5 hold, and* $\{\Omega_e\}_{e=1}^E$ *be a regular family of finite elements. Then there is a constant c such that for any element* Ω_e *in the family and all functions* $v \in H^{k+1}(\Omega_e)$,

$$\|v - \Pi_e v\|_{m,\Omega_e} \le c\, h_e^{k+1-m} |v|_{k+1,\Omega_e}, \quad m \le k+1. \tag{9.30}$$

It is not difficult to deduce this result from Theorem 9.5. The property (i) of a regular family of finite elements is used to replace the ratio h_e^{k+1}/ρ_e^m in (9.25) by $c\, h_e^{k+1-m}$.

EXAMPLE 9.7. Let $\Omega \subset \mathbb{R}^2$ be partitioned by a regular family $\{\Omega_e\}_{e=1}^E$

of triangular elements. A typical element Ω_e is a triangle, and its three vertices are used as the nodal points. The space X_e spanned by the local interpolation functions is $P_1(\Omega_e)$, so that $k = 1$. Assuming that v belongs to $H^2(\Omega_e)$, the estimate (9.30) gives

$$\|v - \Pi_e v\|_{m,\Omega_e} \leq c\, h_e^{2-m} |v|_{2,\Omega_e} \tag{9.31}$$

for $m = 0, 1, 2$. □

9.4 Global Interpolation Error Estimates

Having established properties of finite element interpolations over individual elements, we turn now to estimating the error of the global interpolation of a function defined on the entire domain Ω. Specifically, we have a function $v \in C(\overline{\Omega})$, and we construct its interpolant $\Pi^h v$ in the finite element space V^h according to

$$\Pi^h v = \sum_{i=1}^{N} v(\boldsymbol{x}_i) N_i, \tag{9.32}$$

where N_i, $i = 1, \ldots, N$, are the global basis functions that span V^h. Here N_i is the global basis function associated with the node $\boldsymbol{x}_i$, i.e., N_i is a piecewise polynomial of degree less than or equal to k, uniquely determined by the property $N_i(\boldsymbol{x}_j) = \delta_{ij}$. If the node $\boldsymbol{x}_i$ is a vertex $\boldsymbol{x}_j^{(e)}$ of the element Ω^e, then $N_i|_{\Omega_e} = N_j^e$. If $\boldsymbol{x}_i$ is not a node of Ω_e, then $N_i|_{\Omega_e} = 0$. Thus the functions N_i are constructed from local basis functions N_i^e, and it is clear that the restriction of $\Pi^h v$ to any element Ω_e is in fact $\Pi_e v$.

Since we will be primarily interested in obtaining error estimates for finite element solutions of second-order problems, we need to estimate the interpolation error $\|u - \Pi^h u\|_{1,\Omega}$ (recall that $m = 1$ for second-order problems). In the same way as $\|u - \Pi_e u\|_{m,\Omega_e}$ is estimated in terms of the parameter h_e, a suitable parameter is required for the global estimate. For this purpose, suppose that we are dealing with a *regular* family of finite elements, and set

$$h = \max_{1 \leq e \leq E} h_e. \tag{9.33}$$

The quantity h is called the *mesh parameter* and is a measure of how refined the mesh is. Hence, if it is possible to obtain an interpolation error estimate of the form

$$\|u - \Pi^h u\|_{1,\Omega} \leq c\, h^{\beta} |u|_{k+1,\Omega},$$

then we are assured of convergence as $h \to 0$, provided that $\beta > 0$.

The mesh parameter provides a natural way of quantifying the dimension of the space V^h that occurs in finite element approximations. For each value of h the approximate solution u^h is sought in a finite-dimensional space V^h, with the hope that the error $\|u-u^h\|$ approaches zero as $h \to 0$. The smaller h is, the finer the subdivision, and hence the larger the dimension of V^h will be. The family of finite element spaces V^h is said to approximate V as $h \to 0$ if for any $v \in V$, $\inf\{\|v - v^h\| : v^h \in V^h\} \to 0$ as $h \to 0$.

The following global *interpolation* error estimate establishes the precise sense in which V^h approximates V. Our discussion here is valid for $V = H^1(\Omega)$, $V = H_0^1(\Omega)$, or $V = H^1_{\Gamma_0}(\Omega)$, $\Gamma_0 \subset \Gamma$, depending on the form of the boundary condition. When we have homogeneous Dirichlet condition on a part or the whole boundary, we agree that those global basis functions N_i associated with the nodes $\boldsymbol{x}_i$ lying on that part of the boundary are removed from the expression (9.32) as well as the finite element space V^h. In case $\Gamma_0 \neq \Gamma$, we assume that if the intersection of a side of some element with the boundary is not empty, then either no Dirichlet condition is specified on that side or the Dirichlet condition is specified on the whole side. In this way, $V^h \subset V$.

THEOREM 9.8. *Assume that all the conditions of Corollary 9.6 hold. Then there exists a constant c independent of h such that for $m = 0, 1$,*

$$\|v - \Pi^h v\|_{m,\Omega} \le c\, h^{k+1-m} |v|_{k+1,\Omega} \quad \forall\, v \in H^{k+1}(\Omega). \tag{9.34}$$

PROOF. We notice that $\hat{X} \subset H^1(\hat{\Omega})$ and $V^h \subset C(\overline{\Omega})$ imply $V^h \subset H^1(\Omega)$. Hence $\Pi^h u \in H^1(\Omega)$ with $\Pi^h u|_{\Omega_e} = \Pi_e u$. Applying Corollary 9.6 with $m = 0$ or 1, we have

$$\begin{aligned} \|u - \Pi^h u\|_{m,\Omega} &= \left(\sum_{e=1}^{E} \|u - \Pi_e u\|^2_{m,\Omega_e} \right)^{1/2} \\ &\le \left(\sum_{e=1}^{E} c\, h_e^{2(k+1-m)} |u|^2_{k+1,\Omega_e} \right)^{1/2} \\ &\le c\, h^{k+1-m} \left(\sum_{e=1}^{E} |u|^2_{k+1,\Omega_e} \right)^{1/2} \\ &= c\, h^{k+1-m} |u|_{k+1,\Omega}. \end{aligned}$$

□

We make a remark on finite element interpolation of possibly discontinuous functions. In the above discussion of the finite element interpolation error analysis, we assume that the function being interpolated is continuous, so that it is meaningful to use its finite element interpolation defined in (9.32). The continuity condition is guaranteed by the assumption $v \in H^{k+1}(\Omega)$ and $k + 1 > d/2$. In case $k + 1 \le d/2$, an $H^{k+1}(\Omega)$-function is no longer necessarily continuous. Instead of (9.32), we can choose $\Pi^h v$

to be Clément's interpolant, which is well-defined even when $k+1 \leq d/2$, and the interpolation error estimates stated in Theorem 9.8 are still valid. For detail, see Clément [26]. We will use the same symbol $\Pi^h v$ to denote the "regular" finite element interpolant (9.32) when v is continuous, and in case v is discontinuous, $\Pi^h v$ is Clément's interpolant. In either case, we have the error estimates (9.34).

Orthogonal projections are another possibility in case the function to be interpolated is not continuous. Let $\Pi_0^h : L^2(\Omega) \to V^h$ and $\Pi_1^h : V \to V^h$ be the $L^2(\Omega)$- and $H^1(\Omega)$-orthogonal projection operators, defined by

$$\Pi_0^h u \in V^h, \quad (\Pi_0^h u, v^h)_{0,\Omega} = (u, v^h)_{0,\Omega} \quad \forall\, v^h \in V^h, \tag{9.35}$$
$$\Pi_1^h u \in V^h, \quad (\Pi_1^h u, v^h)_{1,\Omega} = (u, v^h)_{1,\Omega} \quad \forall\, v^h \in V^h, \tag{9.36}$$

respectively. The orthogonal projections $\Pi_0^h u$ and $\Pi_1^h u$ are uniquely defined for $u \in L^2(\Omega)$ and $u \in V$. We have the following estimates for the projection errors (cf. [103]).

THEOREM 9.9. *Assume that all the conditions, except that $k+1 > d/2$, of Corollary 9.6 hold. Suppose the family of triangulations $\{\Omega_e\}_{e=1}^E$ is quasi-uniform. Then if $u \in H^{l+1}(\Omega)$, $0 \leq l \leq k$, we have the estimates*

$$\|\Pi_0^h u - u\|_{0,\Omega} + h\, \|\Pi_0^h u - u\|_{1,\Omega} \leq c\, h^{l+1} |u|_{l+1,\Omega}, \tag{9.37}$$
$$\|\Pi_1^h u - u\|_{0,\Omega} + h\, \|\Pi_1^h u - u\|_{1,\Omega} \leq c\, h^{l+1} |u|_{l+1,\Omega}. \tag{9.38}$$

A family of partitions $\{\Omega_e\}_{e=1}^E$ is called *quasi-uniform* if the family is regular and there exists a constant $\tau > 0$ such that

$$\frac{\min_{1 \leq e \leq E} h_e}{\max_{1 \leq e \leq E} h_e} \geq \tau.$$

10

Approximation of Variational Problems

In this chapter we consider the approximation by the finite element method of variational equations and inequalities. In Chapter 6 we have reviewed some standard results for the well-posedness of variational equations and inequalities. The results can also be applied to the corresponding discrete problems over finite-dimensional spaces; in this way, we can then conclude the well-posedness of the discretized variational equations and inequalities. As we will see, Céa's lemma (Theorem 10.1) reduces the task of estimating finite element solution errors for an elliptic variational equation problem to that of estimating approximation errors. For approximations of variational inequalities, we will show results of the type of Céa's lemma. Then an application of the theory of finite element interpolation error estimates reviewed in Chapter 9 provides order error estimates for finite element solutions of variational equations and inequalities. Some references on finite element approximations of variational equations have been mentioned in Chapter 9. For detailed accounts on numerical solutions of variational inequalities, the reader may consult, among others, Glowinski, Lions, and Trémolières [45], Glowinski [44], Kikuchi and Oden [70], Hlaváček, Haslinger, Nečas, and Lovíšek [61], and, more recently, Haslinger, Hlaváček, and Nečas [57].

10.1 Approximation of Elliptic Variational Equations

Our discussion is given in the abstract framework found in the statement of the Lax–Milgram lemma (Theorem 5.9). Let V be a real Hilbert space with the norm $\|\cdot\|$. Let $a(\cdot,\cdot)$ be a bilinear form on V and ℓ a linear functional on V. The general form of a variational equation is then

$$u \in V, \quad a(u, v) = \langle \ell, v \rangle \quad \forall\, v \in V. \tag{10.1}$$

Let $V^h \subset V$ be a finite element space. Then the corresponding finite element problem is

$$u^h \in V^h, \quad a(u^h, v^h) = \langle \ell, v^h \rangle \quad \forall\, v^h \in V^h. \tag{10.2}$$

We have the following basic result about the discrete problem (10.2).

THEOREM 10.1. *Assume that the bilinear form $a(\cdot,\cdot)$ is V-elliptic,*

$$\exists\, \alpha > 0, \quad a(v, v) \ge \alpha \, \|v\|^2 \quad \forall\, v \in V, \tag{10.3}$$

and is bounded,

$$\exists\, M < \infty, \quad |a(u, v)| \le M \, \|u\| \, \|v\| \quad \forall\, u, v \in V. \tag{10.4}$$

Assume that the linear functional ℓ is continuous on V. Then both problems (10.1) and (10.2) have unique solutions u and u^h. Furthermore, for the error $u - u^h$, we have the inequality

$$\|u - u^h\| \le c \inf_{v^h \in V^h} \|u - v^h\|. \tag{10.5}$$

PROOF. The existence and uniqueness for the problems (10.1) and (10.2) follow from the Lax–Milgram lemma. We need only to prove the inequality (10.5). Since $V^h \subset V$, we have the following error relation from (10.1) and (10.2):

$$a(u - u^h, v^h) = 0 \quad \forall\, v^h \in V^h.$$

Using the assumption (10.3), the error relation, and the assumption (10.4), we have for any $v^h \in V^h$,

$$\begin{aligned} \alpha \, \|u - u^h\|^2 &\le a(u - u^h, u - u^h) \\ &= a(u - u^h, u - v^h) \\ &\le M \, \|u - u^h\| \, \|u - v^h\|. \end{aligned}$$

Thus,

$$\|u - u^h\| \le \frac{M}{\alpha} \, \|u - v^h\| \quad \forall\, v^h \in V^h.$$

Therefore, (10.5) holds. □

The inequality (10.5) is known as Céa's lemma in the literature. Such an inequality was first proved by Céa [20] for the case where the bilinear form is symmetric, and it was extended to the nonsymmetric case in Birkhoff, Schultz, and Varga [12]. The inequality (10.5) shows that to estimate the finite element solution error, it suffices to estimate the approximation error $\inf_{v^h \in V^h} \|u - v^h\|$.

For the rest of the section we combine Theorem 10.1 and the results on finite element interpolation theory presented in Chapter 9 to derive error estimates for finite element approximations of linear second-order elliptic problems. Let V be a subspace of $H^1(\Omega)$ (possibly $V = H^1(\Omega)$) and let $V^h \subset V$ be a finite element space.

THEOREM 10.2. *Consider the variational form of a linear second-order elliptic boundary value problem: Find $u \in V$ such that*

$$a(u, v) = \langle \ell, v \rangle \quad \forall\, v \in V, \tag{10.6}$$

where the bilinear form $a(\cdot, \cdot)$ is continuous and V-elliptic, and ℓ is a linear continuous form on V. Let $V^h \subset V$ be a finite element space consisting of piecewise polynomials of degree less than or equal to k, and suppose that all the assumptions of Corollary 9.6 hold. Then for the finite element solution u^h of (10.6), we have the following error estimate:

$$\|u - u^h\|_{1,\Omega} \le c\, h^k |u|_{k+1,\Omega}.$$

PROOF. From (10.5) with $v^h = \Pi^h u$ and (9.34) with $m = 1$ we obtain

$$\|u - u^h\|_{1,\Omega} \le c\, \|u - \Pi^h u\|_{1,\Omega} \le c\, h^k |u|_{k+1,\Omega}.$$ □

It may happen in practice that the solution u is not smooth enough to belong to $H^{k+1}(\Omega)$. For example, if we know from the theory of elliptic boundary value problems that u is in $H^2(\Omega)$ but not in $H^3(\Omega)$, then the use of piecewise quadratic finite element functions for a problem in $\mathbb{R}^2$ will mean that $k = 2$, or $k + 1 = 3$, and the seminorm $|v|_{3,\Omega}$ in (9.34) does not necessarily make sense. We overcome this problem by going back to Section 9.3 and noting that the entire theory developed there still holds if we replace $k + 1$ by r, and hence also k by $r - 1$, where $r \le k + 1$ is any positive integer. Specifically, we do this in Theorems 9.3 and 9.5 and in Corollary 9.6. Of course, r must be such that $H^r(\hat{\Omega}) \hookrightarrow C(\hat{\Omega})$ (that is, $r > d/2$ and $r \ge m$). The estimate (9.30) then reads, for $v \in H^r(\Omega_e)$,

$$\|v - \Pi_e v\|_{m,\Omega_e} \le c\, h_e^{\mu - m} |v|_{\mu,\Omega_e}, \quad \mu = \min\{k + 1, r\}.$$

Correspondingly, the global interpolation error estimate (9.34) is changed to

$$\|v - \Pi^h v\|_{m,\Omega} \le c\, h^{\mu - m} |v|_{\mu,\Omega}.$$

In the context of the finite element approximation of a linear elliptic boundary value problem of second order, if the solution u is in $H^r(\Omega)$ for some $r > 1$, then

$$\|u - u^h\|_{1,\Omega} \le c\, h^{\mu-1} |u|_{\mu,\Omega}, \quad \mu = \min\{k+1, r\}. \tag{10.7}$$

We mention that an estimate of the form (10.7) is still valid even if we no longer require r (and hence μ) to be an integer. This extension is made possible by employing the theory of interpolation of Banach spaces and operators; for a comprehensive presentation of such an interpolation theory, one may consult Bergh and Löfström [9].

We take one more step from the error estimate (10.7). As it stands, the error bound involves the *unknown* quantity $|u|_{\mu,\Omega}$ on the right-hand side. This dependence on u can be removed if regularity estimates for the solution are available. Let $l \ge 0$. If the boundary of the domain Ω and the coefficients of the differential operator are smooth, and the boundary condition does not change its type (from Dirichlet condition to Neumann condition, or vice versa), then the solution u of a second-order elliptic boundary value problem lies in $H^{l+2}(\Omega)$, provided that the right-hand side of the differential equation satisfies $f \in H^l(\Omega)$ and the boundary condition function satisfies $g \in H^{l-1/2}(\partial\Omega)$ for the Dirichlet type boundary condition or $g \in H^{l-3/2}(\partial\Omega)$ for the Dirichlet type boundary condition. Furthermore, for some constant $c > 0$,

$$\|u\|_{l+2,\Omega} \le c\,(\|f\|_{l,\Omega} + \|g\|_{\bar{l},\partial\Omega}), \tag{10.8}$$

where, $\bar{l} = l - \frac{1}{2}$ for a Dirichlet boundary value problem and $\bar{l} = l - \frac{3}{2}$ for a Neumann boundary value problem. As an example, we consider the boundary value problem with the homogeneous Dirichlet boundary condition $u = 0$. The finite element theory developed here is applicable only to polygonal domains (in $\mathbb{R}^2$), but it is known that the estimate (10.8) holds for a range of the values l, depending on the largest internal angle of the polygon; for detail, see Grisvard [46]. We may set $r = l + 2$, and with $\mu = \min\{k+1, r\}$, since

$$|u|_\mu \le \|u\|_{l+2} \le c\,\|f\|_l,$$

the dependence on $|u|_{\mu,\Omega}$ in (10.7) may be removed. One sample result is the following.

COROLLARY 10.3. *Let the conditions for Theorem 10.2 hold. Assume that the coefficients of the partial differential equation are smooth. Assume that Ω is a polygonal domain for which the regularity estimate* (10.8) *holds. Let the right-hand side of the equation $f \in H^l(\Omega)$ be given and the boundary condition be $u = 0$. Then for some constant c,*

$$\|u - u^h\|_{1,\Omega} \le c\, h^\beta \|f\|_l, \tag{10.9}$$

where $\beta = \min(k, l+1)$.

According to Theorem 10.2 and Corollary 10.3, since the order of convergence β is governed by the smaller of k and $l+1$, when $l \le k-1$, the convergence order is governed by the smoothness of f. For example, if f is only in $L^2(\Omega) = H^0(\Omega)$, then it suffices to use linear finite elements.

EXAMPLE 10.4. Consider the problem

$$\begin{aligned} -\Delta u &= f \quad \text{in } \Omega, \\ u &= 0 \quad \text{on } \Gamma, \end{aligned}$$

where $\Omega \subset \mathbb{R}^d$ is a polygonal domain. The corresponding variational formulation is

$$u \in H_0^1(\Omega), \quad \int_\Omega \nabla u \cdot \nabla u \, dx = \int_\Omega f \, v \, dx \quad \forall \, v \in H_0^1(\Omega),$$

and this problem has a unique solution. The discrete problem

$$u^h \in V^h, \quad \int_\Omega \nabla u^h \cdot \nabla v^h \, dx = \int_\Omega f \, v^h \, dx \quad \forall \, v^h \in V^h$$

also has a unique solution. Here $V^h \subset H_0^1(\Omega)$ consists of piecewise polynomials of degree less than or equal to k, corresponding to a regular triangulation of the domain Ω. If $f \in H^l(\Omega)$, $l \ge 0$, and the regularity estimate (10.8) holds (with $g = 0$), then the error is estimated by

$$\|u - u^h\|_{1,\Omega} \le c \, h^\beta \|f\|_{l,\Omega},$$

where $\beta = \min(k, l+1)$. Thus if linear elements ($k = 1$) are used, the error is of order 1, since $l+1$ will not be less than 1. □

10.2 Approximation of EVI of the First Kind

We first recall the general framework for elliptic variational inequalities of the first kind. Let V be a real Hilbert space, and $K \subset V$ nonempty, convex, and closed. Assume that $a(\cdot, \cdot)$ is a V-elliptic and bounded bilinear form on V and ℓ a continuous linear functional on V. Then according to Theorem 6.4, the elliptic variational inequality of the first kind

$$u \in K, \quad a(u, v - u) \ge \langle \ell, v - u \rangle \quad \forall \, v \in K \tag{10.10}$$

has a unique solution. Let $V^h \subset V$ be a finite element space, and let $K^h \subset V^h$ be nonempty, convex, and closed. Then the finite element approximation of the problem (10.10) is

$$u^h \in K^h, \quad a(u^h, v^h - u^h) \ge \langle \ell, v^h - u^h \rangle \quad \forall \, v^h \in K^h. \tag{10.11}$$

Another application of Theorem 6.4 shows that the discrete problem (10.11) has a unique solution under the stated assumptions on the given data.

A general convergence result of the finite element method can be found in [44]. Here we are interested in order error estimation for the finite element solution u^h. We follow Falk [39] and first give an abstract error estimate.

THEOREM 10.5. *There is a constant $c > 0$ independent of h and u such that*

$$\begin{aligned} \|u - u^h\| & \\ \leq c \Big\{ & \inf_{v^h \in K^h} \Big[\|u - v^h\| + |a(u, v^h - u) - \langle \ell, v^h - u \rangle|^{1/2} \Big] \\ & + \inf_{v \in K} |a(u, v - u^h) - \langle \ell, v - u^h \rangle|^{1/2} \Big\}. \end{aligned} \tag{10.12}$$

PROOF. From (10.10) and (10.11), we find that

$$\begin{aligned} a(u, u) &\leq a(u, v) - \langle \ell, v - u \rangle \quad \forall\, v \in K, \\ a(u^h, u^h) &\leq a(u^h, v^h) - \langle \ell, v^h - u^h \rangle \quad \forall\, v^h \in K^h. \end{aligned}$$

Using these relations, together with the V-ellipticity and boundedness of the bilinear form $a(\cdot, \cdot)$, we have for any $v \in K$ and $v^h \in K^h$,

$$\begin{aligned} \alpha \, \|u - u^h\|^2 &\leq a(u - u^h, u - u^h) \\ &= a(u, u) + a(u^h, u^h) - a(u, u^h) - a(u^h, u) \\ &\leq a(u, v - u^h) - \langle \ell, v - u^h \rangle + a(u, v^h - u) - \langle \ell, v^h - u \rangle \\ &\quad + a(u^h - u, v^h - u) \\ &\leq a(u, v - u^h) - \langle \ell, v - u^h \rangle + a(u, v^h - u) - \langle \ell, v^h - u \rangle \\ &\quad + \tfrac{1}{2} \alpha \, \|u - u^h\|^2 + c \, \|v^h - u\|^2. \end{aligned}$$

Thus the inequality (10.12) holds. □

Theorem 10.5 is a generalization of Céa's lemma to the finite element approximation of elliptic variational inequalities of the first kind. Indeed, the inequality (10.12) reduces to Céa's inequality in the case of finite element approximation of a variational equation problem, because in this case, $K = V$, $K^h = V^h$, and $V^h \subset V$, so

$$a(u, v^h - u) - \langle \ell, v^h - u \rangle = 0$$

and

$$\inf_{v \in K} |a(u, v - u^h) - \langle \ell, v - u^h \rangle| = 0.$$

When $K^h \subset K$, we have the so-called internal approximation of the elliptic variational inequality of the first kind. Since now $u^h \in K$, the second term on the right-hand side of (10.12) vanishes, and the error inequality (10.12) reduces to

$$\|u - u^h\| \le c \inf_{v^h \in K^h} \left[\|u - v^h\| + |a(u, v^h - u) - \langle \ell, v^h - u \rangle|^{1/2} \right].$$

Based on the inequality (10.12), it is then possible to derive order error estimates for the approximation of some elliptic variational inequalities of the first kind. For instance, for the obstacle problem mentioned in Section 6.2, it is shown in Falk [39] that if $u, \psi \in H^2(\Omega)$ and if linear elements on a regular mesh are used, then one has the optimal order error estimate

$$\|u - u^h\|_{H^1(\Omega)} \le c\,h$$

for some constant $c > 0$ independent of h.

10.3 Approximation of EVI of the Second Kind

As in Section 6.2, let V be a real Hilbert space, $a(\cdot, \cdot)$ a V-elliptic, bounded bilinear form, ℓ a linear continuous functional on V. Also, let $j(\cdot)$ be a proper, convex, and l.s.c. functional on V. Under these assumptions, by Theorem 6.6 there exists a unique solution of the elliptic variational inequality of the second kind,

$$u \in V, \quad a(u, v - u) + j(v) - j(u) \ge \langle \ell, v - u \rangle \quad \forall\, v \in V. \tag{10.13}$$

Let $V^h \subset V$ be a finite element space. Then the finite element approximation of the problem (10.13) is to find $u^h \in V^h$ such that

$$a(u^h, v^h - u^h) + j(v^h) - j(u^h) \ge \langle \ell, v^h - u^h \rangle \quad \forall\, v^h \in V^h. \tag{10.14}$$

Assuming additionally that $j(\cdot)$ is proper also on V^h, as is always the case in applications, we can use Theorem 6.6 to conclude that the discrete problem (10.14) has a unique solution u^h and $j(u^h) \subset \mathbb{R}$. We will now derive an abstract error estimate for $u - u^h$.

THEOREM 10.6. *There is a constant $c > 0$ independent of h and u such that*

$$\|u - u^h\| \le c \inf_{v^h \in V^h} \left\{ \|u - v^h\| + |a(u, v^h - u) + j(v^h) - j(u) - \langle \ell, v^h - u \rangle|^{1/2} \right\}. \tag{10.15}$$

PROOF. We let $v = u^h$ in (10.13) and add the resulting inequality to the inequality (10.10) to obtain an error relation

$$a(u, u^h - u) + a(u^h, v^h - u^h) + j(v^h) - j(u) \ge \langle \ell, v^h - u \rangle \quad \forall\, v^h \in V^h.$$

Using this error relation, together with the V-ellipticity and boundedness of the bilinear form, we have for any $v^h \in V^h$,

$$\begin{aligned}
&\alpha \|u - u^h\|^2 \\
&\quad \le a(u - u^h, u - u^h) \\
&\quad = -a(u, u^h - u) - a(u^h, v^h - u^h) + a(u^h - u, v^h - u) \\
&\qquad + a(u, v^h - u) \\
&\quad \le a(u - u^h, u - v^h) + a(u, v^h - u) + j(v^h) - j(u) - \langle \ell, v^h - u \rangle \\
&\quad \le M \|u - u^h\| \, \|u - v^h\| + a(u, v^h - u) + j(v^h) - j(u) - \langle \ell, v^h - u \rangle \\
&\quad \le \tfrac{1}{2} \alpha \|u - u^h\|^2 + c \|u - v^h\|^2 \\
&\qquad + a(u, v^h - u) + j(v^h) - j(u) - \langle \ell, v^h - u \rangle,
\end{aligned}$$

from which it is easy to see that (10.15) holds. □

We observe that Theorem 10.6 is a generalization of Céa's lemma to the finite element approximation of elliptic variational inequalities of the second kind. The inequality (10.15) is the basis for order error estimates of finite element solutions of various application problems.

Let us apply the inequality (10.15) to derive an error estimate for some finite element solution of a model problem. Let $\Omega \subset \mathbb{R}^2$ be an open bounded set, with a Lipschitz domain $\partial\Omega$. We take

$$\begin{aligned}
V &= H^1(\Omega), \\
a(u, v) &= \int_\Omega (\nabla u \, \nabla v + u \, v) \, dx, \\
\langle \ell, v \rangle &= \int_\Omega f \, v \, dx, \\
j(v) &= g \int_{\partial\Omega} |v| \, ds.
\end{aligned}$$

Here $f \in L^2(\Omega)$ and $g > 0$ are given. This problem is a simplified version of the friction problem in elasticity. We choose this model problem for its simplicity, while at the same time it contains the main feature of an elliptic variational inequality of the second kind. Applying Theorem 6.6, we see that the corresponding variational inequality problem

$$u \in V, \quad a(u, v - u) + j(v) - j(u) \ge \langle \ell, v - u \rangle \quad \forall \, v \in V \tag{10.16}$$

has a unique solution. Given a finite element space V^h, let u^h denote the corresponding finite element solution defined in (10.14). To simplify the exposition, we will assume below that Ω is a polygonal domain and write $\partial\Omega = \cup_{i=1}^{i_0} \Gamma_i$, where each Γ_i is a line segment. For an error estimation, we have the following result.

THEOREM 10.7. *Assume, for the model problem, that* $u \in H^2(\Omega)$, *and for* *each i, $u|_{\Gamma_i} \in H^2(\Gamma_i)$. Let V^h be a piecewise linear finite element space constructed from a regular partition of the domain Ω. Let $u^h \in V^h$ be the finite element solution defined by (10.14). Then we have the optimal order error estimate*

$$\|u - u^h\|_{H^1(\Omega)} \le c(u)\, h. \tag{10.17}$$

PROOF. We apply the result of Theorem 10.6:

$$\begin{aligned}
& a(u, v^h - u) + j(v^h) - j(u) - \langle \ell, v^h - u\rangle \\
&\quad = \int_{\partial\Omega} \left[\frac{\partial u}{\partial n}\,(v^h - u) + g\,\left(|v^h| - |u|\right)\right] ds \\
&\qquad + \int_\Omega (-\Delta u + u - f)\,(v^h - u)\, dx \\
&\quad \le \left(\left\|\frac{\partial u}{\partial n}\right\|_{L^2(\partial\Omega)} + g\sqrt{\operatorname{meas}(\partial\Omega)}\right) \|v^h - u\|_{L^2(\partial\Omega)} \\
&\qquad + \|-\Delta u + u - f\|_{L^2(\Omega)} \|v^h - u\|_{L^2(\Omega)},
\end{aligned}$$

where, $\partial/\partial n$ is the normal derivative operator on $\partial\Omega$. Using (10.15), we get

$$\begin{aligned}
& \|u - u^h\|_{H^1(\Omega)} \\
&\quad \le c(u) \inf_{v^h \in V^h} \left\{\|u - v^h\|_{H^1(\Omega)} + \|u - v^h\|^{1/2}_{L^2(\partial\Omega)} + \|u - v^h\|^{1/2}_{L^2(\Omega)}\right\} \\
&\quad \le c(u) \left\{\|u - \Pi^h u\|_{H^1(\Omega)} + \|u - \Pi^h u\|^{1/2}_{L^2(\partial\Omega)} + \|u - \Pi^h u\|^{1/2}_{L^2(\Omega)}\right\},
\end{aligned}$$

where $\Pi^h u \in V^h$ is the piecewise linear interpolant of u. From the regularity assumptions on u, we have

$$\begin{aligned}
\|u - \Pi^h u\|_{H^1(\Omega)} &\le c\, h\, |u|_{H^2(\Omega)}, \\
\|u - \Pi^h u\|_{L^2(\partial\Omega)} &\le c\, h^2\, |u|_{H^2(\partial\Omega)}, \\
\|u - \Pi^h u\|_{L^2(\Omega)} &\le c\, h^2\, |u|_{H^2(\Omega)},
\end{aligned}$$

using Theorem 9.8. Therefore, the error estimate (10.17) holds. □

Let us return to the general case. A major issue in solving the discrete system (10.14) is the treatment of the nondifferentiable term. In practice, several approaches can be used, e.g., regularization technique, method of Lagrangian multipliers. Here we study the approach by approximating $j(v^h)$ with $j_h(v^h)$, obtained through numerical integrations. Then the numerical method is to find $u^h \in V^h$ such that

$$a(u^h, v^h - u^h) + j_h(v^h) - j_h(u^h) \ge \langle \ell, v^h - u^h\rangle \quad \forall\, v^h \in V^h. \tag{10.18}$$

The following convergence theorem is proved in [44, 45].

THEOREM 10.8. *Assume that* $\{V^h\}_h \subset V$ *is a family of finite-dimensional subspaces such that for a dense subset* U *of* V *one can define mappings* $r_h : U \to V^h$ *with* $\lim_{h\to 0} r_h v = v$ *in* V, *for any* $v \in U$. *Assume that* j_h *is convex, l.s.c., and uniformly proper in* h, *and if* $v^h \rightharpoonup v$ *in* V, *then* $\liminf_{h\to 0} j_h(v^h) \geq j(v)$. *Finally, assume* $\lim_{h\to 0} j_h(r_h v) = j(v)$ *for any* $v \in U$. *Then for the solution of* (10.18), *we have the convergence*

$$\lim_{h\to 0} \|u - u^h\| = 0.$$

In the above theorem, the functional family $\{j_h\}_h$ is said to be uniformly proper in h if there exist $\ell_0 \in V'$ and $c_0 \in \mathbb{R}$ such that

$$j_h(v^h) \geq \langle \ell_0, v^h \rangle + c_0 \quad \forall\, v^h \in V^h,\ \forall\, h.$$

This theorem gives some rather general assumptions under which one has the convergence of the finite element solutions. However, it does not provide information on the convergence order of the approximations. In the following, we prove an inequality of the form (10.15).

THEOREM 10.9. *Assume that*

$$j(v^h) \leq j_h(v^h) \quad \forall\, v^h \in V^h. \tag{10.19}$$

Let u^h *be defined by* (10.18). *Then*

$$\|u - u^h\| \leq c \inf_{v^h \in V^h} \left\{ \|u - v^h\| + |a(u, v^h - u) + j_h(v^h) - j(u) - \langle \ell, v^h - u \rangle|^{1/2} \right\}. \tag{10.20}$$

PROOF. Choosing $v = u^h$ in (10.13) and adding the resulting inequality to (10.18), we obtain, for any $v^h \in V^h$,

$$a(u, u^h - u) + a(u^h, v^h - u^h) + j(u^h) - j_h(u^h) + j_h(v^h) - j(u) \geq \langle \ell, v^h - u \rangle.$$

Using the assumption (10.19) for $v^h = u^h$, we then have

$$a(u, u^h - u) + a(u^h, v^h - u^h) + j_h(v^h) - j(u) \geq \langle \ell, v^h - u \rangle \quad \forall\, v^h \in V^h.$$

The rest of the argument is similar to that in the proof of Theorem 10.6 and is hence omitted. □

Let us now comment on the assumption (10.19). In some applications, the functional $j(\cdot)$ is of the form $j(v) = I(g\,|v|)$ in which I is an integration operator and $g \geq 0$ a given nonnegative function. One method for constructing practically useful approximate functionals j_h is through numerical quadrature, where $j_h(v^h) = I_h(g\,|v^h|)$, I_h denoting a numerical integration operator. Let $\{\phi_i\}_i$ be the set of functions chosen from a basis of the space V^h, which defines the functions v^h over the integration region. Assume that the basis functions $\{\phi_i\}_i$ are nonnegative. Writing

$$v^h = \sum_i v_i \phi_i$$

on the integration region, we define

$$j_h(v^h) = \sum_i |v_i| \, I(g\,\phi_i). \tag{10.21}$$

Obviously, j_h constructed in this way enjoys the property (10.19). We will see next in the analysis for solving the model problem that a certain polynomial invariance property is preserved through a construction of the form (10.21). Such a property is needed in proving optimal order error estimates.

Let us again consider the model problem. Assume that we use linear elements to construct the finite element space V^h. Denote by $\{P_i\}$ the nodes of the triangulation that lie on the boundary, numbered consecutively. Let $\{\phi_i\}$ be the canonical basis functions of the space V^h, corresponding to the nodes $\{P_i\}$. Then $\phi_i \geq 0$. Thus we define

$$j_h(v^h) = g \sum_i |\overline{P_i P_{i+1}}| \, \frac{1}{2} \left(|v^h(P_i)| + |v^h(P_{i+1})| \right). \tag{10.22}$$

Assume $u \in H^2(\Omega)$. By Theorem 10.9, the finite element solution error satisfies

$$\begin{aligned} \|u - u^h\|_{H^1(\Omega)} \leq c \Big\{ & \|u - \Pi^h u\|_{H^1(\Omega)} \\ & + |a(u, \Pi^h u - u) + j_h(\Pi^h u) - j(u) - \langle \ell, \Pi^h u - u \rangle|^{1/2} \Big\}, \end{aligned} \tag{10.23}$$

where $\Pi^h u \in V^h$ is the piecewise linear interpolant of the solution u. Let us first estimate the difference $j_h(\Pi^h u) - j(u)$. We have

$$\begin{aligned} & j_h(\Pi^h u) - j(u) \\ & \quad = g \sum_i \Big\{ \frac{1}{2} |\overline{P_i P_{i+1}}| \left(|u(P_i)| + |u(P_{i+1})| \right) - \int_{\overline{P_i P_{i+1}}} |u| \, ds \Big\}. \end{aligned} \tag{10.24}$$

Now, if $u|_{\overline{P_iP_{i+1}}}$ keeps the same sign, then

$$\begin{aligned}
&\left|\frac{1}{2}\,|\overline{P_iP_{i+1}}|\,\left(|u(P_i)|+|u(P_{i+1})|\right)-\int_{\overline{P_iP_{i+1}}}|u|\,ds\right|\\
&\quad=\left|\frac{1}{2}\,|\overline{P_iP_{i+1}}|\,\left(u(P_i)+u(P_{i+1})\right)-\int_{\overline{P_iP_{i+1}}}u\,ds\right|\\
&\quad=\left|\int_{\overline{P_iP_{i+1}}}(u-\Pi^h u)\,ds\right|\\
&\quad\le\int_{\overline{P_iP_{i+1}}}|u-\Pi^h u|\,ds.
\end{aligned}$$

Assume that $u|_{\overline{P_iP_{i+1}}}$ changes its sign. It is easy to see that

$$\sup_{\overline{P_iP_{i+1}}}|u|\le h\,\|u\|_{W^{1,\infty}(P_iP_{i+1})}$$

if $u|_{\overline{P_iP_{i+1}}}\in W^{1,\infty}(P_iP_{i+1})$, which is implied by $u|_{\Gamma_i}\in H^2(\Gamma_i)$, $i=1,\dots,i_0$, an assumption made in Theorem 10.7. Thus,

$$\left|\frac{1}{2}\,|\overline{P_iP_{i+1}}|\,\left(|u(P_i)|+|u(P_{i+1})|\right)-\int_{\overline{P_iP_{i+1}}}|u|\,ds\right|\le c\,h^2\|u\|_{W^{1,\infty}(P_iP_{i+1})}.$$

Therefore, if the exact solution u changes its sign only finitely many times on $\partial\Omega$, then from (10.24) we find that

$$|j_h(\Pi^h u)-j(u)|\le c\,h^2\sum_{i=1}^{i_0}\|u\|_{W^{1,\infty}(\Gamma_i)}+c\,\|u-\Pi^h u\|_{L^1(\partial\Omega)}.$$

Using (10.23), we then get

$$\begin{aligned}
\|u-u^h\|_{H^1(\Omega)}\le c\Big\{&\|u-\Pi^h u\|_{H^1(\Omega)}+\left\|\frac{\partial u}{\partial n}\right\|_{L^2(\partial\Omega)}\|u-\Pi^h u\|_{L^2(\partial\Omega)}\\
&+h\,[\sum_{i=1}^{i_0}\|u\|_{W^{1,\infty}(\Gamma_i)}]^{1/2}+\|u-\Pi^h u\|_{L^1(\partial\Omega)}^{1/2}\\
&+\|-\Delta u+u-f\|_{L^2(\Omega)}\|u-\Pi^h u\|_{L^2(\Omega)}\Big\}.
\end{aligned}$$

In conclusion, if $u\in H^2(\Omega)$, $u|_{\Gamma_i}\in W^{1,\infty}(\Gamma_i)$ for $i=1,\dots,i_0$, and if $u|_{\partial\Omega}$ changes its sign only finitely many times, then we have the error estimate

$$\|u-u^h\|_{H^1(\Omega)}\le c(u)\,h;$$

that is, the approximation of j by j_h does not cause a degradation in the convergence order of the finite element method.

REMARK. If $f \in L^2(\Omega)$, then $u \in H^{1+\alpha}(\Omega)$ for some $\alpha \in [\frac{1}{2}, 1]$. It is shown in [45] that for the finite element solution defined by (10.18) and (10.22), the estimate

$$\|u - u^h\|_{H^1(\Omega)} \le c(\|f\|_{L^2(\Omega)}, g, \varepsilon)\, h^{\min\{1/2, (\alpha-\varepsilon)/(1-\varepsilon)\}}$$

holds for arbitrarily small $\varepsilon > 0$. □

If quadratic elements are used, one can construct basis functions by using nodal shape functions and side modes (cf. [121]). Then the basis functions are nonnegative, and an error analysis similar to the above one can be done.

10.4 Approximation of Parabolic Variational Inequalities

The work on numerical analysis of parabolic variational inequalities is not as abundant as that for solving elliptic variational inequalities. In [45], one can find a detailed convergence analysis for some standard fully discrete approximations (finite difference discretization in time and finite element discretization in space) of the parabolic variational inequality (6.45). It is a delicate matter to derive order error estimates for numerical solutions if only proved solution regularity is used in derivation. To describe such an example, let us turn to the problem setting that appears in Theorem 6.9 and to order error estimates for some numerical approximations of the problem.

First, we give a result for fully discrete approximations, due to Johnson [68]. We discretize the time interval $I = [0, T]$ into N equal parts and denote the step-size by $k = T/N$ and the nodal points by $t_n = nk$, $n = 0, 1, \ldots, N$. Let $h \in (0, 1]$ denote the mesh parameter in a finite element triangulation of the domain Ω. Let $\{V^h\}$ be a family of finite-dimensional subspaces of V, and assume that $K^h = V^h \cap K$ is nonempty. Then K^h is a nonempty, closed, convex subset of V^h. A fully discrete approximation of the problem in Theorem 6.9 based on a backward difference approximation of the time derivative is the following: Find $u^{hk} = \{u_n^{hk}\}_{n=0}^N \subset V^h$ such that u_0^{hk} is an approximation of the initial value u_0 in the sense that

$$\|u_0^{hk} - u_0\|_0 \le c\, h,$$

and such that for $n = 1, 2, \ldots, N$,

$$(\delta u_n^{hk}, v^h - u_n^{hk}) + a(u_n^{hk}, v^h - u_n^{hk}) \ge \langle f_n, v^h - u_n^{hk}\rangle \quad \forall\, v^h \in K^h. \tag{10.25}$$

Recall that the solution of the continuous problem has the regularity given in Theorem 6.9. Under the assumption that the solution u does not change too frequently from zero to positive values or vice versa (for the exact meaning of this, cf. [68]), it is proved in [68] that there exists a constant c independent of k and h such that

$$\max_n \|u_n - u_n^{hk}\|_0 + \left(\sum_{n=1}^{N} \|u_n - u_n^{hk}\|_V^2 k \right)^{1/2} \leq c\,[(\log k^{-1})^{1/4} k^{3/4} + h]. \tag{10.26}$$

An extension of this result to the case in which a generalized midpoint approximation is considered is given by Vuik [126]. Instead of (10.25), one solves the following problem for u_n^{hk}:

$$(\delta u_n^{hk}, v^h - u_n^{hk}) + a(u_{n-1+\theta}^{hk}, v^h - u_n^{hk}) \geq \langle f_n, v^h - u_n^{hk} \rangle \quad \forall\, v^h \in K^h. \tag{10.27}$$

Here $v_{n-1+\theta} = v_{n-1} + \theta(v_n - v_{n-1})$ with $\theta \in (0, 1]$. The error estimation is rather technical and will not be repeated in full detail; instead, we point out that the error estimate essentially differs from (10.26) in that the constant c is replaced by $c(g(\mu))^{-1/2}$, in which $g(\mu) = \min\{4\mu - 1 + 2\theta, 2\mu - \frac{1}{2} + \frac{3}{2}\theta\}$ and μ is a constant that lies in $((1 - 2\theta)/4, (1 - \theta)/2] \cap [0, c\,h^2/k]$. Under stronger regularity conditions on the data, the estimate can be improved to one of $O(k + h)$.

One distinction between the convergence results for parabolic variational inequalities and equations is evident if one compares the results for parabolic equations obtained by Douglas and Dupont [31] with those of Vuik [126]; this concerns the order of convergence associated with the Crank–Nicolson scheme, which corresponds to (10.27) with $\theta = \frac{1}{2}$. It is well known that the Crank–Nicolson scheme leads to convergence of $O(k^2)$ when parabolic equations are approximated. In the case of variational inequalities, however, the lack of regularity of the solution precludes a similar result, so that the estimate for the case $\theta = \frac{1}{2}$ cannot be improved beyond one of the type (10.26).

11

Approximations of the Abstract Problem

As a prelude to the error analysis of various numerical schemes for solving the primal variational problem, we will first give a convergence analysis and derive error estimates for numerical solutions of the abstract problem, introduced in Chapter 7, which includes the primal variational problem as a special case. In the next chapter, we will apply the results presented here to perform an error analysis for various numerical approximation schemes for solving the primal problem. For convenience, let us recall the abstract problem.

PROBLEM ABS. *Find* $w : [0,T] \to H$, $w(0) = 0$, *such that for almost all* $t \in (0,T)$, $\dot{w}(t) \in K$ *and*

$$a(w(t), z - \dot{w}(t)) + j(z) - j(\dot{w}(t)) \geq \langle \ell(t), z - \dot{w}(t) \rangle \quad \forall\, z \in K. \tag{11.1}$$

Under the assumptions that

- H is a Hilbert space
- $K \subset H$ is a nonempty, closed, convex cone
- $a : H \times H \to \mathbb{R}$ is a bilinear form on H, symmetric, bounded and H-elliptic
- $\ell \in H^1(0,T;H')$, $\ell(0) = 0$
- $j : K \to \mathbb{R}$ is nonnegative, convex, positively homogeneous, and Lipschitz continuous

we have the existence of a unique solution $w \in H^1(0,T;H)$ of the problem ABS. For convenience, later on we will refer to these assumptions as the *standard assumptions* for the problem ABS. In this chapter we will always assume that these standard assumptions hold. We also recall that there exists $w^* \in H^1(0,T;H')$ such that

$$a(w(t), z) + \langle w^*(t), z\rangle = \langle \ell(t), z\rangle \quad \forall\, z \in H. \tag{11.2}$$

In the first three sections of the chapter we will derive error estimates for spatially discrete, time-discrete, and fully discrete approximations of the abstract problem. Order error estimates are obtained under the assumption that the solution of the problem is sufficiently regular. Notice that a regularity theory for the elastoplasticity problem is largely not available at the moment, and it is likely that the solution of the abstract variational problem does not have the high regularity required for the various order error estimates. Thus it is of interest to see whether we still have convergence of the numerical solutions under the basic solution regularity proved in Chapter 7. Such a convergence analysis is carried out in Section 11.4.

The following elementary result result will be used repeatedly:

$$a, b, x \geq 0 \text{ and } x^2 \leq a\,x + b \Longrightarrow x^2 \leq a^2 + 2\,b. \tag{11.3}$$

11.1 Spatially Discrete Approximations

We consider discrete internal approximations of the abstract problem ABS, in which H is replaced by a family of finite-dimensional subspaces $\{H^h\}$, and correspondingly, the set K is replaced by a family of finite-dimensional subsets $\{K^h\}$. The approximations are referred to as internal because of the properties $H^h \subset H$ and $K^h \subset K$. The subspaces $\{H^h\}$ are intended to be finite element spaces, though much of the analysis applies to more general situations.

Let $\{H^h\}$ be a family of finite-dimensional subspaces of H, with the property that

$$\lim_{h\to 0} \inf_{z^h \in H^h} \|z - z^h\|_H = 0 \quad \forall\, z \in H. \tag{11.4}$$

Set $K^h = H^h \cap K$, which is nonempty, since $0 \in K^h$. Then a spatially discrete internal approximation of Problem ABS is

PROBLEM ABS^h. *Find* $w^h : [0,T] \to H^h$, $w^h(0) = 0$, *such that for almost all* $t \in (0,T)$, $\dot{w}^h(t) \in K^h$ *and*

$$a(w^h(t), z^h - \dot{w}^h(t)) + j(z^h) - j(\dot{w}^h(t)) \geq \langle \ell(t), z^h - \dot{w}^h(t)\rangle \quad \forall\, z^h \in K^h. \tag{11.5}$$

We note that for any given h, K^h is a nonempty, closed, convex cone in H^h. Thus, the existence of a unique solution w^h to Problem ABS^h follows from Theorem 7.3 with H and K replaced by H^h and K^h, respectively. We also note from Theorem 7.3 that $w^h \in H^1(0,T;H)$. This regularity result implies that $w^h \in C([0,T];H)$; in particular, the value $w^h(0)$ is well-defined. From Theorem 7.4 we have the stability estimate

$$\|w^{(1)h} - w^{(2)h}\|_{L^\infty(0,T;H)} \le c\,\|\dot{\ell}^{(1)} - \dot{\ell}^{(2)}\|_{L^1(0,T;H')},$$

for semidiscrete solutions $w^{(1)h}$ and $w^{(2)h}$ corresponding to two right-hand sides $\ell^{(1)}, \ell^{(2)} \in H^1(0,T;H')$ with $\ell^{(1)}(0) = \ell^{(2)}(0) = 0$.

The main purpose of the section is to give an estimate for the semidiscrete approximation error $w - w^h$. For convenience we will use the notation

$$\|w\|_a^2 = a(w,w).$$

Note that $\|\cdot\|_a$ is a norm on H, equivalent to $\|\cdot\|_H$.

We begin by setting $z = \dot{w}^h(t) \in K$ in (11.1) to obtain

$$a(w(t), \dot{w}^h(t) - \dot{w}(t)) + j(\dot{w}^h(t)) - j(\dot{w}(t)) \ge \langle \ell(t), \dot{w}^h(t) - \dot{w}(t)\rangle. \quad (11.6)$$

We now add (11.6) to (11.5) and obtain

$$\begin{aligned} &a(w(t), \dot{w}^h(t) - \dot{w}(t)) + a(w^h(t), z^h - \dot{w}^h(t)) + j(z^h) - j(\dot{w}(t)) \\ &\quad \ge \langle \ell(t), z^h - \dot{w}(t)\rangle. \end{aligned} \quad (11.7)$$

Using (11.7), Theorem 7.3, (7.34), and (7.36), we have, for any $z^h \in K^h$,

$$\begin{aligned} &\frac{1}{2}\frac{d}{dt}\|w(t) - w^h(t)\|_a^2 \\ &\quad = a(w(t) - w^h(t), \dot{w}(t) - \dot{w}^h(t)) \\ &\quad = a(w(t) - w^h(t), \dot{w}(t) - z^h) + a(w(t) - w^h(t), z^h - \dot{w}^h(t)) \\ &\quad \le a(w(t) - w^h(t), \dot{w}(t) - z^h) + a(w(t), z^h - \dot{w}^h(t)) \\ &\qquad + a(w(t), \dot{w}^h(t) - \dot{w}(t)) + j(z^h) - j(\dot{w}(t)) - \langle \ell(t), z^h - \dot{w}(t)\rangle \\ &\quad = a(w(t) - w^h(t), \dot{w}(t) - z^h) + j(z^h) - j(\dot{w}(t)) \\ &\qquad - \langle w^*(t), z^h - \dot{w}(t)\rangle \\ &\quad \le a(w(t) - w^h(t), \dot{w}(t) - z^h) + j(z^h) - j(\dot{w}(t)) + j(\dot{w}(t) - z^h), \end{aligned}$$

where in the last step we used (7.36), which in turn is derived using the positive homogeneity of $j(\cdot)$. Alternatively, we have the regularity estimate (7.48). Thus, we have $w^* \in C([0,T];H')$, and

$$-\langle w^*(t), z^h - \dot{w}(t)\rangle \le c\,\|z^h - \dot{w}(t)\|_H,$$

which can be used in deriving (11.8) below. Now, using the Lipschitz continuity of $j(\cdot)$, we find that for any $z^h \in K^h$,

$$\frac{1}{2}\frac{d}{dt}\|w(t) - w^h(t)\|_a^2 \le a(w(t) - w^h(t), \dot{w}(t) - z^h) + c\,\|z^h - \dot{w}(t)\|_H. \quad (11.8)$$

Since

$$
\begin{aligned}
a(w(t) &- w^h(t), \dot{w}(t) - z^h) \\
&\leq a(w(t) - w^h(t), w(t) - w^h(t))^{1/2} a(\dot{w}(t) - z^h, \dot{w}(t) - z^h)^{1/2} \\
&\leq c \left(\|w(t) - w^h(t)\|_a^2 + \|\dot{w}(t) - z^h\|_H^2 \right),
\end{aligned}
$$

from (11.8) we find that for any $z^h = z^h(t) \in K^h$,

$$
\begin{aligned}
&\frac{d}{dt}\|w(t) - w^h(t)\|_a^2 \\
&\quad \leq c \left(\|w(t) - w^h(t)\|_a^2 + \|\dot{w}(t) - z^h(t)\|_H^2 + \|\dot{w}(t) - z^h(t)\|_H \right) \quad (11.9)
\end{aligned}
$$

We multiply the inequality (11.9) by e^{-ct} and integrate from 0 to t to obtain

$$
\|w(t) - w^h(t)\|_a^2 \leq c\, e^{ct} \int_0^t e^{-cs} \left(\|\dot{w}(s) - z^h(s)\|_H^2 + \|\dot{w}(s) - z^h(s)\|_H \right) ds.
$$

This in turn leads to the Céa-type inequality

$$
\|w - w^h\|_{L^\infty(0,T;H)} \leq c \inf_{z^h \in L^2(0,T;K^h)} \|\dot{w} - z^h\|_{L^2(0,T;H)}^{1/2}. \tag{11.10}
$$

THEOREM 11.1. *Suppose the standard assumptions on H, K, a, ℓ, and j are satisfied. Assume that H^h is a finite-dimensional subspace of H, and $K^h = H^h \cap K$. Let $w \in H^1(0,T;H)$ and $w^h \in H^1(0,T;H)$ be the solutions of the problems* ABS *and* ABS^h, *respectively. Then the estimate* (11.10) *holds.*

The inequality (11.10) is the basis for various asymptotic error estimates.

11.2 Time-Discrete Approximations

We now turn to the analysis of another type of semidiscrete scheme, which is obtained by discretizing the time domain. One such scheme has already been seen in Section 7.1, where time-discretization was used as a first stage in proving the existence result. Our aim in this section is to derive error estimates for such semidiscrete approximate solutions, and this will be done for a family of time-discrete schemes that result from approximating the time derivative by generalized midpoint rules. Extensive work has been carried out, both in the context of plasticity and in more general settings, on the *stability* of generalized midpoint and related schemes. Further details may be found in the survey works Simo [114] (in the context of plasticity) and Stuart and Humphries [120] (in a more general context, but pertaining only to ordinary differential equations).

As in Section 7.1, we divide the time interval $[0,T]$ into N equal subintervals with node points $t_n = nk$, $0 \le n \le N$, where $k = T/N$ is the step-size. For the given linear functional $\ell \in H^1(0,T;H')$ and the solution $w \in H^1(0,T;H)$, we use the notation $\ell_n = \ell(t_n)$ and $w_n = w(t_n)$, which are well-defined. The symbol Δw_n is used to denote the backward difference $w_n - w_{n-1}$, and $\delta w_n = \Delta w_n / k$ for the backward divided difference. In this and later sections, no summation is implied over the repeated index n.

Let $\theta \in \left[\frac{1}{2}, 1\right]$ be a parameter. The reason we restrict the value of θ to be in $\left[\frac{1}{2}, 1\right]$ is explained in the remark at the end of the section. A family of generalized midpoint time-discrete approximations of the problem ABS is now introduced.

PROBLEM ABS^k. Find $w^k = \{w_n^k\}_{n=0}^N \subset H$, $w_0^k = 0$, such that for $n = 1, \dots, N$, $\delta w_n^k \in K$ and

$$\begin{aligned} a(\theta\, w_n^k + (1-\theta)\, w_{n-1}^k, z - \delta w_n^k) + j(z) - j(\delta w_n^k) \\ \ge \langle \ell_{n-1+\theta}, z - \delta w_n^k\rangle \qquad \forall\, z \in K. \end{aligned} \tag{11.11}$$

Here, $\ell_{n-1+\theta} = \ell(t_{n-1+\theta})$, and $t_{n-1+\theta} = (n-1+\theta)\,k = \theta\, t_n + (1-\theta)\, t_{n-1}$.

For simplicity in writing, we will not explicitly exhibit the dependence on θ of the solution w^k.

In (11.11) we can replace δw_n^k by Δw_n^k using the positive homogeneity of j. An equivalent way of writing (11.11) is therefore

$$\begin{aligned} \theta\, a(\Delta w_n^k, z - \Delta w_n^k) + j(z) - j(\Delta w_n^k) \\ \ge \langle \ell_{n-1+\theta}, z - \Delta w_n^k\rangle - a(w_{n-1}^k, z - \Delta w_n^k) \quad \forall\, z \in K. \end{aligned} \tag{11.12}$$

Then we can prove the existence of a unique solution of the problem ABS^k using a procedure similar to that employed in the proof of Lemma 7.1.

We now derive an error estimate for the approximation ABS^k. Set $e_n = w_n - w_n^k$, $0 \le n \le N$, for the approximation errors. We recall that $\|w\|_a = a(w,w)^{1/2}$ defines a norm on H, equivalent to $\|w\|_H$. Consider the quantity

$$A_n = a(\theta\, e_n + (1-\theta)\, e_{n-1}, \delta e_n). \tag{11.13}$$

First we have

$$\begin{aligned} A_n &= \frac{1}{k}\left[\theta\, a(e_n, e_n) - (2\theta-1)\, a(e_n, e_{n-1}) - (1-\theta)\, a(e_{n-1}, e_{n-1})\right] \\ &\ge \frac{1}{k}\left[\theta\, \|e_n\|_a^2 - (2\theta-1)\, \|e_n\|_a \|e_{n-1}\|_a - (1-\theta)\, \|e_{n-1}\|_a^2\right] \\ &\ge \frac{1}{k}\left[\theta\, \|e_n\|_a^2 - (2\theta-1)\, \tfrac{1}{2}\left(\|e_n\|_a^2 + \|e_{n-1}\|_a^2\right) - (1-\theta)\, \|e_{n-1}\|_a^2\right]. \end{aligned}$$

So we have the lower bound

$$A_n \ge \frac{1}{2k}\left(\|e_n\|_a^2 - \|e_{n-1}\|_a^2\right). \tag{11.14}$$

Next, we derive an upper bound for A_n. We notice that A_n can be expressed as

$$a(\theta\, w_n + (1-\theta)\, w_{n-1}, \delta w_n - \delta w_n^k) - a(\theta\, w_n^k + (1-\theta)\, w_{n-1}^k, \delta w_n - \delta w_n^k).$$

We use (11.11) with $z = \delta w_n$ for the second term on the right-hand side of the above inequality to obtain

$$\begin{aligned} A_n \le\ & a(\theta\, w_n + (1-\theta)\, w_{n-1}, \delta w_n - \delta w_n^k) \\ & + j(\delta w_n) - j(\delta w_n^k) - \langle \ell_{n-1+\theta}, \delta w_n - \delta w_n^k \rangle. \end{aligned} \tag{11.15}$$

From (11.1) with $z = \delta w_n^k$ at $t = t_{n-1+\theta}$, we get

$$\begin{aligned} 0 \le\ & a(w_{n-1+\theta}, \delta w_n^k - \dot{w}_{n-1+\theta}) \\ & + j(\delta w_n^k) - j(\dot{w}_{n-1+\theta}) - \langle \ell_{n-1+\theta}, \delta w_n^k - \dot{w}_{n-1+\theta} \rangle, \end{aligned}$$

which is added to (11.15) to yield

$$\begin{aligned} A_n \le\ & a(\theta\, w_n + (1-\theta)\, w_{n-1}, \delta w_n - \delta w_n^k) + a(w_{n-1+\theta}, \delta w_n^k - \dot{w}_{n-1+\theta}) \\ & + j(\delta w_n) - j(\dot{w}_{n-1+\theta}) - \langle \ell_{n-1+\theta}, \delta w_n - \dot{w}_{n-1+\theta} \rangle. \end{aligned}$$

After some elementary manipulation we arrive at the upper bound

$$\begin{aligned} A_n \le\ & a(E_{n,\theta}(w), \delta e_n) + a(w_{n-1+\theta}, \delta w_n - \dot{w}_{n-1+\theta}) \\ & + j(\delta w_n) - j(\dot{w}_{n-1+\theta}) - \langle \ell_{n-1+\theta}, \delta w_n - \dot{w}_{n-1+\theta} \rangle \end{aligned} \tag{11.16}$$

for A_n, where

$$E_{n,\theta}(w) = \theta\, w_n + (1-\theta)\, w_{n-1} - w_{n-1+\theta}. \tag{11.17}$$

Combining the bounds (11.14) and (11.16), and using the Lipschitz continuity of $j(\cdot)$, we get

$$\frac{1}{2k}\left(\|e_n\|_a^2 - \|e_{n-1}\|_a^2\right) \le \frac{1}{k}\, a(E_{n,\theta}(w), e_n - e_{n-1}) + c\, \|\delta w_n - \dot{w}_{n-1+\theta}\|_H,$$

that is,

$$\|e_n\|_a^2 - \|e_{n-1}\|_a^2 \le 2\, a(E_{n,\theta}(w), e_n - e_{n-1}) + c\, k\, \|\delta w_n - \dot{w}_{n-1+\theta}\|_H. \tag{11.18}$$

Set

$$M = \max_n \|e_n\|_a.$$

Since $e_0 = 0$, a mathematical induction based on (11.18) reveals that for $n = 1, \ldots, N$,

$$
\begin{aligned}
\|e_n\|_a^2 &\le 2\sum_{j=1}^{n} a(E_{j,\theta}(w), e_j - e_{j-1}) + c\,k \sum_{j=1}^{n} \|\delta w_j - \dot{w}_{j-1+\theta}\|_H \\
&= 2\,a(E_{n,\theta}(w), e_n) + 2\sum_{j=1}^{n-1} a(E_{j,\theta}(w) - E_{j+1,\theta}(w), e_j) \\
&\quad + c\,k \sum_{j=1}^{n} \|\delta w_j - \dot{w}_{j-1+\theta}\|_H \\
&\le c\,\|E_{n,\theta}(w)\|_H M + c\sum_{j=1}^{n-1} \|E_{j,\theta}(w) - E_{j+1,\theta}(w)\|_H M \\
&\quad + c\,k \sum_{j=1}^{n} \|\delta w_j - \dot{w}_{j-1+\theta}\|_H .
\end{aligned}
$$

Hence,

$$
\begin{aligned}
M^2 &\le c\Big(\|E_{N,\theta}(w)\|_H + \sum_{j=1}^{N-1} \|E_{j,\theta}(w) - E_{j+1,\theta}(w)\|_H\Big) M \\
&\quad + c\,k \sum_{j=1}^{N} \|\delta w_j - \dot{w}_{j-1+\theta}\|_H .
\end{aligned} \tag{11.19}
$$

We then apply (11.3) to find that

$$
\begin{aligned}
M^2 &\le c\Big(\|E_{N,\theta}(w)\|_H + \sum_{j=1}^{N-1} \|E_{j,\theta}(w) - E_{j+1,\theta}(w)\|_H\Big)^2 \\
&\quad + c\,k \sum_{j=1}^{N} \|\delta w_j - \dot{w}_{j-1+\theta}\|_H ,
\end{aligned}
$$

or

$$
\begin{aligned}
&\max_n \|w_n - w_n^k\|_H \\
&\quad \le c\Big(\|E_{N,\theta}(w)\|_H + \sum_{j=1}^{N-1} \|E_{j,\theta}(w) - E_{j+1,\theta}(w)\|_H\Big) \\
&\quad + c\Big\{k \sum_{j=1}^{N} \|\delta w_j - \dot{w}_{j-1+\theta}\|_H\Big\}^{1/2} .
\end{aligned} \tag{11.20}
$$

To proceed further, we need to estimate each term on the right-hand side of (11.20). We have the following lemmas.

LEMMA 11.2. *Under the assumption $\ddot{w} \in L^1(0,T;H)$, we have*

$$\|E_{n,\theta}(w)\|_H \le 2\,\theta\,(1-\theta)\,k\,\|\ddot{w}\|_{L^1(t_{n-1},t_n;H)}.$$

If we further assume that $\ddot{w} \in L^\infty(0,T;H)$, then

$$\|E_{n,\theta}(w)\|_H \le \frac{\theta\,(1-\theta)}{2}\,k^2\|\ddot{w}\|_{L^\infty(0,T;H)}.$$

PROOF. We use the Taylor expansions of w about $t_{n-1+\theta}$:

$$w_n = w_{n-1+\theta} + (1-\theta)\,k\,\dot{w}_{n-1+\theta} + \int_{t_{n-1+\theta}}^{t_n} (t_n - t)\,\ddot{w}(t)\,dt, \quad (11.21)$$

$$w_{n-1} = w_{n-1+\theta} - \theta\,k\,\dot{w}_{n-1+\theta} + \int_{t_{n-1+\theta}}^{t_{n-1}} (t_{n-1} - t)\,\ddot{w}(t)\,dt. \quad (11.22)$$

Hence

$$E_{n,\theta}(w) = \int_{t_{n-1+\theta}}^{t_n} \theta\,(t_n - t)\,\ddot{w}(t)\,dt + \int_{t_{n-1}}^{t_{n-1+\theta}} (1-\theta)\,(t - t_{n-1})\,\ddot{w}(t)\,dt,$$

and the results follow. □

LEMMA 11.3. *Under the assumption $\ddot{w} \in L^1(0,T;H)$, we have*

$$\|E_{n,\theta}(w) - E_{n+1,\theta}(w)\|_H \le c\,k\,\|\ddot{w}\|_{L^1(t_{n-1},t_{n+1};H)}.$$

If we further assume that $w^{(3)} \in L^1(0,T;H)$, then

$$\|E_{n,\theta}(w) - E_{n+1,\theta}(w)\|_H \le c\,k^2\|w^{(3)}\|_{L^1(t_{n-1},t_{n+1};H)}.$$

PROOF. The first result follows from Lemma 11.2. To prove the second result, we again use Taylor expansions of w about $t_{n-1+\theta}$, this time with one more term, that is,

$$\begin{aligned} w_n &= w_{n-1+\theta} + (1-\theta)\,k\,\dot{w}_{n-1+\theta} + \frac{(1-\theta)^2}{2}\,k^2\ddot{w}_{n-1+\theta} \\ &\quad + \frac{1}{2}\int_{t_{n-1+\theta}}^{t_n} (t_n - t)^2\,w^{(3)}(t)\,dt \end{aligned} \quad (11.23)$$

and

$$\begin{aligned} w_{n-1} &= w_{n-1+\theta} - \theta\,k\,\dot{w}_{n-1+\theta} + \frac{\theta^2}{2}\,k^2\ddot{w}_{n-1+\theta} \\ &\quad + \frac{1}{2}\int_{t_{n-1+\theta}}^{t_{n-1}} (t_{n-1} - t)^2\,w^{(3)}(t)\,dt. \end{aligned} \quad (11.24)$$

Then

$$E_{n,\theta}(w) = \frac{\theta\,(1-\theta)}{2}\,k^2 \ddot{w}_{n-1+\theta} + \frac{\theta}{2}\int_{t_{n-1+\theta}}^{t_n} (t_n - t)^2\, w^{(3)}(t)\,dt$$
$$+ \frac{1-\theta}{2}\int_{t_{n-1+\theta}}^{t_{n-1}} (t_{n-1} - t)^2\, w^{(3)}(t)\,dt.$$

Thus

$$\begin{aligned}
&E_{n,\theta}(w) - E_{n+1,\theta}(w) \\
&\quad= \frac{\theta\,(1-\theta)}{2}\,k^2\left(\ddot{w}_{n-1+\theta} - \ddot{w}_{n+\theta}\right) \\
&\qquad+ \frac{1}{2}\left[\int_{t_{n-1+\theta}}^{t_n} \theta\,(t_n - t)^2\, w^{(3)}(t)\,dt \right. \\
&\qquad\qquad \left. - \int_{t_{n-1}}^{t_{n-1+\theta}} (1-\theta)\,(t - t_{n-1})^2\, w^{(3)}(t)\,dt\right] \\
&\qquad- \frac{1}{2}\left[\int_{t_{n+\theta}}^{t_{n+1}} \theta\,(t_{n+1} - t)^2\, w^{(3)}(t)\,dt \right. \\
&\qquad\qquad \left. - \int_{t_n}^{t_{n+\theta}} (1-\theta)\,(t - t_n)^2\, w^{(3)}(t)\,dt\right],
\end{aligned}$$

and the estimate follows. □

LEMMA 11.4. *Assume that $\ddot{w} \in L^1(0,T;H)$. Then*

$$\|\delta w_n - \dot{w}_{n-1+\theta}\|_H \le \|\ddot{w}\|_{L^1(t_{n-1},t_n;H)}.$$

Furthermore, if $w^{(3)} \in L^1(0,T;H)$, then

$$\|\delta w_n - \dot{w}_{n-1/2}\|_H \le \frac{k}{8}\,\|w^{(3)}\|_{L^1(t_{n-1},t_n;H)}.$$

PROOF. The first inequality follows from (11.21) and (11.22), while the second follows from (11.23) and (11.24). □

From (11.20) and Lemmas 11.2, 11.3, and 11.4, we obtain the following result.

THEOREM 11.5. *Suppose the standard assumptions on H, K, a, ℓ, and j are satisfied. Let $w \in H^1(0,T;H)$ and w^k be the solutions of the problems* ABS *and* ABS^k, *respectively. Then if $\ddot{w} \in L^\infty(0,T;H)$, we have*

$$\max_n \|w_n - w_n^k\|_H \le c\sqrt{k}, \tag{11.25}$$

and, if $\theta = \frac{1}{2}$ and $w^{(3)} \in L^1(0,T;H)$, we have

$$\max_n \|w_n - w_n^k\|_H \le c\,k. \tag{11.26}$$

We note that the estimates (11.25) and (11.26) are probably not of optimal order, in that the expected error bounds should be $c\,k$ and $c\,k^2$, respectively, for the two cases.

REMARK. In the foregoing discussion, the parameter θ is restricted to be in $\left[\frac{1}{2}, 1\right]$. Let us see whether it is feasible to use other values of θ in the scheme (11.11). The choice $\theta > 1$ is not good, for one will have to use a value of ℓ outside the time interval $[0, T]$. Obviously, $\theta = 0$ cannot be used, for then the scheme (11.11) is meaningless. The scheme corresponding to $0 \neq \theta < \frac{1}{2}$ should never be used, for it is then unstable and would lead to meaningless numerical results in practical computations. To see this, let us consider the extreme case when $j = 0$ and $K = H$. Then the continuous problem is to find w, $w(0) = 0$, such that at any time $t \in [0, T]$,

$$a(w(t), z) = \langle \ell(t), z\rangle \quad \forall\, z \in H,$$

and the time-discrete approximation is $w_0^k = 0$, and for $n = 1, \dots, N$,

$$a(\theta\, w_n^k + (1-\theta)\, w_{n-1}^k, z) = \langle \ell_{n-1+\theta}, z\rangle \quad \forall\, z \in H.$$

Now consider a perturbed problem for w^k: $\hat{w}_0^k = \varepsilon$, and for $n = 1, \dots, N$,

$$a(\theta\, \hat{w}_n^k + (1-\theta)\, \hat{w}_{n-1}^k, z) = \langle \ell_{n-1+\theta}, z\rangle \quad \forall\, z \in H.$$

For the difference $e_n = \hat{w}_n^k - w_n^k$ we have $e_0 = \varepsilon$ and for $n = 1, \dots, N$,

$$a(\theta\, e_n^k + (1-\theta)\, e_{n-1}^k, z) = 0 \quad \forall\, z \in H.$$

Thus, for $n \geq 1$,

$$e_n = -\frac{1-\theta}{\theta}\, e_{n-1},$$

and as a consequence,

$$e_n = (-1)^n \left(\frac{1-\theta}{\theta}\right)^n e_0 = (-1)^n \left(\frac{1-\theta}{\theta}\right)^n \varepsilon.$$

Since $0 \neq \theta < \frac{1}{2}$, it follows that $|(1-\theta)/\theta| > 1$. Therefore, a small perturbation in the initial value may cause arbitrarily large errors in the numerical approximation.

11.3 Fully Discrete Approximations

From the point of view of applications, it is more important to consider fully discrete approximations, where the temporal and spatial variables are simultaneously discretized. As before, we divide the time interval $I =$

$[0,T]$ into N equal parts and denote the step-size by $k = T/N$, the nodal points by $t_n = nk$, $n = 0, 1, \ldots, N$, and subintervals by $I_n = [t_{n-1}, t_n]$, $n = 1, 2, \ldots, N$. Again we use $h \in (0,1]$ for the mesh parameter of a triangulation of the domain Ω. Let $\{H^h\}$ be a family of finite-dimensional subspaces of H, and let $K^h = H^h \cap K$. As is noted in Section 11.1, K^h is a nonempty, closed, convex cone in H^h, and in H as well. Let $\theta \in [\frac{1}{2}, 1]$ be a parameter. The family of fully discrete approximation schemes that we will analyze in this section is the following.

PROBLEM ABS^{hk}. *Find $w^{hk} = \{w_n^{hk}\}_{n=0}^N$, where $w_n^{hk} \in H^h$, $0 \le n \le N$, with $w_0^{hk} = 0$, such that for $n = 1, 2, \ldots, N$, $\delta w_n^{hk} \in K^h$ and*

$$
\begin{aligned}
& a(\theta\, w_n^{hk} + (1-\theta)\, w_{n-1}^{hk}, z^h - \delta w_n^{hk}) + j(z^h) - j(\delta w_n^{hk}) \\
& \quad \ge \langle \ell_{n-1+\theta}, z^h - \delta w_n^{hk} \rangle \quad \forall\, z^h \in K^h.
\end{aligned} \tag{11.27}
$$

The remark at the end of the previous subsection applies also to the fully discrete schemes. Hence, we do not consider the case where $\theta < \frac{1}{2}$ or $\theta > 1$.

The first thing we do is to show the well-posedness of the problem ABS^{hk}.

THEOREM 11.6. *The problem $\mathbf{P}^{hk}$ admits a unique solution w^{hk}. The solution is stable in the sense that for $\ell^{(1)}, \ell^{(2)} \in H^1(0,T;H')$ with $\ell^{(1)}(0) = \ell^{(2)}(0) = 0$, the corresponding solutions $w_n^{(1)hk}$ and $w_n^{(2)hk}$, $0 \le n \le N$, satisfy the inequality*

$$
\max_{0 \le n \le N} \|w_n^{(1)hk} - w_n^{(2)hk}\|_H \le c\, \|\ell^{(1)} - \ell^{(2)}\|_{L^\infty(0,T;H')}. \tag{11.28}
$$

PROOF. Once again, because of the positive homogeneity of j and the cone property of K^h, the inequality (11.27) can be rewritten as

$$
\begin{aligned}
& a(\theta\, w_n^{hk} + (1-\theta)\, w_{n-1}^{hk}, z^h - \Delta w_n^{hk}) + j(z^h) - j(\Delta w_n^{hk}) \\
& \quad \ge \langle \ell_{n-1+\theta}, z^h - \Delta w_n^{hk} \rangle \quad \forall\, z^h \in K^h,
\end{aligned}
$$

or

$$
\begin{aligned}
& \theta\, a(\Delta w_n^{hk}, z^h - \Delta w_n^{hk}) + j(z^h) - j(\Delta w_n^{hk}) \\
& \quad \ge \langle l_{n-1+\theta}, z^h - \Delta w_n^{hk} \rangle - a(w_{n-1}^{hk}, z^h - \Delta w_n^{hk}) \quad \forall\, z^h \in K^h.
\end{aligned} \tag{11.29}
$$

Now the existence and uniqueness results can be obtained following the arguments used in proving Lemma 7.1.

We then derive the stability inequality (11.28). For given $\ell^{(1)}$ and $\ell^{(2)}$, let $w_n^{1,hk}$ and $w_n^{2,hk}$, $0 \le n \le N$, be the corresponding fully discrete solutions. Then for $n = 1, 2, \ldots, N$, we have $\delta w_n^{(1)hk}, \delta w_n^{(2)hk} \in K^h$, and

$$
\begin{aligned}
& a(\theta\, w_n^{(1)hk} + (1-\theta)\, w_{n-1}^{(1)hk}, z^h - \delta w_n^{(1)hk}) + j(z^h) - j(\delta w_n^{(1)hk}) \\
& \quad \ge \langle \ell_{n-1+\theta}^{(1)}, z^h - \delta w_n^{(1)hk} \rangle \qquad \forall\, z^h \in K^h,
\end{aligned} \tag{11.30}
$$

$$\begin{aligned}
&a(\theta\, w_n^{(2)hk} + (1-\theta)\, w_{n-1}^{(2)hk}, z^h - \delta w_n^{(2)hk}) + j(z^h) - j(\delta w_n^{(2)hk}) \\
&\quad \geq \langle \ell_{n-1+\theta}^{(2)}, z^h - \delta w_n^{(2)hk} \rangle \qquad \forall\, z^h \in K^h.
\end{aligned} \tag{11.31}$$

Let $e_n = w_n^{(1)hk} - w_n^{(2)hk}$ denote the difference between the two solutions. Taking $z^h = \delta w_n^{(2)hk}$ in (11.30) and $z^h = \delta w_n^{(1)hk}$ in (11.31), and adding the two resultant inequalities, we obtain

$$a(\theta\, e_n + (1-\theta)\, e_{n-1}, \delta e_n) \leq \langle \ell_{n-1+\theta}^{(1)} - \ell_{n-1+\theta}^{(2)}, \delta e_n \rangle.$$

Using (11.14) for a lower bound of the left-hand side of the above inequality, we find that for $n = 1, \ldots, N$,

$$\|e_n\|_a^2 - \|e_{n-1}\|_a^2 \leq 2\, \langle \ell_{n-1+\theta}^{(1)} - \ell_{n-1+\theta}^{(2)}, e_n - e_{n-1} \rangle.$$

Since $e_0 = 0$, a simple induction shows that

$$\begin{aligned}
\|e_n\|_a^2 &\leq 2 \sum_{j=1}^{n} \langle \ell_{j-1+\theta}^{(1)} - \ell_{j-1+\theta}^{(2)}, e_j - e_{j-1} \rangle \\
&= 2\, \langle \ell_{n-1+\theta}^{(1)} - \ell_{n-1+\theta}^{(2)}, e_n \rangle \\
&\quad + 2 \sum_{j=1}^{n-1} \langle (\ell_{j-1+\theta}^{(1)} - \ell_{j+\theta}^{(1)}) - (\ell_{j-1+\theta}^{(2)} - \ell_{j+\theta}^{(2)}), e_j \rangle.
\end{aligned}$$

With $M = \max_n \|e_n\|_a$, we then find that for $n = 1, \ldots, N$,

$$\begin{aligned}
\|e_n\|_a^2 \leq c \Big(&\|\ell_{n-1+\theta}^{(1)} - \ell_{n-1+\theta}^{(2)}\|_{H'} \\
&+ \sum_{j=1}^{n-1} \|(\ell_{j-1+\theta}^{(1)} - \ell_{j+\theta}^{(1)}) - (\ell_{j-1+\theta}^{(2)} - \ell_{j+\theta}^{(2)})\|_{H'} \Big) M.
\end{aligned}$$

Therefore,

$$\begin{aligned}
M^2 \leq c \Big(&\|\ell_{N-1+\theta}^{(1)} - \ell_{N-1+\theta}^{(2)}\|_{H'} \\
&+ \sum_{j=1}^{N-1} \|(\ell_{j-1+\theta}^{(1)} - \ell_{j+\theta}^{(1)}) - (\ell_{j-1+\theta}^{(2)} - \ell_{j+\theta}^{(2)})\|_{H'} \Big) M,
\end{aligned}$$

that is,

$$\begin{aligned}
&\max_n \|e_n\|_a \\
&\quad \leq c \Big(\|\ell_{N-1+\theta}^{(1)} - \ell_{N-1+\theta}^{(2)}\|_{H'} \\
&\qquad + \sum_{j=1}^{N-1} \|(\ell_{j-1+\theta}^{(1)} - \ell_{j+\theta}^{(1)}) - (\ell_{j-1+\theta}^{(2)} - \ell_{j+\theta}^{(2)})\|_{H'} \Big).
\end{aligned} \tag{11.32}$$

We then apply the inequality (5.25) to get the estimate (11.28). □

Now we turn our attention to an error analysis for the fully discrete scheme. The quantity of interest is the error $e_n = w_n - w_n^{hk}$, $0 \le n \le N$. We consider the quantities

$$A_n = a(\theta\, e_n + (1-\theta)\, e_{n-1}, \delta e_n), \quad n = 1, \ldots, N. \tag{11.33}$$

As in (11.14), we have a lower bound for A_n:

$$A_n \ge \frac{1}{2k} \left(\|e_n\|_a^2 - \|e_{n-1}\|_a^2 \right). \tag{11.34}$$

To derive an upper bound for A_n, we write

$$\begin{aligned} A_n &= a(\theta\, w_n + (1-\theta)\, w_{n-1}, \delta w_n - \delta w_n^{hk}) \\ &\quad - a(\theta\, w_n^{hk} + (1-\theta)\, w_{n-1}^{hk}, \delta w_n - z_n^h) \\ &\quad - a(\theta\, w_n^{hk} + (1-\theta)\, w_{n-1}^{hk}, z_n^h - \delta w_n^{hk}), \end{aligned}$$

where $z_n^h \in K^h$ is arbitrary. Using (11.27) to handle the last term, we obtain

$$\begin{aligned} A_n &\le a(\theta\, w_n + (1-\theta)\, w_{n-1}, \delta w_n - \delta w_n^{hk}) \\ &\quad - a(\theta\, w_n^{hk} + (1-\theta)\, w_{n-1}^{hk}, \delta w_n - z_n^h) \\ &\quad + j(z_n^h) - j(\delta w_n^{hk}) - \langle \ell_{n-1+\theta}, z_n^h - \delta w_n^{hk} \rangle. \end{aligned} \tag{11.35}$$

Now we take $z = \delta w_n^{hk} \in K$ in (11.1) at $t = t_{n-1+\theta}$ to get

$$\begin{aligned} 0 &\le a(w_{n-1+\theta}, \delta w_n^{hk} - \dot{w}_{n-1+\theta}) + j(\delta w_n^{hk}) - j(\dot{w}_{n-1+\theta}) \\ &\quad - \langle \ell_{n-1+\theta}, \delta w_n^{hk} - \dot{w}_{n-1+\theta} \rangle. \end{aligned} \tag{11.36}$$

Adding (11.35) and (11.36), we have

$$\begin{aligned} A_n &\le a(\theta\, w_n + (1-\theta)\, w_{n-1}, \delta w_n - \delta w_n^{hk}) \\ &\quad - a(\theta\, w_n^{hk} + (1-\theta)\, w_{n-1}^{hk}, \delta w_n - z_n^h) \\ &\quad + a(w_{n-1+\theta}, \delta w_n^{hk} - \dot{w}_{n-1+\theta}) \\ &\quad + j(z_n^h) - j(\dot{w}_{n-1+\theta}) - \langle \ell_{n-1+\theta}, z_n^h - \dot{w}_{n-1+\theta} \rangle, \end{aligned}$$

which is rewritten as

$$\begin{aligned} A_n &\le \frac{1}{k}\, a(E_{n,\theta}(w), e_n - e_{n-1}) \\ &\quad + a(\theta\, e_n + (1-\theta)\, e_{n-1}, \delta w_n - z_n^h) \\ &\quad - a(\theta\, w_n + (1-\theta)\, w_{n-1}, \delta w_n - z_n^h) \\ &\quad + a(w_{n-1+\theta}, \delta w_n - \dot{w}_{n-1+\theta}) + j(z_n^h) - j(\dot{w}_{n-1+\theta}) \\ &\quad - \langle \ell_{n-1+\theta}, z_n^h - \dot{w}_{n-1+\theta} \rangle, \end{aligned}$$

where $E_{n,\theta}(w)$ is the quantity defined in (11.17). Taking $z = \dot{w}_{n-1+\theta} - z_n^h$ in (11.2) at $t = t_{n-1+\theta}$, we get

$$\begin{aligned} a(w_{n-1+\theta}, \dot{w}_{n-1+\theta} - z_n^h) + \langle w^*_{n-1+\theta}, \dot{w}_{n-1+\theta} - z_n^h \rangle \\ = \langle \ell_{n-1+\theta}, \dot{w}_{n-1+\theta} - z_n^h \rangle. \end{aligned}$$

We thus have the upper bound

$$\begin{aligned} A_n \le\, & \frac{1}{k}\, a(E_{n,\theta}(w), e_n - e_{n-1}) \\ & + a(\theta\, e_n + (1-\theta)\, e_{n-1}, \delta w_n - z_n^h) - a(E_{n,\theta}(w), \delta w_n - z_n^h) \\ & + j(z_n^h) - j(\dot{w}_{n-1+\theta}) + \langle w^*_{n-1+\theta}, \dot{w}_{n-1+\theta} - z_n^h \rangle \end{aligned}$$

for A_n. From this upper bound and the lower bound (11.34), we obtain the inequality

$$\begin{aligned} & \frac{1}{2k} \left(\|e_n\|_a^2 - \|e_{n-1}\|_a^2 \right) \\ & \quad \le \frac{1}{k}\, a(E_{n,\theta}(w), e_n - e_{n-1}) + c\, M\, \|\delta w_n - z_n^h\|_H \\ & \qquad + c\, \|E_{n,\theta}(w)\|_H\, \|\delta w_n - z_n^h\|_H + c\, \|\dot{w}_{n-1+\theta} - z_n^h\|_H \end{aligned} \tag{11.37}$$

where $M = \max_n \|e_n\|_a$. Thus

$$\begin{aligned} & \|e_n\|_a^2 - \|e_{n-1}\|_a^2 \\ & \quad \le 2\, a(E_{n,\theta}(w), e_n - e_{n-1}) + c\, M\, k\, \|\delta w_n - z_n^h\|_H \\ & \qquad + c\, k\, \|E_{n,\theta}(w)\|_H\, \|\delta w_n - z_n^h\|_H + c\, k\, \|\dot{w}_{n-1+\theta} - z_n^h\|_H. \end{aligned}$$

A simple induction argument yields (noting that $e_0 = 0$), for $1 \le n \le N$,

$$\begin{aligned} \|e_n\|_a^2 \le\, & 2 \sum_{j=1}^n a(E_{j,\theta}(w), e_j - e_{j-1}) + c\, M\, k \sum_{j=1}^n \|\delta w_j - z_j^h\|_H \\ & + c\, k\, (\max_n \|E_{n,\theta}(w)\|_H) \sum_{j=1}^n \|\delta w_j - z_j^h\|_H \\ & + c\, k \sum_{j=1}^n \|\dot{w}_{j-1+\theta} - z_j^h\|_H. \end{aligned}$$

Notice that

$$\begin{aligned} & \sum_{j=1}^n a(E_{j,\theta}(w), e_j - e_{j-1}) \\ & \qquad = a(E_{n,\theta}(w), e_n) + \sum_{j=1}^{n-1} a(E_{j,\theta}(w) - E_{j+1,\theta}(w), e_j). \end{aligned}$$

Hence

$$M^2 \le c\,M\left(\|E_{N,\theta}(w)\|_H + \sum_{j=1}^{N-1} \|E_{j,\theta}(w) - E_{j+1,\theta}(w)\|_H \right.$$
$$\left. + \,k \sum_{j=1}^{N} \|\delta w_j - z_j^h\|_H\right)$$
$$+ \,c\,k \max_n \|E_{n,\theta}(w)\|_H \sum_{j=1}^{N} \|\delta w_j - z_j^h\|_H$$
$$+ \,c\,k \sum_{j=1}^{N} \|\dot{w}_{j-1+\theta} - z_j^h\|_H. \tag{11.38}$$

Using the relation (11.3), we find from (11.38) that

$$\max_n \|w_n - w_n^{hk}\|_a$$
$$\le c\left(\|E_{N,\theta}(w)\|_H + \sum_{j=1}^{N-1} \|E_{j,\theta}(w) - E_{j+1,\theta}(w)\|_H \right.$$
$$\left. + \,k \sum_{j=1}^{N} \|\delta w_j - z_j^h\|_H\right)$$
$$+ \,c\left\{k \max_n \|E_{n,\theta}(w)\|_H \sum_{j=1}^{N} \|\delta w_j - z_j^h\|_H\right\}^{1/2}$$
$$+ \,c\left\{k \sum_{j=1}^{N} \|\dot{w}_{j-1+\theta} - z_j^h\|_H\right\}^{1/2}. \tag{11.39}$$

Concrete error estimates follow from (11.39) if we apply the results given in Lemmas 11.2, 11.3, and 11.4. Let us focus on the orders of the schemes by assuming that the solution is smooth. Specifically, we assume that $w \in W^{3,1}(0,T;H)$. Then we also have $\ddot{w} \in L^\infty(0,T;H)$. From Lemma 11.2,

$$\max_n \|E_{n,\theta}(w)\|_H \le c\,k^2.$$

From Lemma 11.3,

$$\sum_{j=1}^{N-1} \|E_{j,\theta}(w) - E_{j+1,\theta}(w)\|_H \le c\,k^2,$$

and from Lemma 11.4,

$$\sum_{j=1}^{N} \|\delta w_j - \dot{w}_{j-1+\theta}\|_H \leq c \quad \text{if } \theta \neq \frac{1}{2},$$

$$\sum_{j=1}^{N} \|\delta w_j - \dot{w}_{j-1/2}\|_H \leq c\,k.$$

With these inequalities, together with the triangle inequality

$$\|\delta w_j - z_j^h\|_H \leq \|\delta w_j - \dot{w}_{j-1+\theta}\|_H + \|\dot{w}_{j-1+\theta} - z_j^h\|_H,$$

we obtain the following error estimates assuming $w^{(3)} \in L^1(0,T;H)$:

$$\begin{aligned}\max_n \|w_n - w_n^{hk}\|_a &\\ \leq c\,k + c\,k \sum_{j=1}^{N} \|\dot{w}_{j-1+\theta} - z_j^h\|_H &\\ + c\left\{k \sum_{j=1}^{N} \|\dot{w}_{j-1+\theta} - z_j^h\|_H\right\}^{1/2} & \quad \text{if } \theta \neq \frac{1}{2}\end{aligned} \tag{11.40}$$

and

$$\begin{aligned}\max_n \|w_n - w_n^{hk}\|_a &\\ \leq c\,k^2 + c\,k \sum_{j=1}^{N} \|\dot{w}_{j-1/2} - z_j^h\|_H &\\ + c\left\{k \sum_{j=1}^{N} \|\dot{w}_{j-1/2} - z_j^h\|_H\right\}^{1/2} & \quad \text{if } \theta = \frac{1}{2}.\end{aligned} \tag{11.41}$$

Since $z_j^h \in K^h$, $1 \leq j \leq N$, are arbitrary, and since the finite-dimensional subspaces satisfy the relation (11.4), we can rewrite the error estimates (11.40) and (11.41) in the more concise forms

$$\begin{aligned}\max_n \|w_n - w_n^{hk}\|_a &\\ \leq c\,k + c\left\{k \sum_{j=1}^{N} \inf_{z_j^h \in K^h} \|\dot{w}_{j-1+\theta} - z_j^h\|_H\right\}^{1/2} & \quad \text{if } \theta \neq \frac{1}{2}\end{aligned} \tag{11.42}$$

and

$$\begin{aligned}\max_n \|w_n - w_n^{hk}\|_a &\\ \leq c\,k^2 + c\left\{k \sum_{j=1}^{N} \inf_{z_j^h \in K^h} \|\dot{w}_{j-1/2} - z_j^h\|_H\right\}^{1/2} & \quad \text{if } \theta = \frac{1}{2}.\end{aligned} \tag{11.43}$$

We now summarize the results of the section in the form of a theorem.

THEOREM 11.7. *Suppose the standard assumptions on H, K, a, ℓ, and j are satisfied. Let $w \in H^1(0,T;H)$ and w^{hk} be the solutions of the problems* ABS *and* ABS^{hk}, *respectively. Then if $w \in W^{3,1}(0,T;H)$, we have the estimates* (11.42) *and* (11.43).

We observe that the orders are optimal with respect to the time step-size in the error estimates (11.42) and (11.43). In particular, when $\theta = 1$, we have a backward Euler scheme, and it is a first-order method with respect to the temporal step-size. When $\theta = \frac{1}{2}$, we have the second-order accurate Crank–Nicolson-type scheme.

11.4 Convergence Under Minimal Regularity

We assumed a certain degree of regularity of the solution in the error analysis presented in the last several sections. Since a regularity theory is still to be developed for the elastoplasticity problem as well as the abstract problem, it is of interest to examine whether we can show convergence of the various numerical methods under the minimal regularity condition of the solution provided by the existence theorem.

Recall that the problem ABS has a unique solution $w \in H^1(0,T;H)$. From the density result (5.27) we see that for any $\varepsilon > 0$ there is a function $\overline{w} \in C^\infty([0,T];H)$ such that

$$\|w - \overline{w}\|_{H^1(0,T;H)} \le \varepsilon, \tag{11.44}$$

i.e.,

$$\int_0^T \|w(t) - \overline{w}(t)\|_H^2 dt + \int_0^T \|\dot{w}(t) - \dot{\overline{w}}(t)\|_H^2 dt \le \varepsilon^2.$$

Since

$$\|w - \overline{w}\|_{C([0,T];H)} = \max_{0 \le t \le T} \|w(t) - \overline{w}(t)\|_H \le c\,\|w - \overline{w}\|_{H^1(0,T;H)},$$

we also have

$$\|w - \overline{w}\|_{C([0,T];H)} \le c\,\varepsilon. \tag{11.45}$$

We will prove the convergence of the time-discrete solutions and fully discrete solutions to the exact solution w of the problem ABS. A convergence analysis for the spatially discrete schemes can be easily done based on the inequality (11.10); the detailed argument is omitted, and here we mention only that for the convergence of the spatially discrete solutions we need to assume the hypotheses (H_1) and (H_2) later as in the convergence analysis for the fully discrete solutions.

The convergence argument below looks long, but the main idea is simple. The solution is approximated arbitrarily closely by smooth functions, and for smooth functions we can use Taylor expansions and derive various estimates.

Convergence of the time-discrete scheme. We first observe that the inequality (11.20) is useful for deriving order error estimates under various regularity assumptions on the solution, yet we cannot use it for a convergence analysis under the basic solution regularity condition, because then the pointwise value $\dot{w}_{j-1+\theta}$ occurring in (11.20) is not well-defined. This suggests that we modify the derivation and prove some result similar to (11.20) without the appearance of the $\dot{w}_{j-1+\theta}$ terms. We use the same notations as in Section 11.2. Additionally, we use $I_n = [t_{n-1}, t_n]$, $n = 1, \ldots, N$, to denote the time subintervals. From (11.15), we have the inequality

$$\begin{aligned} \frac{1}{2k} & \left(\|e_n\|_a^2 - \|e_{n-1}\|_a^2\right) \\ & \le a(\theta\, w_n + (1-\theta)\, w_{n-1}, \delta w_n - \delta w_n^k) \\ & \quad + j(\delta w_n) - j(\delta w_n^k) - \langle \ell_{n-1+\theta}, \delta w_n - \delta w_n^k \rangle. \end{aligned} \tag{11.46}$$

We now take $z = \delta w_n^k$ in (11.1),

$$a(w(t), \delta w_n^k - \dot{w}(t)) + j(\delta w_n^k) - j(\dot{w}(t)) \ge \langle \ell(t), \delta w_n^k - \dot{w}(t) \rangle,$$

and then integrate over I_n to obtain

$$\begin{aligned} 0 \le & \frac{1}{k} \int_{I_n} a(w(t), \delta w_n^k - \dot{w}(t))\, dt + j(\delta w_n^k) - \frac{1}{k} \int_{I_n} j(\dot{w}(t))\, dt \\ & - \frac{1}{k} \int_{I_n} \langle \ell(t), \delta w_n^k - \dot{w}(t) \rangle\, dt. \end{aligned} \tag{11.47}$$

Adding the inequalities (11.46) and (11.47), we find that

$$\frac{1}{2k} \left(\|e_n\|_a^2 - \|e_{n-1}\|_a^2\right) \le Q_1 + Q_2 + Q_3, \tag{11.48}$$

where

$$\begin{aligned} Q_1 &= a(\theta\, w_n + (1-\theta)\, w_{n-1}, \delta w_n - \delta w_n^k) \\ & \quad + \frac{1}{k} \int_{I_n} a(w(t), \delta w_n^k - \dot{w}(t))\, dt, \\ Q_2 &= \frac{1}{k} \int_{I_n} [j(\delta w_n) - j(\dot{w}(t))]\, dt, \\ Q_3 &= -\langle \ell_{n-1+\theta}, \delta w_n - \delta w_n^k \rangle - \frac{1}{k} \int_{I_n} \langle \ell(t), \delta w_n^k - \dot{w}(t) \rangle\, dt. \end{aligned}$$

Let us estimate each of these three terms. We define

$$w_n^a = \frac{1}{k} \int_{I_n} w(t)\,dt \in H, \quad n = 1, \ldots, N, \tag{11.49}$$

for local averages of $w(t)$, and similar to (11.17), we introduce the quantities

$$E_{n,\theta}^a(w) = \theta\, w_n + (1-\theta)\, w_{n-1} - w_n^a. \tag{11.50}$$

We have

$$\begin{aligned}
Q_1 &= a(\theta\, w_n + (1-\theta)\, w_{n-1} - w_n^a, \delta w_n - \delta w_n^k) \\
&\quad + \frac{1}{k} \int_{I_n} a(w(t), \delta w_n)\,dt - \frac{1}{k} \int_{I_n} a(w(t), \dot{w}(t))\,dt \\
&= \frac{1}{k}\, a(E_{n,\theta}^a(w), e_n - e_{n-1}) \\
&\quad + \frac{1}{2k} \left[2\, a(w_n^a, w_n - w_{n-1}) - a(w_n, w_n) + a(w_{n-1}, w_{n-1})\right].
\end{aligned}$$

Since

$$\begin{aligned}
&2\, a(w_n^a, w_n - w_{n-1}) - a(w_n, w_n) + a(w_{n-1}, w_{n-1}) \\
&\quad = 2\, a(w_n^a, w_n - w_{n-1}) - a(w_n, w_n - w_{n-1}) + a(w_{n-1}, w_n - w_{n-1}) \\
&\quad = a(2\, w_n^a - w_n - w_{n-1}, w_n - w_{n-1}),
\end{aligned}$$

we see that

$$Q_1 = \frac{1}{k}\, a(E_{n,\theta}^a(w), e_n - e_{n-1}) + \frac{1}{k}\, a(w_n^a - \tfrac{1}{2}\,(w_n + w_{n-1}), w_n - w_{n-1}). \tag{11.51}$$

For the second term Q_2, we use the Lipschitz continuity of $j(\cdot)$ on K,

$$|Q_2| \le \frac{c}{k} \int_{I_n} \|\delta w_n - \dot{w}(t)\|_H\,dt.$$

Since

$$\delta w_n - \dot{w}(t) = \frac{1}{k} \int_{I_n} (\dot{w}(s) - \dot{w}(t))\,ds,$$

we have

$$\begin{aligned}
|Q_2| &\le \frac{c}{k^2} \int_{I_n} \int_{I_n} \|\dot{w}(s) - \dot{w}(t)\|_H\,ds\,dt \\
&\le \frac{c}{k^2} \int_{I_n \times I_n} \big[\|\dot{w}(s) - \dot{\overline{w}}(s)\|_H + \|\dot{w}(t) - \dot{\overline{w}}(t)\|_H \\
&\qquad\qquad + \|\dot{\overline{w}}(s) - \dot{\overline{w}}(t)\|_H\big]\,ds\,dt \\
&= \frac{c}{k} \int_{I_n} \|\dot{w}(t) - \dot{\overline{w}}(t)\|_H\,dt + \frac{c}{k^2} \int_{I_n} \int_{I_n} \left\| \int_s^t \ddot{\overline{w}}(\tau)\,d\tau \right\|_H ds\,dt.
\end{aligned}$$

Now

$$\int_{I_n}\int_{I_n}\left\|\int_s^t \dddot{w}(\tau)\,d\tau\right\|_H ds\,dt \le k^2\int_{I_n}\|\dddot{w}(t)\|_H dt.$$

Therefore,

$$|Q_2| \le \frac{c}{k}\int_{I_n}\|\dot{w}(t)-\dot{\overline{w}}(t)\|_H dt + c\int_{I_n}\|\dddot{w}(t)\|_H dt. \tag{11.52}$$

Analogous to (11.49), we use

$$\ell_n^a = \frac{1}{k}\int_{I_n}\ell(t)\,dt \in H', \quad n=1,\ldots,N \tag{11.53}$$

for local averages of ℓ. Then

$$Q_3 = \langle \ell_n^a - \ell_{n-1+\theta}, \delta w_n - \delta w_n^k\rangle + \frac{1}{k}\int_{I_n}\langle \ell(t), \dot{w}(t)-\delta w_n\rangle\,dt.$$

Now,

$$\int_{I_n}\langle \ell(t),\delta w_n\rangle\,dt = \left\langle \frac{1}{k}\int_{I_n}\ell(t)\,dt, \int_{I_n}\dot{w}(s)\,ds\right\rangle = \int_{I_n}\langle \ell_n^a, \dot{w}(t)\rangle\,dt.$$

Hence

$$Q_3 = \frac{1}{k}\langle \ell_n^a - \ell_{n-1+\theta}, e_n - e_{n-1}\rangle + \frac{1}{k}\int_{I_n}\langle \ell(t)-\ell_n^a, \dot{w}(t)\rangle\,dt. \tag{11.54}$$

Combining (11.48), (11.51), (11.52), and (11.54), we have

$$\begin{aligned}
&\frac{1}{2k}\left(\|e_n\|_a^2 - \|e_{n-1}\|_a^2\right)\\
&\quad\le \frac{1}{k}\,a(E_{n,\theta}^a(w), e_n - e_{n-1}) + \frac{1}{k}\,a(w_n^a - \tfrac{1}{2}(w_n + w_{n-1}), w_n - w_{n-1})\\
&\qquad+ \frac{c}{k}\int_{I_n}\|\dot{w}(t)-\dot{\overline{w}}(t)\|_H dt + c\int_{I_n}\|\dddot{w}(t)\|_H dt\\
&\qquad+ \frac{1}{k}\langle \ell_n^a - \ell_{n-1+\theta}, e_n - e_{n-1}\rangle + \frac{1}{k}\int_{I_n}\langle \ell(t)-\ell_n^a, \dot{w}(t)\rangle\,dt.
\end{aligned}$$

Then

$$\begin{aligned}
&\|e_n\|_a^2 - \|e_{n-1}\|_a^2\\
&\quad\le 2\,a(E_{n,\theta}^a(w), e_n - e_{n-1}) + 2\,a(w_n^a - \tfrac{1}{2}(w_n + w_{n-1}), w_n - w_{n-1})\\
&\qquad+ c\int_{I_n}\|\dot{w}(t)-\dot{\overline{w}}(t)\|_H dt + c\,k\int_{I_n}\|\dddot{w}(t)\|_H dt\\
&\qquad+ 2\,\langle \ell_n^a - \ell_{n-1+\theta}, e_n - e_{n-1}\rangle + 2\int_{I_n}\langle \ell(t)-\ell_n^a, \dot{w}(t)\rangle\,dt.
\end{aligned}$$

Since $e_0 = 0$, a mathematical induction argument shows that

$$\begin{aligned}
\|e_n\|_a^2 \le 2 \sum_{j=1}^{n} a(E_{j,\theta}^a(w), e_j - e_{j-1}) \\
+ 2 \sum_{j=1}^{n} a(w_j^a - \tfrac{1}{2}(w_j + w_{j-1}), w_j - w_{j-1}) \\
+ c \int_0^{t_n} \|\dot{w}(t) - \dot{\overline{w}}(t)\|_H dt + c\,k \int_0^{t_n} \|\dddot{\overline{w}}(t)\|_H dt \\
+ 2 \sum_{j=1}^{n} \langle \ell_j^a - \ell_{j-1+\theta}, e_j - e_{j-1} \rangle \\
+ 2 \sum_{j=1}^{n} \int_{I_j} \langle \ell(t) - \ell_j^a, \dot{w}(t) \rangle \, dt
\end{aligned}$$

for $n = 1, \ldots, N$. Let $M = \max_{1 \le n \le N} \|e_n\|_a$. We use the identities

$$\begin{aligned}
&\sum_{j=1}^{n} a(E_{j,\theta}^a(w), e_j - e_{j-1}) \\
&\qquad = a(E_{n,\theta}^a(w), e_n) + \sum_{j=1}^{n-1} a(E_{j,\theta}^a(w) - E_{j+1,\theta}^a(w), e_j), \\
&\sum_{j=1}^{n} \langle \ell_j^a - \ell_{j-1+\theta}, e_j - e_{j-1} \rangle \\
&\qquad = \langle \ell_n^a - \ell_{n-1+\theta}, e_n \rangle + \sum_{j=1}^{n-1} \langle (\ell_j^a - \ell_{j-1+\theta}) - (\ell_{j+1}^a - \ell_{j+\theta}), e_j \rangle
\end{aligned}$$

and get from the above inequality

$$\begin{aligned}
M^2 \le c\,M \Big\{ \|E_{N,\theta}^a(w)\|_H + \sum_{n=1}^{N-1} \|E_{n,\theta}^a(w) - E_{n+1,\theta}^a(w)\|_H \\
+ \|\ell_N^a - \ell_{N-1+\theta}\|_{H'} \\
+ \sum_{n=1}^{N-1} \|(\ell_n^a - \ell_{n-1+\theta}) - (\ell_{n+1}^a - \ell_{n+\theta})\|_{H'} \Big\} \\
+ c \sum_{n=1}^{N} \|w_n^a - \tfrac{1}{2}(w_n + w_{n-1})\|_H \|w_n - w_{n-1}\|_H \\
+ c \sum_{n=1}^{N} \int_{I_n} \|\ell(t) - \ell_n^a\|_{H'} \|\dot{w}(t)\|_H dt \\
+ c \int_0^{T} \|\dot{w}(t) - \dot{\overline{w}}(t)\|_H dt + c\,k \int_0^{T} \|\dddot{\overline{w}}(t)\|_H dt.
\end{aligned}$$

Applying the result (11.3), we then have

$$\begin{aligned}
\max_{1\le n\le N} \|e_n\|_a
&\le c\Big\{\|E^a_{N,\theta}(w)\|_H + \sum_{n=1}^{N-1} \|E^a_{n,\theta}(w) - E^a_{n+1,\theta}(w)\|_H \\
&\quad + \|\ell^a_N - \ell_{N-1+\theta}\|_{H'} + \sum_{n=1}^{N-1} \|(\ell^a_n - \ell_{n-1+\theta}) - (\ell^a_{n+1} - \ell_{n+\theta})\|_{H'}\Big\} \\
&\quad + c\Big\{\sum_{n=1}^{N} \|w^a_n - \tfrac{1}{2}(w_n + w_{n-1})\|_H \|w_n - w_{n-1}\|_H \\
&\quad + \int_0^T \|\dot{w}(t) - \dot{\overline{w}}(t)\|_H dt + k\int_0^T \|\dddot{\overline{w}}(t)\|_H dt \\
&\quad + \sum_{n=1}^{N} \int_{I_n} \|\ell(t) - \ell^a_n\|_{H'} \|\dot{w}(t)\|_H dt\Big\}^{1/2}.
\end{aligned} \tag{11.55}$$

Let us analyze each term on the right-hand side of (11.55). First, for the terms involving $E^a_{n,\theta}(w)$, we have

$$\begin{aligned}
&\|E^a_{n,\theta}(w) - E^a_{n+1,\theta}(w)\|_H \\
&\quad \le \|E^a_{n,\theta}(\overline{w}) - E^a_{n+1,\theta}(\overline{w})\|_H \\
&\qquad + \|E^a_{n,\theta}(w - \overline{w})\|_H + \|E^a_{n+1,\theta}(w - \overline{w})\|_H.
\end{aligned} \tag{11.56}$$

Since $w(t) - \overline{w}(t)$ is continuous in t,

$$w^a_n - \overline{w}^a_n = \frac{1}{k}\int_{I_n} [w(t) - \overline{w}(t)]\, dt = w(\tau_n) - \overline{w}(\tau_n), \quad \text{for some } \tau_n \in I_n.$$

Hence

$$\begin{aligned}
&E^a_{n,\theta}(w - \overline{w}) \\
&\quad = \theta\,(w_n - \overline{w}_n) + (1-\theta)\,(w_{n-1} - \overline{w}_{n-1}) - (w(\tau_n) - \overline{w}(\tau_n)) \\
&\quad = \theta \int_{\tau_n}^{t_n} [\dot{w}(t) - \dot{\overline{w}}(t)]\, dt + (1-\theta) \int_{\tau_n}^{t_{n-1}} [\dot{w}(t) - \dot{\overline{w}}(t)]\, dt,
\end{aligned}$$

and then

$$\|E^a_{n,\theta}(w - \overline{w})\|_H \le c\int_{t_{n-1}}^{t_n} \|\dot{w}(t) - \dot{\overline{w}}(t)\|_H dt.$$

Similarly,

$$\|E^a_{n+1,\theta}(w - \overline{w})\|_H \le c\int_{t_n}^{t_{n+1}} \|\dot{w}(t) - \dot{\overline{w}}(t)\|_H dt.$$

Noticing that

$$E^a_{n,\theta}(\overline{w}) = E_{n,\theta}(\overline{w}) + \overline{w}_{n-1+\theta} - \overline{w}^a_n,$$

we have

$$\begin{aligned} &E^a_{n,\theta}(\overline{w}) - E^a_{n+1,\theta}(\overline{w}) \\ &\quad = (E_{n,\theta}(\overline{w}) - E_{n+1,\theta}(\overline{w})) + (\overline{w}_{n-1+\theta} - \overline{w}^a_n) - (\overline{w}_{n+\theta} - \overline{w}^a_{n+1}). \end{aligned}$$

By Lemma 11.3,

$$\|E_{n,\theta}(\overline{w}) - E_{n+1,\theta}(\overline{w})\|_H \le c\,k\,\|\dddot{\overline{w}}\|_{L^1(t_{n-1},t_{n+1};H)}.$$

We use the Taylor expansion

$$\overline{w}(t) = \overline{w}_{n-1+\theta} + \dot{\overline{w}}_{n-1+\theta}(t - t_{n-1+\theta}) + \int_{t_{n-1+\theta}}^{t} (t-s)\,\ddot{\overline{w}}(s)\,ds$$

to get

$$\overline{w}_{n-1+\theta} - \overline{w}^a_n = -\frac{1-2\,\theta}{2}\,k\,\dot{\overline{w}}_{n-1+\theta} - \frac{1}{k}\int_{I_n}\int_{t_{n-1+\theta}}^{t} (t-s)\,\ddot{\overline{w}}(s)\,ds\,dt. \tag{11.57}$$

So

$$\begin{aligned} &(\overline{w}_{n-1+\theta} - \overline{w}^a_n) - (\overline{w}_{n+\theta} - \overline{w}^a_{n+1}) \\ &\quad = \frac{1-2\,\theta}{2}\,k\,\left(\dot{\overline{w}}_{n+\theta} - \dot{\overline{w}}_{n-1+\theta}\right) - \frac{1}{k}\int_{I_n}\int_{t_{n-1+\theta}}^{t} (t-s)\,\ddot{\overline{w}}(s)\,ds\,dt \\ &\qquad + \frac{1}{k}\int_{I_{n+1}}\int_{t_{n+\theta}}^{t} (t-s)\,\ddot{\overline{w}}(s)\,ds\,dt, \end{aligned}$$

in which $\dot{\overline{w}}_{n+\theta} - \dot{\overline{w}}_{n-1+\theta} = \int_{t_{n-1+\theta}}^{t_{n+\theta}} \ddot{\overline{w}}(t)\,dt$, and then

$$\begin{aligned} &\|E^a_{n,\theta}(\overline{w}) - E^a_{n+1,\theta}(\overline{w})\|_H \\ &\quad \le \|E_{n,\theta}(\overline{w}) - E_{n+1,\theta}(\overline{w})\|_H \\ &\qquad + \|(\overline{w}_{n-1+\theta} - \overline{w}^a_n) - (\overline{w}_{n+\theta} - \overline{w}^a_{n+1})\|_H \end{aligned}$$

implies

$$\|E^a_{n,\theta}(\overline{w}) - E^a_{n+1,\theta}(\overline{w})\|_H \le c\,k\,\|\dddot{\overline{w}}\|_{L^1(t_{n-1},t_{n+1};H)}.$$

Therefore (cf. (11.56)),

$$\begin{aligned} &\|E^a_{n,\theta}(w) - E^a_{n+1,\theta}(w)\|_H \\ &\quad \le c\,k\,\|\dddot{\overline{w}}\|_{L^1(t_{n-1},t_{n+1};H)} + c\int_{t_{n-1}}^{t_{n+1}} \|\dot{w}(t) - \dot{\overline{w}}(t)\|_H dt, \end{aligned}$$

and thus,

$$\sum_{n=1}^{N-1} \|E_{n,\theta}^a(w) - E_{n+1,\theta}^a(w)\|_H \\ \leq c\,k\,\|\ddot{\overline{w}}\|_{L^1(0,T;H)} + c\,\|\dot{w} - \dot{\overline{w}}\|_{L^1(0,T;H)} \\ \leq c\,k\,\|\ddot{\overline{w}}\|_{L^1(0,T;H)} + c\,\varepsilon. \tag{11.58}$$

The formula (11.57) with $n = N$ implies

$$\|\overline{w}_{N-1+\theta} - \overline{w}_N^a\|_H \leq c\,k\,\left[\|\dot{\overline{w}}\|_{L^\infty(t_{N-1},t_N;H)} + \|\ddot{\overline{w}}\|_{L^1(t_{N-1},t_N;H)}\right].$$

By Lemma 11.2,

$$\|E_{N,\theta}(\overline{w})\|_H \leq c\,k\,\|\ddot{\overline{w}}\|_{L^1(t_{N-1},t_N;H)}.$$

It is not difficult to see that

$$\|E_{N,\theta}^a(w) - E_{N,\theta}^a(\overline{w})\|_H \leq c \sup_{t_{N-1}\leq t\leq t_N} \|w(t) - \overline{w}(t)\|_H.$$

Applying (11.45), we then have

$$\|E_{N,\theta}^a(w) - E_{N,\theta}^a(\overline{w})\|_H \leq c\,\varepsilon.$$

Using the last several bounds in the inequality

$$\|E_{N,\theta}^a(w)\|_H \leq \|E_{N,\theta}^a(w) - E_{N,\theta}^a(\overline{w})\|_H \\ + \|E_{N,\theta}(\overline{w})\|_H + \|\overline{w}_{N-1+\theta} - \overline{w}_N^a\|_H,$$

we obtain

$$\|E_{N,\theta}^a(w)\|_H \leq c\,\varepsilon + c\,k\,\left[\|\dot{\overline{w}}\|_{L^\infty(t_{N-1},t_N;H)} + \|\ddot{\overline{w}}\|_{L^1(t_{N-1},t_N;H)}\right]. \tag{11.59}$$

We will now estimate the terms involving approximations of ℓ. First we have an $\overline{\ell} \in C^\infty([0,T];H')$ such that

$$\|\ell - \overline{\ell}\|_{H^1(0,T;H')} \leq \varepsilon, \tag{11.60}$$

and then

$$\|\ell - \overline{\ell}\|_{C([0,T];H')} \leq c\,\|\ell - \overline{\ell}\|_{H^1(0,T;H')} \leq c\,\varepsilon. \tag{11.61}$$

We have

$$\|\ell_N^a - \ell_{N-1+\theta}\|_{H'} \\ \leq \|\overline{\ell}_N^a - \overline{\ell}_{N-1+\theta}\|_{H'} + \|\ell_N^a - \overline{\ell}_N^a\|_{H'} + \|\ell_{N-1+\theta} - \overline{\ell}_{N-1+\theta}\|_{H'} \\ \leq \|\overline{\ell}_N^a - \overline{\ell}_{N-1+\theta}\|_{H'} + c\,\|\ell - \overline{\ell}\|_{L^\infty(0,T;H')} \\ \leq \|\overline{\ell}_N^a - \overline{\ell}_{N-1+\theta}\|_{H'} + c\,\varepsilon.$$

Using the formula (11.57) for $\overline{\ell}$, we see that

$$\|\overline{\ell}_N^a - \overline{\ell}_{N-1+\theta}\|_{H'} \leq c\,k\,\Big[\|\dot{\overline{\ell}}\|_{L^\infty(t_{N-1},t_N;H')} + \|\ddot{\overline{\ell}}\|_{L^1(t_{N-1},t_N;H')}\Big].$$

Hence,

$$\|\ell_N^a - \ell_{N-1+\theta}\|_{H'} \leq c\,k\,\Big[\|\dot{\overline{\ell}}\|_{L^\infty(0,T;H')} + \|\ddot{\overline{\ell}}\|_{L^1(0,T;H')}\Big] + c\,\varepsilon. \qquad (11.62)$$

Similarly,

$$\begin{aligned}
&\|(\ell_n^a - \ell_{n-1+\theta}) - (\ell_{n+1}^a - \ell_{n+\theta})\|_{H'} \\
&\quad \leq \|(\overline{\ell}_n^a - \overline{\ell}_{n-1+\theta}) - (\overline{\ell}_{n+1}^a - \overline{\ell}_{n+\theta})\|_{H'} \\
&\qquad + \|(\ell_n^a - \ell_{n+1}^a) - (\overline{\ell}_n^a - \overline{\ell}_{n+1}^a)\|_{H'} \\
&\qquad + \|(\ell_{n-1+\theta} - \ell_{n+\theta}) - (\overline{\ell}_{n-1+\theta} - \overline{\ell}_{n+\theta})\|_{H'} \\
&\quad \leq \|(\overline{\ell}_n^a - \overline{\ell}_{n-1+\theta}) - (\overline{\ell}_{n+1}^a - \overline{\ell}_{n+\theta})\|_{H'} + c\,\|\dot{\ell} - \dot{\overline{\ell}}\|_{L^1(t_{n-1},t_{n+1};H')}.
\end{aligned}$$

Use the formula (11.57) for $\overline{\ell}$ once more:

$$\|(\overline{\ell}_n^a - \overline{\ell}_{n-1+\theta}) - (\overline{\ell}_{n+1}^a - \overline{\ell}_{n+\theta})\|_{H'} \leq c\,k\,\|\ddot{\overline{\ell}}\|_{L^1(t_{n-1},t_{n+1};H')}.$$

Therefore,

$$\begin{aligned}
&\sum_{n=1}^{N-1} \|(\ell_n^a - \ell_{n-1+\theta}) - (\ell_{n+1}^a - \ell_{n+\theta})\|_{H'} \\
&\qquad \leq c\,k\,\|\ddot{\overline{\ell}}\|_{L^1(0,T;H')} + c\,\|\dot{\ell} - \dot{\overline{\ell}}\|_{L^1(0,T;H')}.
\end{aligned}$$

Hence,

$$\sum_{n=1}^{N-1} \|(\ell_n^a - \ell_{n-1+\theta}) - (\ell_{n+1}^a - \ell_{n+\theta})\|_{H'} \leq c\,k\,\|\ddot{\overline{\ell}}\|_{L^1(0,T;H')} + c\,\varepsilon. \qquad (11.63)$$

For the last sum in (11.55), we first use the Cauchy–Schwarz inequality:

$$\begin{aligned}
&\sum_{n=1}^{N} \int_{I_n} \|\ell(t) - \ell_n^a\|_{H'} \|\dot{w}(t)\|_H dt \\
&\qquad \leq \|\dot{w}\|_{L^2(0,T;H)} \Big[\sum_{n=1}^{N} \int_{I_n} \|\ell(t) - \ell_n^a\|_{H'}^2 dt\Big]^{1/2}.
\end{aligned}$$

Then using the inequality

$$\|\ell(t) - \ell_n^a\|_{H'}^2 \leq c\,\Big[\|\overline{\ell}(t) - \overline{\ell}_n^a\|_{H'}^2 + \|\ell_n^a - \overline{\ell}_n^a\|_{H'}^2 + \|\ell(t) - \overline{\ell}(t)\|_{H'}^2\Big],$$

we find that

$$\sum_{n=1}^{N}\int_{I_n}\|\ell(t)-\ell_n^a\|_{H'}^2dt$$
$$\leq c\sum_{n=1}^{N}\int_{I_n}\|\overline{\ell}(t)-\overline{\ell}_n^a\|_{H'}^2\,dt+c\,k\sum_{n=1}^{N}\|\ell_n^a-\overline{\ell}_n^a\|_{H'}^2$$
$$+\,c\,\|\ell-\overline{\ell}\|_{L^2(0,T;H')}^2.$$

Now,

$$\overline{\ell}(t)-\overline{\ell}_n^a=\frac{1}{k}\int_{I_n}[\overline{\ell}(t)-\overline{\ell}(s)]\,ds=\frac{1}{k^2}\int_{I_n}\int_s^t\dot{\overline{\ell}}(\tau)\,d\tau\,ds,$$

so

$$\sum_{n=1}^{N}\int_{I_n}\|\overline{\ell}(t)-\overline{\ell}_n^a\|_{H'}^2\,dt\leq c\,k^2\sum_{n=1}^{N}\int_{I_n}\|\dot{\overline{\ell}}(\tau)\|_{H'}^2d\tau=c\,k^2\|\dot{\overline{\ell}}\|_{L^2(0,T;H')}^2.$$

Similarly,

$$\sum_{n=1}^{N}\|\ell_n^a-\overline{\ell}_n^a\|_{H'}^2\leq\frac{1}{k}\sum_{n=1}^{N}\int_{I_n}\|\ell(t)-\overline{\ell}(t)\|_{H'}^2\,dt\leq\frac{1}{k}\,\|\ell-\overline{\ell}\|_{L^2(0,T;H')}^2.$$

Therefore,

$$\sum_{n=1}^{N}\int_{I_n}\|\ell(t)-\ell_n^a\|_{H'}^2dt\leq c\,k^2\|\dot{\overline{\ell}}\|_{L^2(0,T;H')}^2+c\,\|\ell-\overline{\ell}\|_{L^2(0,T;H')}^2,$$

and then

$$\sum_{n=1}^{N}\int_{I_n}\|\ell(t)-\ell_n^a\|_{H'}\|\dot{w}(t)\|_H dt\leq c\,\|\dot{w}\|_{L^2(0,T;H)}\left[k\,\|\dot{\overline{\ell}}\|_{L^2(0,T;H')}+\varepsilon\right].\tag{11.64}$$

Finally, we estimate the terms $\|w_n^a-\frac{1}{2}\,(w_n+w_{n-1})\|_H$ and $\|w_n-w_{n-1}\|_H$. We have

$$w_n^a-\tfrac{1}{2}\,(w_n+w_{n-1})$$
$$=\frac{1}{k}\int_{I_n}\left[w(t)-\tfrac{1}{2}\,(w_n+w_{n-1})\right]\,dt$$
$$=-\frac{1}{2\,k}\int_{I_n}\left[\int_t^{t_n}\dot{w}(s)\,ds+\int_t^{t_{n-1}}\dot{w}(s)\,ds\right]dt,$$

and thus

$$\|w_n^a - \tfrac{1}{2}(w_n + w_{n-1})\|_H \le c \int_{I_n} \|\dot{w}(t)\|_H dt.$$

Also,

$$\|w_n - w_{n-1}\|_H = \left\| \int_{I_n} \dot{w}(t)\, dt \right\|_H \le \int_{I_n} \|\dot{w}(t)\|_H\, dt.$$

Then

$$\begin{aligned}
&\sum_{n=1}^{N} \|w_n^a - \tfrac{1}{2}(w_n + w_{n-1})\|_H \|w_n - w_{n-1}\|_H \\
&\qquad \le c \sum_{n=1}^{N} \Big(\int_{I_n} \|\dot{w}(t)\|_H\, dt \Big)^2 \\
&\qquad \le c\, k\, \|\dot{w}\|_{L^2(0,T;H)}.
\end{aligned} \tag{11.65}$$

Summarizing, using (11.55), (11.58), (11.59), (11.62)–(11.65), and the inequality

$$\|\dot{w} - \dot{\overline{w}}\|_{L^1(0,T;H)} \le c\, \|\dot{w} - \dot{\overline{w}}\|_{L^2(0,T;H)} \le c\, \varepsilon,$$

we find the following error estimate

$$\begin{aligned}
&\max_{1\le n\le N} \|w_n^k - w_n\|_H \\
&\quad \le c \Big\{ \varepsilon + k\, (\|\dot{\overline{w}}\|_{L^\infty(0,T;H)} + \|\ddot{\overline{w}}\|_{L^1(0,T;H)} \\
&\qquad\qquad + \|\dot{\overline{\ell}}\|_{L^\infty(0,T;H')} + \|\ddot{\overline{\ell}}\|_{L^1(0,T;H')}) \Big\} \\
&\qquad + c \Big\{ \varepsilon + k\, (\|\dot{w}\|_{L^2(0,T;H)} + \|\ddot{\overline{w}}\|_{L^1(0,T;H)}) \\
&\qquad\qquad + \|\dot{w}\|_{L^2(0,T;H)}\, (\varepsilon + k\, \|\dot{\overline{\ell}}\|_{L^2(0,T;H')}) \Big\}^{1/2}.
\end{aligned} \tag{11.66}$$

The estimate (11.66) implies convergence under the basic solution regularity condition.

THEOREM 11.8. *Suppose the standard assumptions on H, K, a, ℓ, and j are satisfied. Let $w \in H^1(0,T;H)$ and w^k be the solutions of the problems* ABS *and* ABS^k*, respectively. Then the time-discrete solution w^k converges to w in the sense that*

$$\max_{1\le n\le N} \|w_n^k - w_n\|_H \to 0 \quad \text{as } k \to 0. \tag{11.67}$$

Convergence of the fully discrete scheme. Under the basic regularity condition $w \in H^1(0,T;H)$, we cannot use the estimate (11.39) to show the convergence of the fully discrete method ABS^{hk} because the pointwise values $\dot{w}_{j-1+\theta}$ in the estimate are not defined. Thus we need to derive an estimate similar to (11.39) without the appearance of pointwise values of $\dot{w}$. It will not be enough for the purpose of showing convergence if we have only the property (11.4) for the finite element space. What we need is the property (11.4) in certain uniform manner for a set of elements z. We make the following additional assumptions about the function space and the finite element space.

(H$_1$) There exists a subspace $H_0 \subset H$, such that $H^1(0,T;H_0 \cap K)$ is dense in $H^1(0,T;K)$ in the norm of $H^1(0,T;H)$.

(H$_2$) For some constants c and $\alpha > 0$, we have the estimate

$$\inf_{z^h \in K^h} \|z - z^h\|_H \le c\,\|z\|_{H_0}\, h^\alpha \quad \forall\, z \in H_0 \cap K. \tag{11.68}$$

In the next chapter, when we apply the general result proved here for the convergence of the fully discrete solutions for solving the primal variational problem, we will need to verify both hypotheses (H$_1$) and (H$_2$).

By the assumption (H$_1$), for any $\varepsilon > 0$ we have $\tilde{w} \in H^1(0,T;H_0)$ such that

$$\|w - \tilde{w}\|_{H^1(0,T;H)} \le \varepsilon. \tag{11.69}$$

We still use $e_n = w_n - w_n^{hk}$, $0 \le n \le N$, to denote the errors, and we consider the quantity A_n defined in (11.33). A lower bound of A_n is given by (11.34) and an upper bound by (11.35). Instead of (11.36), we integrate (11.1) with $z = \delta w_n^{hk} \in K$ from $t = t_{n-1}$ to $t = t_n$,

$$\begin{aligned} 0 \le {} & \frac{1}{k}\int_{I_n} a(w(t), \delta w_n^{hk} - \dot{w}(t))\,dt + j(\delta w_n^{hk}) \\ & - \frac{1}{k}\int_{I_n} j(\dot{w}(t))\,dt - \frac{1}{k}\int_{I_n} \langle \ell(t), \delta w_n^{hk} - \dot{w}(t)\rangle\,dt. \end{aligned} \tag{11.70}$$

We then add (11.70) to (11.35) to obtain

$$A_n \le R_1 + R_2 + R_3, \tag{11.71}$$

where

$$\begin{aligned} R_1 &= a(\theta\, w_n + (1-\theta)\, w_{n-1}, \delta e_n) + \frac{1}{k}\int_{I_n} a(w(t), \delta w_n^{hk} - \dot{w}(t))\,dt \\ &\quad - a(\theta\, w_n^{hk} + (1-\theta)\, w_{n-1}^{hk}, \delta w_n - z_n^h), \\ R_2 &= j(z_n^h) - \frac{1}{k}\int_{I_n} j(\dot{w}(t))\,dt, \\ R_3 &= -\langle \ell_{n-1+\theta}, z_n^h - \delta w_n^{hk}\rangle - \frac{1}{k}\int_{I_n} \langle \ell(t), \delta w_n^{hk} - \dot{w}(t)\rangle\,dt. \end{aligned}$$

As before, we use

$$w_n^a = \frac{1}{k} \int_{I_n} w(t)\, dt, \quad n = 1, \ldots, N,$$

for local averages and let

$$E_{n,\theta}^a(w) = \theta\, w_n + (1-\theta)\, w_{n-1} - w_n^a, \quad n = 1, \ldots, N.$$

Then

$$\begin{aligned} R_1 &= a(E_{n,\theta}^a(w), \delta e_n) + \frac{1}{k} \int_{I_n} a(w(t), \delta w_n - \dot{w}(t))\, dt \\ &\quad + a(\theta\, e_n + (1-\theta)\, e_{n-1}, \delta w_n - z_n^h) \\ &\quad - a(\theta\, w_n + (1-\theta)\, w_{n-1}, \delta w_n - z_n^h). \end{aligned} \tag{11.72}$$

Since

$$R_2 = \frac{1}{k} \int_{I_n} \left[j(z_n^h) - j(\dot{w}(t)) \right] dt,$$

using the Lipschitz continuity of $j(\cdot)$ on K, we have

$$|R_2| \le \frac{c}{k} \int_{I_n} \|z_n^h - \dot{w}(t)\|_H dt \le \frac{c}{k} \int_{I_n} \|z_n^h - \dot{w}(t)\|_H dt. \tag{11.73}$$

Finally, R_3 can be rewritten as

$$\begin{aligned} R_3 &= \langle \ell_n^a - \ell_{n-1+\theta}, \delta e_n \rangle + \frac{1}{k} \int_{I_n} \langle \ell(t), \dot{w}(t) - \delta w_n \rangle\, dt \\ &\quad - \langle \ell_{n-1+\theta}, z_n^h - \delta w_n \rangle. \end{aligned} \tag{11.74}$$

Combine (11.34) with (11.71)–(11.74) and multiply the resulting inequality by $2k$,

$$\begin{aligned} &\|e_n\|_a^2 - \|e_{n-1}\|_a^2 \\ &\quad \le 2\, a(E_{n,\theta}^a(w), e_n - e_{n-1}) + 2 \int_{I_n} a(w(t), \delta w_n - \dot{w}(t))\, dt \\ &\qquad + 2\, k\, a(\theta\, e_n + (1-\theta)\, e_{n-1}, \delta w_n - z_n^h) \\ &\qquad - 2\, k\, a(\theta\, w_n + (1-\theta)\, w_{n-1}, \delta w_n - z_n^h) \\ &\qquad + c \int_{I_n} \|z_n^h - \dot{w}(t)\|_H dt \\ &\qquad + 2\, \langle \ell_n^a - \ell_{n-1+\theta}, e_n - e_{n-1} \rangle + 2 \int_{I_n} \langle \ell(t), \dot{w}(t) - \delta w_n \rangle\, dt \\ &\qquad - 2\, k\, \langle \ell_{n-1+\theta}, z_n^h - \delta w_n \rangle. \end{aligned}$$

Again we set

$$M = \max_{1 \le n \le N} \|e_n\|_a.$$

Since $e_0 = 0$, mathematical induction based on the above inequality reveals that for $n = 1, \ldots, N$,

$$\begin{aligned}
\|e_n\|_a^2 &\\
\le 2 &\sum_{j=1}^n a(E_{j,\theta}^a(w), e_j - e_{j-1}) + 2 \sum_{j=1}^n \int_{I_j} a(w(t), \delta w_j - \dot{w}(t))\, dt \\
&+ 2 \sum_{j=1}^n \int_{I_j} \langle \ell(t), \dot{w}(t) - \delta w_j \rangle\, dt + 2 \sum_{j=1}^n \langle \ell_j^a - \ell_{j-1+\theta}, e_j - e_{j-1} \rangle \\
&+ c\,k\,(M + \|w\|_{L^\infty(0,T;H)}) \sum_{j=1}^n \|\delta w_j - z_j^h\|_H \\
&+ c \sum_{j=1}^n \int_{I_j} \|z_j^h - \dot{w}(t)\|_H dt + c\,k\,\|\ell\|_{L^\infty(0,T;H')} \sum_{j=1}^n \|\delta w_j - z_j^h\|_H.
\end{aligned}$$

Since

$$\begin{aligned}
&\sum_{j=1}^n a(E_{j,\theta}^a(w), e_j - e_{j-1}) \\
&\qquad = a(E_{n,\theta}^a(w), e_n) + \sum_{j=1}^{n-1} a(E_{j,\theta}^a(w) - E_{j+1,\theta}^a(w), e_j)
\end{aligned}$$

and

$$\begin{aligned}
&\sum_{j=1}^n \langle \ell_j^a - \ell_{j-1+\theta}, e_j - e_{j-1} \rangle \\
&\qquad = \langle \ell_n^a - \ell_{n-1+\theta}, e_n \rangle + \sum_{j=1}^{n-1} \langle (\ell_j^a - \ell_{j-1+\theta}) - (\ell_{j+1}^a - \ell_{j+\theta}), e_j \rangle,
\end{aligned}$$

we see that

$$
\begin{aligned}
&\|e_n\|_a^2 \\
&\quad \le c\,M\left(\|E^a_{n,\theta}(w)\|_H + \sum_{j=1}^{n-1}\|E^a_{j,\theta}(w) - E^a_{j+1,\theta}(w)\|_H\right) \\
&\quad + 2\sum_{j=1}^{n}\int_{I_j} a(w(t),\delta w_j - \dot{w}(t))\,dt + 2\sum_{j=1}^{n}\int_{I_j}\langle \ell(t),\dot{w}(t) - \delta w_j\rangle\,dt \\
&\quad + c\,M\left(\|\ell^a_n - \ell_{n-1+\theta}\|_{H'} + \sum_{j=1}^{n-1}\|(\ell^a_j - \ell_{j-1+\theta}) - (\ell^a_{j+1} - \ell_{j+\theta})\|_{H'}\right) \\
&\quad + c\,k\,(M + \|w\|_{L^\infty(0,T;H)})\sum_{j=1}^{n}\|\delta w_j - z^h_j\|_H \\
&\quad + c\sum_{j=1}^{n}\int_{I_j}\|z^h_j - \dot{w}(t)\|_H dt + c\,k\,\|\ell\|_{L^\infty(0,T;H')}\sum_{j=1}^{n}\|\delta w_j - z^h_j\|_H.
\end{aligned}
$$

Then

$$
\begin{aligned}
M^2 \le c\,M\Bigg\{&\|E^a_{N,\theta}(w)\|_H + \sum_{n=1}^{N-1}\|E^a_{n,\theta}(w) - E^a_{n+1,\theta}(w)\|_H \\
&+ k\sum_{n=1}^{N}\|\delta w_n - z^h_n\|_H + \|\ell^a_N - \ell_{N-1+\theta}\|_{H'} \\
&+ \sum_{n=1}^{N-1}\|(\ell^a_n - \ell_{n-1+\theta}) - (\ell^a_{n+1} - \ell_{n+\theta})\|_{H'}\Bigg\} \\
&+ c\,(\|w\|_{L^\infty(0,T;H)} + \|\ell\|_{L^\infty(0,T;H')})\sum_{n=1}^{N}\int_{I_n}\|\delta w_n - \dot{w}(t)\|_H\,dt \\
&+ c\,k\,(\|w\|_{L^\infty(0,T;H)} + \|\ell\|_{L^\infty(0,T;H')})\sum_{n=1}^{N}\|\delta w_n - z^h_n\|_H \\
&+ c\sum_{n=1}^{N}\int_{I_n}\|\dot{w}(t) - z^h_n\|_H\,dt.
\end{aligned}
$$

Applying (11.3), we see that

$$
\begin{aligned}
M \le c\Big\{ & \|E^a_{N,\theta}(w)\|_H + \sum_{n=1}^{N-1} \|E^a_{n,\theta}(w) - E^a_{n+1,\theta}(w)\|_H \\
& + k \sum_{n=1}^{N} \|\delta w_n - z^h_n\|_H + \|\ell^a_N - \ell_{N-1+\theta}\|_{H'} \\
& + \sum_{n=1}^{N-1} \|(\ell^a_n - \ell_{n-1+\theta}) - (\ell^a_{n+1} - \ell_{n+\theta})\|_{H'} \Big\} \\
& + c\Big\{ \big(\|w\|_{L^\infty(0,T;H)} + \|\ell\|_{L^\infty(0,T;H')}\big) \sum_{n=1}^{N} \int_{I_n} \|\delta w_n - \dot{w}(t)\|_H \, dt \\
& \quad + k\,\big(\|w\|_{L^\infty(0,T;H)} + \|\ell\|_{L^\infty(0,T;H')}\big) \sum_{n=1}^{N} \|\delta w_n - z^h_n\|_H \\
& \quad + \sum_{n=1}^{N} \int_{I_n} \|\dot{w}(t) - z^h_n\|_H \, dt \Big\}^{1/2}.
\end{aligned}
\tag{11.75}
$$

We now estimate the quantity

$$\sum_{n=1}^{N} \int_{I_n} \|\delta w_n - \dot{w}(t)\|_H \, dt.$$

We write

$$\delta w_n - \dot{w}(t) = \frac{1}{k} \int_{I_n} [\dot{w}(s) - \dot{w}(t)] \, ds.$$

Hence,

$$
\begin{aligned}
& \int_{I_n} \|\delta w_n - \dot{w}(t)\|_H \, dt \\
& \quad \le \frac{1}{k} \int_{I_n \times I_n} \big[\|\dot{w}(s) - \dot{\tilde{w}}(s)\|_H + \|\dot{w}(t) - \dot{\tilde{w}}(t)\|_H \\
& \qquad\qquad + \|\dot{\tilde{w}}(s) - \dot{\tilde{w}}(t)\|_H \big] \, ds \, dt \\
& \quad = c \int_{I_n} \|\dot{w}(t) - \dot{\tilde{w}}(t)\|_H \, dt + \frac{1}{k} \int_{I_n} \int_{I_n} \Big\| \int_t^s \ddot{\tilde{w}}(\tau) \, d\tau \Big\|_H \, ds \, dt \\
& \quad \le c \int_{I_n} \|\dot{w}(t) - \dot{\tilde{w}}(t)\|_H \, dt + c\, k \int_{I_n} \|\ddot{\tilde{w}}(t)\|_H \, dt.
\end{aligned}
$$

Therefore,

$$
\begin{aligned}
& \sum_{n=1}^{N} \int_{I_n} \|\delta w_n - \dot{w}(t)\|_H \, dt \\
& \qquad \le c\, \|\dot{w} - \dot{\tilde{w}}\|_{L^1(0,T;H)} + c\, k\, \|\ddot{\tilde{w}}\|_{L^1(0,T;H)} \\
& \qquad \le c\, \varepsilon + c\, k\, \|\ddot{\tilde{w}}\|_{L^1(0,T;H)}.
\end{aligned}
\tag{11.76}
$$

Next, from

$$\|\dot{w}(t) - z_n^h\|_H \le \|\delta w_n - \dot{w}(t)\|_H + \|\delta w_n - z_n^h\|_H,$$

we see that

$$\sum_{n=1}^N \int_{I_n} \|\dot{w}(t) - z_n^h\|_H \, dt \le \sum_{n=1}^N \int_{I_n} \|\delta w_n - \dot{w}(t)\|_H \, dt + k \sum_{n=1}^N \|\delta w_n - z_n^h\|_H. \quad (11.77)$$

It remains for us to estimate the term

$$k \sum_{n=1}^N \|\delta w_n - z_n^h\|_H.$$

We have

$$\delta w_n - z_n^h = \frac{1}{k} \int_{I_n} \dot{w}(t) \, dt - z_n^h = \frac{1}{k} \int_{I_n} \left[\dot{w}(t) - \dot{\tilde{w}}(t)\right] \, dt + \delta \tilde{w}_n - z_n^h,$$

and so

$$\|\delta w_n - z_n^h\|_H \le \frac{1}{k} \int_{I_n} \|\dot{w}(t) - \dot{\tilde{w}}(t)\|_H \, dt + \|\delta \tilde{w}_n - z_n^h\|_H,$$

$$k \sum_{n=1}^N \|\delta w_n - z_n^h\|_H \le \|\dot{w} - \dot{\tilde{w}}\|_{L^1(0,T;H)} + k \sum_{n=1}^N \|\delta \tilde{w}_n - z_n^h\|_H$$
$$\le c\,\varepsilon + k \sum_{n=1}^N \|\delta \tilde{w}_n - z_n^h\|_H. \quad (11.78)$$

Using the bounds (11.58), (11.59), (11.62), (11.63), and (11.76)–(11.78) in the inequality (11.75) and noticing the arbitrariness of $z_n^h \in K^h$, we obtain the estimate

$$\max_{1 \le n \le N} \|w_n^{hk} - w_n\|_H \le c \Big\{ \varepsilon + D_h(\tilde{w}) + k \Big(\|\dot{\tilde{w}}\|_{L^\infty(0,T;H)} + \|\ddot{\tilde{w}}\|_{L^1(0,T;H)} + \|\dot{\tilde{\ell}}\|_{L^\infty(0,T;H')} + \|\ddot{\tilde{\ell}}\|_{L^1(0,T;H')} \Big) \Big\}$$
$$+ c \Big\{ (\varepsilon + k\, \|\ddot{\tilde{w}}\|_{L^1(0,T;H)} + D_h(\tilde{w})) \times (\|\dot{w}\|_{L^\infty(0,T;H)} + \|\dot{\ell}\|_{L^\infty(0,T;H')} + 1) \Big\}^{1/2}, \quad (11.79)$$

where

$$D_{hk}(\tilde{w}) = k \sum_{n=1}^{N} \inf_{z_n^h \in K^h} \|\delta \tilde{w}_n - z_n^h\|_H. \tag{11.80}$$

By the assumption (H$_2$), we see that

$$\inf_{z_n^h \in K^h} \|\delta \tilde{w}_n - z_n^h\|_H \le c\, h^\alpha \|\delta \tilde{w}_n\|_{H_0} \le \frac{c\, h^\alpha}{k} \int_{I_n} \|\dot{\tilde{w}}(t)\|_{H_0} dt,$$

and thus

$$D_{hk}(\tilde{w}) = k \sum_{n=1}^{N} \inf_{z_n^h \in K^h} \|\delta \tilde{w}_n - z_n^h\|_H \le c\, h^\alpha \|\dot{\tilde{w}}\|_{L^1(0,T;H_0)}.$$

Use this inequality in the estimate (11.79):

$$\begin{aligned}
&\max_{1 \le n \le N} \|w_n^{hk} - w_n\|_H \\
&\quad \le c \Big\{ \varepsilon + h^\alpha \|\dot{\tilde{w}}\|_{L^1(0,T;H_0)} + k \Big(\|\dot{\tilde{w}}\|_{L^\infty(0,T;H)} + \|\ddot{\tilde{w}}\|_{L^1(0,T;H)} \\
&\qquad + \|\dot{\tilde{\ell}}\|_{L^\infty(0,T;H')} + \|\ddot{\tilde{\ell}}\|_{L^1(0,T;H')} \Big) \Big\} \\
&\quad + c \Big\{ \big(\varepsilon + k\, \|\ddot{\tilde{w}}\|_{L^1(0,T;H)} + h^\alpha \|\dot{\tilde{w}}\|_{L^1(0,T;H_0)} \big) \\
&\qquad \times \big(\|\dot{w}\|_{L^\infty(0,T;H)} + \|\dot{\ell}\|_{L^\infty(0,T;H')} + 1 \big) \Big\}^{1/2}.
\end{aligned} \tag{11.81}$$

The estimate (11.81) implies the convergence under the basic solution regularity condition.

THEOREM 11.9. *Suppose the standard assumptions on H, K, a, ℓ, and j are satisfied. Let $w \in H^1(0,T;H)$ and w^{hk} be the solutions of the problems* ABS *and* ABS^{hk}*, respectively. Then the fully discrete solution w^{hk} converges to w in the sense that*

$$\max_{1 \le n \le N} \|w_n^{hk} - w_n\|_H \to 0 \quad \text{as } h, k \to 0. \tag{11.82}$$

12
Numerical Analysis of the Primal Problem

In this chapter we consider numerical approximations for the primal variational problem of elastoplasticity. We start with the derivation of error estimates for various numerical schemes approximating the solution of the primal variational problem by applying the results for the abstract variational problem proved in the last chapter. We also discuss the convergence property for various schemes under the basic solution regularity condition.

Then we consider the practically important issue of the implementation of numerical schemes and, in particular, the algorithms that are employed in such schemes. The algorithms considered here are of predictor–corrector type. Detailed derivation of the solution algorithms is given in Section 12.2. Convergence of the solution algorithms is discussed in Section 12.3.

A major difficulty in solving the primal variational problem numerically (and similarly, the inequality problem in a corrector step in the solution algorithms discussed in Section 12.2) is the treatment of the nondifferentiable functional $j(\cdot)$. Several approaches can be used to circumvent the difficulty in practice. One approach is the regularization method, where the nondifferentiable term is approximated by a sequence of differentiable ones. Convergence and error estimations for the regularization method are the main topics of Section 12.4. A practically efficient approach is discretizing the inequality for the continuous variables involving the nondifferentiable term to give a set of uncoupled inequalities at integration points. We give a detailed error analysis for one such method in Section 12.5.

12.1 Error Analysis of Discrete Approximations of the Primal Problem

In this section we apply the general results presented in the last chapter on numerical approximations for the abstract problem to derive various error estimates in the context of the primal form of the elastoplasticity problem for concrete selections of the finite elements under suitable regularity assumptions on the solution and to show the convergence of the methods under the basic regularity condition of the solution. We continue to restrict attention to the problems with combined linear kinematic-isotropic hardening or with linear kinematic hardening only, so that the results of the last chapter are directly applicable.

The problem with combined linear kinematic and isotropic hardening. The continuous problem PRIM1 is stated in Section 7.1 and analyzed in Section 7.3.

Let $Z^h = V^h \times Q_0^h \times M^h$ be a finite-dimensional subspace of Z. Let $K^h = Z^h \cap K = V^h \times K_0^h$, where

$$K_0^h = \{(\boldsymbol{q}^h, \mu^h) \in Q_0^h \times M^h : |\boldsymbol{q}^h| \le \mu^h \text{ in } \Omega\}.$$

In the spatially discrete internal approximation of the problem, we seek $\boldsymbol{w}^h = (\boldsymbol{u}^h, \boldsymbol{p}^h, \gamma^h) : [0,T] \to Z^h$, $\boldsymbol{w}^h(0) = \boldsymbol{0}$, such that for almost all $t \in (0,T)$, $\dot{\boldsymbol{w}}^h(t) \in K^h$ and

$$a(\boldsymbol{w}^h(t), \boldsymbol{z}^h - \dot{\boldsymbol{w}}^h(t)) + j(\boldsymbol{z}^h) - j(\dot{\boldsymbol{w}}^h(t)) \ge \langle \boldsymbol{\ell}_n, \boldsymbol{z}^h - \dot{\boldsymbol{w}}^h(t)\rangle \quad \forall\, \boldsymbol{z}^h \in K^h. \tag{12.1}$$

From the discussion in the last chapter we know that the discrete problem has a unique solution $\boldsymbol{w}^h$. Since $j(\boldsymbol{z})$ depends on $\boldsymbol{q}$ only, a careful examination of the argument in Section 11.1 shows that we may modify the error estimate (11.10) to read

$$\begin{aligned} &\|\boldsymbol{w} - \boldsymbol{w}^h\|_{L^\infty(0,T;Z)} \\ &\quad \le c\Big[\inf_{(\boldsymbol{q}^h,\mu^h)\in L^2(0,T;K_0^h)} \Big(\|\dot{\boldsymbol{p}} - \boldsymbol{q}^h\|^{1/2}_{L^2(0,T;Q)} + \|\dot{\gamma} - \mu^h\|_{L^2(0,T;M)}\Big) \\ &\qquad + \inf_{\boldsymbol{v}^h\in L^2(0,T;V^h)} \|\dot{\boldsymbol{u}} - \boldsymbol{v}^h\|_{L^2(0,T;V)}\Big]. \end{aligned} \tag{12.2}$$

The inequality (12.2) is the basis for various order error estimates. For example, suppose that we use linear elements for V^h and piecewise constants for both Q_0^h and M^h. Assume that $\dot{\boldsymbol{u}} \in L^2(0,T;(H^2(\Omega))^3)$, $\dot{\boldsymbol{p}} \in L^2(0,T;(H^1(\Omega))^{3\times 3})$, and $\dot{\gamma} \in L^2(0,T;H^1(\Omega))$. Then using the standard interpolation error estimates for finite elements reviewed in Chapter 9, we have

$$\inf_{\boldsymbol{v}^h\in L^2(0,T;V^h)} \|\dot{\boldsymbol{u}} - \boldsymbol{v}^h\|_{L^2(0,T;V)} \le c\,h.$$

Let $\boldsymbol{q}^h = \Pi^h \dot{\boldsymbol{p}}$ be the orthogonal projection of $\dot{\boldsymbol{p}}$ onto Q_0^h with respect to the inner product of Q. We observe that on each element, $\Pi^h \dot{\boldsymbol{p}}$ is the average value of $\dot{\boldsymbol{p}}$ on the element. Similarly, we take $\mu^h = \Pi^h \dot{\gamma}$ to be the orthogonal projection of $\dot{\gamma}$ onto M^h with respect to the inner product of M. Since $\dot{\boldsymbol{w}} \in K$ and K is convex, we have $(\Pi^h \dot{\boldsymbol{p}}, \Pi^h \dot{\gamma}) \in K_0^h$. Thus,

$$\|\dot{\boldsymbol{p}} - \Pi^h \dot{\boldsymbol{p}}\|_{L^2(0,T;Q)} \le c\,h,$$
$$\|\dot{\gamma} - \Pi^h \dot{\gamma}\|_{L^2(0,T;M)} \le c\,h,$$

and from (12.2) we get the error estimate

$$\|\boldsymbol{w} - \boldsymbol{w}^h\|_{L^\infty(0,T;Z)} \le c\,h^{1/2}. \tag{12.3}$$

If $\dot{\boldsymbol{p}} \in L^2(0,T;(H^2(\Omega))^{3\times 3})$ and $\dot{\gamma} \in L^2(0,T;H^2(\Omega))$, we can use either discontinuous or continuous piecewise linear functions for both Q_0^h and M^h. By choosing $\Pi^h \dot{\boldsymbol{p}}$ and $\Pi^h \dot{\gamma}$ to be the piecewise linear interpolants of $\dot{\boldsymbol{p}}$ and $\dot{\gamma}$, we have $(\Pi^h \dot{\boldsymbol{p}}, \Pi^h \dot{\gamma}) \in K_0^h$, and

$$\|\dot{\boldsymbol{p}} - \Pi^h \dot{\boldsymbol{p}}\|_{L^2(0,T;Q)} \le c\,h^2,$$
$$\|\dot{\gamma} - \Pi^h \dot{\gamma}\|_{L^2(0,T;M)} \le c\,h^2.$$

Then the error estimate for this case becomes

$$\|\boldsymbol{w} - \boldsymbol{w}^h\|_{L^\infty(0,T;Z)} \le c\,h. \tag{12.4}$$

Results for time-discrete approximations can be deduced in a similar way to those in Section 11.2 and are omitted.

Now let us consider fully discrete approximations. As in the last chapter, we divide the time interval $[0,T]$ by evenly spaced nodes $t_n = nk$, $n = 0,1,\ldots,N$, with $k = T/N$ the step-size. The most useful schemes from the family of fully discrete approximations considered in Section 11.3 are the backward Euler scheme (corresponding to $\theta = 1$) and the Crank–Nicolson scheme (corresponding to $\theta = \frac{1}{2}$). Therefore, in the discussion below we will mention only the results pertaining to these two schemes.

In the backward Euler approximation of the problem, $\boldsymbol{w}_0^{hk} = \boldsymbol{0}$, and we compute $\boldsymbol{w}_n^{hk} = (\boldsymbol{u}_n^{hk}, \boldsymbol{p}_n^{hk}, \gamma_n^{hk}) : [0,T] \to Z^h$, $n = 1,2,\ldots,N$, such that $\delta \boldsymbol{w}_n^{hk} \in K^h$ and

$$a(\boldsymbol{w}_n^{hk}, \boldsymbol{z}^h - \delta \boldsymbol{w}_n^{hk}) + j(\boldsymbol{z}^h) - j(\delta \boldsymbol{w}_n^{hk}) \ge \langle \boldsymbol{\ell}_n, \boldsymbol{z}^h - \delta \boldsymbol{w}_n^{hk} \rangle \quad \forall\, \boldsymbol{z}^h \in Z^h. \tag{12.5}$$

We have a unique solution for the backward Euler scheme. By the estimate (11.42), again noting that $j(\boldsymbol{z})$ depends only on $\boldsymbol{q}$, we find that if $\ddot{\boldsymbol{w}} \in L^2(0,T;Z)$, then

$$\max_{0\le n\le N} \|\boldsymbol{w}_n - \boldsymbol{w}_n^{hk}\|_Z^2 \le c\,k \sum_{n=1}^N \Big[\inf_{\boldsymbol{v}^h \in V^h} \|\dot{\boldsymbol{u}}_n - \boldsymbol{v}^h\|_V^2$$
$$+ \inf_{(\boldsymbol{q}^h,\mu^h)\in K_0^h} \left(\|\dot{\boldsymbol{p}}_n - \boldsymbol{q}^h\|_Q + \|\dot{\gamma}_n - \mu^h\|_M^2 \right) \Big] + c\,k^2. \tag{12.6}$$

Assume that $\dot{\boldsymbol{u}} \in C([0,T];(H^2(\Omega))^3)$, $\dot{\boldsymbol{p}} \in C([0,T];(H^1(\Omega))^{3\times 3})$, and $\dot{\gamma} \in C([0,T];H^1(\Omega))$. If we use linear elements for V^h and piecewise constants for both Q_0^h and M^h, then as is noted earlier, $(\Pi^h \dot{\boldsymbol{p}}, \Pi^h \dot{\gamma}) \in K_0^h$, and for $n = 1, \ldots, N$,

$$\inf_{\boldsymbol{v}^h \in V^h} \|\dot{\boldsymbol{u}}_n - \boldsymbol{v}^h\|_V \le c\,h,$$
$$\|\dot{\boldsymbol{p}}_n - \Pi^h \dot{\boldsymbol{p}}_n\|_Q \le c\,h,$$
$$\|\dot{\gamma}_n - \Pi^h \dot{\gamma}_n\|_M \le c\,h.$$

Therefore, we have the error estimate

$$\max_{0 \le n \le N} \|\boldsymbol{w}_n - \boldsymbol{w}_n^{hk}\|_Z \le c\,(h^{1/2} + k). \tag{12.7}$$

If $\dot{\boldsymbol{p}} \in C([0,T];(H^2(\Omega))^{3\times 3})$, $\dot{\gamma} \in C([0,T];H^2(\Omega))$, and we use either discontinuous or continuous piecewise linear functions for both Q_0^h and M^h, then the error estimate for this case becomes

$$\max_{0 \le n \le N} \|\boldsymbol{w}_n - \boldsymbol{w}_n^{hk}\|_Z \le c\,(h + k). \tag{12.8}$$

Similarly, the Crank–Nicolson scheme for the primal problem has a unique solution, and for the two different choices of the finite element spaces, under suitable smoothness assumptions on the solution of the original problems, we have the error estimates

$$\max_{0 \le n \le N} \|\boldsymbol{w}_n - \boldsymbol{w}_n^{hk}\|_Z \le c\,(h^{1/2} + k^2), \tag{12.9}$$
$$\max_{0 \le n \le N} \|\boldsymbol{w}_n - \boldsymbol{w}_n^{hk}\|_Z \le c\,(h + k^2) \tag{12.10}$$

replacing (12.7) and (12.8), respectively.

If we do not make any regularity assumptions on the solution $\boldsymbol{w}$ of the primal problem, the order error estimates (12.3), (12.4), and (12.7)–(12.10) no longer hold. Nevertheless, we still have convergence of the numerical solutions:

$$\|\boldsymbol{w} - \boldsymbol{w}^h\|_{L^\infty(0,T;Z)} \to 0 \quad \text{as } h \to 0$$

for the spatially discrete schemes,

$$\max_{0 \le n \le N} \|\boldsymbol{w}_n - \boldsymbol{w}_n^k\|_Z \to 0 \quad \text{as } k \to 0$$

for the time-discrete schemes, and

$$\max_{0 \le n \le N} \|\boldsymbol{w}_n - \boldsymbol{w}_n^{hk}\|_Z \to 0 \quad \text{as } h, k \to 0$$

for the fully discrete schemes. Of course, for the convergence of the fully discrete schemes, we need to verify the hypotheses (H_1) and (H_2). This is done next.

Verification of the hypotheses ($\mathbf{H}_1$) and ($\mathbf{H}_2$). In the context of the problem PRIM1 (we give the following discussion for a d-dimensional domain $\Omega \subset \mathbb{R}^d$),

$$\begin{aligned} H &= Z = (H_0^1(\Omega))^d \times Q_0 \times L^2(\Omega), \\ K &= Z_p = \{\boldsymbol{z} = (\boldsymbol{v}, \boldsymbol{q}, \mu) \in Z : |\boldsymbol{q}| \le \mu \text{ a.e. in } \Omega\}. \end{aligned}$$

We will show that we can take

$$H_0 = (H_0^1(\Omega) \cap C^\infty(\overline{\Omega}))^d \times (Q_0 \cap C^\infty(\overline{\Omega})) \times (L^2(\Omega) \cap C^\infty(\overline{\Omega})) \quad (12.11)$$

in (H_1) and (H_2). For this purpose we will make some preparations.

The following result is found in Zeidler [129] (Proposition 23.2).

PROPOSITION 12.1. *Assume that X is a Banach space, $1 \le q < \infty$. Then the space $C([0,T]; X)$ is dense in $L^q(0,T;X)$.*

Using this proposition, we can prove the next result.

PROPOSITION 12.2. *Assume that X is a Banach space, $1 \le q < \infty$, and l is a nonnegative integer. Then the space $C^l([0,T]; X)$ is dense in $W^{l,q}(0,T;X)$.*

PROOF. We prove the result for $l = 1$. A similar argument applies for other values of l.

Let $u \in W^{1,q}(0,T;X)$. Then $u' \in L^q(0,T;X)$. By Proposition 12.1, we can find a sequence $\{v_n\} \subset C([0,T]; X)$ such that

$$v_n \to u' \quad \text{in } L^q(0,T;X).$$

Define

$$u_n(t) = u(0) + \int_0^t v_n(t)\, dt.$$

Then $\{u_n\} \subset C^1([0,T], X)$, and $u_n \to u$ in $W^{1,q}(0,T;X)$. □

Define

$$P(0,T;X) = \{p : p(t) = \sum_{i=0}^m a_i t^i,\ a_i \in X,\ 0 \le i \le m,\ m = 0, 1, \dots\},$$

the space of the polynomials with values in X. Obviously, $P(0,T;X) \subset C^\infty([0,T]; X)$. The following result is found in [129] (page 442).

PROPOSITION 12.3. *Assume that X is a Banach space, $X_0 \subset X$ is dense in X, $1 \le q < \infty$, and l is a nonnegative integer. Then $P(0,T;X_0)$ is dense in $C^l([0,T]; X)$.*

Combining Propositions 12.2 and 12.3, we have the next result.

PROPOSITION 12.4. *Assume that X is a Banach space, $X_0 \subset X$ is dense in X, $1 \le q < \infty$, and l is a nonnegative integer. Then $P(0,T;X_0)$ is dense in $W^{l,q}(0,T;X)$.*

Now we recall the following two smooth density results. Let $k \ge 0$, $1 \le p < \infty$. Then

$$C_0^\infty(\Omega) \text{ is dense in } W_0^{k,p}(\Omega). \tag{12.12}$$

If the boundary $\partial\Omega$ is Lipschitz continuous, then

$$C^\infty(\overline{\Omega}) \text{ is dense in } W^{k,p}(\Omega). \tag{12.13}$$

From Proposition 12.4, (12.12), and (12.13), we see that $H^1(0,T;C_0^\infty(\Omega))$ is dense in $H^1(0,T;H_0^1(\Omega))$ and $H^1(0,T;C^\infty(\overline{\Omega}))$ is dense in $H^1(0,T;L^2(\Omega))$. Thus given $\boldsymbol{w} = (\boldsymbol{u},\boldsymbol{p},\gamma) \in H^1(0,T;K)$, we can find a sequence $\boldsymbol{w}_n = (\boldsymbol{u}_n,\boldsymbol{p}_n,\gamma_n) \in H^1(0,T;(C_0^\infty(\Omega))^d \times (C^\infty(\overline{\Omega}))^{d\times d} \times C^\infty(\overline{\Omega}))$ converging to $\boldsymbol{w}$ in $H^1(0,T;Z)$. In order to see that the space H_0 defined in (12.11) has the property

$$H^1(0,T;H_0 \cap K) \text{ is dense in } H^1(0,T;K), \tag{12.14}$$

we need

$$\boldsymbol{p}_n \in Q_0, \quad |\boldsymbol{p}_n| \le \gamma_n \text{ in } \Omega. \tag{12.15}$$

To make sure (12.15) is valid, let us briefly review a typical proof of the density result (12.13) (cf. Evans [35]).

We first introduce some notation. For $\boldsymbol{x}_0 \in \mathbb{R}^d$ and $r > 0$, we use

$$B(\boldsymbol{x}_0,r) = \{\boldsymbol{x} \in \mathbb{R}^d : \|\boldsymbol{x}-\boldsymbol{x}_0\| < r\}$$

and

$$\overline{B}(\boldsymbol{x}_0,r) = \{\boldsymbol{x} \in \mathbb{R}^d : \|\boldsymbol{x}-\boldsymbol{x}_0\| \le r\}$$

to denote an open and a closed ball centered at $\boldsymbol{x}_0$ with radius r. Here and below the vector norm in $\mathbb{R}^d$ is the Euclidean norm. Define

$$J(\boldsymbol{x}) = \begin{cases} c_0 e^{1/(\|\boldsymbol{x}\|^2-1)}, & \|\boldsymbol{x}\| < 1, \\ 0, & \|\boldsymbol{x}\| \ge 1, \end{cases}$$

where $c_0 > 0$ is chosen such that

$$\int_{\mathbb{R}^d} J(\boldsymbol{x})\,dx = 1.$$

The function $J(\cdot)$ is infinitely smooth. Then we define the standard mollifier

$$J_\epsilon(\boldsymbol{x}) = \frac{1}{\epsilon^d} J(\boldsymbol{x}/\epsilon),$$

which has the properties that

$$J_\epsilon \in C^\infty(\mathbb{R}^d), \quad \int_{\mathbb{R}^d} J_\epsilon(\boldsymbol{x})\, dx = 1, \quad J_\epsilon(\boldsymbol{x}) = 0 \text{ for } \|\boldsymbol{x}\| \geq \epsilon.$$

Now we sketch the main steps for the proof of (12.13).

STEP 1. LOCALIZATION. Since $\partial\Omega$ is compact and is Lipschitz continuous, there exist finitely many points $\boldsymbol{x}_m \in \partial\Omega$, positive numbers $r_m > 0$, and Lipschitz continuous functions $\gamma_m : \mathbb{R}^{d-1} \to \mathbb{R}$, $1 \leq m \leq M$, such that

$$\partial\Omega \subset \bigcup_{m=1}^{M} B(\boldsymbol{x}_m, r_m/2),$$

and for each m, after relabeling the coordinate axes if necessary,

$$\Omega \cap \overline{B}(\boldsymbol{x}_m, r_m) = \{\boldsymbol{x} \in \overline{B}(\boldsymbol{x}_m, r_m) : x_d > \gamma_m(x_1, \ldots, x_{d-1})\}.$$

Define

$$\Omega_m = \Omega \cap \overline{B}(\boldsymbol{x}_m, r_m/2), \quad 1 \leq m \leq M,$$

and choose an open set $\Omega_0 \subset\subset \Omega$ with the property

$$\Omega \subset \bigcup_{m=0}^{M} \Omega_m.$$

Let $\{\zeta_m\}_{m=0}^M$ be a smooth partition of unity subordinate to $\{\Omega_m\}_{m=0}^M$; i.e., for each m, $\zeta_m \in C^\infty(\mathbb{R}^d)$, $\operatorname{supp}(\zeta_m) \subset \Omega_m$, and

$$\sum_{m=0}^{M} \zeta_m(\boldsymbol{x}) \equiv 1, \quad \boldsymbol{x} \in \Omega.$$

Then we have the decomposition

$$u = u \sum_{m=0}^{M} \zeta_m = \sum_{m=0}^{M} u_m,$$

where

$$u_m = u\, \zeta_m.$$

STEP 2. SMOOTH APPROXIMATION OF u_0. Take an $\epsilon_0 \in (0, \text{dist}\,(\partial\Omega_0, \partial\Omega))$, and consider only those ϵ with $\epsilon \le \epsilon_0/2$. Define

$$\Omega_0' = \{\boldsymbol{x} \in \Omega : \text{dist}\,(\boldsymbol{x}, \partial\Omega) > \epsilon_0/2\}.$$

Define the mollification of u_0 with respect to the spatial variable

$$u_0^\epsilon(\boldsymbol{x}) = (J_\epsilon * u_0)(\boldsymbol{x}) = \int_{B(O,\epsilon)} J_\epsilon(\boldsymbol{y})\, u_0(\boldsymbol{x} - \boldsymbol{y})\, dy.$$

We have, for ϵ sufficiently small,

$$u_0^\epsilon \in C_0^\infty(\overline{\Omega_0'})$$

and

$$\lim_{\epsilon \to 0} \|u_0^\epsilon - u_0\|_{W^{k,p}(\Omega)} = 0.$$

STEP 3. SMOOTH APPROXIMATION OF u_m, $1 \le m \le M$. Fix an $m = 1, \ldots, M$ and consider the smooth approximation of u_m. Recall that there exist an $r_m > 0$ and a Lipschitz continuous function γ_m such that upon relabeling the coordinate axes if necessary,

$$\Omega \cap \overline{B}(\boldsymbol{x}_m, r_m) = \{\boldsymbol{x} \in \overline{B}(\boldsymbol{x}_m, r_m) : x_d > \gamma_m(x_1, \ldots, x_{d-1})\}.$$

Let $\Omega_m = \Omega \cap B(\boldsymbol{x}_m, r_m/2)$. Denote by $\text{Lip}(\gamma_m)$ the Lipschitz constant of the function γ_m, and write $\boldsymbol{e}_n = (0, \ldots, 0, 1)^T$ in the local coordinates. For any $\boldsymbol{x} \in \Omega_m$, we define $\boldsymbol{x}^\epsilon = \boldsymbol{x} + \alpha\,\epsilon\,\boldsymbol{e}_n$, where we choose $\alpha = \sqrt{2}\max\{\text{Lip}\,(\gamma_m), 1\}$. It can be verified that if ϵ is small enough,

$$\overline{B}(\boldsymbol{x}^\epsilon, \epsilon) \subset \Omega \cap \overline{B}(\boldsymbol{x}, r_m).$$

Now we let

$$v_m^\epsilon(\boldsymbol{x}) = u_m(\boldsymbol{x}^\epsilon), \quad \boldsymbol{x} \in \Omega_m,$$

and define

$$u_m^\epsilon(\boldsymbol{x}) = (J_\epsilon * v_m^\epsilon)(\boldsymbol{x}) = \int_{B(O,\epsilon)} J_\epsilon(\boldsymbol{y})\, v_m^\epsilon(\boldsymbol{x} - \boldsymbol{y})\, dy.$$

We have

$$u_m^\epsilon \in C^\infty(\overline{\Omega}_m)$$

and

$$\lim_{\epsilon \to 0} \|u_m^\epsilon - u_m\|_{W^{k,p}(\Omega)} = 0.$$

STEP 4. GLOBAL SMOOTH APPROXIMATION WITH RESPECT TO $\boldsymbol{x}$. Define

$$u^\epsilon = \sum_{m=0}^{M} u_m^\epsilon.$$

Then

$$u^\epsilon \in C^\infty(\overline{\Omega})$$

and

$$u^\epsilon \to u \quad \text{in } W^{k,p}(\Omega) \text{ as } \epsilon \to 0.$$

It is evident from the proof that

$$\boldsymbol{p} \in Q_0 \Longrightarrow \boldsymbol{p}^\epsilon \in Q_0$$

and

$$|\boldsymbol{p}| \le \gamma \text{ a.e. in } \Omega \Longrightarrow |\boldsymbol{p}^\epsilon| \le \gamma^\epsilon \text{ in } \Omega.$$

Thus (12.14), and therefore (H_1), is satisfied.

As for (H_2), we assume that the finite element space V^h for approximating $V = (H_0^1(\Omega))^d$ contains piecewise linear functions and that the finite element spaces Q_0^h and M^h for Q_0 and M contain piecewise constants. Then for any $\boldsymbol{z} = (\boldsymbol{v}, \boldsymbol{q}, \mu) \in H_0 \cap K$, we define $\boldsymbol{z} = (\Pi^h \boldsymbol{v}, \Pi^h \boldsymbol{q}, \Pi^h \mu)$, where $\Pi^h \boldsymbol{v}$ is the piecewise linear interpolant of $\boldsymbol{v}$, and $\Pi^h \boldsymbol{q}$ and $\Pi^h \mu$ are elementwise averages of $\boldsymbol{q}$ and μ. It is not difficult to see that $\boldsymbol{z}^h \in K^h$. By the standard finite element interpolation theory,

$$\|\boldsymbol{z} - \boldsymbol{z}^h\|_Z \le c\, h\, (\|\boldsymbol{v}\|_{H^2(\Omega)} + \|\boldsymbol{q}\|_{H^1(\Omega)} + \|\mu\|_{H^1(\Omega)}).$$

Hence, (H_2) is satisfied with $\alpha = 1$.

The problem with linear kinematic hardening. Now we consider discrete approximations to the solution $\boldsymbol{w}$ of the problem with linear kinematic hardening only. The continuous problem PRIM2 is stated in Section 7.1 and analyzed in Section 7.3.

Let V^h and Q_0^h be finite element subspaces of V and Q_0, and set $Z^h = V^h \times Q_0^h$. Then a spatially discrete approximation of the problem is to find $\boldsymbol{w}^h = (\boldsymbol{u}^h, \boldsymbol{p}^h) \in Z^h$, $\boldsymbol{w}^h(0) = \boldsymbol{0}$, such that

$$\begin{aligned} a(\boldsymbol{w}^h(t), \boldsymbol{z}^h - \dot{\boldsymbol{w}}^h(t)) + j(\boldsymbol{z}^h) - j(\dot{\boldsymbol{w}}^h(t)) \\ \ge \langle \boldsymbol{\ell}(t), \boldsymbol{z}^h - \dot{\boldsymbol{w}}^h(t) \rangle \qquad \forall\, \boldsymbol{z}^h = (\boldsymbol{v}^h, \boldsymbol{q}^h) \in Z^h. \end{aligned} \tag{12.16}$$

The semidiscrete approximation problem has a unique solution $\boldsymbol{w}^h(t)$, $t \in [0, T]$. Since the functional $j(\boldsymbol{z})$ depends on the second component $\boldsymbol{q}$ of $\boldsymbol{z}$

only, the term $c\,\|\boldsymbol{z}^h - \dot{\boldsymbol{w}}(t)\|_H$ on the right-hand side of (11.10) may be replaced by $c\,\|\boldsymbol{q}^h - \dot{\boldsymbol{p}}(t)\|_Q$. Thus, the error estimate (11.10) becomes, for this case,

$$\begin{aligned}\sup_{0\le t\le T} &\|\boldsymbol{w}(t) - \boldsymbol{w}^h(t)\|_Z^2 \\ &\le c\Big\{\inf_{\boldsymbol{v}^h\in L^2(0,T;V^h)} \|\dot{\boldsymbol{u}} - \boldsymbol{v}^h\|_{L^2(0,T;V)}^2 \\ &\qquad + \inf_{\boldsymbol{q}^h\in L^2(0,T;Q_0^h)} \|\dot{\boldsymbol{p}} - \boldsymbol{q}^h\|_{L^1(0,T;Q)}\Big\}. \end{aligned} \tag{12.17}$$

Once again we omit the discussion on the time-discrete approximations.

Now we consider fully discrete approximations of the problem and present some error estimates for the backward Euler Crank–Nicolson schemes. We divide $[0,T]$ into N equal parts and use $k = T/N$ for the step-size. The backward Euler method amounts to finding $\boldsymbol{w}^{hk} = \{\boldsymbol{w}_n^{hk}\}_{n=0}^N$, where $\boldsymbol{w}_n^{hk} = (\boldsymbol{u}_n^{hk}, \boldsymbol{p}_n^{hk}) \in Z^h, 0 \le n \le N$, $\boldsymbol{w}_0^{hk} = \boldsymbol{0}$, such that for $n = 1, 2, \ldots, N$,

$$\begin{aligned} a(\boldsymbol{w}_n^{hk}, \boldsymbol{z}^h - \delta\boldsymbol{w}_n^{hk}) &+ j(\boldsymbol{z}^h) - j(\delta\boldsymbol{w}_n^{hk}) \\ &\ge \langle \boldsymbol{\ell}_n, \boldsymbol{z}^h - \delta\boldsymbol{w}_n^{hk}\rangle \qquad \forall\, \boldsymbol{z}^h = (\boldsymbol{v}^h, \boldsymbol{q}^h) \in Z^h. \end{aligned} \tag{12.18}$$

The discrete problem has a unique solution. Again we observe that the term $\|z^h - \dot{w}_n\|_H$ on the right-hand side of the inequality (11.37) may be replaced by $\|\boldsymbol{q}^h - \dot{\boldsymbol{p}}_n\|_Q$. Therefore, the error estimate (11.42) for the case of the problem (12.18) becomes

$$\begin{aligned}\max_{0\le n\le N} &\|\boldsymbol{w}_n - \boldsymbol{w}_n^{hk}\|_Z^2 \\ &\le c\,k \sum_{n=1}^N \Big(\inf_{\boldsymbol{q}^h\in Q_0^h} \|\dot{\boldsymbol{p}}_n - \boldsymbol{q}^h\|_Q + \inf_{\boldsymbol{v}^h\in V^h} \|\dot{\boldsymbol{u}}_n - \boldsymbol{v}^h\|_V^2\Big) \\ &\quad + c\,k^2 \|\ddot{\boldsymbol{w}}\|_{L^2(0,T;Z)}^2. \end{aligned} \tag{12.19}$$

For the Crank–Nicolson scheme we compute $\boldsymbol{w}^{hk} = \{\boldsymbol{w}_n^{hk}\}_{n=0}^N$, where $\boldsymbol{w}_n^{hk} = (\boldsymbol{u}_n^{hk}, \boldsymbol{p}_n^{hk}) \in Z^h, 0 \le n \le N$, $\boldsymbol{w}_0^{hk} = \boldsymbol{0}$, such that for $n = 1, 2, \ldots, N$,

$$\begin{aligned} a(\tfrac{1}{2}\,(\boldsymbol{w}_n^{hk} + \boldsymbol{w}_{n-1}^{hk}), \boldsymbol{z}^h - \delta\boldsymbol{w}_n^{hk}) &+ j(\boldsymbol{z}^h) - j(\delta\boldsymbol{w}_n^{hk}) \\ &\ge \langle \boldsymbol{\ell}_{n-1/2}, \boldsymbol{z}^h - \delta\boldsymbol{w}_n^{hk}\rangle \qquad \forall\, \boldsymbol{z}^h = (\boldsymbol{v}^h, \boldsymbol{q}^h) \in Z^h. \end{aligned} \tag{12.20}$$

The discrete problem has a unique solution. Assuming $\boldsymbol{w} \in W^{2,\infty}(0,T;Z)$ and $\boldsymbol{w}^{(3)} \in L^1(0,T;Z)$, we have the error estimate

$$\begin{aligned}\max_{0\le n\le N} &\|\boldsymbol{w}_n - \boldsymbol{w}_n^{hk}\|_Z^2 \\ &\le c\,k \sum_{n=1}^N \Big(\inf_{\boldsymbol{q}^h\in Q_0^h} \|\dot{\boldsymbol{p}}_{n-1/2} - \boldsymbol{q}^h\|_Q + \inf_{\boldsymbol{v}^h\in V^h} \|\dot{\boldsymbol{u}}_{n-1/2} - \boldsymbol{v}^h\|_V^2\Big) \\ &\quad + c\,k^4. \end{aligned} \tag{12.21}$$

The inequalities (12.19) and (12.21) are the basis for deriving various convergence order estimates, which can be obtained as in the case of combined linear isotropic–kinematic hardening.

If we do not make regularity assumptions on the solution $\boldsymbol{w}$, we no longer have order error estimates for the numerical solutions. However, we can still apply the results from Section 11.4 to conclude the convergence of the numerical solutions, as for the case in solving the primal problem with combined linear kinematic–isotropic hardening. Here the hypotheses (H_1) and (H_2) are easier to verify because the problem is posed on the whole space, $K = H = V \times Q_0$. The reader can verify the hypotheses for kinematic hardening by modifying the argument presented for the more complicated case of combined linear kinematic–isotropic hardening.

12.2 Solution Algorithms

In this section we discuss details of solution algorithms for the primal elastoplasticity problems. We will be concerned with a particular class of algorithms that have been employed in computational approaches to these problems (see, for example, Reddy and Martin [107], Simo [114]). While algorithms of this kind are often developed and implemented in the context of fully discrete problems, in which the discretization is carried out using finite elements for the spatial domain, we will develop the necessary algorithms for the spatially continuous case, that is, for time-discrete schemes. A parallel treatment can be given for the fully discrete schemes. Furthermore, the questions of whether to discretize spatially, and how, are ones that are essentially independent of the details of the time-discretization, and may be posed subsequently.

For simplicity of presentation, let us consider the particular elastoplasticity problem with linear kinematic hardening and von Mises yield condition. We recall from Section 7.1 that the problem is to find $\boldsymbol{w} = (\boldsymbol{u}, \boldsymbol{p}) : [0, T] \to Z$ with $\boldsymbol{w}(0) = \boldsymbol{0}$ such that for almost all $t \in (0, T)$,

$$a(\boldsymbol{w}(t), \boldsymbol{z} - \dot{\boldsymbol{w}}(t)) + j(\boldsymbol{z}) - j(\dot{\boldsymbol{w}}(t)) \geq \langle \boldsymbol{\ell}(t), \boldsymbol{z} - \dot{\boldsymbol{w}}(t) \rangle \quad \forall\, \boldsymbol{z} = (\boldsymbol{v}, \boldsymbol{q}) \in Z, \tag{12.22}$$

where $Z = V \times Q_0$ with

$$\begin{aligned} V &= (H_0^1(\Omega))^3, \\ Q_0 &= \{\boldsymbol{q} = (q_{ij}) : q_{ij} = q_{ji},\ q_{ij} \in L^2(\Omega),\ \operatorname{tr} \boldsymbol{q} = 0\}, \end{aligned}$$

and

$$a(\boldsymbol{w}, \boldsymbol{z}) = \int_\Omega [\boldsymbol{C}(\boldsymbol{\epsilon}(\boldsymbol{u}) - \boldsymbol{p}) : (\boldsymbol{\epsilon}(\boldsymbol{v}) - \boldsymbol{q}) + k_1 \boldsymbol{p} : \boldsymbol{q}] \, dx, \tag{12.23}$$

$$j(\boldsymbol{z}) = \int_\Omega c_0 |\boldsymbol{q}(\boldsymbol{x})| \, dx, \tag{12.24}$$

$$\langle \boldsymbol{\ell}(t), \boldsymbol{z} \rangle = \int_\Omega \boldsymbol{f}(t) \cdot \boldsymbol{v} \, dx. \tag{12.25}$$

We take as an example the backward Euler time-discrete scheme, the spatially continuous version of (12.18), to show how various iterative solution algorithms are employed to solve the discrete problem. A similar discussion applies to the more general generalized midpoint schemes.

The time-discrete problem involves the computation of a sequence $\boldsymbol{w}^k = (\boldsymbol{u}^k, \boldsymbol{p}^k) = \{(\boldsymbol{u}_n^k, \boldsymbol{p}_n^k)\}_{n=1}^N \subset Z$, $\boldsymbol{w}_0^k = \boldsymbol{0}$, such that for $n = 1, \dots, N$,

$$a(\boldsymbol{w}_n^k, \boldsymbol{z} - \delta \boldsymbol{w}_n^k) + j(\boldsymbol{z}) - j(\delta \boldsymbol{w}_n^k) \ge \langle \boldsymbol{\ell}_n, \boldsymbol{z} - \delta \boldsymbol{w}_n^k \rangle \quad \forall\, \boldsymbol{z} \in Z. \tag{12.26}$$

The algorithms of interest are all of predictor–corrector type. Investigations of convergence have been carried out in the context of the fully discrete (finite element) formulation by Martin and Caddemi [85] and, at a more general level, by Reddy and Martin [108]. These investigations have been concerned primarily with the question of whether the algorithms generate minimizing sequences, rather than with the consequential question of whether such sequences in fact converge. In the course of developing the solution algorithms we will pay close attention to whether they produce minimizing sequences. Then in next section we will examine rigorously the question of convergence for some of the algorithms.

We first rewrite (12.26) in a form such that the increment $\Delta \boldsymbol{w}_n^k$ is the primary unknown. Attention is focused on a particular time t_n, and with $\boldsymbol{w}_{n-1}^k$ known, the problem is one of finding $\boldsymbol{w}_n^k \in Z$ such that

$$a(\Delta \boldsymbol{w}_n^k, \boldsymbol{z} - \Delta \boldsymbol{w}_n^k) + j(\boldsymbol{z}) - j(\Delta \boldsymbol{w}_n^k) \ge \langle \boldsymbol{L}_n, \boldsymbol{z} - \Delta \boldsymbol{w}_n^k \rangle \quad \forall\, \boldsymbol{z} \in Z, \tag{12.27}$$

where the functional $\boldsymbol{L}_n$ is defined by

$$\langle \boldsymbol{L}_n, \boldsymbol{z} \rangle = \langle \boldsymbol{\ell}_n, \boldsymbol{z} \rangle - a(\boldsymbol{w}_{n-1}^k, \boldsymbol{z}). \tag{12.28}$$

Crucial to the algorithm will be the consideration of this variational inequality as a combination of an equation and an inequality. Indeed, if we return to the definition (12.23) of $a(\cdot, \cdot)$, we see that this bilinear form may be written, with $\boldsymbol{w} = (\boldsymbol{u}, \boldsymbol{p})$ and $\boldsymbol{z} = (\boldsymbol{v}, \boldsymbol{q})$, as

$$a(\boldsymbol{w}, \boldsymbol{z}) = b(\boldsymbol{u}, \boldsymbol{v}) - c(\boldsymbol{p}, \boldsymbol{v}) - c(\boldsymbol{q}, \boldsymbol{u}) + d(\boldsymbol{p}, \boldsymbol{q}), \tag{12.29}$$

where

$$b : V \times V \to \mathbb{R}, \quad b(\boldsymbol{u}, \boldsymbol{v}) = \int_\Omega \boldsymbol{C}\,\boldsymbol{\epsilon}(\boldsymbol{u}) : \boldsymbol{\epsilon}(\boldsymbol{v})\,dx,$$

$$c : Q_0 \times V \to \mathbb{R}, \quad c(\boldsymbol{q}, \boldsymbol{v}) = \int_\Omega \boldsymbol{C}\,\boldsymbol{q} : \boldsymbol{\epsilon}(\boldsymbol{v})\,dx,$$

$$d : Q_0 \times Q_0 \to \mathbb{R}, \quad d(\boldsymbol{p}, \boldsymbol{q}) = \int_\Omega (\boldsymbol{C}\,\boldsymbol{p} : \boldsymbol{q} + k_1 \boldsymbol{p} : \boldsymbol{q})\,dx.$$

The linear functional $\boldsymbol{L}_n(\cdot)$ may likewise be decomposed by writing it in the form

$$\langle \boldsymbol{L}_n, \boldsymbol{z} \rangle = \langle \boldsymbol{L}_{n,1}, \boldsymbol{v} \rangle + \langle \boldsymbol{L}_{n,2}, \boldsymbol{q} \rangle,$$

in which

$$\boldsymbol{L}_{n,1} : V \to \mathbb{R}, \quad \langle \boldsymbol{L}_{n,1}, \boldsymbol{v} \rangle = \int_\Omega [\boldsymbol{f}_n \cdot \boldsymbol{v} - \boldsymbol{\sigma}_{n-1}^k \cdot \boldsymbol{\epsilon}(\boldsymbol{v})]\,dx,$$

and

$$\boldsymbol{L}_{n,2} : Q_0 \to \mathbb{R}, \quad \langle \boldsymbol{L}_{n,2}, \boldsymbol{q} \rangle = \int_\Omega \boldsymbol{\chi}_{n-1}^k : \boldsymbol{q}\,dx,$$

where

$$\begin{aligned} \boldsymbol{\sigma}_{n-1}^k &= \boldsymbol{C}\,(\boldsymbol{\epsilon}(\boldsymbol{u}_{n-1}^k) - \boldsymbol{p}_{n-1}^k), \\ \boldsymbol{\chi}_{n-1}^k &= \boldsymbol{\sigma}_{n-1}^k + k_1 \boldsymbol{p}_{n-1}^k \end{aligned}$$

are known from the previous step of the computation. Thus (12.27) can be written in the form

$$b(\Delta \boldsymbol{u}_n^k, \boldsymbol{v}) - c(\Delta \boldsymbol{p}_n^k, \boldsymbol{v}) = \langle \boldsymbol{L}_{n,1}, \boldsymbol{v} \rangle \qquad \forall\, \boldsymbol{v} \in V, \tag{12.30}$$

$$\begin{aligned} &j(\boldsymbol{q}) - j(\Delta \boldsymbol{p}_n^k) - c(\boldsymbol{q} - \Delta \boldsymbol{p}_n^k, \Delta \boldsymbol{u}_n^k) + d(\Delta \boldsymbol{p}_n^k, \boldsymbol{q} - \Delta \boldsymbol{p}_n^k) \\ &\qquad \geq \langle \boldsymbol{L}_{n,2}, \boldsymbol{q} - \Delta \boldsymbol{p}_n^k \rangle \qquad \forall\, \boldsymbol{q} \in Q_0. \end{aligned} \tag{12.31}$$

Here, we identify $j(\boldsymbol{z})$ with $j(\boldsymbol{q})$ in view of the expression of $j(\boldsymbol{z})$ defined in (12.24).

This variational inequality can be reformulated as a minimization problem. Such a formulation follows directly from the (equivalent) formulation (12.27), and the problem may be considered as one of finding $\Delta \boldsymbol{w}_n^k \in Z$ such that

$$\mathcal{L}_n(\Delta \boldsymbol{w}_n^k) \leq \mathcal{L}_n(\boldsymbol{z}) \quad \forall\, \boldsymbol{z} \in Z, \tag{12.32}$$

where the functional $\mathcal{L}_n$ is defined by

$$\mathcal{L}_n(\boldsymbol{z}) = \frac{1}{2} a(\boldsymbol{z}, \boldsymbol{z}) + j(\boldsymbol{q}) - \langle \boldsymbol{L}_n, \boldsymbol{z} \rangle. \tag{12.33}$$

In order to simplify the notation in the derivation and convergence analysis of solution algorithms, we will view the problem (12.30)–(12.31) as a special case of an abstract problem to be defined next.

First we introduce various spaces, functionals, and assumptions. Let V and Λ be two Hilbert spaces. Let there be given three continuous bilinear forms, $b : V \times V \to \mathbb{R}$, $c : \Lambda \times V \to \mathbb{R}$, and $d : \Lambda \times \Lambda \to \mathbb{R}$; two continuous linear forms, $\ell_1 : V \to \mathbb{R}$ and $\ell_2 : \Lambda \to \mathbb{R}$; and one functional, $j : \Lambda \to \mathbb{R}$. Assume that b and d are symmetric. Also assume that j is nonnegative, convex, and Lipschitz continuous, and is of the form

$$j(\mu) = \int_\Omega D(\mu(\boldsymbol{x}))\, dx,$$

where the function $D(\mu)$ is not differentiable at $\mu = 0$ and is two times differentiable everywhere else. For $w = (u, \lambda)$, $z = (v, \mu) \in V \times \Lambda$, define

$$a(w, z) = b(u, v) - c(\lambda, v) - c(\mu, u) + d(\lambda, \mu). \tag{12.34}$$

We further assume that $a : (V \times \Lambda) \times (V \times \Lambda) \to \mathbb{R}$ is $(V \times \Lambda)$-elliptic, that is, for some constant $c_0 > 0$,

$$a(z, z) \geq c_0(\|v\|_V^2 + \|\mu\|_\Lambda^2) \quad \forall\, z = (v, \mu) \in V \times \Lambda. \tag{12.35}$$

PROBLEM ABS^d. Find $u \in V$ and $\lambda \in \Lambda$ such that

$$b(u, v) - c(\lambda, v) = \langle \ell_1, v \rangle \quad \forall\, v \in V, \tag{12.36}$$

$$j(\mu) - j(\lambda) - c(\mu - \lambda, u) + d(\lambda, \mu - \lambda) \geq \langle \ell_2, \mu - \lambda \rangle \quad \forall\, \mu \in \Lambda. \tag{12.37}$$

The smoothness assumptions on j essentially restrict applications to problems involving smooth yield surfaces. The implications for the algorithm of a nonsmooth yield surface such as that corresponding to the Tresca yield condition are treated in Rencontré, Bird, and Martin [110], and Simo, Kennedy, and Govindjee [117].

We will develop solution algorithms for the problem ABS^d, instead of the concrete problem formulated in (12.30) and (12.31). It will prove convenient to reformulate this problem as a minimization problem. Let us introduce an "energy" functional

$$L(z) = \tfrac{1}{2} a(z, z) + j(\mu) - \langle \ell_1, v \rangle - \langle \ell_2, \mu \rangle, \quad z = (v, \mu) \in V \times \Lambda. \tag{12.38}$$

By a standard approach, it can be shown that the variational inequality problem (12.36)–(12.37) is equivalent to the minimization problem

$$w \in V \times \Lambda, \quad L(w) \leq L(z) \;\; \forall\, z \in V \times \Lambda. \tag{12.39}$$

From the assumptions made on the data, it is readily seen that the functional $L(\cdot)$ is strictly convex and coercive over $V \times \Lambda$, and hence the minimization problem (12.39) has a unique solution. Thus, the variational inequality problem (12.36)–(12.37) also has a unique solution $(u, \lambda) \in V \times \Lambda$. The framework given in the problem ABS^d is fairly general; by a slight modification of the space setting (replacing the space Λ by a nonempty closed convex cone of a Hilbert space), we can also include in this frame the elastoplasticity problem with combined linear kinematic–isotropic hardening, using a generalized midpoint rule.

The solution algorithm. We make use of a two-step predictor–corrector strategy for solving the variational problem ABS^d. In a general iteration step we have estimates u^{i-1} and λ^{i-1}, and we seek new, improved estimates u^i and λ^i. Let u^0 and λ^0 be some initial guess, e.g., we may take $u^0 = 0$ and $\lambda^0 = 0$. We consider first conditions under which the iterative scheme generates a minimizing sequence for the "energy" functional L. This requires that

$$\Delta L^i \equiv L(u^i, \lambda^i) - L(u^{i-1}, \lambda^{i-1}) < 0. \tag{12.40}$$

The two steps of the iteration scheme will be referred to as the *predictor step* and the *corrector step*. In the predictor step we replace L by a *quadratic* functional $L^{(i)}$, chosen in such a way that

$$L^{(i)}(u^{i-1}, \lambda^{i-1}) = L(u^{i-1}, \lambda^{i-1}). \tag{12.41}$$

The minimization of the unconstrained quadratic functional $L^{(i)}$ is a linear problem, and leads to the estimate (u^i, λ^{*i}). In the corrector step, we minimize $L(u^i, \mu)$ over all μ to find λ^i. We shall discuss the implementation of each of these steps in further detail. At the moment, we note that it is possible to set conditions under which each of the predictor and corrector steps leads to a decrease in the functional L. In other words, the sequence generated in this manner is a *minimizing sequence*. The question of whether this sequence is actually convergent will be investigated in the next section.

We define

$$\Delta L_P^i = L(u^i, \lambda^{*i}) - L(u^{i-1}, \lambda^{i-1}) \tag{12.42}$$

and

$$\Delta L_C^i = L(u^i, \lambda^i) - L(u^i, \lambda^{*i}). \tag{12.43}$$

A sufficient condition for monotonic behavior is clearly

$$\Delta L_P^i < 0 \quad \text{and} \quad \Delta L_C^i \le 0. \tag{12.44}$$

The definition of the corrector step leads immediately to condition $(12.44)_2$, so that our sufficient condition for monotonic behavior is met if condition

$(12.44)_1$ is satisfied. We first discuss briefly the corrector step.

The corrector step. In the corrector step we start with a new estimate u^i obtained from the predictor step, and seek to minimize $L(u^i, \mu)$ over all $\mu \in \Lambda$. In this minimization u^i is held unchanged, and $L(u^i, \mu)$ achieves its minimum at $\mu = \lambda^i$. From (12.38) we see that

$$\begin{aligned} L(u^i, \mu) &= -c(\mu, u^i) + \tfrac{1}{2} d(\mu, \mu) + j(\mu) - \langle \ell_2, \mu \rangle \\ &\qquad + \text{terms independent of } \mu. \end{aligned} \tag{12.45}$$

With $u^i \in V$ given, we are therefore required to find $\lambda^i \in \Lambda$ such that

$$j(\mu) - j(\lambda^i) + d(\lambda^i, \mu - \lambda^i) \geq \langle \ell_2, \mu - \lambda^i \rangle + c(\mu - \lambda^i, u^i) \quad \forall\, \mu \in \Lambda. \tag{12.46}$$

Comparison with (12.37) shows that the problem of minimizing the functional (12.45) is equivalent to solving the variational inequality (12.37), with $u = u^i$ there.

If the variables have the required degree of smoothness, it is possible to numerically solve the problem (12.46) pointwise. Indeed, this amounts to solving a discrete version of (12.46) pointwise. For a fully discrete approximation of the problem (12.36)–(12.37), this procedure is implemented by solving for the variable λ at integration points. Thus, rather than having to solve simultaneously for the variable λ at all integration points, it is necessary only to solve a sequence of small uncoupled problems at each integration point. This collocation-type approach is made possible by the fact that spatial derivatives of the variable λ (which corresponds to the internal variables in elastoplasticity problems) do not appear anywhere. Further details may be found, for example, in Martin and Reddy [87], Reddy and Martin [107, 108] for the small strain case, and in Eve and Reddy [37] for finite strain problems.

The predictor step. We will consider several predictor steps discussed in the literature. It is convenient to introduce a change of variables

$$\hat{v} = v - u^{i-1}, \quad \hat{\mu} = \mu - \lambda^{i-1},$$

and to define $\hat{z} = (\hat{v}, \hat{\mu})$. Now we rewrite the functional L in terms of $\hat{z}$, thanks to the expression (12.34). We find, after some straightforward algebraic manipulation, that

$$\begin{aligned} L(v, \mu) &= L(u^{i-1} + \hat{v}, \lambda^{i-1} + \hat{\mu}) \\ &= \tfrac{1}{2} a(\hat{z}, \hat{z}) + j(\lambda^{i-1} + \hat{\mu}) - j(\lambda^{i-1}) - \langle \chi^{i-1}, \hat{\mu} \rangle \\ &\quad - \langle R^i, \hat{v} \rangle + \text{terms independent of } \hat{v}, \hat{\mu} \end{aligned} \tag{12.47}$$

where

$$\begin{aligned} \langle \chi^{i-1}, \hat{\mu} \rangle &= c(\hat{\mu}, u^{i-1}) - d(\lambda^{i-1}, \hat{\mu}) + \langle \ell_2, \hat{\mu} \rangle, \\ \langle R^i, \hat{v} \rangle &= -b(u^{i-1}, \hat{v}) + c(\lambda^{i-1}, \hat{v}) + \langle \ell_1, \hat{v} \rangle. \end{aligned}$$

In the context of the plasticity problem, the linear functional R^i is the residual, and represents the out-of-balance forces at the end of the $(i-1)$th iteration (cf. (12.36)). Clearly, if $R^i = 0$, then the iterative process is complete, since the equilibrium equation (12.36) is satisfied at (u^{i-1}, λ^{i-1}). At this point L achieves its minimum value. This follows from the fact that in (12.47), the first term is nonnegative from the positive definiteness of the bilinear form $a(\cdot,\cdot)$, and $j(\lambda^{i-1} + \hat{\mu}) - j(\lambda^{i-1}) - \langle \chi^{i-1}, \hat{\mu} \rangle \geq 0$ from the fact that (u^{i-1}, λ^{i-1}) satisfies the inequality (12.37). Thus, if $R^i = 0$, then L achieves its least value at $\hat{z} = (\hat{u}, \hat{\lambda}) = (0, 0)$.

The first choice of a predictor that we shall introduce is the *elastic* predictor. In the elastic predictor we set $\hat{\mu} = 0$ in (12.47), and instead of minimizing L, we minimize the functional

$$L^{(i)}(\hat{v}) = \tfrac{1}{2} b(\hat{v}, \hat{v}) - \langle R^i, \hat{v} \rangle. \tag{12.48}$$

Here, we use the notation $L^{(i)}(\hat{v})$ rather than $L^{(i)}(\hat{v}, 0)$. The minimum of $L^{(i)}$ is achieved at $\hat{u}^i$, which satisfies

$$b(\hat{u}^i, \hat{v}) = \langle R^i, \hat{v} \rangle \quad \forall\, \hat{v} \in V. \tag{12.49}$$

This is simply the elastic boundary value problem with the loading given by R^i. We set $u^i = \hat{u}^i + u^{i-1}$, and since there is no change in λ,

$$\lambda^{*i} = \lambda^{i-1}.$$

Use of the elastic predictor results in a nonpositive ΔL_P^i. To see this, we observe that

$$\begin{aligned} \Delta L_P^i &= L(u^i, \lambda^{*i}) - L(u^{i-1}, \lambda^{i-1}) \\ &= L(u^i, \lambda^{i-1}) - L(u^{i-1}, \lambda^{i-1}) \\ &= L^{(i)}(\hat{u}^i) - L^{(i)}(0) \\ &\leq 0 \end{aligned}$$

with equality if and only if $R^i = 0$.

In practice, however, the decrease of the energy functional $L(\cdot)$ is slow when the elastic predictor is used. A modified elastic predictor was given by Comi and Maier [28], within the context of the spatially discrete (finite element) formulation of the elastoplasticity problem, and was shown to improve the speed of monotonic decrease. We briefly describe their extension in the present variational framework.

First we introduce another bilinear form $b^*(\cdot,\cdot)$ on $V \times V$. Then we replace (12.49) by the problem of finding $\hat{u}^{i*} \in V$ such that

$$b^*(\hat{u}^{i*}, \hat{v}) = \langle R^i, \hat{v} \rangle \quad \forall\, \hat{v} \in V. \tag{12.50}$$

The solution $\hat{u}^{i*}$ differs from the solution $\hat{u}^i$ of the problem (12.49), and we still set $\lambda^{*i} = \lambda^{i-1}$. It follows that

$$\begin{aligned}\Delta L_P^i &= L(u^{i*}, \lambda^{*i}) - L(u^{i-1}, \lambda^{i-1}) \\ &= L(u^{i*}, \lambda^{i-1}) - L(u^{i-1}, \lambda^{i-1}) \\ &= \tfrac{1}{2} b(u^{i*}, u^{i*}) - \langle R^i, u^{i*} \rangle \\ &= \tfrac{1}{2} b(u^{i*}, u^{i*}) - b^*(u^{i*}, u^{i*}).\end{aligned}$$

It is evident that if the bilinear form b^* strictly dominates $\frac{1}{2}b$ in the sense that

$$\tfrac{1}{2} b(v, v) < b^*(v, v) \quad \forall v \in V, \ v \neq 0, \tag{12.51}$$

then we have

$$\Delta L_P^i \leq 0$$

with equality if and only if $u^{i*} = 0$, or equivalently, $R^i = 0$. In the context of the elastoplasticity problem (12.30)–(12.31), the bilinear form $b^*(\cdot, \cdot)$ can be constructed from $b(\cdot, \cdot)$ by replacing the elasticity tensor $\boldsymbol{C}$ with a tensor $\boldsymbol{C}^*$. Then the requirement (12.51) amounts to the condition that the tensor $\boldsymbol{C}^* - \frac{1}{2}\boldsymbol{C}$ be positive definite. This permits a variety of choices for $\boldsymbol{C}^*$ (and in the finite element framework, for the modified stiffness matrix $\boldsymbol{K}^*$) that lead to the desired monotonically decreasing behavior. Some possibilities for the construction of the tensor $\boldsymbol{C}^*$ are discussed for the spatially discrete case by Comi and Maier [28].

In proceeding further, we differentiate between active and inactive regions of the domain. At the end of the $(i-1)$th iteration the domain Ω is partitioned into two disjoint parts: the active region $\Omega^{p(i)}$, which is defined to be the subset of Ω comprising points at which $\lambda^{i-1} \neq 0$, and the inactive region $\Omega^{e(i)}$, which is the subset comprising points at which $\lambda^{i-1} = 0$. Note that these two subsets can change after each iteration. In setting up the quadratic approximation $L^{(i)}$ of L in the following discussion, we shall assume that $\hat{\lambda}^{*i} = 0$ in the inactive region. We note further that for the first iteration the entire domain is regarded as inactive, and hence, by setting $\hat{\lambda}^0 = 0$ throughout Ω, we are forced to use the elastic predictor. This is logical in the sense that the load increment may lead to global unloading, for which the elastic predictor is the only acceptable predictor for convergence. The effect of permitting changes $\hat{\lambda}^{*i}$ only in the active region is that the dissipation function D (and hence also the functional j) is differentiable at $\hat{\lambda}^{i-1}$.

In setting up the quadratic approximation of L we restrict all integrals involving the variable μ to the active region. The solution $\hat{\lambda}^{*i}$ will thus be a function that is identically zero in the inactive region.

For completeness, we shall refer briefly to the concept of a secant predictor used in Bird and Martin [10]. Here we replace D in the active region

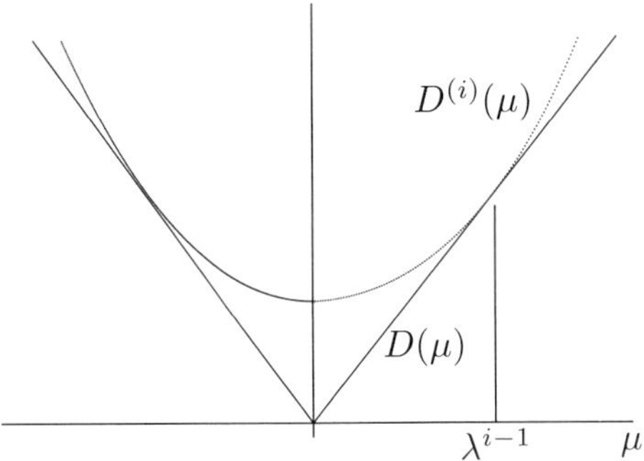

Figure 12.1: Approximation $D^{(i)}$ of D for the secant predictor

by a quadratic function $D^{(i)}$, defined in such a way that

$$D^{(i)}(\lambda^{i-1}) = D(\lambda^{i-1}), \tag{12.52}$$

$$\nabla D^{(i)}(\lambda^{i-1}) = \nabla D(\lambda^{i-1}) = \chi^{i-1} \tag{12.53}$$

and

$$D(\mu) \leq D^{(i)}(\mu) \quad \forall \, \mu \in \Lambda. \tag{12.54}$$

The quadratic function $D^{(i)}$ thus fits inside the cone D, touching at the point λ^{i-1} (see Figure 12.1). The exact form of $D^{(i)}$ will depend on the nature of D; the von Mises case is discussed in [10].

Let

$$j^{(i)}(\mu) = \int_\Omega D^{(i)}(\mu)\, dx, \quad \mu \in \Lambda.$$

Here, as stated earlier, the integration region is implicitly understood to be the active region of λ^{i-1}. The quadratic approximating function for L is now

$$L^{(i)}(z) = \tfrac{1}{2}a(z,z) + j^{(i)}(\mu) - \langle \ell_1, v\rangle - \langle \ell_2, \mu\rangle \tag{12.55}$$

for $z = (v,\mu) \in V \times \Lambda$.

The functional $L^{(i)}$ achieves its minimum at $(u^i, \lambda^{*i}) = (u^{i-1} + \hat{u}^i, \lambda^{i-1} + \hat{\lambda}^{*i})$, and $(\hat{u}^i, \hat{\lambda}^{*i})$ satisfies

$$b(\hat{u}^i, \hat{v}) - c(\hat{\lambda}^{*i}, \hat{v}) = \langle R^i, \hat{v}\rangle \quad \forall \, \hat{v} \in V, \tag{12.56}$$

$$j^{(i)}(\lambda^{i-1} + \hat{\mu}) - j^{(i)}(\lambda^{i-1} + \hat{\lambda}^{*i}) - c(\hat{\mu} - \hat{\lambda}^{*i}, \hat{u}^i) + d(\hat{\lambda}^{*i}, \hat{\mu} - \hat{\lambda}^{*i}) \geq \langle \chi^{i-1}, \hat{\mu} - \hat{\lambda}^{*i}\rangle \quad \forall \, \hat{\mu} \in \Lambda, \tag{12.57}$$

where R^i and χ^{i-1} are defined as before. We note that since $j^{(i)}$ is a differentiable quadratic form, the inequality (12.57) is actually equivalent to the linear equation

$$\langle \nabla j^{(i)}(\lambda^{i-1} + \hat{\lambda}^{*i}), \hat{\mu} \rangle - c(\hat{\mu}, \hat{u}^i) + d(\hat{\lambda}^{*i}, \hat{\mu}) = \langle \chi^{i-1}, \hat{\mu} \rangle \quad \forall\, \hat{\mu} \in \Lambda. \quad (12.58)$$

Now let us check whether ΔL_P^i is decreasing. First we note that

$$L(u^{i-1}, \lambda^{i-1}) = L^{(i)}(u^{i-1}, \lambda^{i-1})$$

by the condition (12.52). Then by using the condition (12.54), we have

$$\begin{aligned}
\Delta L_P^i &= L(u^i, \lambda^{*i}) - L(u^{i-1}, \lambda^{i-1}) \\
&= L^{(i)}(u^i, \lambda^{*i}) + j(\lambda^{*i}) - j^{(i)}(\lambda^{*i}) - L^{(i)}(u^{i-1}, \lambda^{i-1}) \\
&\leq L^{(i)}(u^i, \lambda^{*i}) - L^{(i)}(u^{i-1}, \lambda^{i-1}) \\
&\leq 0,
\end{aligned}$$

where the last inequality becomes an equality if and only if $R^i = 0$. Thus $(12.44)_1$ is satisfied for the secant method.

However, usually the rate of decrease for the secant predictor is still slow. The case of real interest is the tangent predictor, which we shall discuss next.

The tangent predictor. In the tangent predictor, we define $D^{(i)}$ as the second order Taylor expansion of D about λ^{i-1}. Again, we need only to define $D^{(i)}$ in the active region. We put, with $\hat{\mu} = \mu - \lambda^{i-1}$,

$$D^{(i)}(\mu) = D(\lambda^{i-1}) + \chi^{i-1} \cdot \hat{\mu} + \tfrac{1}{2} \hat{\mu} \cdot B \hat{\mu}, \quad (12.59)$$

where

$$\begin{aligned}
\chi^{i-1} &= \nabla D(\lambda^{i-1}), \\
B &= \nabla^2 D(\lambda^{i-1}).
\end{aligned}$$

From the convexity of D, we infer that the function $D^{(i)}$ is convex.

As in the case of the secant predictor, we define

$$j^{(i)}(\mu) = \int_\Omega D^{(i)}(\mu)\, dx, \quad \mu \in \Lambda.$$

The quadratic approximation for L is

$$L^{(i)}(z) = \tfrac{1}{2} a(z, z) + j^{(i)}(\mu) - \langle \ell_1, v \rangle - \langle \ell_2, \mu \rangle \quad (12.60)$$

for $z = (v, \mu) \in V \times \Lambda$. The functional $L^{(i)}$ achieves its minimum at

$$(u^i, \lambda^{*i}) = (u^{i-1} + \hat{u}^i, \lambda^{i-1} + \hat{\lambda}^{*i}),$$

and $(\hat{u}^i, \hat{\lambda}^{*i})$ satisfies

$$b(\hat{u}^i, \hat{v}) - c(\hat{\lambda}^{*i}, \hat{v}) = \langle R^i, \hat{v} \rangle \quad \forall\, \hat{v} \in V, \tag{12.61}$$

$$- c(\hat{\mu}, \hat{u}^i) + d(\hat{\lambda}^{*i}, \hat{\mu}) + \langle B\hat{\lambda}^{*i}, \hat{\mu} \rangle = 0 \quad \forall\, \hat{\mu} \in \Lambda. \tag{12.62}$$

Again, R^i is defined as before.

This formulation leads, in the spatially discrete case, to the consistent tangent predictor of Simo and Taylor [118]. To see this, note that (12.61) and (12.62) will yield, after the introduction of a finite element basis, the set of equations

$$\begin{aligned} \boldsymbol{K}\boldsymbol{a} - \boldsymbol{L}\boldsymbol{\alpha} &= \boldsymbol{R}^i, \\ -\boldsymbol{L}^T\boldsymbol{a} + (\boldsymbol{M} + \boldsymbol{Q})\boldsymbol{\alpha} &= \boldsymbol{0}, \end{aligned} \tag{12.63}$$

in which $\boldsymbol{a}$ is the vector of nodal displacements, $\boldsymbol{\alpha}$ is the vector of internal variables at Gauss points (see Martin and Caddemi [85] for further details), $\boldsymbol{K}$ is the conventional stiffness matrix, and $\boldsymbol{Q}$ is the matrix that arises from the term containing B. Elimination of the vector $\boldsymbol{\alpha}$ from this set of equations leads to the equation

$$\boldsymbol{K}^C\boldsymbol{a} = \left\{\boldsymbol{K} - \boldsymbol{L}(\boldsymbol{M} + \boldsymbol{Q})^{-1}\boldsymbol{L}^T\right\}\boldsymbol{a} = \boldsymbol{R}^i, \tag{12.64}$$

thus defining the consistent tangent predictor modulus $\boldsymbol{K}^C$. An alternative formulation, in which we expand to only first order in (12.59), or alternatively set $\boldsymbol{Q} = \boldsymbol{0}$, leads to the conventional tangent predictor still used in much finite element software. The consistent predictor is associated with a quadratic rate of convergence, and we shall confine our attention to this case. However, results for the conventional tangent predictor can be inferred by setting $\boldsymbol{Q} = \boldsymbol{0}$.

Now let us check whether it is possible to have monotonic convergence for the tangent predictor. As before, we write

$$\Delta L_P^i = \left[L^{(i)}(u^i, \lambda^{*i}) - L^{(i)}(u^{i-1}, \lambda^{i-1})\right] + \left[j(\lambda^{*i}) - j^{(i)}(\lambda^{*i})\right]. \tag{12.65}$$

From the definition of the solution (u^i, λ^{*i}), we see that the first term on the right-hand side of (12.65) is nonpositive. However, we have no control over the sign of the second term on the right-hand side of (12.65). Indeed, since $D^{(i)}$ is the second-order Taylor approximation of D, the difference $j(\mu) - j^{(i)}(\mu)$ could be positive for certain μ and negative for some other μ. Hence it seems that the tangent predictor in its pure form will not produce a minimizing sequence.

Thus we are led to ask for what set of sufficient conditions it is possible to show convergence for the tangent predictor. Here we give details to show the line search procedure (see, for example, [29, 41, 89]) as a means of resolving this problem.

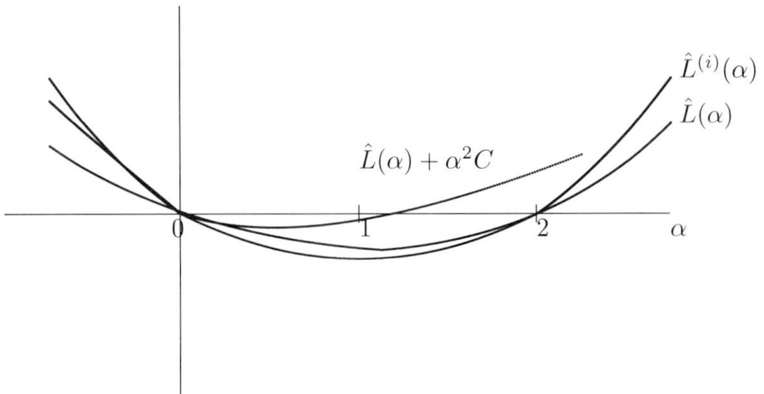

Figure 12.2: The functions arising from the line search procedure

We consider a modified prediction of the form $(\alpha\hat{u}^i, \alpha\hat{\lambda}^{*i})$, where $\alpha > 0$ is a scalar to be determined and $(\hat{u}^i, \hat{\lambda}^{*i})$ is the solution of (12.61)–(12.62).

Referring back to the formula (12.47) for the functional $L(v, \mu)$, we discard the terms that are independent of $\hat{v}$ and $\hat{\mu}$, and define

$$\begin{aligned}\overline{L}(\hat{z}) &= \tfrac{1}{2}a(\hat{z}, \hat{z}) + j(\lambda^{i-1} + \hat{\mu}) - j(\lambda^{i-1}) - \langle \chi^{i-1}, \hat{\mu}\rangle - \langle R^i, \hat{v}\rangle,\\ \overline{L}^{(i)}(\hat{z}) &= \tfrac{1}{2}a(\hat{z}, \hat{z}) + j^{(i)}(\lambda^{i-1} + \hat{\mu}) - j(\lambda^{i-1}) - \langle \chi^{i-1}, \hat{\mu}\rangle - \langle R^i, \hat{v}\rangle.\end{aligned}$$

We note that $j^{(i)}(\lambda^{i-1}) = j(\lambda^{i-1})$. Now define two functions of α according to

$$\begin{aligned}\hat{L}(\alpha) &= \overline{L}(\alpha\hat{u}^i, \alpha\hat{\lambda}^{*i}),\\ \hat{L}^{(i)}(\alpha) &= \overline{L}^{(i)}(\alpha\hat{u}^i, \alpha\hat{\lambda}^{*i}).\end{aligned}$$

Prototype graphs of the functions are shown in Figure 12.2. Since the function $\hat{L}^{(i)}$ is quadratic in α and has its minimum value at $\alpha = 1$, we see that $\hat{L}^{(i)}|_{\alpha=2} = \hat{L}^{(i)}|_{\alpha=0}$. Now consider the function

$$\hat{L} + \alpha^2 C, \qquad C = \tfrac{1}{2}\langle \hat{\lambda}^{*i}, B\,\hat{\lambda}^{*i}\rangle \geq 0. \tag{12.66}$$

Since D is a convex function, it is evident that

$$\hat{L}(\alpha) + \alpha^2 C \geq \hat{L}^{(i)}(\alpha). \tag{12.67}$$

The function is plotted in Figure 12.2.

We note that $\hat{L}(\alpha) + \alpha^2 C$ and $\hat{L}^{(i)}(\alpha)$ have the same value and gradient at $\alpha = 0$. It follows then that

$$\hat{L}(\alpha) + \alpha^2 C < \hat{L}(0)$$

for α in some range $0 < \alpha < \alpha'$, where $\alpha' \leq 1$. Finally,

$$\hat{L}(\alpha) \leq \hat{L}(\alpha) + \alpha^2 C.$$

At this level of generality it is not clear what value of α will provide the least value of $\hat{L}$; this indeed is the merit of the line search algorithm in which the optimal value of α is found approximately. The optimal value of α may be less than or greater than unity, and clearly if it is significantly different from unity, its adoption holds promise for faster convergence in the predictor step.

While we cannot forecast the optimal value of α, we are assured that the least value of $\hat{L}(\alpha)$ will be less than $\hat{L}(0)$, in view of the fact that the slope of $\hat{L}(\alpha)$ at $\alpha = 0$ is negative. Further, if the optimal value of α is larger than unity, then $\hat{L}|_{\alpha=1} < \hat{L}|_{\alpha=0}$. Thus the conclusion given above holds: there exists α' with $0 < \alpha' \leq 1$ such that $\hat{L}(\alpha) < \hat{L}(0)$ for $0 < \alpha < \alpha'$. For a choice of α in this range, a monotonic decrease is assured.

The essential result that follows from the discussion above is that the condition $\Delta L_P^i \leq 0$, $\Delta L_C^i < 0$ is not assured when the consistent tangent predictor is used directly. However, if for some iteration this condition does not hold, there will exist a value α in the range $0 < \alpha \leq 1$ for which it holds. In practice, the rate of decrease can be judged by comparing the residual at the end of the iteration with the previous residual; our analysis suggests that if the rate has the wrong sign in any iteration, the corrector step of the iteration should be repeated with decreasing values of α until it is again of the correct sign.

This comment strengthens the argument for the adoption of the line search algorithm; if the optimal or a near optimal value of α is chosen in each iteration, then the desired monotonic decrease is assured.

12.3 Convergence Analysis of the Solution Algorithms

In the last section we presented and analyzed some solution algorithms of the predictor–corrector type. The solution algorithms with the first three predictors are shown to lead to minimizing sequences of approximations. Whereas the solution algorithm with the tangent predictor does not, in general, enjoy this property, it can be adapted, by introducing a line search technique, so that the resulting algorithm leads to the desired monotonically decreasing behavior.

While the monotonic property guarantees that the sequence of approximations is a minimizing sequence, it does not imply the convergence of the sequence itself. Theoretically, it is possible that the limit of the minimizing sequence may fail to be the minimizer of the function. Therefore, it is important to know whether the solution algorithms presented in the last section do produce convergent results. We will rigorously prove the convergence of the solution algorithms for the first three predictors in this section.

We will perform the convergence analysis in the framework of the problem ABS^d. In particular, we need to solve (12.36)–(12.37). In this section we assume that the conditions stated in Problem ABS^d are satisfied. Thus, for example, the bilinear form a is $(V \times \Lambda)$-elliptic in the sense of (12.35). By taking $\mu = 0$ and $v = 0$ in turn in (12.35), we find that the bilinear form b is V-elliptic, the bilinear form d is Λ-elliptic, and with the same constant c_0 in (12.35),

$$b(v,v) \geq c_0 \|v\|_V^2 \quad \forall\, v \in V, \tag{12.68}$$

$$d(\mu,\mu) \geq c_0 \|\mu\|_\Lambda^2 \quad \forall\, \mu \in \Lambda. \tag{12.69}$$

Convergence of the elastic predictor. For convenience, we restate the algorithm for the elastic predictor in a more compact form.

ALGORITHM 1. Choose $w^0 = (u^0, \lambda^0) \in V \times \Lambda$ as the initial guess.

For $i = 1, 2, \ldots,$

Predictor: Compute $u^i \in V$ such that

$$b(u^i, v) = \langle \ell_1, v\rangle + c(\lambda^{i-1}, v) \quad \forall\, v \in V. \tag{12.70}$$

Corrector: Compute $\lambda^i \in \Lambda$ such that

$$j(\mu) - j(\lambda^i) + d(\lambda^i, \mu - \lambda^i) \geq \langle \ell_2, \mu - \lambda^i\rangle + c(\mu - \lambda^i, u^i) \quad \forall\, \mu \in \Lambda. \tag{12.71}$$

THEOREM 12.5. *Under the assumptions stated in Problem* ABS^d*, Algorithm 1 converges:*

$$u^i \to u \quad \text{in } V \quad \textit{and} \quad \lambda^i \to \lambda \quad \text{in } \Lambda \qquad \textit{as } i \to \infty,$$

where (u, λ) *is the solution of Problem* ABS^d.

PROOF. First we notice that the sequence $\{(u^i, \lambda^i)\}_{i\geq 1}$ is well-defined.

We take $\mu = \lambda^{i-1}$ in (12.71) to obtain

$$j(\lambda^{i-1}) - j(\lambda^i) + d(\lambda^i, \lambda^{i-1} - \lambda^i) \geq \langle \ell_2, \lambda^{i-1} - \lambda^i\rangle + c(\lambda^{i-1} - \lambda^i, u^i), \tag{12.72}$$

and take $v = u^{i-1} - u^i$ in (12.70) to obtain

$$b(u^i, u^{i-1} - u^i) = \langle \ell_1, u^{i-1} - u^i\rangle + c(\lambda^{i-1}, u^{i-1} - u^i). \tag{12.73}$$

Now we consider the energy difference (recall the definition (12.38) of the energy L):

$$\begin{aligned} L(w^{i-1}) &- L(w^i) \\ &= \tfrac{1}{2}\left[b(u^{i-1}, u^{i-1}) - b(u^i, u^i)\right] - \left[c(\lambda^{i-1}, u^{i-1}) - c(\lambda^i, u^i)\right] \\ &\quad + \tfrac{1}{2}\left[d(\lambda^{i-1}, \lambda^{i-1}) - d(\lambda^i, \lambda^i)\right] + j(\lambda^{i-1}) - j(\lambda^i) \\ &\quad - \langle \ell_1, u^{i-1} - u^i\rangle - \langle \ell_2, \lambda^{i-1} - \lambda^i\rangle, \end{aligned}$$

where $w^i = (u^i, \lambda^i)$. Using the relations (12.72) and (12.73), we have

$$\begin{aligned}
&L(w^{i-1}) - L(w^i) \\
&\quad \geq \tfrac{1}{2} \left[b(u^{i-1}, u^{i-1}) - b(u^i, u^i) - 2\, b(u^i, u^{i-1} - u^i) \right] \\
&\qquad - \left[c(\lambda^{i-1}, u^{i-1}) - c(\lambda^i, u^i) \right. \\
&\qquad\quad \left. - c(\lambda^{i-1} - \lambda^i, u^i) - c(\lambda^{i-1}, u^{i-1} - u^i) \right] \\
&\qquad + \tfrac{1}{2} \left[d(\lambda^{i-1}, \lambda^{i-1}) - d(\lambda^i, \lambda^i) - 2\, d(\lambda^i, \lambda^{i-1} - \lambda^i) \right] \\
&\quad = \tfrac{1}{2} \left[b(u^{i-1} - u^i, u^{i-1} - u^i) + d(\lambda^{i-1} - \lambda^i, \lambda^{i-1} - \lambda^i) \right].
\end{aligned}$$

Using (12.68) and (12.69) we thus have a useful inequality:

$$\begin{aligned}
L(w^{i-1}) - L(w^i) &\geq \frac{c_0}{2} \left(\|u^{i-1} - u^i\|_V^2 + \|\lambda^{i-1} - \lambda^i\|_\Lambda^2 \right) \\
&= \frac{c_0}{2} \|w^{i-1} - w^i\|_Z^2. \qquad (12.74)
\end{aligned}$$

A first consequence of the above inequality is that the sequence $\{L(w^i)\}_i$ is decreasing. Since the sequence $\{L(w^i)\}_i$ is bounded below by $L(w)$, with w the solution of the problem ABS^d or equivalently the minimizer of $L(\cdot)$, we see that the sequence $\{L(w^i)\}_i$ has a limit. We may use the inequality (12.74) again to find that

$$\|w^{i-1} - w^i\|_Z \to 0 \quad \text{as } i \to \infty, \qquad (12.75)$$

since $\{L(w^i)\}_i$ is convergent and hence is a Cauchy sequence.

The next step is to show that the limit is attained at w. We choose $\mu = \lambda^i$ in (12.37) to obtain

$$j(\lambda^i) - j(\lambda) - c(\lambda^i - \lambda, u) + d(\lambda, \lambda^i - \lambda) \geq \langle \ell_2, \lambda^i - \lambda \rangle, \qquad (12.76)$$

and choose $v = u^i \quad u$ in (12.36) to obtain

$$b(u, u^i - u) - c(\lambda, u^i - u) = \langle \ell_1, u^i - u \rangle. \qquad (12.77)$$

Next we consider the energy difference $L(w^i) - L(w)$. This time, we will derive both a lower bound and an upper bound for the energy difference. For a lower bound, we use the inequalities (12.76) and (12.77) to obtain

$$\begin{aligned}
&L(w^i) - L(w) \\
&\quad = \tfrac{1}{2} \left[b(u^i, u^i) - b(u, u) \right] - \left[c(\lambda^i, u^i) - c(\lambda, u) \right] \\
&\qquad + \tfrac{1}{2} \left[d(\lambda^i, \lambda^i) - d(\lambda, \lambda) \right] + j(\lambda^i) - j(\lambda) \\
&\qquad - \langle \ell_1, u^i - u \rangle - \langle \ell_2, \lambda^i - \lambda \rangle \\
&\quad \geq \tfrac{1}{2} \left[b(u^i, u^i) - b(u, u) - 2\, b(u, u^i - u) \right] \\
&\qquad - \left[c(\lambda^i, u^i) - c(\lambda, u) - c(\lambda^i - \lambda, u) - c(\lambda, u^i - u) \right] \\
&\qquad + \tfrac{1}{2} \left[d(\lambda^i, \lambda^i) - d(\lambda, \lambda) - 2\, d(\lambda, \lambda^i - \lambda) \right] \\
&\quad = \tfrac{1}{2}\, b(u^i - u, u^i - u) - c(\lambda^i - \lambda, u^i - u) + \tfrac{1}{2}\, d(\lambda^i - \lambda, \lambda^i - \lambda) \\
&\quad = \tfrac{1}{2}\, a(w^i - w, w^i - w).
\end{aligned}$$

Using the Z-ellipticity of $a(\cdot,\cdot)$, (12.35), we then have the lower bound

$$L(w^i) - L(w) \geq \frac{c_0}{2} \|w^i - w\|_Z^2. \tag{12.78}$$

To derive an upper bound for the energy difference, we first take $\mu = \lambda$ in (12.71) and $v = u^i - u$ in (12.70) to obtain

$$j(\lambda) - j(\lambda^i) + d(\lambda^i, \lambda - \lambda^i) \geq \langle \ell_2, \lambda - \lambda^i \rangle + c(\lambda - \lambda^i, u^i), \tag{12.79}$$

$$b(u^i, u^i - u) = \langle \ell_1, u^i - u \rangle + c(\lambda^{i-1}, u^i - u). \tag{12.80}$$

Using these two inequalities, we have

$$\begin{aligned}
&L(w^i) - L(w) \\
&\quad = \tfrac{1}{2}\, \big[b(u^i, u^i) - b(u, u)\big] - \big[c(\lambda^i, u^i) - c(\lambda, u)\big] \\
&\qquad + \tfrac{1}{2}\, \big[d(\lambda^i, \lambda^i) - d(\lambda, \lambda)\big] + j(\lambda^i) - j(\lambda) \\
&\qquad - \langle \ell_1, u^i - u \rangle - \langle \ell_2, \lambda^i - \lambda \rangle \\
&\quad \leq \tfrac{1}{2}\, \big[b(u^i, u^i) - b(u, u) - 2\, b(u^i, u^i - u)\big] \\
&\qquad - \big[c(\lambda^i, u^i) - c(\lambda, u) + c(\lambda - \lambda^i, u^i) - c(\lambda^{i-1}, u^i - u)\big] \\
&\qquad + \tfrac{1}{2}\, \big[d(\lambda^i, \lambda^i) - d(\lambda, \lambda) + 2\, d(\lambda^i, \lambda - \lambda^i)\big] \\
&\quad = -\tfrac{1}{2}\, b(u^i - u, u^i - u) + c(\lambda^{i-1} - \lambda, u^i - u) \\
&\qquad - \tfrac{1}{2}\, d(\lambda^i - \lambda, \lambda^i - \lambda) \\
&\quad = -\tfrac{1}{2}\, a(w^i - w, w^i - w) + c(\lambda^{i-1} - \lambda^i, u^i - u) \\
&\quad \leq c(\lambda^{i-1} - \lambda^i, u^i - u).
\end{aligned}$$

Using the continuity of the bilinear form c and the well-known inequality

$$x\, y \leq \epsilon\, x^2 + \frac{1}{4\,\epsilon}\, y^2 \quad \forall\, x, y \in \mathbb{R},\ \forall\, \epsilon > 0,$$

we then get

$$L(w^i) - L(w) \leq \frac{c_0}{4} \|u^i - u\|_V^2 + c\, \|\lambda^{i-1} - \lambda^i\|_\Lambda^2 \quad \text{for some } c > 0. \tag{12.81}$$

Combining the inequalities (12.78), (12.81) and the obvious inequality

$$\|u^i - u\|_V \leq \|w^i - w\|_Z,$$

we have

$$\|w^i - w\|_Z \leq c\, \|\lambda^i - \lambda^{i-1}\|_\Lambda \to 0 \quad \text{as } i \to \infty,$$

by virtue of (12.75). □

A more careful examination of the argument leading to the inequality (12.71) shows that together with (12.78), actually we have

$$a(w^i - w, w^i - w) \le c(\lambda^{i-1} - \lambda^i, u^i - u).$$

Now assume that we know the continuity constant c_2 for the bilinear form $c(\cdot,\cdot)$, that is, the constant appearing in the inequality

$$c(\mu, v) \le c_2 \|\mu\|_\Lambda \|v\|_V.$$

Then

$$\begin{aligned} c_0 \|w^i - w\|_Z^2 &\le a(w^i - w, w^i - w) \\ &\le c(\lambda^{i-1} - \lambda^i, u^i - u) \\ &\le c_2 \|\lambda^{i-1} - \lambda^i\|_\Lambda \|u^i - u\|_V \\ &\le c_2 \|\lambda^{i-1} - \lambda^i\|_\Lambda \|w^i - w\|_Z. \end{aligned}$$

Therefore, we have a computable error estimate

$$\|w^i - w\|_Z \le \frac{c_2}{c_0} \|\lambda^{i-1} - \lambda^i\|_\Lambda,$$

which is useful in estimating the error associated with an iterate w^i.

Convergence of the modified elastic predictor. For the second predictor considered in the last section, we need an auxiliary bilinear form $b^* : V \times V \to \mathbb{R}$. Then we state the second solution algorithm in compact form as follows:

ALGORITHM 2. Choose $w^0 = (u^0, \lambda^0) \in V \times \Lambda$ as the initial guess.
For $i = 1, 2, \ldots,$
Predictor: Compute $u^i \in V$ such that

$$b^*(u^i, v) = b^*(u^{i-1}, v) - b(u^{i-1}, v) + c(\lambda^{i-1}, v) + \langle \ell_1, v\rangle \quad \forall\, v \in V. \tag{12.82}$$

Corrector: Compute $\lambda^i \in \Lambda$ such that

$$j(\mu) - j(\lambda^i) + d(\lambda^i, \mu - \lambda^i) \ge \langle \ell_2, \mu - \lambda^i\rangle + c(\mu - \lambda^i, u^i) \quad \forall\, \mu \in \Lambda. \tag{12.83}$$

We observe that the corrector step is the same as that in Algorithm 1. For the convergence of Algorithm 2, we need to make some assumptions on the bilinear form b^*.

THEOREM 12.6. *We keep the assumptions stated in Problem* ABS^d. *Furthermore, we assume that the bilinear form $b^* : V \times V \to \mathbb{R}$ is continuous, symmetric, and that there exists a constant $c_1 > 0$ such that*

$$b^*(v, v) - \tfrac{1}{2}\, b(v, v) \ge c_1 \|v\|_V^2 \quad \forall\, v \in V. \tag{12.84}$$

Then Algorithm 2 converges:

$$u^i \to u \quad \text{in } V, \qquad \lambda^i \to \lambda \quad \text{in } \Lambda \qquad \text{as } i \to \infty.$$

PROOF. From the assumptions, we see that the sequence $\{(u^i, \lambda^i)\}$ is well-defined.

The idea of the convergence proof is the same as that for Theorem 12.5. First we consider the energy difference $L(w^{i-1}) - L(w^i)$. Since the corrector steps are the same for the two algorithms, we still have the inequality (12.72). We take $v = u^{i-1} - u^i$ in (12.82) to obtain

$$\begin{aligned} b^*(u^i, u^{i-1} - u^i) &= b^*(u^{i-1}, u^{i-1} - u^i) - b(u^{i-1}, u^{i-1} - u^i) \\ &\quad + c(\lambda^{i-1}, u^{i-1} - u^i) + \langle \ell_1, u^{i-1} - u^i \rangle. \end{aligned} \tag{12.85}$$

Using the inequalities (12.72) and (12.85), we can find a lower bound for the energy difference:

$$\begin{aligned} L(w^{i-1}) - L(w^i) &\geq b^*(u^{i-1} - u^i, u^{i-1} - u^i) - \tfrac{1}{2}\, b(u^{i-1} - u^i, u^{i-1} - u^i) \\ &\quad + \tfrac{1}{2}\, d(\lambda^{i-1} - \lambda^i, \lambda^{i-1} - \lambda^i). \end{aligned}$$

Using the assumption (12.84) and the Λ-ellipticity of the bilinear form d, we then find that

$$L(w^{i-1}) - L(w^i) \geq \min\{c_1, c_0/2\} \left(\|u^{i-1} - u^i\|_V^2 + \|\lambda^{i-1} - \lambda^i\|_\Lambda^2 \right). \tag{12.86}$$

Once again, from (12.86) we infer that

$$\|w^{i-1} - w^i\|_Z \to 0 \quad \text{as } i \to \infty. \tag{12.87}$$

Now we consider the energy difference $L(w^i) - L(w)$. We note that in deriving the lower bound (12.78) for the difference $L(w^i) - L(w)$ in the proof of the last theorem, we used only the relations (12.36) and (12.37). Thus, for Algorithm 2, we still have the inequality (12.78). To have an upper bound for the energy difference, we note that the inequality (12.79) remains unchanged, since it is derived from the corrector step. We choose $v = u^i - u$ in (12.82) to obtain

$$\begin{aligned} b^*(u^i, u^i - u) &= b^*(u^{i-1}, u^i - u) - b(u^{i-1}, u^i - u) \\ &\quad + c(\lambda^{i-1}, u^i - u) + \langle \ell_1, u^i - u \rangle. \end{aligned} \tag{12.88}$$

Using the inequalities (12.79) and (12.88), after some straightforward algebraic manipulations we have

$$
\begin{aligned}
L(w^i) &- L(w) \\
&\le -\tfrac{1}{2}\, b(u^i - u, u^i - u) + b(u^i - u^{i-1}, u^i - u) - b^*(u^i - u^{i-1}, u^i - u) \\
&\quad + c(\lambda^{i-1} - \lambda, u^i - u) - \tfrac{1}{2}\, d(\lambda^i - \lambda, \lambda^i - \lambda) \\
&= -\tfrac{1}{2}\, a(w^i - w, w^i - w) + b(u^i - u^{i-1}, u^i - u) \\
&\quad - b^*(u^i - u^{i-1}, u^i - u) + c(\lambda^{i-1} - \lambda^i, u^i - u) \\
&\le b(u^i - u^{i-1}, u^i - u) - b^*(u^i - u^{i-1}, u^i - u) + c(\lambda^{i-1} - \lambda^i, u^i - u) \\
&\le \frac{c_0}{4}\, \|u^i - u\|_V^2 + c\, \left(\|u^i - u^{i-1}\|_V^2 + \|\lambda^i - \lambda^{i-1}\|_\Lambda^2 \right) \\
&\le \frac{c_0}{4}\, \|w^i - w\|_Z^2 + c\, \left(\|u^i - u^{i-1}\|_V^2 + \|\lambda^i - \lambda^{i-1}\|_\Lambda^2 \right),
\end{aligned}
$$

which, combined with (12.78), implies that

$$\|w^i - w\|_Z \le c\, \|w^i - w^{i-1}\|_Z \to 0 \quad \text{as } i \to \infty.$$

Thus the convergence of the iterates is proved. □

Convergence of the secant predictor. Now we turn to the convergence analysis for the solution algorithm with the secant predictor. We observe that for the ith iteration in the predictor step, we minimize the functional

$$\tfrac{1}{2}\, b(v, v) - c(\mu, v) + \tfrac{1}{2}\, d(\mu, \mu) + j^{(i)}(\mu) - \langle \ell_1, v \rangle - \langle \ell_2, \mu \rangle,$$

where

$$j^{(i)}(\mu) = \int_\Omega D^{(i)}(\mu)\, dx$$

for a quadratic function $D^{(i)}$ satisfying the conditions (12.52)–(12.54). The minimizer is denoted by (u^i, λ^{*i}), where u^i is an updated solution component, while the intermediate value λ^{*i} is not needed further. The third solution algorithm can now be stated as follows.

ALGORITHM 3. Choose a $w^0 = (u^0, \lambda^0) \in V \times \Lambda$ as the initial guess.

For $i = 1, 2, \ldots,$

Predictor: Compute $u^i \in V$ such that u^i, together with $\lambda^{*i} \in \Lambda$, satisfies

$$b(u^i, v) - c(\lambda^{*i}, v) = \langle \ell_1, v \rangle \quad \forall\, v \in V, \tag{12.89}$$

$$
\begin{aligned}
j^{(i)}(\mu) - j^{(i)}(\lambda^{*i}) &+ d(\lambda^{*i}, \mu - \lambda^{*i}) \\
&\ge \langle \ell_2, \mu - \lambda^{*i} \rangle + c(\mu - \lambda^{*i}, u^i) \quad \forall\, \mu \in \Lambda.
\end{aligned} \tag{12.90}
$$

Corrector: Compute $\lambda^i \in \Lambda$ such that

$$j(\mu) - j(\lambda^i) + d(\lambda^i, \mu - \lambda^i) \ge \langle \ell_2, \mu - \lambda^i \rangle + c(\mu - \lambda^i, u^i) \quad \forall\, \mu \in \Lambda. \tag{12.91}$$

As was noted in the last section, the inequality (12.90) is actually a linear equation (cf. (12.58)). For the convergence analysis, however, it is more convenient to leave the relation in the form of an inequality. Once again, we observe that the corrector step is the same as in Algorithm 1, so any relations derived from the corrector step in the proof of Theorem 12.5 are still valid for Algorithm 3.

THEOREM 12.7. *Under the assumptions stated in Problem* ABS^d, *Algorithm 3 converges:*

$$u^i \to u \quad \text{in } V \quad \text{and} \quad \lambda^i \to \lambda \quad \text{in } \Lambda \qquad \text{as } i \to \infty.$$

PROOF. Denote by $w^{*i} = (u^i, \lambda^{*i})$ the intermediate solution from the predictor step. Let us first consider the difference

$$\begin{aligned} &L(w^{i-1}) - L(w^{*i}) \\ &\quad = \tfrac{1}{2}\left[b(u^{i-1}, u^{i-1}) - b(u^i, u^i)\right] - \left[c(\lambda^{i-1}, u^{i-1}) - c(\lambda^{*i}, u^i)\right] \\ &\quad\quad + \tfrac{1}{2}\left[d(\lambda^{i-1}, \lambda^{i-1}) - d(\lambda^{*i}, \lambda^{*i})\right] + j(\lambda^{i-1}) - j(\lambda^{*i}) \\ &\quad\quad - \langle \ell_1, u^{i-1} - u^i \rangle - \langle \ell_2, \lambda^{i-1} - \lambda^{*i} \rangle. \end{aligned}$$

We take $v = u^{i-1} - u^i$ in (12.89) to obtain

$$b(u^i, u^{i-1} - u^i) - c(\lambda^{*i}, u^{i-1} - u^i) = \langle \ell_1, u^{i-1} - u^i \rangle, \tag{12.92}$$

and take $\mu = \lambda^{i-1}$ in (12.90) to obtain

$$\begin{aligned} &j^{(i)}(\lambda^{i-1}) - j^{(i)}(\lambda^{*i}) + d(\lambda^{*i}, \lambda^{i-1} - \lambda^{*i}) \\ &\quad \geq \langle \ell_2, \lambda^{i-1} - \lambda^{*i} \rangle + c(\lambda^{i-1} - \lambda^{*i}, u^i). \end{aligned} \tag{12.93}$$

Using (12.92) and (12.93), we find that

$$\begin{aligned} &L(w^{i-1}) - L(w^{*i}) \\ &\quad \geq \tfrac{1}{2}\left[b(u^{i-1}, u^{i-1}) - b(u^i, u^i) - 2\, b(u^i, u^{i-1} - u^i)\right] \\ &\quad\quad - \left[c(\lambda^{i-1}, u^{i-1}) - c(\lambda^{*i}, u^i)\right. \\ &\quad\quad\quad \left. - c(\lambda^{*i}, u^{i-1} - u^i) - c(\lambda^{i-1} - \lambda^{*i}, u^i)\right] \\ &\quad\quad + \tfrac{1}{2}\left[d(\lambda^{i-1}, \lambda^{i-1}) - d(\lambda^{*i}, \lambda^{*i}) - 2\, d(\lambda^{*i}, \lambda^{i-1} - \lambda^{*i})\right] \\ &\quad\quad + j(\lambda^{i-1}) - j(\lambda^{*i}) - j^{(i)}(\lambda^{i-1}) + j^{(i)}(\lambda^{*i}) \\ &\quad = \tfrac{1}{2}\, b(u^{i-1} - u^i, u^{i-1} - u^i) - c(\lambda^{i-1} - \lambda^{*i}, \lambda^{i-1} - \lambda^{*i}) \\ &\quad\quad + \tfrac{1}{2}\, d(\lambda^{i-1} - \lambda^{*i}, \lambda^{i-1} - \lambda^{*i}) + j^{(i)}(\lambda^{*i}) - j(\lambda^{*i}) \\ &\quad \geq \tfrac{1}{2}\, a(w^{i-1} - w^{*i}, w^{i-1} - w^{*i}), \end{aligned}$$

where we have used the relations $j^{(i)}(\lambda^{i-1}) = j(\lambda^{i-1})$ and $j(\mu) \leq j^{(i)}(\mu)$ for any $\mu \in \Lambda$. Thus, we have the inequality

$$L(w^{i-1}) - L(w^{*i}) \geq \frac{c_0}{2}\, \|w^{i-1} - w^{*i}\|_Z^2. \tag{12.94}$$

From the definition of the corrector step, we have $L(w^i) \le L(w^{*i})$. Hence, from (12.94) we get

$$L(w^{i-1}) - L(w^i) \ge \frac{c_0}{2} \|w^{i-1} - w^{*i}\|_Z^2.$$

With the same argument used before, we then conclude that

$$w^{i-1} - w^{*i} \to 0 \quad \text{as } i \to \infty. \tag{12.95}$$

As will be seen below, we will need a result of the form (12.95), with the superscript $i-1$ in w^{i-1} replaced by i. To obtain such a result, we choose $\mu = \lambda^{i-1}$ in (12.91) to obtain

$$j(\lambda^{i-1}) - j(\lambda^i) + d(\lambda^i, \lambda^{i-1} - \lambda^i) \ge \langle \ell_2, \lambda^{i-1} - \lambda^i \rangle + c(\lambda^{i-1} - \lambda^i, u^i). \tag{12.96}$$

Then in (12.91) again, we replace the index i by $i-1$, and choose $\mu = \lambda^i$ to obtain

$$j(\lambda^i) - j(\lambda^{i-1}) + d(\lambda^{i-1}, \lambda^i - \lambda^{i-1}) \ge \langle \ell_2, \lambda^i - \lambda^{i-1} \rangle + c(\lambda^i - \lambda^{i-1}, u^{i-1}). \tag{12.97}$$

The inequalities (12.96) and (12.97) are added to give

$$-d(\lambda^{i-1} - \lambda^i, \lambda^{i-1} - \lambda^i) \ge -c(\lambda^{i-1} - \lambda^i, u^{i-1} - u^i),$$

or

$$d(\lambda^{i-1} - \lambda^i, \lambda^{i-1} - \lambda^i) \le c(\lambda^{i-1} - \lambda^i, u^{i-1} - u^i).$$

Using the Λ-ellipticity of d and the continuity of c, we have

$$\|\lambda^{i-1} - \lambda^i\|_\Lambda \le c \, \|u^{i-1} - u^i\|_V \to 0 \quad \text{as } i \to \infty,$$

by (12.95). This inequality, together with (12.95), implies

$$\lambda^{*i} - \lambda^i \to 0 \quad \text{as } i \to \infty. \tag{12.98}$$

We now proceed to consider the energy difference

$$\begin{aligned} L(w^i) - L(w) = {} & \tfrac{1}{2} \left[b(u^i, u^i) - b(u, u) \right] - \left[c(\lambda^i, u^i) - c(\lambda, u) \right] \\ & + \tfrac{1}{2} \left[d(\lambda^i, \lambda^i) - d(\lambda, \lambda) \right] \\ & + j(\lambda^i) - j(\lambda) - \langle \ell_1, u^i - u \rangle - \langle \ell_2, \lambda^i - \lambda \rangle. \end{aligned}$$

As before, we have the lower bound

$$L(w^i) - L(w) \ge \tfrac{1}{2} \, a(w^i - w, w^i - w) \ge \frac{c_0}{2} \|w^i - w\|_Z^2. \tag{12.99}$$

To derive an upper bound for the energy difference, we take $v = u^i - u$ in (12.89), $\mu = \lambda$ in (12.91), to obtain

$$b(u^i, u^i - u) - c(\lambda^{*i}, u^i - u) = \langle \ell_1, u^i - u \rangle$$

and

$$j(\lambda) - j(\lambda^i) + d(\lambda^i, \lambda - \lambda^i) \geq \langle \ell_2, \lambda - \lambda^i \rangle + c(\lambda - \lambda^i, u^i).$$

Using these two inequalities, we have

$$\begin{aligned} L(w^i) - L(w) \leq\ & \tfrac{1}{2}\left[b(u^i, u^i) - b(u, u) - 2\, b(u^i, u^i - u)\right] \\ & - \left[c(\lambda^i, u^i) - c(\lambda, u) - c(\lambda^{*i}, u^i - u) + c(\lambda - \lambda^i, u^i)\right] \\ & + \tfrac{1}{2}\left[d(\lambda^i, \lambda^i) - d(\lambda, \lambda) + 2\, d(\lambda^i, \lambda - \lambda^i)\right] \\ =\ & -\tfrac{1}{2}\, a(w^i - w, w^i - w) + c(\lambda^{*i} - \lambda^i, u^i - u) \\ \leq\ & c(\lambda^{*i} - \lambda^i, u^i - u). \end{aligned}$$

This inequality and (12.99) imply, with the Z-ellipticity of a and the continuity of c, that

$$\|w^i - w\|_Z^2 \leq c\, \|\lambda^{*i} - \lambda^i\|_\Lambda \|u^i - u\|_V \leq c\, \|\lambda^{*i} - \lambda^i\|_\Lambda \|w^i - w\|_Z.$$

Therefore,

$$\|w^i - w\|_Z \leq c\, \|\lambda^{*i} - \lambda^i\|_\Lambda \to 0 \quad \text{as } i \to \infty,$$

by virtue of (12.98). □

We observe that in the convergence proof we did not use the condition (12.53) in the construction of the quadratic functional $j^{(i)}$.

12.4 Regularization Technique and A Posteriori Error Analysis

In this section we analyze the regularization technique for solving variational inequalities involving nondifferentiable terms. As a sample problem, we consider the solution of the backward Euler time-discrete scheme (12.27) for the elastoplasticity problem (12.22). Similar results hold for other numerical schemes (in particular, the fully discrete schemes) for solving (12.22), and for more general plasticity models such as those involving combined kinematic and isotropic hardening.

In addition, we will restrict attention to the case of the von Mises yield condition, for which the dissipation function is given by $D(\boldsymbol{q}) = c_0|\boldsymbol{q}|$; the generalization to arbitrary yield conditions can be achieved with a little modification on the following arguments.

We will first discuss the idea of the regularization technique and present some general convergence theorems for the technique. For the commonly used regularization methods (resulting from commonly used regularization functions), we prove an a priori error estimate, which shows directly the convergence of the regularization sequences. Besides the a priori error estimate, we will also provide an a posteriori error analysis for the regularization technique. An a posteriori error estimate gives a computable error bound once the solution of a regularized problem is computed. From the application viewpoint, an a posteriori error estimate is more useful for actual implementation of a regularization method.

To simplify the notation in this section, we will use $\boldsymbol{w}$ and $\boldsymbol{\ell}$ to stand for the quantities $\Delta \boldsymbol{w}_n^k$ and $\boldsymbol{L}_n$ in (12.27). Thus, the problem to be solved is

$$\boldsymbol{w} \in Z, \quad a(\boldsymbol{w}, \boldsymbol{z}-\boldsymbol{w}) + j(\boldsymbol{z}) - j(\boldsymbol{w}) \geq \langle \boldsymbol{\ell}, \boldsymbol{z}-\boldsymbol{w} \rangle \quad \forall\, \boldsymbol{z} \in Z. \tag{12.100}$$

Here, as before, $Z = V \times Q_0$, $\boldsymbol{w} = (\boldsymbol{u}, \boldsymbol{p})$, $\boldsymbol{z} = (\boldsymbol{v}, \boldsymbol{q})$, and

$$a(\boldsymbol{w}, \boldsymbol{z}) = \int_\Omega \left[\boldsymbol{C}(\boldsymbol{\epsilon}(\boldsymbol{u}) - \boldsymbol{p}) : (\boldsymbol{\epsilon}(\boldsymbol{v}) - \boldsymbol{q}) + k_1 \boldsymbol{p} : \boldsymbol{q}\right] dx, \tag{12.101}$$

$$j(\boldsymbol{z}) = \int_\Omega c_0 |\boldsymbol{q}(x)|\, dx, \tag{12.102}$$

$$\langle \boldsymbol{\ell}, \boldsymbol{z} \rangle = \langle \boldsymbol{\ell}_1, \boldsymbol{v} \rangle + \langle \boldsymbol{\ell}_2, \boldsymbol{q} \rangle. \tag{12.103}$$

Here, $\boldsymbol{\ell}_1$ is a continuous linear form on V, and $\boldsymbol{\ell}_2$ is a continuous linear form on Q_0. Thus, $\boldsymbol{\ell}$ is a continuous linear form on $Z = V \times Q_0$.

In this section we will additionally assume $\boldsymbol{f}(t) \in (L^2(\Omega))^3$ for a.a. $t \in [0, T]$. It will be convenient for us to identify the functionals $\boldsymbol{\ell}_1$ and $\boldsymbol{\ell}_2$ with functions $\boldsymbol{\ell}_1 \in (L^2(\Omega))^3$ and $\boldsymbol{\ell}_2 \in Q$ such that

$$\langle \boldsymbol{\ell}_1, \boldsymbol{v} \rangle = \int_\Omega \boldsymbol{\ell}_1 \cdot \boldsymbol{v}\, dx,$$

$$\langle \boldsymbol{\ell}_2, \boldsymbol{q} \rangle = \int_\Omega \boldsymbol{\ell}_2 : \boldsymbol{q}\, dx.$$

Then

$$\langle \boldsymbol{\ell}, \boldsymbol{z} \rangle = \int_\Omega (\boldsymbol{\ell}_1 \cdot \boldsymbol{v} + \boldsymbol{\ell}_2 : \boldsymbol{q})\, dx \quad \forall\, \boldsymbol{z} = (\boldsymbol{v}, \boldsymbol{q}) \in Z.$$

The regularization technique. In a regularization method, the nondifferentiable term $j(\boldsymbol{z})$ is approximated by a family of differentiable functionals $j_\varepsilon(\boldsymbol{z}) = \int_\Omega \phi_\varepsilon(|\boldsymbol{q}|)\, dx$, where $\phi_\varepsilon(|\boldsymbol{q}|)$ is differentiable with respect to $\boldsymbol{q}$. The regularized approximation of (12.100) is

$$\boldsymbol{w}_\varepsilon \in Z, \quad a(\boldsymbol{w}_\varepsilon, \boldsymbol{z}-\boldsymbol{w}_\varepsilon) + j_\varepsilon(\boldsymbol{z}) - j_\varepsilon(\boldsymbol{w}_\varepsilon) \geq \langle \boldsymbol{\ell}, \boldsymbol{z}-\boldsymbol{w}_\varepsilon \rangle \quad \forall\, \boldsymbol{z} \in Z. \tag{12.104}$$

Since j_ε is differentiable, this problem is actually an equation, namely,

$$\boldsymbol{w}_\varepsilon \in Z, \quad a(\boldsymbol{w}_\varepsilon, \boldsymbol{z}) + \langle j'_\varepsilon(\boldsymbol{w}_\varepsilon), \boldsymbol{z}\rangle = \langle \boldsymbol{\ell}, \boldsymbol{z}\rangle \quad \forall\, \boldsymbol{z} \in Z. \tag{12.105}$$

For a given nondifferentiable term, there are many ways to construct sequences of differentiable approximations. We list three possible choices of a regularizing sequence for the nondifferentiable functional j of (12.102).

CHOICE 1. $j_\varepsilon(\boldsymbol{z}) = \displaystyle\int_\Omega c_0 \phi^1_\varepsilon(|\boldsymbol{q}|)\, dx$, where

$$\phi^1_\varepsilon(t) = \sqrt{t^2 + \varepsilon^2}.$$

CHOICE 2. $j_\varepsilon(\boldsymbol{z}) = \displaystyle\int_\Omega c_0 \phi^2_\varepsilon(|\boldsymbol{q}|)\, dx$, where

$$\phi^2_\varepsilon(t) = \begin{cases} t - \dfrac{\varepsilon}{2} & \text{if } t \ge \varepsilon, \\ \dfrac{1}{2\varepsilon}\, t^2 & \text{if } |t| \le \varepsilon, \\ -t - \dfrac{\varepsilon}{2} & \text{if } t \le -\varepsilon. \end{cases}$$

CHOICE 3. $j_\varepsilon(\boldsymbol{z}) = \displaystyle\int_\Omega c_0 \phi^3_\varepsilon(|\boldsymbol{q}|)\, dx$, where

$$\phi^3_\varepsilon(t) = \begin{cases} t & \text{if } t \ge \varepsilon, \\ \dfrac{1}{2}\left(\dfrac{t^2}{\varepsilon} + \varepsilon\right) & \text{if } |t| \le \varepsilon, \\ -t & \text{if } t \le -\varepsilon. \end{cases}$$

Convergence, a priori error estimate. We first consider the convergence of the regularization technique (cf. [45, 62]).

THEOREM 12.8. *Let V be a Hilbert space; $a : V \times V \to \mathbb{R}$ a continuous, V-elliptic bilinear form; $j : V \to \overline{\mathbb{R}}$ a proper, nonnegative, convex, and weakly continuous functional; and ℓ a linear continuous form on V. Assume that $j_\varepsilon : V \to \mathbb{R}$ is proper, nonnegative, convex, and weakly l.s.c. Assume further that*

$$j_\varepsilon(v) \to j(v) \quad \forall\, v \in V,$$

$$u_\varepsilon \to u \text{ weakly in } V \Longrightarrow j(u) \le \liminf_{\varepsilon\to 0} j_\varepsilon(u_\varepsilon).$$

Let $u, u_\varepsilon \in V$ be the solutions of the variational inequalities

$$a(u, v-u) + j(v) - j(u) \ge \langle \ell, v-u\rangle \quad \forall\, v \in V$$

and

$$a(u_\varepsilon, v - u_\varepsilon) + j_\varepsilon(v) - j_\varepsilon(u_\varepsilon) \geq \langle \ell, v - u_\varepsilon \rangle \quad \forall\, v \in V,$$

respectively. Then, $u_\varepsilon \to u$ *in* V *as* $\varepsilon \to 0$.

In the context of the problem (12.100) and its regularization (12.104), the conditions stated in Theorem 12.8 are satisfied for all three choices of the regularization function. So with any of these choices the regularization method (12.104) produces a convergent sequence:

$$\boldsymbol{w}_\varepsilon \to \boldsymbol{w} \quad \text{as } \varepsilon \to 0. \tag{12.106}$$

Actually, with the chosen regularization functions it is easy to verify that

$$\big|\phi_\varepsilon(|\boldsymbol{q}|) - |\boldsymbol{q}|\big| \leq c\,\varepsilon \quad \forall\, \boldsymbol{q}. \tag{12.107}$$

Using (12.107), we can derive an a priori error estimate for the regularization methods.

THEOREM 12.9. *With any of the three choices for the regularization function, the regularization method* (12.104) *converges, and*

$$\|\boldsymbol{w}_\varepsilon - \boldsymbol{w}\|_Z \leq c\,\sqrt{\varepsilon}. \tag{12.108}$$

PROOF. We take $\boldsymbol{z} = \boldsymbol{w}_\varepsilon$ in (12.100), $\boldsymbol{z} = \boldsymbol{w}$ in (12.104), add the two inequalities, and use the Z-ellipticity of a to obtain

$$\begin{aligned} c_0\, \|\boldsymbol{w}_\varepsilon - \boldsymbol{w}\|_Z^2 &\leq a(\boldsymbol{w}_\varepsilon - \boldsymbol{w}, \boldsymbol{w}_\varepsilon - \boldsymbol{w}) \\ &\leq [j(\boldsymbol{w}_\varepsilon) - j_\varepsilon(\boldsymbol{w}_\varepsilon)] + [j_\varepsilon(\boldsymbol{w}) - j(\boldsymbol{w})] \\ &\leq c\,\varepsilon. \end{aligned}$$

Then (12.108) follows by applying (12.107). □

A posteriori error analysis. To derive a posteriori error estimates for the regularization technique, we employ the duality theory reviewed in Section 4.1. Specifically, in our analysis of the regularization technique, we reformulate the problem (12.100) in the form of (4.32). Then, in applying Theorem 4.6, we take y there to be the solution $\boldsymbol{w}$ of the problem (12.100), and we will take z in (4.34) to be the solution $\boldsymbol{w}_\varepsilon$ of the regularized problem (12.104). The procedure of deriving an estimate for the error $\boldsymbol{w} - \boldsymbol{w}_\varepsilon$ then consists of two steps:

STEP 1. Find a suitable lower bound for the difference D defined in (4.33). We will show that this difference D can be bounded below by the quantity $c_1\|\boldsymbol{w} - \boldsymbol{w}_\varepsilon\|_Z^2$, with $c_1 = c_0/2$, and c_0 is the constant in the Z-ellipticity inequality for the bilinear form a.

STEP 2. Take an appropriate s^* such that the right-hand side of (4.34) will provide a close upper bound for the quantity D. The auxiliary quantity s^*

in (4.34) must be easily computable from the solution $\boldsymbol{w}_\varepsilon$ of the regularized problem. We will make such a selection and discuss the efficiency of the resulting a posteriori error estimate.

To apply Theorem 4.6 we use the following problem setting:

$$Z = V \times Q_0 \text{ with the norm of } V \times Q,$$
$$S = Q \text{ with the norm of } Q,$$

and with $\boldsymbol{z} = (\boldsymbol{v}, \boldsymbol{q}) \in Z$, $\boldsymbol{s} \in S$,

$$F\boldsymbol{z} = \boldsymbol{\epsilon}(\boldsymbol{v}),$$
$$J(\boldsymbol{z}, \boldsymbol{s}) = \int_\Omega \left[\tfrac{1}{2}\,\boldsymbol{C}\,(\boldsymbol{s}-\boldsymbol{q}) : (\boldsymbol{s}-\boldsymbol{q}) + \frac{k_1}{2}\,|\boldsymbol{q}|^2 + c_0|\boldsymbol{q}|\right] dx - \langle \boldsymbol{\ell}, \boldsymbol{z}\rangle.$$

We identify Q' with Q, and use $\boldsymbol{s}^*$ to denote a generic element in Q'. It is easily seen that the problem (12.100) is equivalent to the minimization problem (4.32) with the above identification. Now let us use the definition (4.8) to compute the conjugate function

$$\begin{aligned} &J^*(F^*\boldsymbol{s}^*, -\boldsymbol{s}^*) \\ &\quad= \sup_{\boldsymbol{z}\in Z, \boldsymbol{s}\in S} \left[\langle \boldsymbol{s}^*, F\boldsymbol{z}\rangle - \langle \boldsymbol{s}^*, \boldsymbol{s}\rangle - J(\boldsymbol{z}, \boldsymbol{s})\right] \\ &\quad= \sup_{\boldsymbol{z}\in Z, \boldsymbol{s}\in S} \int_\Omega \big[\boldsymbol{s}^* : (\boldsymbol{\epsilon}(\boldsymbol{v}) - \boldsymbol{s}) - \tfrac{1}{2}\boldsymbol{C}(\boldsymbol{s}-\boldsymbol{q}) : (\boldsymbol{s}-\boldsymbol{q}) - \tfrac{1}{2}\,k_1|\boldsymbol{q}|^2 \\ &\qquad\qquad - c_0|\boldsymbol{q}| + \boldsymbol{\ell}_1\cdot\boldsymbol{v} + \boldsymbol{\ell}_2 : \boldsymbol{q}\big]\, dx. \end{aligned}$$

We have

$$\begin{aligned} &J^*(F^*\boldsymbol{s}^*, -\boldsymbol{s}^*) \\ &\quad= \int_\Omega \tfrac{1}{2}\,\boldsymbol{C}^{-1}\boldsymbol{s}^* : \boldsymbol{s}^*\, dx \\ &\qquad+ \sup_{\boldsymbol{z}\in Z} \int_\Omega \big[\boldsymbol{s}^* : \boldsymbol{\epsilon}(\boldsymbol{v}) + \boldsymbol{\ell}_1\cdot\boldsymbol{v} - \tfrac{1}{2}\,k_1|\boldsymbol{q}|^2 - c_0|\boldsymbol{q}| + (\boldsymbol{\ell}_2 - \boldsymbol{s}^*) : \boldsymbol{q}\big]\, dx \\ &\quad= \int_\Omega \tfrac{1}{2}\,\boldsymbol{C}^{-1}\boldsymbol{s}^* : \boldsymbol{s}^*\, dx \\ &\qquad+ \sup_{\boldsymbol{z}\in Z} \int_\Omega \left[\boldsymbol{s}^* : \boldsymbol{\epsilon}(\boldsymbol{v}) + \boldsymbol{\ell}_1\cdot\boldsymbol{v} - \tfrac{1}{2}\,k_1|\boldsymbol{q}|^2 + \left(|\boldsymbol{s}^{*D} - \boldsymbol{\ell}_2^D| - c_0\right)|\boldsymbol{q}|\right] dx, \end{aligned}$$

where $\boldsymbol{s}^{*D}$ is the deviatoric part of the tensor $\boldsymbol{s}^*$. Thus we have

$$J^*(F^*\boldsymbol{s}^*, -\boldsymbol{s}^*) = \begin{cases} \displaystyle\int_\Omega \left[\frac{1}{2\,k_1}\left(|\boldsymbol{s}^{*D} - \boldsymbol{\ell}_2^D| - c_0\right)_+^2 + \frac{1}{2}\,\boldsymbol{C}^{-1}\boldsymbol{s}^* : \boldsymbol{s}^*\right] dx \\ \qquad\qquad \text{if } \displaystyle\int_\Omega \left[\boldsymbol{s}^* : \boldsymbol{\epsilon}(\boldsymbol{v}) + \boldsymbol{\ell}_1\cdot\boldsymbol{v}\right] dx = 0 \ \ \forall\, \boldsymbol{v} \in V, \\ +\infty \quad \text{otherwise,} \end{cases} \tag{12.109}$$

where $x_+ = \max\{x, 0\}$.

Now for any $\boldsymbol{z} \in Z$, consider the difference

$$\begin{aligned} D(\boldsymbol{w}, \boldsymbol{z}) &= J(\boldsymbol{z}, F\boldsymbol{z}) - J(\boldsymbol{w}, F\boldsymbol{w}) \\ &= \int_\Omega \Big[\tfrac{1}{2}\, \boldsymbol{C}(\boldsymbol{\epsilon}(\boldsymbol{v}) - \boldsymbol{q}) : (\boldsymbol{\epsilon}(\boldsymbol{v}) - \boldsymbol{q}) \\ &\qquad - \tfrac{1}{2}\, \boldsymbol{C}(\boldsymbol{\epsilon}(\boldsymbol{u}) - \boldsymbol{p}) : (\boldsymbol{\epsilon}(\boldsymbol{u}) - \boldsymbol{p}) \\ &\qquad + \tfrac{1}{2}\, k_1(|\boldsymbol{q}|^2 - |\boldsymbol{p}|^2) + c_0(|\boldsymbol{q}| - |\boldsymbol{p}|) \Big]\, dx - \langle \boldsymbol{\ell}, \boldsymbol{z} - \boldsymbol{w} \rangle. \end{aligned}$$

First we derive a lower bound for $D(\boldsymbol{w}, \boldsymbol{z})$. Using (12.100), we find that

$$\begin{aligned} \langle \boldsymbol{\ell}, \boldsymbol{z} - \boldsymbol{w} \rangle \le \int_\Omega \Big[& \boldsymbol{C}(\boldsymbol{\epsilon}(\boldsymbol{u}) - \boldsymbol{p}) : (\boldsymbol{\epsilon}(\boldsymbol{v} - \boldsymbol{u}) - (\boldsymbol{q} - \boldsymbol{p})) \\ & + k_1 \boldsymbol{p} : (\boldsymbol{q} - \boldsymbol{p}) + c_0(|\boldsymbol{q}| - |\boldsymbol{p}|) \Big]\, dx. \end{aligned}$$

Thus

$$\begin{aligned} D(\boldsymbol{w}, \boldsymbol{z}) &\ge \int_\Omega \Big[\tfrac{1}{2}\, \boldsymbol{C}(\boldsymbol{\epsilon}(\boldsymbol{v}) - \boldsymbol{q}) : (\boldsymbol{\epsilon}(\boldsymbol{v}) - \boldsymbol{q}) \\ &\qquad - \tfrac{1}{2}\, \boldsymbol{C}(\boldsymbol{\epsilon}(\boldsymbol{u}) - \boldsymbol{p}) : (\boldsymbol{\epsilon}(\boldsymbol{u}) - \boldsymbol{p}) \\ &\qquad - \boldsymbol{C}(\boldsymbol{\epsilon}(\boldsymbol{u}) - \boldsymbol{p}) : (\boldsymbol{\epsilon}(\boldsymbol{v} - \boldsymbol{u}) - (\boldsymbol{q} - \boldsymbol{p})) \\ &\qquad + \tfrac{1}{2}\, k_1(|\boldsymbol{q}|^2 - |\boldsymbol{p}|^2) - k_1 \boldsymbol{p} : (\boldsymbol{q} - \boldsymbol{p}) \Big]\, dx \\ &= \tfrac{1}{2}\, a(\boldsymbol{z} - \boldsymbol{w}, \boldsymbol{z} - \boldsymbol{w}). \end{aligned}$$

By the Z-ellipticity of a, we then obtain

$$D(\boldsymbol{w}, \boldsymbol{z}) \ge c_1 \|\boldsymbol{z} - \boldsymbol{w}\|_Z^2 \quad \forall\, \boldsymbol{z} \in Z. \tag{12.110}$$

In particular, with $\boldsymbol{z} = \boldsymbol{w}_\varepsilon$ in (12.110), we have

$$D(\boldsymbol{w}, \boldsymbol{w}_\varepsilon) \ge c_1 \|\boldsymbol{w}_\varepsilon - \boldsymbol{w}\|_Z^2. \tag{12.111}$$

An upper bound for $D(\boldsymbol{w}, \boldsymbol{w}_\varepsilon)$ comes from (4.34). Examining this expression for $J^*(F^*\boldsymbol{s}^*, \ \boldsymbol{s}^*)$, we will say that an auxiliary $\boldsymbol{s}^* \in S'(= S)$ is admissible if

$$\int_\Omega [\boldsymbol{s}^* : \boldsymbol{\epsilon}(\boldsymbol{v}) + \boldsymbol{\ell}_1 \cdot \boldsymbol{v}]\, dx = 0 \quad \forall\, \boldsymbol{v} \in V.$$

We observe that the value of $J^*(F^*\boldsymbol{s}^*, -\boldsymbol{s}^*)$ is finite if and only if $\boldsymbol{s}^*$ is admissible. For an admissible $\boldsymbol{s}^*$, we have

$$\begin{aligned} D(\boldsymbol{w}, \boldsymbol{w}_\varepsilon) \le \int_\Omega \Big[& \frac{1}{2}\, \boldsymbol{C}(\boldsymbol{\epsilon}(\boldsymbol{u}_\varepsilon) - \boldsymbol{p}_\varepsilon) : (\boldsymbol{\epsilon}(\boldsymbol{u}_\varepsilon) - \boldsymbol{p}_\varepsilon) \\ & + \frac{1}{2\,k_1} \left(|\boldsymbol{s}^{*D} - \boldsymbol{\ell}_2^D| - c_0 \right)_+^2 + \frac{1}{2}\, \boldsymbol{C}^{-1}\boldsymbol{s}^* : \boldsymbol{s}^* \\ & + \frac{k_1}{2}\, |\boldsymbol{p}_\varepsilon|^2 + c_0 |\boldsymbol{p}_\varepsilon| \Big]\, dx - \langle \boldsymbol{\ell}, \boldsymbol{w}_\varepsilon \rangle. \end{aligned}$$

The regularized problem (12.105) can be decomposed into two relations,

$$\int_\Omega [\boldsymbol{C}(\boldsymbol{\epsilon}(\boldsymbol{u}_\varepsilon) - \boldsymbol{p}_\varepsilon) : \boldsymbol{\epsilon}(\boldsymbol{v}) - \boldsymbol{\ell}_1 \cdot \boldsymbol{v}]\, dx = 0 \quad \forall\, \boldsymbol{v} \in V \tag{12.112}$$

and

$$\int_\Omega [-\boldsymbol{C}(\boldsymbol{\epsilon}(\boldsymbol{u}_\varepsilon) - \boldsymbol{p}_\varepsilon) + k_1 \boldsymbol{p}_\varepsilon + c_0 \phi'_\varepsilon(|\boldsymbol{p}_\varepsilon|) - \boldsymbol{\ell}_2] : \boldsymbol{q}\, dx = 0 \quad \forall\, \boldsymbol{q} \in Q_0. \tag{12.113}$$

By (12.112), the quantity

$$\boldsymbol{s}^* = -\boldsymbol{C}(\boldsymbol{\epsilon}(\boldsymbol{u}_\varepsilon) - \boldsymbol{p}_\varepsilon) \tag{12.114}$$

is admissible. Then an equivalent way of writing (12.113) is

$$\boldsymbol{s}^{*D} + k_1 \boldsymbol{p}_\varepsilon + c_0 \phi'_\varepsilon(|\boldsymbol{p}_\varepsilon|)^D - \boldsymbol{\ell}_2^D = 0,$$

from which we easily obtain

$$|\boldsymbol{s}^{*D} - \boldsymbol{\ell}_2^D| = |k_1 \boldsymbol{p}_\varepsilon + c_0 \phi'_\varepsilon(|\boldsymbol{p}_\varepsilon|)^D|. \tag{12.115}$$

From (12.109) with $\boldsymbol{v} = \boldsymbol{u}_\varepsilon$, we have

$$\langle \boldsymbol{\ell}_1, \boldsymbol{u}_\varepsilon \rangle = \int_\Omega \boldsymbol{C}(\boldsymbol{\epsilon}(\boldsymbol{u}_\varepsilon) - \boldsymbol{p}_\varepsilon) : \boldsymbol{\epsilon}(\boldsymbol{u}_\varepsilon).$$

From (12.113) with $\boldsymbol{q} = \boldsymbol{p}_\varepsilon$, we have

$$\langle \boldsymbol{\ell}_2, \boldsymbol{p}_\varepsilon \rangle = \int_\Omega [-\boldsymbol{C}(\boldsymbol{\epsilon}(\boldsymbol{u}_\varepsilon) - \boldsymbol{p}_\varepsilon) + k_1 \boldsymbol{p}_\varepsilon + c_0 \phi'_\varepsilon(|\boldsymbol{p}_\varepsilon|)] : \boldsymbol{p}_\varepsilon\, dx.$$

Using these two relations and (12.115), we can simplify the upper bound for $D(\boldsymbol{w}, \boldsymbol{w}_\varepsilon)$ to get

$$\begin{aligned} D(\boldsymbol{w}, \boldsymbol{w}_\varepsilon) \le \int_\Omega \Big[& c_0 \left(|\boldsymbol{p}_\varepsilon| - \phi'_\varepsilon(|\boldsymbol{p}_\varepsilon|) : \boldsymbol{p}_\varepsilon\right) - \frac{k_1}{2} |\boldsymbol{p}_\varepsilon|^2 \\ & + \frac{1}{2\, k_1} \left(|k_1 \boldsymbol{p}_\varepsilon + c_0 \phi'_\varepsilon(|\boldsymbol{p}_\varepsilon|)^D| - c_0\right)_+^2 \Big]\, dx, \end{aligned}$$

which together with (12.111) yields an a posteriori error estimate for the solution of the regularized problem (12.104), that is,

$$\begin{aligned} & c_1 \|\boldsymbol{w}_\varepsilon - \boldsymbol{w}\|_Z^2 \\ & \le \int_\Omega \Big[c_0 \left(|\boldsymbol{p}_\varepsilon| - \phi'_\varepsilon(|\boldsymbol{p}_\varepsilon|) : \boldsymbol{p}_\varepsilon\right) - \frac{k_1}{2} |\boldsymbol{p}_\varepsilon|^2 \\ & \qquad + \frac{1}{2\, k_1} \left(|k_1 \boldsymbol{p}_\varepsilon + c_0 \phi'_\varepsilon(|\boldsymbol{p}_\varepsilon|)^D| - c_0\right)_+^2 \Big]\, dx. \end{aligned} \tag{12.116}$$

Now we derive the concrete a posteriori error bound for each of the three choices of the regularization function. For the first choice,

$$(\phi_\varepsilon^1)'(t) = \frac{t}{\sqrt{t^2+\varepsilon^2}}.$$

Hence the error estimate (12.116) in this case reduces to

$$\begin{aligned} c_1 \|\boldsymbol{w}_\varepsilon - \boldsymbol{w}\|_Z^2 &\le \int_\Omega \left[\frac{c_0 |\boldsymbol{p}_\varepsilon|\, \varepsilon^2}{\sqrt{|\boldsymbol{p}_\varepsilon|^2+\varepsilon^2}\,(\sqrt{|\boldsymbol{p}_\varepsilon|^2+\varepsilon^2} + |\boldsymbol{p}_\varepsilon|)} - \frac{k_1}{2}\, |\boldsymbol{p}_\varepsilon|^2 \right. \\ &\quad \left. + \frac{1}{2\,k_1} \left(\left| k_1 + \frac{c_0}{\sqrt{|\boldsymbol{p}_\varepsilon|^2+\varepsilon^2}} \right| |\boldsymbol{p}_\varepsilon| - c_0 \right)_+^2 \right] dx. \end{aligned} \tag{12.117}$$

For the second choice of the regularization function,

$$(\phi_\varepsilon^2)'(t) = \begin{cases} 1, & \text{if } t \ge \varepsilon, \\ \dfrac{1}{\varepsilon}\, t, & \text{if } |t| \le \varepsilon, \\ -1, & \text{if } t \le -\varepsilon, \end{cases}$$

the a posteriori error estimate is

$$\begin{aligned} c_1 \|\boldsymbol{w}_\varepsilon - \boldsymbol{w}\|_Z^2 &\le \int_{\Omega_\varepsilon} \left[c_0 |\boldsymbol{p}_\varepsilon| \left(1 - \frac{|\boldsymbol{p}_\varepsilon|}{\varepsilon}\right) - \frac{k_1}{2}\, |\boldsymbol{p}_\varepsilon|^2 \right. \\ &\quad \left. + \frac{1}{2\,k_1} \left(k_1 |\boldsymbol{p}_\varepsilon| + c_0 \Big(\frac{|\boldsymbol{p}_\varepsilon|^2}{\varepsilon} - 1 \Big) \right)_+^2 \right] dx, \end{aligned} \tag{12.118}$$

where $\Omega_\varepsilon = \{x \in \Omega :\ |\boldsymbol{p}_\varepsilon(x)| \le \varepsilon\}$.

For the third choice of the regularization function,

$$(\phi_\varepsilon^3)'(t) = \begin{cases} 1, & \text{if } t \ge \varepsilon, \\ \dfrac{1}{\varepsilon}\, t, & \text{if } |t| \le \varepsilon, \\ -1, & \text{if } t \le -\varepsilon. \end{cases}$$

Hence, the resulting a posteriori error estimate has the same form as that of the estimate (12.118).

To see the efficiency of the a posteriori error estimates we observe that in (12.117), the summation of the second and third terms in the integrand on the right-hand side of (12.117) is nonpositive. Hence, a simple consequence of (12.117) is

$$c_1 \|\boldsymbol{w}_\varepsilon - \boldsymbol{w}\|_Z^2 \le \int_\Omega \frac{c_0 |\boldsymbol{p}_\varepsilon|\, \varepsilon^2}{\sqrt{|\boldsymbol{p}_\varepsilon|^2+\varepsilon^2}\,(\sqrt{|\boldsymbol{p}_\varepsilon|^2+\varepsilon^2} + |\boldsymbol{p}_\varepsilon|)}\, dx. \tag{12.119}$$

Similarly, for the second and third choices of the regularization function, we have the simple consequence

$$c_1 \|\boldsymbol{w}_\varepsilon - \boldsymbol{w}\|_Z^2 \le \int_{\Omega_\varepsilon} c_0 |\boldsymbol{p}_\varepsilon| \left(1 - \frac{|\boldsymbol{p}_\varepsilon|}{\varepsilon}\right) dx \tag{12.120}$$

of the estimate (12.118). It is worth noting that a similar procedure employing the duality theory can be used for a posteriori error analysis in various processes in applied and computational mathematics. In Han [50, 51], a posteriori error analysis is carried out on effects of mathematical idealizations of models and data. In Han, Jensen, and Shimansky [53], a posteriori error analysis is given for the Kačanov iteration method for solving some nonlinear problems. Numerical experiments in these references show that the derived a posteriori error estimates are efficient.

12.5 Fully Discrete Schemes with Numerical Integration

In this section we analyze another type of method for dealing with the difficulty caused by nondifferentiable terms. We will use numerical quadrature to approximate nondifferentiable terms and as a result get a system of linear equations and uncoupled inequalities at integration points for fully discrete approximations. As we will see, each uncoupled inequality is of small size, and thus can be solved easily. This section follows Han, Jensen, and Reddy [52].

We will present the analysis for the elastoplasticity problem with linear kinematic hardening. In other words, the continuous problem to be solved is the following.

Problem Prim2. Given $\boldsymbol{f} \in H^1(0,T;H^{-1}(\Omega))$ with $\boldsymbol{f}(0) = \boldsymbol{0}$, find $\boldsymbol{w} = (\boldsymbol{u}, \boldsymbol{p}) : [0,T] \to Z$ with $\boldsymbol{w}(0) = \boldsymbol{0}$ such that for almost all $t \in (0,T)$,

$$a(\boldsymbol{w}(t), \boldsymbol{z} - \dot{\boldsymbol{w}}(t)) + j(\boldsymbol{z}) - j(\dot{\boldsymbol{w}}(t)) \ge \langle \boldsymbol{\ell}(t), \boldsymbol{z} - \dot{\boldsymbol{w}}(t)\rangle \quad \forall\, \boldsymbol{z} = (\boldsymbol{v}, \boldsymbol{q}) \in Z. \tag{12.121}$$

Here, as before, the solution space is $Z = V \times Q_0$ with

$$\begin{aligned} V &= [H_0^1(\Omega)]^3, \\ Q_0 &= \{\boldsymbol{q} = (q_{ij})_{3\times 3} : q_{ij} = q_{ji} \in L^2(\Omega),\ \operatorname{tr} \boldsymbol{q} = 0\}. \end{aligned}$$

The bilinear form takes the form

$$a(\boldsymbol{w}, \boldsymbol{z}) = \int_\Omega [\boldsymbol{C}(\boldsymbol{\epsilon}(\boldsymbol{u}) - \boldsymbol{p}) : (\boldsymbol{\epsilon}(\boldsymbol{v}) - \boldsymbol{q}) + k_1\, \boldsymbol{p} : \boldsymbol{q}]\, dx.$$

The linear form $\boldsymbol{\ell}$ is defined by

$$\langle \boldsymbol{\ell}(t), \boldsymbol{z}\rangle = \int_\Omega \boldsymbol{f}(t) \cdot \boldsymbol{v}\, dx.$$

The nondifferentiable functional is

$$j(\boldsymbol{z}) = \int_\Omega c_0 |\boldsymbol{q}|\, dx \quad \text{for } \boldsymbol{z} = (\boldsymbol{v}, \boldsymbol{q}) \in Z.$$

The elasticity tensor $\boldsymbol{C}$ is assumed to be symmetric, bounded, and pointwise stable, and the hardening coefficient k_1 is bounded and uniformly bounded below by 0. As we have seen before in Chapter 7, the problem PRIM2 has a unique solution.

Let us review some fully discrete approximations for solving the problem PRIM2. We divide the time interval $I = [0, T]$ into N equal parts with step size $k = T/N$. The nodal points are denoted by $t_n = nk$ $(n = 0, 1, \ldots, N)$ and subintervals by $I_n = [t_{n-1}, t_n]$ $(n = 1, 2, \ldots, N)$. For a continuous function $\boldsymbol{v}(t)$ with values in one of the spaces Z, V, Q_0, or Z', we use the notation $\boldsymbol{v}_{n-1+\theta} = \boldsymbol{v}(t_{n-1+\theta})$, where $t_{n-1+\theta} = \theta\, t_n + (1-\theta)\, t_{n-1}$ and $\theta \in \left[\frac{1}{2}, 1\right]$. We will also need the notation $\Delta \boldsymbol{v}_n = \boldsymbol{v}_n - \boldsymbol{v}_{n-1}$ and $\delta \boldsymbol{v}_n = \Delta \boldsymbol{v}_n / k$. Our analysis below works also for discrete schemes with nonuniform divisions of I; all the error estimates to be derived are true with k interpreted as the maximal step-size.

Let $\mathcal{T}_h = \{\Omega_e\}_{e=1}^E$ be a regular triangulation of the domain Ω into triangular (tetrahedral) or rectangular (hexahedral) elements. As usual, $h \in (0, 1]$ denotes the maximal side of the elements in the triangulation. Let V^h and Q_0^h be finite element subspaces of V and Q_0, and set $Z^h = V^h \times Q_0^h$. Then a family of fully discrete approximations to the solution of the problem PRIM2 is the following.

PROBLEM PRIM2^{hk}. Find $\boldsymbol{w}^{hk} = \{\boldsymbol{w}_n^{hk}\}_{n=0}^N$, where $\boldsymbol{w}_n^{hk} = (\boldsymbol{u}_n^{hk}, \boldsymbol{p}_n^{hk}) \in Z^h$, $0 \le n \le N$, $\boldsymbol{w}_0^{hk} = \boldsymbol{0}$, such that for $n = 1, 2, \ldots, N$,

$$\begin{aligned} a(\theta\, \boldsymbol{w}_n^{hk} + (1-\theta)\, \boldsymbol{w}_{n-1}^{hk}, \boldsymbol{z}^h - \delta \boldsymbol{w}_n^{hk}) + j(\boldsymbol{z}^h) - j(\delta \boldsymbol{w}_n^{hk}) \\ \ge \langle \boldsymbol{\ell}_{n-1+\theta}, \boldsymbol{z}^h - \delta \boldsymbol{w}_n^{hk} \rangle \qquad \forall\, \boldsymbol{z}^h = (\boldsymbol{v}^h, \boldsymbol{q}^h) \in Z^h. \end{aligned} \tag{12.122}$$

We have shown before that this discrete problem has a unique solution, and furthermore that for the error $\boldsymbol{w}_n - \boldsymbol{w}_n^{hk}$,

$$\begin{aligned} &\max_{0 \le n \le N} \|\boldsymbol{w}_n - \boldsymbol{w}_n^{hk}\|_Z^2 \\ &\quad \le c\,k \sum_{n=1}^N \left(\inf_{\boldsymbol{q}^h \in Q_0^h} \|\dot{\boldsymbol{p}}_{n-1+\theta} - \boldsymbol{q}^h\|_Q + \inf_{\boldsymbol{v}^h \in V^h} \|\dot{\boldsymbol{u}}_{n-1+\theta} - \boldsymbol{v}^h\|_V^2 \right) \\ &\qquad + c\,k^2 \left(\|\dot{\boldsymbol{w}}\|_{L^\infty(0,T;Z)}^2 + \|\ddot{\boldsymbol{w}}\|_{L^1(0,T;Z)}^2 \right), \end{aligned} \tag{12.123}$$

and when $\theta = \frac{1}{2}$,

$$\max_{0\le n\le N} \|\boldsymbol{w}_n - \boldsymbol{w}_n^{hk}\|_Z^2$$

$$\le c\,k \sum_{n=1}^{N} \left(\inf_{\boldsymbol{q}^h\in Q_0^h} \|\dot{\boldsymbol{p}}_{n-1/2} - \boldsymbol{q}^h\|_Q + \inf_{\boldsymbol{v}^h\in V^h} \|\dot{\boldsymbol{u}}_{n-1/2} - \boldsymbol{v}^h\|_V^2 \right)$$

$$+ c\,k^4 \left(\|\ddot{\boldsymbol{w}}\|_{L^\infty(0,T;Z)}^2 + \|\boldsymbol{w}^{(3)}\|_{L^1(0,T;Z)}^2 \right), \tag{12.124}$$

as long as the solution $\boldsymbol{w}$ has the regularity required by the right-hand sides of (12.123) and (12.124). The inequalities (12.123) and (12.124) are the basis for deriving various order error estimates, which can be obtained by applying the theory of finite element interpolation errors.

We assume for definiteness that the elements in the triangulation are either triangles for domains in $\mathbb{R}^2$ or tetrahedra for domains in $\mathbb{R}^3$, and choose

$$V^h = \left\{\boldsymbol{v}^h \in V : \boldsymbol{v}^h|_{\Omega_e} \text{ is linear } \forall\,\Omega_e \in \mathcal{T}_h\right\},$$
$$Q_0^h = \left\{\boldsymbol{q}^h \in Q_0 : \boldsymbol{q}^h|_{\Omega_e} \text{ is linear } \forall\,\Omega_e \in \mathcal{T}_h\right\}.$$

We chose discontinuous piecewise linear functions for the space Q_0^h in order to obtain the first order bound in h. We have

$$\inf_{\boldsymbol{q}^h\in Q_0^h} \|\dot{\boldsymbol{p}}_{n-1+\theta} - \boldsymbol{q}^h\|_Q \le c\,h^2 \left\{ \sum_{\Omega_e} |\dot{\boldsymbol{p}}_{n-1+\theta}|_{H^2(\Omega_e)}^2 \right\}^{1/2},$$

$$\inf_{\boldsymbol{v}^h\in V^h} \|\dot{\boldsymbol{u}}_{n-1+\theta} - \boldsymbol{v}^h\|_V \le c\,h \left\{ \sum_{\Omega_e} |\dot{\boldsymbol{u}}_{n-1+\theta}|_{H^2(\Omega_e)}^2 \right\}^{1/2}.$$

Therefore, if the solution is sufficiently smooth, the error $\max_n \|\boldsymbol{w}_n - \boldsymbol{w}_n^{hk}\|_Z$ is bounded by $O(h+k)$ if $\theta \in (\frac{1}{2}, 1]$, and by $O(h+k^2)$ if $\theta = \frac{1}{2}$.

From a practical point of view it is not convenient to compute the value of $j(\boldsymbol{z}^h)$ for piecewise linear functions $\boldsymbol{z}^h \in Q_0$. One way to overcome the difficulty is to replace $j(\boldsymbol{z}^h)$ by an approximation $j_h(\boldsymbol{z}^h)$, which is achieved through the use of numerical quadratures. We present next one such approximation, in two dimensions; the extension to three dimensions is immediate.

For a typical triangle $\Omega_e \in \mathcal{T}_h$, let A_i, $i = 1, 2, 3$, denote the three vertices of Ω_e. Then for any $\boldsymbol{q}^h \in Q_0^h$, the restriction $\boldsymbol{q}^h|_{\Omega_e}$ is uniquely defined by its values at the vertices, $\boldsymbol{q}^h(A_i)$, $i = 1, 2, 3$. Notice that if a point A is a common vertex of several neighboring triangles, in general, the values of $\boldsymbol{q}^h$ at A computed from the different triangles are different. We approximate the functional j by

$$\int_\Omega c_0|\boldsymbol{q}^h|\,dx = \sum_{\Omega_e} \int_{\Omega_e} c_0|\boldsymbol{q}^h|\,dx \approx c_0 \sum_{\Omega_e} \operatorname{meas}(\Omega_e)\,\frac{1}{3} \sum_{i=1}^{3} |\boldsymbol{q}^h(A_i)|,$$

and define

$$j_h(\boldsymbol{z}^h) = c_0 \sum_{\Omega_e} \text{meas}\,(\Omega_e)\,\frac{1}{3}\sum_{i=1}^{3} |\boldsymbol{q}^h(A_i)|. \tag{12.125}$$

Then the problem $\text{PRIM}2^{hk}$ is replaced by the following problem.

PROBLEM $\text{PRIM}2^{hk}_{\#}$. Find $\boldsymbol{w}^{hk} = \{\boldsymbol{w}_n^{hk}\}_{n=0}^N$, where $\boldsymbol{w}_n^{hk} = (\boldsymbol{u}_n^{hk}, \boldsymbol{p}_n^{hk}) \in Z^h$, $0 \le n \le N$, $\boldsymbol{w}_0^{hk} = \boldsymbol{0}$, such that for $n = 1, 2, \ldots, N$,

$$\begin{aligned} &a(\theta\, \boldsymbol{w}_n^{hk} + (1-\theta)\, \boldsymbol{w}_{n-1}^{hk}, \boldsymbol{z}^h - \delta \boldsymbol{w}_n^{hk}) + j_h(\boldsymbol{z}^h) - j_h(\delta \boldsymbol{w}_n^{hk}) \\ &\quad \ge \langle \boldsymbol{\ell}_{n-1+\theta}, \boldsymbol{z}^h - \delta \boldsymbol{w}_n^{hk}\rangle \quad \forall\, \boldsymbol{z}^h = (\boldsymbol{v}^h, \boldsymbol{q}^h) \in Z^h. \end{aligned} \tag{12.126}$$

The next result gives an error analysis of the numerical solution computed from (12.126), with $j_h(\boldsymbol{z}^h)$ defined through (12.125).

THEOREM 12.10. *For the error of the numerical solution defined by the problem* $\text{PRIM}2^{hk}_{\#}$, *we have the following inequalities, for any* $\boldsymbol{z}_n^h \in Z^h$, $n = 1, \ldots, N$,

$$\begin{aligned} &\max_{0\le n\le N} \|\boldsymbol{w}_n - \boldsymbol{w}_n^{hk}\|_Z \\ &\quad \le c\, k \left(\|\dot{\boldsymbol{w}}\|_{L^\infty(0,T;Z)} + \|\ddot{\boldsymbol{w}}\|_{L^1(0,T;Z)}\right) + c\, k \sum_{n=1}^N \|\dot{\boldsymbol{w}}_{n-1+\theta} - \boldsymbol{z}_n^h\|_Z \\ &\qquad + c \left\{ k \sum_{n=1}^N \left[\|\dot{\boldsymbol{p}}_{n-1+\theta} - \boldsymbol{q}_n^h\|_Q + |j_h(\boldsymbol{z}_n^h) - j(\dot{\boldsymbol{w}}_{n-1+\theta})|\right]\right\}^{1/2} \end{aligned} \tag{12.127}$$

if $\theta \ne \frac{1}{2}$,

$$\begin{aligned} &\max_{0\le n\le N} \|\boldsymbol{w}_n - \boldsymbol{w}_n^{hk}\|_Z \\ &\quad \le c\, k^2 \left(\|\ddot{\boldsymbol{w}}\|_{L^\infty(0,T;Z)} + \|\boldsymbol{w}^{(3)}\|_{L^1(0,T;Z)}\right) + c\, k \sum_{n=1}^N \|\dot{\boldsymbol{w}}_{n-1+\theta} - \boldsymbol{z}_n^h\|_Z \\ &\qquad + c \left\{ k \sum_{n=1}^N \left[\|\dot{\boldsymbol{p}}_{n-1+\theta} - \boldsymbol{q}_n^h\|_Q + |j_h(\boldsymbol{z}_n^h) - j(\dot{\boldsymbol{w}}_{n-1+\theta})|\right]\right\}^{1/2} \end{aligned} \tag{12.128}$$

if $\theta = \frac{1}{2}$.

PROOF. For the approximation defined in (12.125), it is easy to verify that

$$j(\boldsymbol{z}^h) \le j_h(\boldsymbol{z}^h) \quad \forall\, \boldsymbol{z}^h \in Z^h, \tag{12.129}$$

an important property needed in the error analysis below. The practical implications of this inequality are commented at the end of next section for more general finite element spaces and related approximations $j_h(\cdot)$ constructed via numerical quadratures.

The error is denoted by $\boldsymbol{e}_n = \boldsymbol{w}_n - \boldsymbol{w}_n^{hk}$, $0 \le n \le N$. Since the bilinear form $a(\cdot,\cdot)$ is symmetric, bounded, and Z-elliptic, the quantity $\|\boldsymbol{z}\|_a = a(\boldsymbol{z},\boldsymbol{z})^{1/2}$ defines an equivalent norm on Z. Consider next the quantities

$$A_n = a(\theta\, \boldsymbol{e}_n + (1-\theta)\, \boldsymbol{e}_{n-1}, \delta \boldsymbol{e}_n), \quad n = 1, \ldots, N.$$

Recalling $\theta \in \left[\frac{1}{2}, 1\right]$, an elementary argument reveals

$$A_n \ge \frac{1}{2k} \left(\|\boldsymbol{e}_n\|_a^2 - \|\boldsymbol{e}_{n-1}\|_a^2 \right). \tag{12.130}$$

On the other hand, for any $\boldsymbol{z}^h \in Z^h$, we write

$$\begin{aligned} A_n &= a(\theta\, \boldsymbol{w}_n + (1-\theta)\, \boldsymbol{w}_{n-1}, \delta \boldsymbol{w}_n - \delta \boldsymbol{w}_n^{hk}) \\ &\quad - a(\theta\, \boldsymbol{w}_n^{hk} + (1-\theta)\, \boldsymbol{w}_{n-1}^{hk}, \delta \boldsymbol{w}_n - \boldsymbol{z}^h) \\ &\quad - a(\theta\, \boldsymbol{w}_n^{hk} + (1-\theta)\, \boldsymbol{w}_{n-1}^{hk}, \boldsymbol{z}^h - \delta \boldsymbol{w}_n^{hk}). \end{aligned}$$

Using the inequality (12.126), we have

$$\begin{aligned} A_n &\le a(\theta\, \boldsymbol{w}_n + (1-\theta)\, \boldsymbol{w}_{n-1}, \delta \boldsymbol{w}_n - \delta \boldsymbol{w}_n^{hk}) \\ &\quad - a(\theta\, \boldsymbol{w}_n^{hk} + (1-\theta)\, \boldsymbol{w}_{n-1}^{hk}, \delta \boldsymbol{w}_n - \boldsymbol{z}^h) \\ &\quad + j_h(\boldsymbol{z}^h) - j_h(\delta \boldsymbol{w}_n^{hk}) - \langle \boldsymbol{\ell}_{n-1+\theta}, \boldsymbol{z}^h - \delta \boldsymbol{w}_n^{hk} \rangle. \end{aligned}$$

Taking $t = t_{n-1+\theta}$ and $\boldsymbol{z} = \delta \boldsymbol{w}_n^{hk}$ in (12.121), we get

$$\begin{aligned} 0 &\le a(\boldsymbol{w}_{n-1+\theta}, \delta \boldsymbol{w}_n^{hk} - \dot{\boldsymbol{w}}_{n-1+\theta}) + j(\delta \boldsymbol{w}_n^{hk}) - j(\dot{\boldsymbol{w}}_{n-1+\theta}) \\ &\quad - \langle \boldsymbol{\ell}_{n-1+\theta}, \delta \boldsymbol{w}_n^{hk} - \dot{\boldsymbol{w}}_{n-1+\theta} \rangle. \end{aligned}$$

Adding the last two inequalities, we then have

$$\begin{aligned} A_n &\le a(\theta\, \boldsymbol{w}_n + (1-\theta)\, \boldsymbol{w}_{n-1}, \delta \boldsymbol{w}_n - \delta \boldsymbol{w}_n^{hk}) \\ &\quad - a(\theta\, \boldsymbol{w}_n^{hk} + (1-\theta)\, \boldsymbol{w}_{n-1}^{hk}, \delta \boldsymbol{w}_n - \boldsymbol{z}^h) \\ &\quad + a(\boldsymbol{w}_{n-1+\theta}, \delta \boldsymbol{w}_n^{hk} - \dot{\boldsymbol{w}}_{n-1+\theta}) + j_h(\boldsymbol{z}^h) - j(\dot{\boldsymbol{w}}_{n-1+\theta}) \\ &\quad - \langle \boldsymbol{\ell}_{n-1+\theta}, \boldsymbol{z}^h - \dot{\boldsymbol{w}}_{n-1+\theta} \rangle + j(\delta \boldsymbol{w}_n^{hk}) - j_h(\delta \boldsymbol{w}_n^{hk}). \end{aligned}$$

Using the property (12.129) and the inequality (12.130), we conclude that

$$\begin{aligned} &\frac{1}{2k} \left(\|\boldsymbol{e}_n\|_a^2 - \|\boldsymbol{e}_{n-1}\|_a^2 \right) \\ &\quad \le a(\theta\, \boldsymbol{w}_n + (1-\theta)\, \boldsymbol{w}_{n-1}, \delta \boldsymbol{w}_n - \delta \boldsymbol{w}_n^{hk}) \\ &\qquad - a(\theta\, \boldsymbol{w}_n^{hk} + (1-\theta)\, \boldsymbol{w}_{n-1}^{hk}, \delta \boldsymbol{w}_n - \boldsymbol{z}^h) \\ &\qquad + a(\boldsymbol{w}_{n-1+\theta}, \delta \boldsymbol{w}_n^{hk} - \dot{\boldsymbol{w}}_{n-1+\theta}) \\ &\qquad + j_h(\boldsymbol{z}^h) - j(\dot{\boldsymbol{w}}_{n-1+\theta}) - \langle \boldsymbol{\ell}_{n-1+\theta}, \boldsymbol{z}^h - \dot{\boldsymbol{w}}_{n-1+\theta} \rangle. \end{aligned}$$

We can then use the above inequality recursively, and show that for any $\boldsymbol{z}_j^h \in Z^h$, $j = 1, \ldots, N$,

$$
\begin{aligned}
\max_n \|\boldsymbol{w}_n - \boldsymbol{w}_n^{hk}\|_a
&\le c\Big(\|E_{N,\theta}(\boldsymbol{w})\|_Z + \sum_{j=1}^{N-1} \|E_{j,\theta}(\boldsymbol{w}) - E_{j+1,\theta}(\boldsymbol{w})\|_Z\Big) \\
&\quad + c\,k \sum_{j=1}^{N} \|\delta\boldsymbol{w}_j - \boldsymbol{z}_j^h\|_Z \\
&\quad + c\Big\{k \max_n \|E_{n,\theta}(\boldsymbol{w})\|_Z \sum_{j=1}^{N} \|\delta\boldsymbol{w}_j - \boldsymbol{z}_j^h\|_Z\Big\}^{1/2} \\
&\quad + c\Big\{k \sum_{j=1}^{N} \big[\|\dot{\boldsymbol{p}}_{j-1+\theta} - \boldsymbol{q}_j^h\|_Q + |j_h(\boldsymbol{z}_j^h) - j(\dot{\boldsymbol{w}}_{j-1+\theta})|\big]\Big\}^{1/2},
\end{aligned}
$$

where

$$
E_{j,\theta}(\boldsymbol{w}) = \theta\,\boldsymbol{w}_n + (1-\theta)\,\boldsymbol{w}_{n-1} - \boldsymbol{w}_{n-1+\theta}.
$$

This leads in turn to (12.127) and (12.128), using estimates for the terms $\|E_{N,\theta}(\boldsymbol{w})\|_Z$ and $\|E_{j,\theta}(\boldsymbol{w}) - E_{j+1,\theta}(\boldsymbol{w})\|_Z$, $j = 1, \ldots, N-1$. □

The inequalities (12.127) and (12.128) form the basis for order error estimates, under suitable assumptions on the regularity of the solution, as the next theorem shows. We say $\boldsymbol{p}$ changes its sign if one of its components does.

THEOREM 12.11. *Assume that the solution of the problem* PRIM2 *satisfies* $\boldsymbol{w} \in W^{2,1}(0,T;Z)$ *or* $\boldsymbol{w} \in W^{3,1}(0,T;Z)$ *if* $\theta = \frac{1}{2}$, *and for each* n, $\dot{\boldsymbol{u}}_{n-1+\theta} \in \cap_{\Omega_e} H^2(\Omega_e)$, $\dot{\boldsymbol{p}}_{n-1+\theta} \in \cap_{\Omega_e}(W^{1,\infty}(\Omega_e) \cap H^2(\Omega_e))$. *Further, assume that for each* n, $\dot{\boldsymbol{p}}_{n-1+\theta}$ *changes its sign on at most finitely many curves in* Ω. *Therefore, the numerical solution defined by the problem* $\text{PRIM2}_{\#}^{hk}$, *we have the following error estimates:*

$$
\begin{aligned}
\max_{0\le n\le N} \|\boldsymbol{w}_n - \boldsymbol{w}_n^{hk}\|_Z
&\le c\,k\left(\|\dot{\boldsymbol{w}}\|_{L^\infty(0,T;Z)} + \|\ddot{\boldsymbol{w}}\|_{L^1(0,T;Z)}\right) \\
&\quad + c\,h\Big\{k \sum_{n=1}^{N} \Big[\Big(\sum_{\Omega_e} |\dot{\boldsymbol{p}}_{n-1+\theta}|^2_{H^2(\Omega_e)}\Big)^{1/2} + \sup_{\Omega_e} |\dot{\boldsymbol{p}}_{n-1+\theta}|_{W^{1,\infty}(\Omega_e)}\Big]\Big\}^{1/2} \\
&\quad + c\,h\,k \sum_{n=1}^{N} \Big\{\sum_{\Omega_e} \Big[|\dot{\boldsymbol{u}}_{n-1+\theta}|^2_{H^2(\Omega_e)} + |\dot{\boldsymbol{p}}_{n-1+\theta}|^2_{H^1(\Omega_e)}\Big]\Big\}^{1/2}
\end{aligned}
\tag{12.131}
$$

if $\theta \neq \frac{1}{2}$,

$$
\begin{aligned}
\max_{0 \le n \le N} & \|\boldsymbol{w}_n - \boldsymbol{w}_n^{hk}\|_Z \\
& \le c\, k^2 \left(\|\ddot{\boldsymbol{w}}\|_{L^\infty(0,T;Z)} + \|\boldsymbol{w}^{(3)}\|_{L^1(0,T;Z)} \right) \\
& \quad + c\, h \left\{ k \sum_{n=1}^{N} \left[\left(\sum_{\Omega_e} |\dot{\boldsymbol{p}}_{n-1+\theta}|^2_{H^2(\Omega_e)} \right)^{1/2} + \sup_{\Omega_e} |\dot{\boldsymbol{p}}_{n-1+\theta}|_{W^{1,\infty}(\Omega_e)} \right] \right\}^{1/2} \\
& \quad + c\, h\, k \sum_{n=1}^{N} \left\{ \sum_{\Omega_e} \left[|\dot{\boldsymbol{u}}_{n-1+\theta}|^2_{H^2(\Omega_e)} + |\dot{\boldsymbol{p}}_{n-1+\theta}|^2_{H^1(\Omega_e)} \right] \right\}^{1/2} \qquad (12.132)
\end{aligned}
$$

if $\theta = \frac{1}{2}$.

PROOF. In (12.127) and (12.128) we choose $\boldsymbol{z}_n^h = \Pi^h \dot{\boldsymbol{w}}_{n-1+\theta}$, the finite element interpolant of $\dot{\boldsymbol{w}}_{n-1+\theta}$, $n = 1, \ldots, N$. Let us estimate the term

$$
J_{n-1+\theta} = j_h(\Pi^h \dot{\boldsymbol{w}}_{n-1+\theta}) - j(\dot{\boldsymbol{w}}_{n-1+\theta}). \qquad (12.133)
$$

By definition, we have

$$
J_{n-1+\theta} = c_0 \sum_{\Omega_e} \left\{ \operatorname{meas}(\Omega_e) \frac{1}{3} \sum_{i=1}^{3} |\dot{\boldsymbol{p}}_{n-1+\theta}(A_i)| - \int_{\Omega_e} |\dot{\boldsymbol{p}}_{n-1+\theta}|\, dx \right\},
$$

where as before, we use A_i, $i = 1, 2, 3$, to denote the three vertices of a typical triangle Ω_e. We distinguish two cases according to whether or not the function $\dot{\boldsymbol{p}}_{n-1+\theta}$ changes its sign on Ω_e.

If $\dot{\boldsymbol{p}}_{n-1+\theta}$ does not change its sign on Ω_e, then

$$
\begin{aligned}
\operatorname{meas}(\Omega_e) & \frac{1}{3} \sum_{i=1}^{3} |\dot{\boldsymbol{p}}_{n-1+\theta}(A_i)| - \int_{\Omega_e} |\dot{\boldsymbol{p}}_{n-1+\theta}|\, dx \\
& = \int_{\Omega_e} |\Pi^h \dot{\boldsymbol{p}}_{n-1+\theta}|\, dx - \int_{\Omega_e} |\dot{\boldsymbol{p}}_{n-1+\theta}|\, dx,
\end{aligned}
$$

so that

$$
\begin{aligned}
& \left| \operatorname{meas}(\Omega_e) \frac{1}{3} \sum_{i=1}^{3} |\dot{\boldsymbol{p}}_{n-1+\theta}(A_i)| - \int_{\Omega_e} |\dot{\boldsymbol{p}}_{n-1+\theta}|\, dx \right| \\
& \quad \le \int_{\Omega_e} |\Pi^h \dot{\boldsymbol{p}}_{n-1+\theta} - \dot{\boldsymbol{p}}_{n-1+\theta}|\, dx. \qquad (12.134)
\end{aligned}
$$

Now assume that $\dot{\boldsymbol{p}}_{n-1+\theta}$ changes its sign on Ω_e. An application of Taylor's expansion at a zero of $\dot{\boldsymbol{p}}_{n-1+\theta}$ reveals that

$$
|\dot{\boldsymbol{p}}_{n-1+\theta}|_{L^\infty(\Omega_e)} \le c\, h_e\, |\dot{\boldsymbol{p}}_{n-1+\theta}|_{W^{1,\infty}(\Omega_e)},
$$

where h_e is the diameter of Ω_e. Therefore,

$$\left|\text{meas}\,(\Omega_e)\,\frac{1}{3}\sum_{i=1}^{3}|\dot{\boldsymbol{p}}_{n-1+\theta}(A_i)| - \int_{\Omega_e}|\dot{\boldsymbol{p}}_{n-1+\theta}|\,dx\right| \le c\,h_e^3|\dot{\boldsymbol{p}}_{n-1+\theta}|_{W^{1,\infty}(\Omega_e)}. \tag{12.135}$$

Under the assumption that $\dot{\boldsymbol{p}}_{n-1+\theta}$ changes its sign on at most finitely many curves in Ω, we see that from (12.134) and (12.135),

$$|J_{n-1+\theta}| \le c_0\int_\Omega |\Pi^h\dot{\boldsymbol{p}}_{n-1+\theta} - \dot{\boldsymbol{p}}_{n-1+\theta}|\,dx + c\,h^2\sup_{\Omega_e}|\dot{\boldsymbol{p}}_{n-1+\theta}|_{W^{1,\infty}(\Omega_e)}. \tag{12.136}$$

Using (12.136) and the finite element interpolation error estimates, we obtain the estimates (12.131) and (12.132) from (12.127) and (12.128). □

REMARK. Roughly speaking, Theorem 12.11 states that, under the assumption that $\dot{\boldsymbol{p}}_{n-1+\theta}$, $n-1,\ldots,N$, can change their signs on at most finitely many curves, the replacement of the functional $j(\cdot)$ by its numerical quadrature approximation does not cause degradation in the convergence order. If for some n it happens that $\dot{\boldsymbol{p}}_{n-1+\theta}$ changes its sign on infinitely many curves, then the last term in the error bound (12.131) and (12.132) has to be replaced by

$$c\,h\left\{k\sum_{n=1}^{N}\Big[\sum_{\Omega_e}|\dot{\boldsymbol{p}}_{n-1+\theta}|^2_{H^2(\Omega_e)}\Big]^{1/2}\right\}^{1/2} + c\,h^{1/2}\sup_{\Omega_e}|\dot{\boldsymbol{p}}_{n-1+\theta}|^{1/2}_{W^{1,\infty}(\Omega_e)}.$$

Stability of the numerical scheme. We consider the stability of the problem PRIM2$^{hk}_{\#}$. Assume that there is some error associated with the solution $\boldsymbol{w}^{hk}_{n-1}$ at $t = t_{n-1}$, say, caused by rounding errors. We will show that the propagation of the error in the solution at the later time levels is under control. Thus, let $\bar{\boldsymbol{w}}^{hk}_{n-1}$ be an approximation of $\boldsymbol{w}^{hk}_{n-1}$, and let $\bar{\boldsymbol{w}}^{hk}_n$ be the (exact) solution of (12.126) with $\boldsymbol{w}^{hk}_{n-1}$ replaced by $\bar{\boldsymbol{w}}^{hk}_{n-1}$. Therefore, $\bar{\boldsymbol{w}}^{hk}_n$ satisfies

$$\begin{aligned} &a(\theta\,\bar{\boldsymbol{w}}^{hk}_n + (1-\theta)\,\bar{\boldsymbol{w}}^{hk}_{n-1}, \boldsymbol{z}^h - \delta\bar{\boldsymbol{w}}^{hk}_n) + j_h(\boldsymbol{z}^h) - j_h(\delta\bar{\boldsymbol{w}}^{hk}_n) \\ &\quad\ge \langle \boldsymbol{\ell}_{n-1+\theta}, \boldsymbol{z}^h - \delta\bar{\boldsymbol{w}}^{hk}_n\rangle \qquad \forall\,\boldsymbol{z}^h = (\boldsymbol{v}^h,\boldsymbol{q}^h)\in Z^h. \end{aligned} \tag{12.137}$$

We take $\boldsymbol{z}^h = \delta\bar{\boldsymbol{w}}^{hk}_n$ in (12.126), $\boldsymbol{z}^h = \delta\boldsymbol{w}^{hk}_n$ in (12.137), and add the two inequalities to obtain

$$a(\theta\,(\boldsymbol{w}^{hk}_n - \bar{\boldsymbol{w}}^{hk}_n) + (1-\theta)\,(\boldsymbol{w}^{hk}_{n-1} - \bar{\boldsymbol{w}}^{hk}_{n-1}), \delta(\boldsymbol{w}^{hk}_n - \bar{\boldsymbol{w}}^{hk}_n)) \le 0. \tag{12.138}$$

Since $\theta\in[\frac{1}{2},1]$, we have

$$\begin{aligned} &a(\theta\,(\boldsymbol{w}^{hk}_n - \bar{\boldsymbol{w}}^{hk}_n) + (1-\theta)\,(\boldsymbol{w}^{hk}_{n-1} - \bar{\boldsymbol{w}}^{hk}_{n-1}), \\ &\quad(\boldsymbol{w}^{hk}_n - \bar{\boldsymbol{w}}^{hk}_n) - (\boldsymbol{w}^{hk}_{n-1} - \bar{\boldsymbol{w}}^{hk}_{n-1})) \\ &\quad\ge \tfrac{1}{2}\left[a(\boldsymbol{w}^{hk}_n - \bar{\boldsymbol{w}}^{hk}_n, \boldsymbol{w}^{hk}_n - \bar{\boldsymbol{w}}^{hk}_n) - a(\boldsymbol{w}^{hk}_{n-1} - \bar{\boldsymbol{w}}^{hk}_{n-1}, \boldsymbol{w}^{hk}_{n-1} - \bar{\boldsymbol{w}}^{hk}_{n-1})\right]. \end{aligned}$$

Then from (12.138), we see that

$$\|\boldsymbol{w}_n^{hk} - \bar{\boldsymbol{w}}_n^{hk}\|_a \le \|\boldsymbol{w}_{n-1}^{hk} - \bar{\boldsymbol{w}}_{n-1}^{hk}\|_a,$$

i.e., in the norm induced by the bilinear form $a(\cdot,\cdot)$, we have the stability inequality for the propagation of errors.

In the literature, the above type of stability is termed B-stability (cf. Reddy and Martin [107], Simo [114]).

13

Numerical Analysis of the Dual Problem

In this last chapter we present some results on the numerical analysis for the dual formulation of the elastoplasticity problem. For various numerical approximation schemes, we will derive error estimates under sufficient regularity assumptions on the solution and prove the convergence under the basic solution regularity condition. In Section 13.1 we study a family of generalized midpoint schemes for the stress problem. For the dual problem, we analyze several time-discrete schemes in Section 13.2 and fully discrete schemes in Section 13.3.

We then turn our attention to the implementation of numerical methods for solving the dual problem. For simplicity in notation, the discussion will be given in the context of the solution of temporal semidiscrete schemes. The extension of the discussion to fully discrete schemes is straightforward; one needs only to change infinite-dimensional spaces or their subsets to corresponding finite element spaces or their subsets in the argument. At each time level, one needs to solve a variational inequality system for the current state of the generalized stress and the displacement (or velocity). A common practice in engineering is to use an iteration procedure to update the generalized stress and the displacement separately, thus breaking a large-scale problem into two subproblems. Such an iteration procedure is termed a predictor-corrector method. Analysis of some predictor–corrector methods are given in Section 13.4. The main work required to carry out one step of a corrector–predictor method is the solution of a constrained variational inequality for updating the generalized stress. The problem can be equivalently formulated as one of computing the closest-point projection of a trial generalized stress onto a convex set—the admissible set. In the engineering literature, an algorithm for solving the closest-point projection problem is called a *return mapping algorithm* (the algorithm returns a trial generalized stress to the admissible set). We will discuss several return mapping algorithms that are used in actual computations.

For convenience, we recall here the dual variational problem.

PROBLEM DUAL. Given $\boldsymbol{\ell} \in H^1(0,T;V')$ with $\boldsymbol{\ell}(0) = \mathbf{0}$, find $(\boldsymbol{u}, \boldsymbol{\Sigma}) = (\boldsymbol{u}, \boldsymbol{\sigma}, \boldsymbol{\chi}) : [0,T] \to V \times \mathcal{P}$ with $(\boldsymbol{u}(0), \boldsymbol{\Sigma}(0)) = (\mathbf{0}, \mathbf{0})$ such that for almost all $t \in (0,T)$,

$$b(\boldsymbol{v}, \boldsymbol{\sigma}(t)) = \langle \boldsymbol{\ell}(t), \boldsymbol{v} \rangle \quad \forall \boldsymbol{v} \in V, \tag{13.1}$$

$$A(\dot{\boldsymbol{\Sigma}}(t), \boldsymbol{T} - \boldsymbol{\Sigma}(t)) + b(\dot{\boldsymbol{u}}(t), \boldsymbol{\tau} - \boldsymbol{\sigma}(t)) \geq 0 \quad \forall \boldsymbol{T} \in \mathcal{P}. \tag{13.2}$$

Here,

$$\begin{aligned} V &= [H_0^1(\Omega)]^3, \\ \mathcal{P} &= \{\boldsymbol{T} = (\boldsymbol{\tau}, \boldsymbol{\mu}) \in \mathcal{T} : \ (\boldsymbol{\tau}, \boldsymbol{\mu}) \in K \text{ a.e. in } \Omega\} \end{aligned}$$

with

$$\mathcal{T} = S \times M$$

and

$$\begin{aligned} S &= \{\boldsymbol{\tau} = (\tau_{ij}) : \ \tau_{ji} = \tau_{ij}, \ \tau_{ij} \in L^2(\Omega), \ 1 \leq i, j \leq 3\}, \\ M &= \{\boldsymbol{\mu} = (\mu_j) : \ \mu_j \in L^2(\Omega), \ j = 1, \ldots, m\}. \end{aligned}$$

The bilinear forms are

$$\begin{aligned} A : \mathcal{T} \times \mathcal{T} \to \mathbb{R}, \quad & A(\boldsymbol{\Sigma}, \boldsymbol{T}) = \int_\Omega \boldsymbol{\sigma} : \boldsymbol{C}^{-1} \boldsymbol{\tau} \, dx + \int_\Omega \boldsymbol{\chi} : \boldsymbol{H}^{-1} \boldsymbol{\mu} \, dx, \\ b : V \times S \to \mathbb{R}, \quad & b(\boldsymbol{v}, \boldsymbol{\tau}) = -\int_\Omega \boldsymbol{\epsilon}(\boldsymbol{v}) : \boldsymbol{\tau} \, dx. \end{aligned}$$

The linear form is

$$\boldsymbol{\ell}(t) : V \to \mathbb{R}, \qquad \langle \boldsymbol{\ell}(t), \boldsymbol{v} \rangle = -\int_\Omega \boldsymbol{f}(t) \cdot \boldsymbol{v} \, dx.$$

Introducing

$$\mathcal{P}(t) = \{\boldsymbol{T} = (\boldsymbol{\tau}, \boldsymbol{\mu}) \in \mathcal{P} : \ b(\boldsymbol{v}, \boldsymbol{\tau}) = \langle \boldsymbol{\ell}(t), \boldsymbol{v} \rangle \quad \forall \boldsymbol{v} \in V\},$$

we can eliminate the variable $\dot{\boldsymbol{u}}(t)$ from Problem DUAL and obtain the stress problem.

PROBLEM DUAL1. Given $\boldsymbol{\ell} \in H^1(0,T;V')$, $\boldsymbol{\ell}(0) = 0$, find $\boldsymbol{\Sigma} = (\boldsymbol{\sigma}, \boldsymbol{\chi}) : [0,T] \to \mathcal{P}$ with $\boldsymbol{\Sigma}(0) = \mathbf{0}$ such that for almost all $t \in (0,T)$, $\boldsymbol{\Sigma}(t) \in \mathcal{P}(t)$ and

$$A(\dot{\boldsymbol{\Sigma}}(t), \boldsymbol{T} - \boldsymbol{\Sigma}(t)) \geq 0 \quad \forall \boldsymbol{T} = (\boldsymbol{\tau}, \boldsymbol{\mu}) \in \mathcal{P}(t). \tag{13.3}$$

The well-posedness of the problems have been discussed in Chapter 8.

13.1 Time-Discrete Approximations of the Stress Problem

In this section we consider a family of time-discrete approximations that includes as a special case that used in the existence proof in Section 8.2. The goal here is to derive optimal order error estimates for the approximate solutions under sufficient regularity assumptions on the solution and show the convergence under the basic solution condition.

A family of generalized midpoint schemes. Let $\theta \in \left[\frac{1}{2}, 1\right]$ be a parameter. As before, we divide the time interval $[0, T]$ into N equal parts, and denote by $k = T/N$ the step-size. The partition points are $t_n = nk$, $n = 0, 1, \ldots, N$. Let $t_{n-1+\theta} = (n+1-\theta)\, k$, $n = 1, \ldots, N$. We use $\boldsymbol{\Sigma}_n^k$ for an approximate value of $\boldsymbol{\Sigma}(t_n)$. A family of generalized midpoint time-discrete approximations of the problem DUAL1 is as follows.

PROBLEM DUAL1$_\theta^k$. Find a sequence $\{\boldsymbol{\Sigma}_n^k - (\boldsymbol{\sigma}_n^k, \boldsymbol{\chi}_n^k)\}_{n=0}^N \subset \mathcal{T}$ with $\boldsymbol{\Sigma}_0^k = \boldsymbol{0}$ such that for $n = 1, 2, \ldots, N$, $\boldsymbol{\Sigma}_{n-1+\theta}^k = \theta \boldsymbol{\Sigma}_n^k + (1-\theta)\boldsymbol{\Sigma}_{n-1}^k \in \mathcal{P}_{n-1+\theta}$ and

$$A(\Delta\boldsymbol{\Sigma}_n^k, \boldsymbol{T} - \boldsymbol{\Sigma}_{n-1+\theta}^k) \geq 0 \quad \forall \boldsymbol{T} \in \mathcal{P}_{n-1+\theta}. \tag{13.4}$$

The constraint set $\mathcal{P}_{n-1+\theta}$ is defined by

$$\begin{aligned} \mathcal{P}_{n-1+\theta} &\equiv \mathcal{P}(t_{n-1+\theta}) \\ &= \{\boldsymbol{T} = (\boldsymbol{\tau}, \boldsymbol{\mu}) \in \mathcal{P} : b(\boldsymbol{v}, \boldsymbol{\tau}) = \langle \boldsymbol{\ell}(t_{n-1+\theta}), \boldsymbol{v}\rangle \quad \forall \boldsymbol{v} \in V\}. \end{aligned}$$

For the same reason as that in the case of time discretization of the abstract problem (cf. Section 11.1), we do not consider values of θ outside the range $\left[\frac{1}{2}, 1\right]$. For simplicity in writing, we will omit the explicit dependence on θ in the notation $\boldsymbol{\Sigma}_n^k$. When $\theta = 1$, the problem DUAL1$_\theta^k$ is reduced to DUAL1^k, which was used in the existence proof in Section 8.2. By Lemma 8.7, this problem has a unique solution. For other θ in the range $\left[\frac{1}{2}, 1\right]$, the inequality (13.4) can be rewritten in terms of $\boldsymbol{\Sigma}_{n-1+\theta}^k \subset \mathcal{P}_{n-1+\theta}$:

$$A(\boldsymbol{\Sigma}_{n-1+\theta}^k - \boldsymbol{\Sigma}_{n-1}^k, \boldsymbol{T} - \boldsymbol{\Sigma}_{n-1+\theta}^k) \geq 0 \quad \forall \boldsymbol{T} \in \mathcal{P}_{n-1+\theta}.$$

By modifying the proof of Lemma 8.7 in a straightforward way, it can be readily shown that for $\theta \in \left[\frac{1}{2}, 1\right]$, the problem DUAL1$_\theta^k$ also admits a unique solution.

Order error estimates. We now derive error estimates for the time-discrete approximate solutions under sufficient regularity assumptions on the solution $\boldsymbol{\Sigma}$. We consider the relation (13.3) at $t = t_{n-1+\theta}$. Choosing $\boldsymbol{T} = \boldsymbol{\Sigma}_{n-1+\theta}^k \in \mathcal{P}_{n-1+\theta}$, we obtain

$$A(\dot{\boldsymbol{\Sigma}}(t_{n-1+\theta}), \boldsymbol{\Sigma}_{n-1+\theta}^k - \boldsymbol{\Sigma}(t_{n-1+\theta})) \geq 0, \tag{13.5}$$

and choosing $\boldsymbol{T} = \boldsymbol{\Sigma}(t_{n-1+\theta}) \in \mathcal{P}_{n-1+\theta}$ in (13.4), we obtain

$$A(\Delta\boldsymbol{\Sigma}_n^k, \boldsymbol{\Sigma}(t_{n-1+\theta}) - \boldsymbol{\Sigma}_{n-1+\theta}^k) \geq 0. \tag{13.6}$$

The inequality (13.5) is multiplied by k and is added to the inequality (13.6), yielding

$$A\left(\Delta\boldsymbol{\Sigma}_n^k - k\,\dot{\boldsymbol{\Sigma}}(t_{n-1+\theta}), \boldsymbol{\Sigma}(t_{n-1+\theta}) - \boldsymbol{\Sigma}_{n-1+\theta}^k\right) \geq 0. \tag{13.7}$$

If the error is denoted by $\boldsymbol{e}_n = \boldsymbol{\Sigma}(t_n) - \boldsymbol{\Sigma}_n^k$, $n = 1, \ldots, N$, then with

$$E_{n,\theta}(\boldsymbol{\Sigma}) = \theta\,\boldsymbol{\Sigma}(t_n) + (1-\theta)\,\boldsymbol{\Sigma}(t_{n-1}) - \boldsymbol{\Sigma}(t_{n-1+\theta}),$$

the inequality (13.7) can be rewritten as

$$\begin{aligned}A(\boldsymbol{e}_{n-1} - \boldsymbol{e}_n + k\,(\delta\boldsymbol{\Sigma}(t_n) - \dot{\boldsymbol{\Sigma}}(t_{n-1+\theta})), \theta\,\boldsymbol{e}_n + (1-\theta)\,\boldsymbol{e}_{n-1} - E_{n,\theta}(\boldsymbol{\Sigma}))\\ \geq 0,\end{aligned}$$

or

$$\begin{aligned}&A\,(\boldsymbol{e}_n - \boldsymbol{e}_{n-1}, \theta\,\boldsymbol{e}_n + (1-\theta)\,\boldsymbol{e}_{n-1})\\ &\quad\leq A\,(\boldsymbol{e}_n - \boldsymbol{e}_{n-1}, E_{n,\theta}(\boldsymbol{\Sigma}))\\ &\qquad + k\,A\left(\delta\boldsymbol{\Sigma}(t_n) - \dot{\boldsymbol{\Sigma}}(t_{n-1+\theta}), \theta\,\boldsymbol{e}_n + (1-\theta)\,\boldsymbol{e}_{n-1}\right)\\ &\qquad - k\,A\left(\delta\boldsymbol{\Sigma}(t_n) - \dot{\boldsymbol{\Sigma}}(t_{n-1+\theta}), E_{n,\theta}(\boldsymbol{\Sigma})\right).\end{aligned} \tag{13.8}$$

We will use below the equivalent norm $\|\cdot\|_A$ induced by the bilinear form $A(\cdot,\cdot)$:

$$\|\boldsymbol{\Sigma}\|_A^2 = \tfrac{1}{2}\,A(\boldsymbol{\Sigma}, \boldsymbol{\Sigma}).$$

Since $\theta \in \left[\frac{1}{2}, 1\right]$, we have

$$A\,(\boldsymbol{e}_n - \boldsymbol{e}_{n-1}, \theta\,\boldsymbol{e}_n + (1-\theta)\,\boldsymbol{e}_{n-1}) \geq \tfrac{1}{2}\left(\|\boldsymbol{e}_n\|_A^2 - \|\boldsymbol{e}_{n-1}\|_A^2\right). \tag{13.9}$$

Thus, from (13.8), we get

$$\begin{aligned}&\frac{1}{2}\left(\|\boldsymbol{e}_n\|_A^2 - \|\boldsymbol{e}_{n-1}\|_A^2\right)\\ &\quad\leq A\,(\boldsymbol{e}_n - \boldsymbol{e}_{n-1}, E_{n,\theta}(\boldsymbol{\Sigma}))\\ &\qquad + k\,A\left(\delta\boldsymbol{\Sigma}(t_n) - \dot{\boldsymbol{\Sigma}}(t_{n-1+\theta}), \theta\,\boldsymbol{e}_n + (1-\theta)\,\boldsymbol{e}_{n-1}\right)\\ &\qquad - k\,A\left(\delta\boldsymbol{\Sigma}(t_n) - \dot{\boldsymbol{\Sigma}}(t_{n-1+\theta}), E_{n,\theta}(\boldsymbol{\Sigma})\right),\end{aligned}$$

which in turn implies that

$$\begin{aligned}\frac{1}{2}\|\boldsymbol{e}_n\|_A^2 \leq & \sum_{j=1}^{n} A\left(\boldsymbol{e}_j - \boldsymbol{e}_{j-1}, E_{j,\theta}(\boldsymbol{\Sigma})\right) \\ & + k\sum_{j=1}^{n} A\left(\delta\boldsymbol{\Sigma}(t_j) - \dot{\boldsymbol{\Sigma}}(t_{j-1+\theta}), \theta\,\boldsymbol{e}_j + (1-\theta)\,\boldsymbol{e}_{j-1}\right) \\ & - k\sum_{j=1}^{n} A\left(\delta\boldsymbol{\Sigma}(t_j) - \dot{\boldsymbol{\Sigma}}(t_{j-1+\theta}), E_{j,\theta}(\boldsymbol{\Sigma})\right).\end{aligned}$$

Let us use the identity

$$\begin{aligned}& \sum_{j=1}^{n} A\left(\boldsymbol{e}_j - \boldsymbol{e}_{j-1}, E_{j,\theta}(\boldsymbol{\Sigma})\right) \\ & \quad = \sum_{j=1}^{n-1} A\left(\boldsymbol{e}_j, E_{j,\theta}(\boldsymbol{\Sigma}) - E_{j+1,\theta}(\boldsymbol{\Sigma})\right) + A\left(\boldsymbol{e}_n, E_{n,\theta}(\boldsymbol{\Sigma})\right),\end{aligned}$$

and denote by $M = \max_{0\leq n\leq N}\|\boldsymbol{e}_n\|_{\mathcal{T}}$ the maximal error. Then we have, for $n = 1, \ldots, N$,

$$\begin{aligned}\|\boldsymbol{e}_n\|_A^2 \leq c\Big(& \sum_{j=1}^{n-1}\|E_{j,\theta}(\boldsymbol{\Sigma}) - E_{j+1,\theta}(\boldsymbol{\Sigma})\|_{\mathcal{T}} + \|E_{n,\theta}(\boldsymbol{\Sigma})\|_{\mathcal{T}} \\ & + k\sum_{j=1}^{n}\|\delta\boldsymbol{\Sigma}(t_j) - \dot{\boldsymbol{\Sigma}}(t_{j-1+\theta})\|_{\mathcal{T}}\Big)\, M \\ & + c\,k\sum_{j=1}^{n}\|\delta\boldsymbol{\Sigma}(t_j) - \dot{\boldsymbol{\Sigma}}(t_{j-1+\theta})\|_{\mathcal{T}}\|E_{j,\theta}(\boldsymbol{\Sigma})\|_{\mathcal{T}}.\end{aligned}$$

Therefore,

$$\begin{aligned}M^2 \leq c\Big(& \sum_{j=1}^{N-1}\|E_{j,\theta}(\boldsymbol{\Sigma}) - E_{j+1,\theta}(\boldsymbol{\Sigma})\|_{\mathcal{T}} + \|E_{N,\theta}(\boldsymbol{\Sigma})\|_{\mathcal{T}} \\ & + k\sum_{j=1}^{N}\|\delta\boldsymbol{\Sigma}(t_j) - \dot{\boldsymbol{\Sigma}}(t_{j-1+\theta})\|_{\mathcal{T}}\Big)\, M \\ & + c\,k\sum_{j=1}^{N}\|\delta\boldsymbol{\Sigma}(t_j) - \dot{\boldsymbol{\Sigma}}(t_{j-1+\theta})\|_{\mathcal{T}}\|E_{j,\theta}(\boldsymbol{\Sigma})\|_{\mathcal{T}}.\end{aligned}$$

By (11.3), we then have

$$M^2 \le c\Big(\sum_{j=1}^{N-1} \|E_{j,\theta}(\mathbf{\Sigma}) - E_{j+1,\theta}(\mathbf{\Sigma})\|_{\mathcal{T}} + \|E_{N,\theta}(\mathbf{\Sigma})\|_{\mathcal{T}} \\ + k\sum_{j=1}^{N} \|\delta\mathbf{\Sigma}(t_j) - \dot{\mathbf{\Sigma}}(t_{j-1+\theta})\|_{\mathcal{T}}\Big)^2 \\ + c\,k\sum_{j=1}^{N} \|\delta\mathbf{\Sigma}(t_j) - \dot{\mathbf{\Sigma}}(t_{j-1+\theta})\|_{\mathcal{T}} \|E_{j,\theta}(\mathbf{\Sigma})\|_{\mathcal{T}}. \tag{13.10}$$

To proceed further, let us assume that $\mathbf{\Sigma} \in W^{2,1}(0,T;\mathcal{T})$ if $\theta \in \left(\frac{1}{2}, 1\right]$, and $\mathbf{\Sigma} \in W^{3,1}(0,T;\mathcal{T})$ if $\theta = \frac{1}{2}$. Then by Lemma 11.3,

$$\sum_{j=1}^{N-1} \|E_{j,\theta}(\mathbf{\Sigma}) - E_{j+1,\theta}(\mathbf{\Sigma})\|_{\mathcal{T}} \le \begin{cases} c\,k\,\|\ddot{\mathbf{\Sigma}}\|_{L^1(0,T;\mathcal{T})} & \text{if } \theta \in (\frac{1}{2}, 1], \\ c\,k^2\|\mathbf{\Sigma}^{(3)}\|_{L^1(0,T;\mathcal{T})} & \text{if } \theta = \frac{1}{2}; \end{cases}$$

by Lemma 11.2, for $j = 1, \dots, N$,

$$\|E_{j,\theta}(\mathbf{\Sigma})\|_{\mathcal{T}} \le \begin{cases} c\,k\,\|\ddot{\mathbf{\Sigma}}\|_{L^1(0,T;\mathcal{T})} & \text{if } \theta \in (\frac{1}{2}, 1], \\ c\,k^2\|\ddot{\mathbf{\Sigma}}\|_{L^\infty(0,T;\mathcal{T})} & \text{if } \theta = \frac{1}{2}; \end{cases}$$

and by Lemma 11.4,

$$\sum_{j=1}^{N} \|\delta\mathbf{\Sigma}(t_j) - \dot{\mathbf{\Sigma}}(t_{j-1+\theta})\|_{\mathcal{T}} \le \|\ddot{\mathbf{\Sigma}}\|_{L^1(0,T;\mathcal{T})},$$

$$\sum_{j=1}^{N} \|\delta\mathbf{\Sigma}(t_j) - \dot{\mathbf{\Sigma}}(t_{j-1/2})\|_{\mathcal{T}} \le \frac{k}{8}\,\|\mathbf{\Sigma}^{(3)}\|_{L^1(0,T;\mathcal{T})}.$$

Applying these estimates in (13.10), we obtain the following result.

THEOREM 13.1. *Assume that* $\mathbf{\Sigma} \in W^{2,1}(0,T;\mathcal{T})$ *and in the case* $\theta = \frac{1}{2}$, $\mathbf{\Sigma} \in W^{3,1}(0,T;\mathcal{T})$. *For the time-discrete solution* $\{\mathbf{\Sigma}_n^k\}_{n=0}^N$ *of Problem* DUAL1_θ^k, *we have the error estimate*

$$\max_{0\le n\le N} \|\mathbf{\Sigma}(t_n) - \mathbf{\Sigma}_n^k\|_{\mathcal{T}} \le c\,k\,\|\mathbf{\Sigma}\|_{W^{2,1}(0,T;\mathcal{T})},$$

and if $\theta = \frac{1}{2}$,

$$\max_{0\le n\le N} \|\mathbf{\Sigma}(t_n) - \mathbf{\Sigma}_n^k\|_{\mathcal{T}} \le c\,k^2\,\|\mathbf{\Sigma}\|_{W^{3,1}(0,T;\mathcal{T})}.$$

Convergence analysis under minimal regularity conditions. The estimate (13.10) cannot be used to conclude the convergence of the semidiscrete solutions if we have only the basic solution regularity condition $\mathbf{\Sigma} \in$

$H^1(0,T;\mathcal{T})$. Let us modify the estimate (13.10). We still have the relation (13.9). On the other hand,

$$\begin{aligned} A\left(\boldsymbol{e}_n - \boldsymbol{e}_{n-1}, \theta\,\boldsymbol{e}_n + (1-\theta)\,\boldsymbol{e}_{n-1}\right) &= A\left(\Delta\boldsymbol{\Sigma}(t_n), \theta\,\boldsymbol{e}_n + (1-\theta)\,\boldsymbol{e}_{n-1}\right) \\ &\quad - A\left(\Delta\boldsymbol{\Sigma}_n^k, \boldsymbol{\Sigma}(t_{n-1+\theta}) - \boldsymbol{\Sigma}_{n-1+\theta}^k\right) - A\left(\Delta\boldsymbol{\Sigma}_n^k, E_{n,\theta}(\boldsymbol{\Sigma})\right). \end{aligned}$$

Using the inequality (13.6), we obtain

$$\begin{aligned} A\left(\boldsymbol{e}_n - \boldsymbol{e}_{n-1}, \theta\,\boldsymbol{e}_n + (1-\theta)\,\boldsymbol{e}_{n-1}\right) &\le A\left(\Delta\boldsymbol{\Sigma}(t_n), \theta\,\boldsymbol{e}_n + (1-\theta)\,\boldsymbol{e}_{n-1}\right) - A\left(\Delta\boldsymbol{\Sigma}_n^k, E_{n,\theta}(\boldsymbol{\Sigma})\right) \\ &= A\left(\Delta\boldsymbol{\Sigma}(t_n), \theta\,\boldsymbol{e}_n + (1-\theta)\,\boldsymbol{e}_{n-1}\right) \\ &\quad + A\left(\Delta\boldsymbol{e}_n, E_{n,\theta}(\boldsymbol{\Sigma})\right) - A\left(\Delta\boldsymbol{\Sigma}_n, E_{n,\theta}(\boldsymbol{\Sigma})\right). \end{aligned} \tag{13.11}$$

Now, for any $t \in I_n = [t_{n-1}, t_n]$, from Lemma 8.2 we have the existence of a unique $\boldsymbol{\tau}_\delta(t) \in (\operatorname{Ker} B)^\perp$ such that

$$\begin{aligned} b(\boldsymbol{v}, \boldsymbol{\tau}_\delta(t)) &= \langle \boldsymbol{\ell}(t) - \boldsymbol{\ell}_{n-1+\theta}, \boldsymbol{v}\rangle \quad \forall\, \boldsymbol{v} \in V, \\ \|\boldsymbol{\tau}_\delta(t)\|_S &\le c\, \|\boldsymbol{\ell}(t) - \boldsymbol{\ell}_{n-1+\theta}\|_{V'}. \end{aligned}$$

The subscript δ indicates that the quantity is related to a difference. By Assumption 8.4, we have a $\boldsymbol{\mu}_\delta(t) \in M$ such that $\boldsymbol{T}_\delta(t) = (\boldsymbol{\tau}_\delta(t), \boldsymbol{\mu}_\delta(t)) \in \mathcal{P}(t)$ and

$$\|\boldsymbol{T}_\delta(t)\|_{\mathcal{T}} \le c\, \|\boldsymbol{\ell}(t) - \boldsymbol{\ell}_{n-1+\theta}\|_{V'}. \tag{13.12}$$

Let $\boldsymbol{T}(t) = \boldsymbol{\Sigma}_{n-1+\theta}^k + \boldsymbol{T}_\delta(t)$. Obviously, $\boldsymbol{T}(t) \in \mathcal{P}(t)$. We take $\boldsymbol{T} = \boldsymbol{T}(t)$ in (13.3) to obtain

$$A(\dot{\boldsymbol{\Sigma}}(t), \boldsymbol{T}_\delta(t) + \boldsymbol{\Sigma}_{n-1+\theta}^k - \boldsymbol{\Sigma}(t)) \ge 0,$$

i.e.,

$$\begin{aligned} &A(\dot{\boldsymbol{\Sigma}}(t), \theta\,\boldsymbol{e}_n + (1-\theta)\,\boldsymbol{e}_{n-1}) \\ &\quad \le A(\dot{\boldsymbol{\Sigma}}(t), \boldsymbol{T}_\delta(t)) + A(\dot{\boldsymbol{\Sigma}}(t), \theta\,\boldsymbol{\Sigma}(t_n) + (1-\theta)\,\boldsymbol{\Sigma}(t_{n-1}) - \boldsymbol{\Sigma}(t)). \end{aligned}$$

Integrate the above relation over I_n to obtain

$$\begin{aligned} A(\Delta\boldsymbol{\Sigma}(t_n), \theta\,\boldsymbol{e}_n + (1-\theta)\,\boldsymbol{e}_{n-1}) &\le \int_{I_n} A(\dot{\boldsymbol{\Sigma}}(t), \boldsymbol{T}_\delta(t))\, dt \\ &\quad + \int_{I_n} A(\dot{\boldsymbol{\Sigma}}(t), \theta\,\boldsymbol{\Sigma}(t_n) + (1-\theta)\,\boldsymbol{\Sigma}(t_{n-1}) - \boldsymbol{\Sigma}(t))\, dt, \end{aligned}$$

which is then used in (13.11) to yield

$$
\begin{aligned}
A(\boldsymbol{e}_n - \boldsymbol{e}_{n-1}, \theta\, \boldsymbol{e}_n + (1-\theta)\, \boldsymbol{e}_{n-1})
&\le \int_{I_n} A(\dot{\boldsymbol{\Sigma}}(t), \boldsymbol{T}_\delta(t))\, dt + \int_{I_n} A(\dot{\boldsymbol{\Sigma}}(t), \boldsymbol{\Sigma}(t_{n-1+\theta}) - \boldsymbol{\Sigma}(t))\, dt \\
&\quad + A(\Delta \boldsymbol{e}_n, E_{n,\theta}(\boldsymbol{\Sigma})).
\end{aligned} \tag{13.13}
$$

To simplify the writing, we introduce the moduli of continuity

$$
\omega_k(\boldsymbol{\ell}) = \sup\{\|\boldsymbol{\ell}(s) - \boldsymbol{\ell}(t)\|_{V'} :\ 0 \le s, t \le T,\ |t-s| \le k\}, \tag{13.14}
$$

$$
\omega_k(\boldsymbol{\Sigma}) = \sup\{\|\boldsymbol{\Sigma}(s) - \boldsymbol{\Sigma}(t)\|_{\mathcal{T}} :\ 0 \le s, t \le T,\ |t-s| \le k\}. \tag{13.15}
$$

Note that $\boldsymbol{\ell} \in H^1(0,T;V')$ and $\boldsymbol{\Sigma} \in H^1(0,T;\mathcal{T})$ are uniformly continuous with respect to $t \in [0,T]$. Hence $\omega_k(\boldsymbol{\ell}) \to 0$ and $\omega_k(\boldsymbol{\Sigma}) \to 0$ as $k \to 0$.

Combining (13.9), (13.12), and (13.13), we obtain

$$
\begin{aligned}
\|\boldsymbol{e}_n\|_A^2 - \|\boldsymbol{e}_{n-1}\|_A^2 &\le c\,(\omega_k(\boldsymbol{\ell}) + \omega_k(\boldsymbol{\Sigma})) \int_{I_n} \|\dot{\boldsymbol{\Sigma}}(t)\|_{\mathcal{T}}\, dt \\
&\quad + 2\, A(\boldsymbol{e}_n - \boldsymbol{e}_{n-1}, E_{n,\theta}(\boldsymbol{\Sigma})).
\end{aligned}
$$

Applying this inequality recursively and recalling that $\boldsymbol{e}_0 = \boldsymbol{0}$, we see that

$$
\begin{aligned}
&\|\boldsymbol{e}_n\|_A^2 \\
&\quad \le c\,(\omega_k(\boldsymbol{\ell}) + \omega_k(\boldsymbol{\Sigma})) \int_0^{t_n} \|\dot{\boldsymbol{\Sigma}}(t)\|_{\mathcal{T}} dt \\
&\qquad + 2 \sum_{j=1}^{n} A(\boldsymbol{e}_j - \boldsymbol{e}_{j-1}, E_{j,\theta}(\boldsymbol{\Sigma})) \\
&\quad = c\,(\omega_k(\boldsymbol{\ell}) + \omega_k(\boldsymbol{\Sigma})) \int_0^{t_n} \|\dot{\boldsymbol{\Sigma}}(t)\|_{\mathcal{T}} dt \\
&\qquad + 2\, A(\boldsymbol{e}_n, E_{n,\theta}(\boldsymbol{\Sigma})) + 2 \sum_{j=1}^{n-1} A(\boldsymbol{e}_j, E_{j,\theta}(\boldsymbol{\Sigma}) - E_{j+1,\theta}(\boldsymbol{\Sigma})).
\end{aligned}
$$

With $M = \max_{0 \le n \le N} \|\boldsymbol{e}_n\|_{\mathcal{T}}$, we then have

$$
\begin{aligned}
M^2 &\le c\,(\omega_k(\boldsymbol{\ell}) + \omega_k(\boldsymbol{\Sigma}))\, \|\dot{\boldsymbol{\Sigma}}\|_{L^1(0,T;\mathcal{T})} \\
&\quad + c \left(\|E_{N,\theta}(\boldsymbol{\Sigma})\|_{\mathcal{T}} + \sum_{n=1}^{N-1} \|E_{n,\theta}(\boldsymbol{\Sigma}) - E_{n+1,\theta}(\boldsymbol{\Sigma})\|_{\mathcal{T}} \right) M.
\end{aligned}
$$

Now use the inequality (11.3) to obtain

$$
\begin{aligned}
M &\le c \left\{ (\omega_k(\boldsymbol{\ell}) + \omega_k(\boldsymbol{\Sigma}))\, \|\dot{\boldsymbol{\Sigma}}\|_{L^1(0,T;\mathcal{T})} \right\}^{1/2} \\
&\quad + c \left\{ \|E_{N,\theta}(\boldsymbol{\Sigma})\|_{\mathcal{T}} + \sum_{n=1}^{N-1} \|E_{n,\theta}(\boldsymbol{\Sigma}) - E_{n+1,\theta}(\boldsymbol{\Sigma})\|_{\mathcal{T}} \right\}.
\end{aligned} \tag{13.16}
$$

By the density result (5.27), for any $\varepsilon > 0$, we have $\overline{\mathbf{\Sigma}} \in C^\infty([0,T];\mathcal{T})$ such that

$$\|\mathbf{\Sigma} - \overline{\mathbf{\Sigma}}\|_{H^1(0,T;\mathcal{T})} < \varepsilon. \tag{13.17}$$

It is easy to see that

$$\begin{aligned}
&\|E_{N,\theta}(\mathbf{\Sigma})\|_{\mathcal{T}} + \sum_{n=1}^{N-1} \|E_{n,\theta}(\mathbf{\Sigma}) - E_{n+1,\theta}(\mathbf{\Sigma})\|_{\mathcal{T}} \\
&\quad \le \int_{I_N} \|\dot{\mathbf{\Sigma}}(t)\|_{\mathcal{T}}\, dt + \sum_{n=1}^{N-1} \|E_{n,\theta}(\overline{\mathbf{\Sigma}}) - E_{n+1,\theta}(\overline{\mathbf{\Sigma}})\|_{\mathcal{T}} \\
&\quad\quad + c \int_0^T \|\dot{\mathbf{\Sigma}}(t) - \dot{\overline{\mathbf{\Sigma}}}(t)\|_{\mathcal{T}} dt.
\end{aligned}$$

Using Lemma 11.3, we see that

$$\sum_{n=1}^{N-1} \|E_{n,\theta}(\overline{\mathbf{\Sigma}}) - E_{n+1,\theta}(\overline{\mathbf{\Sigma}})\|_{\mathcal{T}} \le c\, k\, \|\ddot{\overline{\mathbf{\Sigma}}}\|_{L^1(0,T;\mathcal{T})}.$$

So finally, from (13.16) we obtain

$$\begin{aligned}
\max_{0\le n\le N} \|\boldsymbol{e}_n\|_{\mathcal{T}} &\le c \left\{ (\omega_k(\boldsymbol{\ell}) + \omega_k(\mathbf{\Sigma}))\, \|\dot{\mathbf{\Sigma}}\|_{L^1(0,T;\mathcal{T})} \right\}^{1/2} \\
&\quad + c \left\{ \|\dot{\mathbf{\Sigma}}\|_{L^1(t_{N-1},t_N;\mathcal{T})} + k\, \|\ddot{\overline{\mathbf{\Sigma}}}\|_{L^1(0,T;\mathcal{T})} + \|\dot{\mathbf{\Sigma}} - \dot{\overline{\mathbf{\Sigma}}}\|_{L^1(0,T;\mathcal{T})} \right\}.
\end{aligned} \tag{13.18}$$

The estimate (13.18) implies the convergence as $k \to 0$. We state this result in the form of a theorem.

THEOREM 13.2. *Under the basic regularity condition $\mathbf{\Sigma} \in H^1(0,T;\mathcal{T})$, the time-discrete solution $\{\mathbf{\Sigma}_n^k\}_{n=0}^N$ of the problem* DUAL1$_\theta^k$ *converges to the exact solution $\mathbf{\Sigma}$ in the sense that*

$$\max_{0\le n\le N} \|\mathbf{\Sigma}(t_n) - \mathbf{\Sigma}_n^k\|_{\mathcal{T}} \to 0 \quad \text{as } k \to 0.$$

13.2 Time-Discrete Approximations of the Dual Problem

Now we study several time-discrete schemes for the dual problem. We are interested in approximating values of the generalized stress $\mathbf{\Sigma}(t_n)$ and the velocity $\boldsymbol{w}(t_n) = \dot{\boldsymbol{u}}(t_n)$, $n = 1, \ldots, N$.

The first temporal semidiscrete scheme is the backward Euler's scheme.

Scheme Dual_1^k. Find $(\boldsymbol{w}^k, \boldsymbol{\Sigma}^k) = \{(\boldsymbol{w}_n^k, \boldsymbol{\Sigma}_n^k)\}_{n=0}^N \subset V \times \mathcal{P}$, $(\boldsymbol{w}_0^k, \boldsymbol{\Sigma}_0^k) = \boldsymbol{0}$ such that for $n = 1, \ldots, N$,

$$b(\boldsymbol{v}, \boldsymbol{\sigma}_n^k) = \langle \boldsymbol{\ell}(t_n), \boldsymbol{v}\rangle \quad \forall\, \boldsymbol{v} \in V, \tag{13.19}$$

$$A(\delta \boldsymbol{\Sigma}_n^k, \boldsymbol{T} - \boldsymbol{\Sigma}_n^k) + b(\boldsymbol{w}_n^k, \boldsymbol{\tau} - \boldsymbol{\sigma}_n^k) \ge 0 \quad \forall\, \boldsymbol{T} \in \mathcal{P}. \tag{13.20}$$

More generally, we can form a family of generalized midpoint schemes. For $\theta \in \left[\frac{1}{2}, 1\right]$ we use $\boldsymbol{\Sigma}_{n-1+\theta}^k = \theta\, \boldsymbol{\Sigma}_n^k + (1-\theta)\, \boldsymbol{\Sigma}_{n-1}^k$ and $\boldsymbol{w}_{n-1+\theta}^k = \theta\, \boldsymbol{w}_n^k + (1-\theta)\, \boldsymbol{w}_{n-1}^k$ to approximate $\boldsymbol{\Sigma}(t_{n-1+\theta})$ and $\boldsymbol{w}(t_{n-1+\theta})$, respectively.

Scheme Dual_2^k. Find $(\boldsymbol{w}^k, \boldsymbol{\Sigma}^k) = \{(\boldsymbol{w}_n^k, \boldsymbol{\Sigma}_n^k)\}_{n=0}^N \subset V \times \mathcal{P}$, $(\boldsymbol{w}_0^k, \boldsymbol{\Sigma}_0^k) = \boldsymbol{0}$ such that for $n = 1, \ldots, N$,

$$b(\boldsymbol{v}, \boldsymbol{\sigma}_{n-1+\theta}^k) = \langle \boldsymbol{\ell}(t_{n-1+\theta}), \boldsymbol{v}\rangle \quad \forall\, \boldsymbol{v} \in V, \tag{13.21}$$

$$\begin{aligned} &A(\delta \boldsymbol{\Sigma}_n^k, \boldsymbol{T} - \boldsymbol{\Sigma}_{n-1+\theta}^k) \\ &\quad + b(\boldsymbol{w}_{n-1+\theta}^k, \boldsymbol{\tau} - \boldsymbol{\sigma}_{n-1+\theta}^k) \ge 0 \quad \forall\, \boldsymbol{T} = (\boldsymbol{\tau}, \boldsymbol{\mu}) \in \mathcal{P}. \end{aligned} \tag{13.22}$$

We notice that since $\mathcal{P}$ is a convex set, from the conditions $\boldsymbol{\Sigma}_n^k, \boldsymbol{\Sigma}_{n-1}^k \in \mathcal{P}$, we have $\boldsymbol{\Sigma}_{n-1+\theta}^k \in \mathcal{P}$. Obviously, in the case $\theta = 1$, the scheme Dual_2^k reduces to the scheme Dual_1^k.

As before, we introduce the constraint set

$$\begin{aligned} \mathcal{P}_{n-1+\theta} &\equiv \mathcal{P}(t_{n-1+\theta}) \\ &= \{\boldsymbol{T} = (\boldsymbol{\tau}, \boldsymbol{\mu}) \in \mathcal{P} : \ b(\boldsymbol{v}, \boldsymbol{\tau}) = \langle \boldsymbol{\ell}(t_{n-1+\theta}), \boldsymbol{v}\rangle \quad \forall\, \boldsymbol{v} \in V\}. \end{aligned}$$

Then $\boldsymbol{\Sigma}_{n-1+\theta}^k \in \mathcal{P}_{n-1+\theta}$ satisfies

$$A(\Delta \boldsymbol{\Sigma}_n^k, \boldsymbol{T} - \boldsymbol{\Sigma}_{n-1+\theta}^k) \ge 0 \quad \forall\, \boldsymbol{T} \in \mathcal{P}_{n-1+\theta},$$

which is exactly (13.4). By the error analysis done in the last section, we see that in terms of the error $\max_n \|\boldsymbol{\Sigma}(t_n) - \boldsymbol{\Sigma}_n^k\|_{\mathcal{T}}$, if the solution is sufficiently smooth, the scheme Dual_2^k is of second-order when $\theta = \frac{1}{2}$, and for $\theta \in \left(\frac{1}{2}, 1\right]$, the scheme Dual_2^k, and hence also the scheme Dual_1^k, is first-order accurate.

The third discretization scheme is a generalized midpoint method of Simo [114].

Scheme Dual_3^k. Let $(\boldsymbol{w}_0^k, \boldsymbol{\Sigma}_0^k) = \boldsymbol{0}$. For $n = 1, \ldots, N$, first compute $(\boldsymbol{w}_{n-1+\theta}^k, \boldsymbol{\Sigma}_{n-1+\theta}^k) \in V \times \mathcal{P}$ satisfying

$$b(\boldsymbol{v}, \boldsymbol{\sigma}_{n-1+\theta}^k) = \langle \boldsymbol{\ell}(t_{n-1+\theta}), \boldsymbol{v}\rangle \quad \forall\, \boldsymbol{v} \in V, \tag{13.23}$$

$$\begin{aligned} &A(\boldsymbol{\Sigma}_{n-1+\theta}^k - \boldsymbol{\Sigma}_{n-1}^k, \boldsymbol{T} - \boldsymbol{\Sigma}_{n-1+\theta}^k) \\ &\quad + \theta\, k\, b(\boldsymbol{w}_{n-1+\theta}^k, \boldsymbol{\tau} - \boldsymbol{\sigma}_{n-1+\theta}^k) \ge 0 \quad \forall\, \boldsymbol{T} = (\boldsymbol{\tau}, \boldsymbol{\mu}) \in \mathcal{P}. \end{aligned} \tag{13.24}$$

Then define

$$\boldsymbol{w}_n^k = \frac{1}{\theta}\, \boldsymbol{w}_{n-1+\theta}^k + \left(1 - \frac{1}{\theta}\right) \boldsymbol{w}_{n-1}^k, \tag{13.25}$$

and finally, find $\mathbf{\Sigma}_n^k \in \mathcal{P}$ such that

$$A(\mathbf{\Sigma}_n^k - \mathbf{\Sigma}_{n-1}^k, \boldsymbol{T} - \mathbf{\Sigma}_n^k) + k\, b(\boldsymbol{w}_n^k, \boldsymbol{\tau} - \boldsymbol{\sigma}_n^k) \ge 0 \quad \forall \boldsymbol{T} \in \mathcal{P}. \tag{13.26}$$

In the analysis of the dual problem in Chapter 8, we have shown the existence of a solution of the problem (13.19)–(13.20). The existence of solutions of the problems (13.21)–(13.22) and (13.23)–(13.24) can be shown similarly. The problem (13.26) is equivalent to a constrained minimization problem,

$$\inf \left\{ \tfrac{1}{2} A(\boldsymbol{T}, \boldsymbol{T}) - A(\mathbf{\Sigma}_{n-1}^k, \boldsymbol{T}) + k\, b(\boldsymbol{w}_n^k, \boldsymbol{\tau}) : \ \boldsymbol{T} \in \mathcal{P} \right\},$$

which has a unique minimizer $\mathbf{\Sigma}_n^k \in \mathcal{P}$. For the rest of the section we give a stability analysis of the above semidiscrete schemes. First we notice that for the continuous problem, we have a result on the contractivity of the solution.

THEOREM 13.3. *Let* $(\boldsymbol{u}_1, \mathbf{\Sigma}_1), (\boldsymbol{u}_2, \mathbf{\Sigma}_2) : [0, T] \to V \times \mathcal{P}$ *satisfy the relations* (13.1) *and* (13.2). *Then*

$$\|\mathbf{\Sigma}_1(t) - \mathbf{\Sigma}_2(t)\|_A \le \|\mathbf{\Sigma}_1(s) - \mathbf{\Sigma}_2(s)\|_A \quad \text{for all } 0 \le s \le t \le T. \tag{13.27}$$

PROOF. The functions $(\boldsymbol{u}_1(t), \mathbf{\Sigma}_1(t))$ and $(\boldsymbol{u}_2(t), \mathbf{\Sigma}_2(t))$ satisfy

$$b(\boldsymbol{v}, \boldsymbol{\sigma}_1(t)) = \langle \boldsymbol{\ell}(t), \boldsymbol{v} \rangle \quad \forall \boldsymbol{v} \in V, \tag{13.28}$$

$$A(\dot{\mathbf{\Sigma}}_1(t), \boldsymbol{T} - \mathbf{\Sigma}_1(t)) + b(\dot{\boldsymbol{u}}_1(t), \boldsymbol{\tau} - \boldsymbol{\sigma}_1(t)) \ge 0 \quad \forall \boldsymbol{T} \in \mathcal{P}, \tag{13.29}$$

and

$$b(\boldsymbol{v}, \boldsymbol{\sigma}_2(t)) = \langle \boldsymbol{\ell}(t), \boldsymbol{v} \rangle \quad \forall \boldsymbol{v} \in V, \tag{13.30}$$

$$A(\dot{\mathbf{\Sigma}}_2(t), \boldsymbol{T} - \mathbf{\Sigma}_2(t)) + b(\dot{\boldsymbol{u}}_2(t), \boldsymbol{\tau} - \boldsymbol{\sigma}_2(t)) \ge 0 \quad \forall \boldsymbol{T} \in \mathcal{P}. \tag{13.31}$$

Subtracting (13.30) from (13.28), we obtain

$$b(\boldsymbol{v}, \boldsymbol{\sigma}_1(t) - \boldsymbol{\sigma}_2(t)) = 0 \quad \forall \boldsymbol{v} \in V. \tag{13.32}$$

Then we take $\boldsymbol{T} = \mathbf{\Sigma}_2(t)$ in (13.29) and $\boldsymbol{T} = \mathbf{\Sigma}_1(t)$ in (13.31), and add the resulting inequalities to obtain

$$-A(\dot{\mathbf{\Sigma}}_1(t) - \dot{\mathbf{\Sigma}}_2(t), \mathbf{\Sigma}_1(t) - \mathbf{\Sigma}_2(t)) - b(\dot{\boldsymbol{u}}_1(t) - \dot{\boldsymbol{u}}_2(t), \boldsymbol{\sigma}_1(t) - \boldsymbol{\sigma}_2(t)) \ge 0.$$

By (13.32),

$$b(\dot{\boldsymbol{u}}_1(t) - \dot{\boldsymbol{u}}_2(t), \boldsymbol{\sigma}_1(t) - \boldsymbol{\sigma}_2(t)) = 0.$$

Hence,

$$A(\dot{\mathbf{\Sigma}}_1(t) - \dot{\mathbf{\Sigma}}_2(t), \mathbf{\Sigma}_1(t) - \mathbf{\Sigma}_2(t)) \le 0,$$

i.e.,

$$\frac{1}{2}\frac{d}{dt}\|\boldsymbol{\Sigma}_1(t) - \boldsymbol{\Sigma}_2(t)\|_A^2 \le 0,$$

from which the contractivity inequality (13.27) follows. □

We say that a numerical scheme for Problem DUAL is stable if its solutions inherit the contractivity property of the solution of the continuous problem. More precisely, we introduce the following definition.

DEFINITION 13.4. A numerical scheme for solving the dual problem DUAL is said to be stable if two numerical solutions $(\boldsymbol{w}_1^k, \boldsymbol{\Sigma}_1^k) = \{(\boldsymbol{w}_{1,n}^k, \boldsymbol{\Sigma}_{1,n}^k)\}_{n=0}^N$ and $(\boldsymbol{w}_2^k, \boldsymbol{\Sigma}_2^k) = \{(\boldsymbol{w}_{2,n}^k, \boldsymbol{\Sigma}_{2,n}^k)\}_{n=0}^N$, generated by two initial values, satisfy the inequality

$$\|\boldsymbol{\Sigma}_{1,n}^k - \boldsymbol{\Sigma}_{2,n}^k\|_A \le \|\boldsymbol{\Sigma}_{1,m}^k - \boldsymbol{\Sigma}_{2,m}^k\|_A \quad \text{for all } 0 \le m \le n \le N. \tag{13.33}$$

Stability is a desirable property for a numerical scheme. The stability estimate (13.33) shows that the propagation of the error at any step is controlled at later steps.

Let us show that all the three schemes are stable.

THEOREM 13.5. *The schemes* DUAL_1^k *and* DUAL_3^k *are stable. If* $\theta \in \left[\frac{1}{2}, 1\right]$, *the scheme* DUAL_2^k *is also stable.*

PROOF. We will prove that when $\theta \in \left[\frac{1}{2}, 1\right]$, the scheme DUAL_2^k is stable. Since the scheme DUAL_1^k is the particular case of the scheme DUAL_2^k with $\theta = 1$, we also get the stability of DUAL_1^k. The stability of the scheme DUAL_3^k can be proved by a similar argument.

Let $(\boldsymbol{w}_1^k, \boldsymbol{\Sigma}_1^k)$ and $(\boldsymbol{w}_2^k, \boldsymbol{\Sigma}_2^k)$ be two solutions computed from the scheme DUAL_2^k with two initial values. Then for $n = 1, \ldots, N$, we have

$$b(\boldsymbol{v}, \boldsymbol{\sigma}_{1,n-1+\theta}^k) = \langle \boldsymbol{\ell}(t_{n-1+\theta}), \boldsymbol{v}\rangle \quad \forall \boldsymbol{v} \in V, \tag{13.34}$$

$$\begin{aligned} &A(\delta\boldsymbol{\Sigma}_{1,n}^k, \boldsymbol{T} - \boldsymbol{\Sigma}_{1,n-1+\theta}^k) \\ &\quad + b(\boldsymbol{w}_{1,n-1+\theta}^k, \boldsymbol{\tau} - \boldsymbol{\sigma}_{1,n-1+\theta}^k) \ge 0 \quad \forall \boldsymbol{T} = (\boldsymbol{\tau}, \boldsymbol{\mu}) \in \mathcal{P}, \end{aligned} \tag{13.35}$$

and

$$b(\boldsymbol{v}, \boldsymbol{\sigma}_{2,n-1+\theta}^k) = \langle \boldsymbol{\ell}(t_{n-1+\theta}), \boldsymbol{v}\rangle \quad \forall \boldsymbol{v} \in V, \tag{13.36}$$

$$\begin{aligned} &A(\delta\boldsymbol{\Sigma}_{2,n}^k, \boldsymbol{T} - \boldsymbol{\Sigma}_{2,n-1+\theta}^k) \\ &\quad + b(\boldsymbol{w}_{2,n-1+\theta}^k, \boldsymbol{\tau} - \boldsymbol{\sigma}_{2,n-1+\theta}^k) \ge 0 \quad \forall \boldsymbol{T} = (\boldsymbol{\tau}, \boldsymbol{\mu}) \in \mathcal{P}. \end{aligned} \tag{13.37}$$

Subtracting (13.36) from (13.34), we obtain

$$b(\boldsymbol{v}, \boldsymbol{\sigma}_{1,n-1+\theta}^k - \boldsymbol{\sigma}_{2,n-1+\theta}^k) = 0 \quad \forall \boldsymbol{v} \in V. \tag{13.38}$$

Notice that $\boldsymbol{\Sigma}^k_{1,n-1+\theta}, \boldsymbol{\Sigma}^k_{2,n-1+\theta} \in \mathcal{P}$. Taking $\boldsymbol{T} = \boldsymbol{\Sigma}^k_{2,n-1+\theta}$ in (13.35) and $\boldsymbol{T} = \boldsymbol{\Sigma}^k_{1,n-1+\theta}$ in (13.37), adding the two resulting inequalities, and using the relation (13.38), we get

$$A\left((\boldsymbol{\Sigma}^k_{1,n} - \boldsymbol{\Sigma}^k_{2,n}) - (\boldsymbol{\Sigma}^k_{1,n-1} - \boldsymbol{\Sigma}^k_{2,n-1}), \boldsymbol{\Sigma}^k_{1,n-1+\theta} - \boldsymbol{\Sigma}^k_{2,n-1+\theta}\right) \le 0. \tag{13.39}$$

By definition, $\boldsymbol{\Sigma}^k_{i,n-1+\theta} = \theta\,\boldsymbol{\Sigma}^k_{i,n} + (1-\theta)\,\boldsymbol{\Sigma}^k_{i,n-1}$, $i = 1, 2$. Since $\theta \in \left[\frac{1}{2}, 1\right]$, from the inequality (13.39) we have

$$\begin{aligned}
\theta\, A(&\boldsymbol{\Sigma}^k_{1,n} - \boldsymbol{\Sigma}^k_{2,n}, \boldsymbol{\Sigma}^k_{1,n} - \boldsymbol{\Sigma}^k_{2,n}) \\
&\le (1-\theta)\, A(\boldsymbol{\Sigma}^k_{1,n-1} - \boldsymbol{\Sigma}^k_{2,n-1}, \boldsymbol{\Sigma}^k_{1,n-1} - \boldsymbol{\Sigma}^k_{2,n-1}) \\
&\quad + (2\theta - 1)\, A(\boldsymbol{\Sigma}^k_{1,n} - \boldsymbol{\Sigma}^k_{2,n}, \boldsymbol{\Sigma}^k_{1,n-1} - \boldsymbol{\Sigma}^k_{2,n-1}) \\
&\le (1-\theta)\, A(\boldsymbol{\Sigma}^k_{1,n-1} - \boldsymbol{\Sigma}^k_{2,n-1}, \boldsymbol{\Sigma}^k_{1,n-1} - \boldsymbol{\Sigma}^k_{2,n-1}) \\
&\quad + \tfrac{1}{2}\,(2\theta - 1)\,\Big[A(\boldsymbol{\Sigma}^k_{1,n} - \boldsymbol{\Sigma}^k_{2,n}, \boldsymbol{\Sigma}^k_{1,n} - \boldsymbol{\Sigma}^k_{2,n}) \\
&\qquad\qquad + A(\boldsymbol{\Sigma}^k_{1,n-1} - \boldsymbol{\Sigma}^k_{2,n-1}, \boldsymbol{\Sigma}^k_{1,n-1} - \boldsymbol{\Sigma}^k_{2,n-1})\Big].
\end{aligned}$$

Therefore,

$$A(\boldsymbol{\Sigma}^k_{1,n} - \boldsymbol{\Sigma}^k_{2,n}, \boldsymbol{\Sigma}^k_{1,n} - \boldsymbol{\Sigma}^k_{2,n}) \le A(\boldsymbol{\Sigma}^k_{1,n-1} - \boldsymbol{\Sigma}^k_{2,n-1}, \boldsymbol{\Sigma}^k_{1,n-1} - \boldsymbol{\Sigma}^k_{2,n-1}),$$

i.e., the scheme is stable. □

We end this section by commenting that the contractivity property implies the uniqueness of a solution. Thus, in particular, for both the continuous problem and discrete problem, the uniqueness of the generalized stress part of the solutions follows immediately.

13.3 Fully Discrete Approximations of the Dual Problem

We now discuss a family of fully discrete approximations to the problem DUAL. We have seen in Chapter 8 that under suitable assumptions, the problem DUAL has a unique solution. The fully discrete schemes discussed here can also be viewed as mixed approximations to the stress problem DUAL1; by "mixed" here we mean that a Lagrange multiplier is introduced as a result of the constraint related to the bilinear form $b(\cdot,\cdot)$.

The schemes. To begin with, we assume that a uniform partition dividing the time interval $[0, T]$ into N subintervals is given, with step-size $k = T/N$. Then we assume that a finite element mesh $\mathcal{T}_h = \{\Omega_e\}_{e=1}^E$ of the spatial domain Ω is constructed in the usual way, with the mesh-size defined by $h = \max h_e$, where h_e is the diameter of the element Ω_e, a general element of the triangulation. Unlike the case of the primal variational problem, where we derive error estimates for any finite element subspaces, here we will consider only a particular choice of finite element subspaces. More precisely, the finite element subspace V^h will consist of piecewise linear functions in $V = [H_0^1(\Omega)]^3$, while S^h and M^h will be the subspaces of S and M, respectively, comprising piecewise constants. Then we define $\mathcal{T}^h = S^h \times M^h$ and

$$\mathcal{P}^h = \{\boldsymbol{T}^h = (\boldsymbol{\tau}^h, \boldsymbol{\mu}^h) \in \mathcal{T}^h : \boldsymbol{T}^h \in K \text{ a.e. in } \Omega\}.$$

Again, let $\theta \in \left[\frac{1}{2}, 1\right]$ be a parameter. The family of fully discrete schemes for the problem DUAL is stated next.

PROBLEM DUAL^{hk}. Find $(\boldsymbol{w}^{hk}, \boldsymbol{\Sigma}^{hk}) = \{(\boldsymbol{w}_n^{hk}, \boldsymbol{\Sigma}_n^{hk})\}_{n=0}^N \subset V^h \times \mathcal{P}^h$ with $(\boldsymbol{w}_0^{hk}, \boldsymbol{\Sigma}_0^{hk}) = \boldsymbol{0}$ such that for $n = 1, \ldots, N$,

$$b(\boldsymbol{v}^h, \boldsymbol{\sigma}_{n-1+\theta}^{hk}) = \langle \boldsymbol{\ell}(t_{n-1+\theta}), \boldsymbol{v}^h \rangle \quad \forall \boldsymbol{v}^h \in V^h, \tag{13.40}$$

$$A_h(\delta \boldsymbol{\Sigma}_n^{hk}, \boldsymbol{T}^h - \boldsymbol{\Sigma}_{n-1+\theta}^{hk}) + b(\boldsymbol{w}_{n-1+\theta}^{hk}, \boldsymbol{\tau}^h - \boldsymbol{\sigma}_{n-1+\theta}^{hk}) \geq 0 \quad \forall \boldsymbol{T}^h = (\boldsymbol{\tau}^h, \boldsymbol{\mu}^h) \in \mathcal{P}^h. \tag{13.41}$$

Here, as before, we use the notation $\boldsymbol{\Sigma}_{n-1+\theta}^{hk} = \theta\, \boldsymbol{\Sigma}_n^{hk} + (1-\theta)\, \boldsymbol{\Sigma}_{n-1}^{hk}$. We use $\boldsymbol{w}_{n-1+\theta}^{hk} \in V^h$ to denote an approximation of the velocity variable $\boldsymbol{w}(t) \equiv \dot{\boldsymbol{u}}(t)$ at $t = t_{n-1+\theta}$. The bilinear form $A_h : \mathcal{T} \times \mathcal{T} \to \mathbb{R}$ is an approximation to $A(\cdot,\cdot)$ and is defined by

$$A_h(\boldsymbol{\Sigma}, \boldsymbol{T}) = \int_\Omega \boldsymbol{\sigma} : \boldsymbol{C}_h^{-1} \boldsymbol{\tau}\, dx + \int_\Omega \boldsymbol{\chi} : \boldsymbol{H}_h^{-1} \boldsymbol{\mu}\, dx, \tag{13.42}$$

in which the approximate moduli $\boldsymbol{C}_h^{-1}$ and $\boldsymbol{H}_h^{-1}$ are piecewise constant approximations of $\boldsymbol{C}^{-1}$ and $\boldsymbol{H}^{-1}$. They can be defined as the piecewise averages of $\boldsymbol{C}^{-1}$ and $\boldsymbol{H}^{-1}$ over each element, for example. The approximations $\boldsymbol{C}_h^{-1}$ and $\boldsymbol{H}_h^{-1}$ are assumed to satisfy the material properties enjoyed by $\boldsymbol{C}^{-1}$ and $\boldsymbol{H}^{-1}$, given in Section 7.1, with the constants there independent of h.

With the same proof technique as that used in Section 8.3, one can show that under assumptions that are the discrete counterparts of Assumptions 8.4 and 8.11, the discrete problem DUAL^{hk} has a solution.

Order error estimates. In the derivation of order error estimates below, we assume that the solution has the regularity required by the expressions at various locations. The precise regularity assumptions needed will be made clear in the statement of Theorem 13.7. We introduce the orthogonal projection operator $\Pi^h : \mathcal{T} \to \mathcal{T}^h$ with respect to the inner product defined by the bilinear form $A_h(\cdot,\cdot)$; that is, for $\boldsymbol{T} \in \mathcal{T}$, $\Pi^h \boldsymbol{T}$ is the unique element in $\mathcal{T}^h$ such that

$$A_h(\boldsymbol{T} - \Pi^h \boldsymbol{T}, \boldsymbol{T}^h) = 0 \quad \forall \boldsymbol{T}^h \in \mathcal{T}^h. \tag{13.43}$$

From the expression (13.42) we see that $\Pi^h \boldsymbol{T} = (\boldsymbol{\tau}_1^h, \boldsymbol{\mu}_1^h)$ with $\boldsymbol{\tau}_1^h \in S^h$ and $\boldsymbol{\mu}_1^h \in M^h$ being orthogonal projections of $\boldsymbol{\tau}$ and $\boldsymbol{\mu}$ onto S^h and M^h in the inner products defined by the bilinear forms $a_h(\cdot,\cdot)$ and $c_h(\cdot,\cdot)$, respectively. Here,

$$\bar{a}_h(\boldsymbol{\sigma}, \boldsymbol{\tau}) = \int_\Omega \boldsymbol{\sigma} : \boldsymbol{C}_h^{-1} \boldsymbol{\tau}\, dx,$$

$$c_h(\boldsymbol{\chi}, \boldsymbol{\mu}) = \int_\Omega \boldsymbol{\chi} \cdot \boldsymbol{H}_h^{-1} \boldsymbol{\mu}\, dx.$$

We will use the same symbol Π^h also to denote these two orthogonal projections. Thus, we will write $\Pi^h \boldsymbol{T} = (\Pi^h \boldsymbol{\tau}, \Pi^h \boldsymbol{\mu})$.

Later, we will need some properties of the orthogonal projections, which are summarized in the following lemma.

LEMMA 13.6. *The orthogonal projections $\Pi^h : S \to S^h$ and $\Pi^h : M \to M^h$ are piecewise averaging operators; that is, for $\boldsymbol{T} = (\boldsymbol{\tau}, \boldsymbol{\mu}) \in \mathcal{T}$, for any element T,*

$$\Pi^h \boldsymbol{\tau}|_{\Omega_e} = \frac{1}{\operatorname{meas}(\Omega_e)} \int_{\Omega_e} \boldsymbol{\tau}\, dx, \quad \Pi^h \boldsymbol{\mu}|_{\Omega_e} = \frac{1}{\operatorname{meas}(\Omega_e)} \int_{\Omega_e} \boldsymbol{\mu}\, dx. \tag{13.44}$$

Consequently, by the convexity of the set $\mathcal{P}$, if $\boldsymbol{T} \in \mathcal{P}$, then $\Pi^h \boldsymbol{T} \in \mathcal{P}^h$. Also, we have

$$b(\boldsymbol{v}^h, \Pi^h \boldsymbol{\tau} - \boldsymbol{\tau}) = 0 \quad \forall \boldsymbol{v}^h \in V^h,\ \boldsymbol{\tau} \in S. \tag{13.45}$$

PROOF. We will prove the first relation in (13.44); the second relation can be proved similarly. From (13.43), we find that

$$\int_\Omega (\Pi^h \boldsymbol{\tau} - \boldsymbol{\tau}) : \boldsymbol{C}_h^{-1} \boldsymbol{\tau}^h\, dx = 0 \quad \forall \boldsymbol{\tau}^h \in S^h,$$

which implies, on each element Ω_e, that

$$\int_{\Omega_e} (\Pi^h \boldsymbol{\tau} - \boldsymbol{\tau})\, dx : \left(\boldsymbol{C}_h^{-1} \boldsymbol{\tau}^h\right)|_{\Omega_e} = 0 \quad \forall \boldsymbol{\tau}^h \in S^h,$$

since $\left(\boldsymbol{C}_h^{-1} \boldsymbol{\tau}^h\right)|_{\Omega_e}$ is constant. Because $\left(\boldsymbol{C}_h^{-1} \boldsymbol{\tau}^h\right)|_{\Omega_e}$ can take on the value of an arbitrary constant, we get

$$\int_{\Omega_e} (\Pi^h \boldsymbol{\tau} - \boldsymbol{\tau})\, dx = \boldsymbol{0}.$$

So the first relation in (13.44) holds. The relation (13.45) is a simple consequence of (13.44) and the fact that $\boldsymbol{\epsilon}(\boldsymbol{v}^h)$ is a piecewise constant for

$\boldsymbol{v}^h \in V^h$. □

Now let $\boldsymbol{e}_n = \boldsymbol{\Sigma}(t_n) - \boldsymbol{\Sigma}_n^{hk}$, $n = 0, 1, \ldots, N$, denote the approximation error, $\boldsymbol{e}_0 = \mathbf{0}$. We consider the expression

$$A_h(\delta \boldsymbol{e}_n, \theta\, \boldsymbol{e}_n + (1-\theta)\, \boldsymbol{e}_{n-1}).$$

Denote by $\|\cdot\|_h$ the norm induced by the discrete bilinear form $A_h(\cdot,\cdot)$, that is,

$$\|\boldsymbol{T}\|_h = A_h(\boldsymbol{T}, \boldsymbol{T})^{\frac{1}{2}}.$$

By the assumptions made on $\boldsymbol{C}_h^{-1}$ and $\boldsymbol{H}_h^{-1}$, the norm $\|\cdot\|_h$ is equivalent to $\|\cdot\|_{\mathcal{T}}$ with the equivalence constants independent of h. Adapting (11.34) to the current case, we have

$$A_h(\delta \boldsymbol{e}_n, \theta\, \boldsymbol{e}_n + (1-\theta)\, \boldsymbol{e}_{n-1}) \geq \frac{1}{2k}\left(\|\boldsymbol{e}_n\|_h^2 - \|\boldsymbol{e}_{n-1}\|_h^2\right). \tag{13.46}$$

To derive an upper bound, we write

$$\begin{aligned} & A_h(\delta \boldsymbol{e}_n, \theta\, \boldsymbol{e}_n + (1-\theta)\, \boldsymbol{e}_{n-1}) \\ & \quad = A_h(\delta \boldsymbol{e}_n, E_{n,\theta}(\boldsymbol{\Sigma})) + A_h(\delta \boldsymbol{e}_n, \boldsymbol{\Sigma}(t_{n-1+\theta}) - \boldsymbol{\Sigma}_{n-1+\theta}^{hk}), \end{aligned} \tag{13.47}$$

where as before, we use the notation

$$E_{n,\theta}(\boldsymbol{\Sigma}) = \theta\, \boldsymbol{\Sigma}(t_n) + (1-\theta)\, \boldsymbol{\Sigma}(t_{n-1}) - \boldsymbol{\Sigma}(t_{n-1+\theta}).$$

For the second term on the right-hand side of (13.47), we have

$$\begin{aligned} & A_h(\delta \boldsymbol{e}_n, \boldsymbol{\Sigma}(t_{n-1+\theta}) - \boldsymbol{\Sigma}_{n-1+\theta}^{hk}) \\ & \quad = A_h(\delta \boldsymbol{\Sigma}(t_n) - \dot{\boldsymbol{\Sigma}}(t_{n-1+\theta}), \boldsymbol{\Sigma}(t_{n-1+\theta}) - \boldsymbol{\Sigma}_{n-1+\theta}^{hk}) \\ & \qquad + A_h(\dot{\boldsymbol{\Sigma}}(t_{n-1+\theta}), \boldsymbol{\Sigma}(t_{n-1+\theta}) - \boldsymbol{\Sigma}_{n-1+\theta}^{hk}) \\ & \qquad - A(\dot{\boldsymbol{\Sigma}}(t_{n-1+\theta}), \boldsymbol{\Sigma}(t_{n-1+\theta}) - \boldsymbol{\Sigma}_{n-1+\theta}^{hk}) \\ & \qquad + A(\dot{\boldsymbol{\Sigma}}(t_{n-1+\theta}), \boldsymbol{\Sigma}(t_{n-1+\theta}) - \boldsymbol{\Sigma}_{n-1+\theta}^{hk}) \\ & \qquad - A_h(\delta \boldsymbol{\Sigma}_n^{hk}, \Pi^h \boldsymbol{\Sigma}(t_{n-1+\theta}) - \boldsymbol{\Sigma}_{n-1+\theta}^{hk}) \\ & \qquad - A_h(\delta \boldsymbol{\Sigma}_n^{hk}, \boldsymbol{\Sigma}(t_{n-1+\theta}) - \Pi^h \boldsymbol{\Sigma}(t_{n-1+\theta})). \end{aligned}$$

The first term on the right-hand side can be rewritten,

$$\begin{aligned} & A_h(\delta \boldsymbol{\Sigma}(t_n) - \dot{\boldsymbol{\Sigma}}(t_{n-1+\theta}), \boldsymbol{\Sigma}(t_{n-1+\theta}) - \boldsymbol{\Sigma}_{n-1+\theta}^{hk}) \\ & \quad = A_h(\delta \boldsymbol{\Sigma}(t_n) - \dot{\boldsymbol{\Sigma}}(t_{n-1+\theta}), -E_{n,\theta}(\boldsymbol{\Sigma})) \\ & \qquad + A_h(\delta \boldsymbol{\Sigma}(t_n) - \dot{\boldsymbol{\Sigma}}(t_{n-1+\theta}), \theta\, \boldsymbol{e}_n + (1-\theta)\, \boldsymbol{e}_{n-1}). \end{aligned}$$

We use (13.2) at $t = t_{n-1+\theta}$ with $\boldsymbol{T} = \boldsymbol{\Sigma}^{hk}_{n-1+\theta}$ (which is in $\mathcal{P}$ by its convexity) to obtain (recalling that we use $\boldsymbol{w}$ to stand for $\dot{\boldsymbol{u}}$)

$$A(\dot{\boldsymbol{\Sigma}}(t_{n-1+\theta}), \boldsymbol{\Sigma}(t_{n-1+\theta}) - \boldsymbol{\Sigma}^{hk}_{n-1+\theta}) \\ \leq b(\boldsymbol{w}(t_{n-1+\theta}), \boldsymbol{\sigma}^{hk}_{n-1+\theta} - \boldsymbol{\sigma}(t_{n-1+\theta})).$$

Taking $\boldsymbol{T}^h = \Pi^h \boldsymbol{\Sigma}(t_{n-1+\theta})$ in (13.41), we get

$$-A_h(\delta \boldsymbol{\Sigma}^{hk}_n, \Pi^h \boldsymbol{\Sigma}(t_{n-1+\theta}) - \boldsymbol{\Sigma}^{hk}_{n-1+\theta}) \\ \leq b(\boldsymbol{w}^{hk}_{n-1+\theta}, \Pi^h \boldsymbol{\sigma}(t_{n-1+\theta}) - \boldsymbol{\sigma}^{hk}_{n-1+\theta}).$$

And finally, because Π^h is the orthogonal projection onto $\mathcal{T}^h$ in the inner product induced by the bilinear form $A_h(\cdot, \cdot)$, we have

$$A_h(\delta \boldsymbol{\Sigma}^{hk}_n, \boldsymbol{\Sigma}(t_{n-1+\theta}) - \Pi^h \boldsymbol{\Sigma}(t_{n-1+\theta})) = 0.$$

Thus, we have

$$\begin{aligned} A_h(\delta \boldsymbol{e}_n, \boldsymbol{\Sigma}(t_{n-1+\theta}) - \boldsymbol{\Sigma}^{hk}_{n-1+\theta}) \\ \leq A_h(\delta \boldsymbol{\Sigma}(t_n) - \dot{\boldsymbol{\Sigma}}(t_{n-1+\theta}), -E_{n,\theta}(\boldsymbol{\Sigma})) \\ + A_h(\delta \boldsymbol{\Sigma}(t_n) - \dot{\boldsymbol{\Sigma}}(t_{n-1+\theta}), \theta\, \boldsymbol{e}_n + (1-\theta)\, \boldsymbol{e}_{n-1}) \\ + A_h(\dot{\boldsymbol{\Sigma}}(t_{n-1+\theta}), \boldsymbol{\Sigma}(t_{n-1+\theta}) - \boldsymbol{\Sigma}^{hk}_{n-1+\theta}) \\ - A(\dot{\boldsymbol{\Sigma}}(t_{n-1+\theta}), \boldsymbol{\Sigma}(t_{n-1+\theta}) - \boldsymbol{\Sigma}^{hk}_{n-1+\theta}) \\ + b(\boldsymbol{w}(t_{n-1+\theta}), \boldsymbol{\sigma}^{hk}_{n-1+\theta} - \boldsymbol{\sigma}(t_{n-1+\theta})) \\ + b(\boldsymbol{w}^{hk}_{n-1+\theta}, \Pi^h \boldsymbol{\sigma}(t_{n-1+\theta}) - \boldsymbol{\sigma}^{hk}_{n-1+\theta}). \end{aligned}$$

From (13.40) and (13.1) we obtain the relation

$$b(\boldsymbol{v}^h, \boldsymbol{\sigma}(t_{n-1+\theta}) - \boldsymbol{\sigma}^{hk}_{n-1+\theta}) = 0 \quad \forall\, \boldsymbol{v}^h \in V^h. \tag{13.48}$$

Using (13.48) and (13.45), we have

$$\begin{aligned} b(\boldsymbol{w}^{hk}_{n-1+\theta}, \Pi^h \boldsymbol{\sigma}(t_{n-1+\theta}) - \boldsymbol{\sigma}^{hk}_{n-1+\theta}) \\ = b(\boldsymbol{w}^{hk}_{n-1+\theta}, \Pi^h \boldsymbol{\sigma}(t_{n-1+\theta}) - \boldsymbol{\sigma}(t_{n-1+\theta})) \\ = 0. \end{aligned} \tag{13.49}$$

Again using (13.48), we get, for any $\boldsymbol{v}^h \in V^h$,

$$\begin{aligned} b(\boldsymbol{w}(t_{n-1+\theta}), \boldsymbol{\sigma}^{hk}_{n-1+\theta} - \boldsymbol{\sigma}(t_{n-1+\theta})) \\ = b(\boldsymbol{w}(t_{n-1+\theta}) - \boldsymbol{v}^h, \boldsymbol{\sigma}^{hk}_{n-1+\theta} - \boldsymbol{\sigma}(t_{n-1+\theta})) \\ = b(\boldsymbol{w}(t_{n-1+\theta}) - \boldsymbol{v}^h, E_{n,\theta}(\boldsymbol{\sigma})) \\ + b(\boldsymbol{w}(t_{n-1+\theta}) - \boldsymbol{v}^h, \\ \theta\, (\boldsymbol{\sigma}^{hk}_n - \boldsymbol{\sigma}(t_n)) + (1-\theta)\, (\boldsymbol{\sigma}^{hk}_{n-1} - \boldsymbol{\sigma}(t_{n-1}))). \end{aligned}$$

Employing all these relations in (13.47), we obtain an upper bound, for any $\boldsymbol{v}^h \in V^h$,

$$
\begin{aligned}
& A_h(\delta \boldsymbol{e}_n, \theta\, \boldsymbol{e}_n + (1-\theta)\, \boldsymbol{e}_{n-1}) \\
& \quad \le A_h(\delta \boldsymbol{e}_n, E_{n,\theta}(\boldsymbol{\Sigma})) - A_h(\delta \boldsymbol{\Sigma}(t_n) - \dot{\boldsymbol{\Sigma}}(t_{n-1+\theta}), E_{n,\theta}(\boldsymbol{\Sigma})) \\
& \qquad + A_h(\delta \boldsymbol{\Sigma}(t_n) - \dot{\boldsymbol{\Sigma}}(t_{n-1+\theta}), \theta\, \boldsymbol{e}_n + (1-\theta)\, \boldsymbol{e}_{n-1}) \\
& \qquad + A_h(\dot{\boldsymbol{\Sigma}}(t_{n-1+\theta}), \boldsymbol{\Sigma}(t_{n-1+\theta}) - \boldsymbol{\Sigma}^{hk}_{n-1+\theta}) \\
& \qquad - A(\dot{\boldsymbol{\Sigma}}(t_{n-1+\theta}), \boldsymbol{\Sigma}(t_{n-1+\theta}) - \boldsymbol{\Sigma}^{hk}_{n-1+\theta}) \\
& \qquad + b(\boldsymbol{w}(t_{n-1+\theta}) - \boldsymbol{v}^h, E_{n,\theta}(\boldsymbol{\sigma})) \\
& \qquad + b(\boldsymbol{w}(t_{n-1+\theta}) - \boldsymbol{v}^h, \\
& \qquad\qquad \theta\, (\boldsymbol{\sigma}^{hk}_n - \boldsymbol{\sigma}(t_n)) + (1-\theta)\, (\boldsymbol{\sigma}^{hk}_{n-1} - \boldsymbol{\sigma}(t_{n-1}))).
\end{aligned}
$$

Set $M = \max_{0 \le n \le N} \|\boldsymbol{e}_n\|_{\mathcal{T}}$, the maximal error. Furthermore, we use the notation

$$
c^h_{n,\theta}(\boldsymbol{w}) = \inf_{\boldsymbol{v}^h \in V^h} \|\boldsymbol{w}(t_{n-1+\theta}) - \boldsymbol{v}^h\|_V \tag{13.50}
$$

and

$$
c^h(\boldsymbol{C}, \boldsymbol{H}) = \max_{\Omega_e} \left\{ \|\boldsymbol{C}_h^{-1} - \boldsymbol{C}^{-1}\|_{L^\infty(\Omega_e)}, \|\boldsymbol{H}_h^{-1} - \boldsymbol{H}^{-1}\|_{L^\infty(\Omega_e)} \right\}. \tag{13.51}
$$

Notice that

$$
\|\boldsymbol{\Sigma}(t_{n-1+\theta}) - \boldsymbol{\Sigma}^{hk}_{n-1+\theta}\|_{\mathcal{T}} \le \|E_{n,\theta}(\boldsymbol{\Sigma})\|_{\mathcal{T}} + M.
$$

Combining the lower and upper bounds for the quantity

$$
A_h(\delta \boldsymbol{e}_n, \theta\, \boldsymbol{e}_n + (1-\theta)\, \boldsymbol{e}_{n-1}),
$$

we obtain

$$
\begin{aligned}
& \frac{1}{2k} \left(\|\boldsymbol{e}_n\|_h^2 - \|\boldsymbol{e}_{n-1}\|_h^2 \right) \\
& \quad \le \frac{1}{k} A_h(\boldsymbol{e}_n - \boldsymbol{e}_{n-1}, E_{n,\theta}(\boldsymbol{\Sigma})) \\
& \qquad + c \left(\|\delta \boldsymbol{\Sigma}(t_n) - \dot{\boldsymbol{\Sigma}}(t_{n-1+\theta})\|_{\mathcal{T}} + c^h_{n,\theta}(\boldsymbol{w}) + c^h(\boldsymbol{C}, \boldsymbol{H})\, \|\dot{\boldsymbol{\Sigma}}\|_{L^\infty(0,T;\mathcal{T})} \right) \\
& \qquad\quad \times (\|E_{n,\theta}(\boldsymbol{\Sigma})\|_{\mathcal{T}} + M).
\end{aligned}
$$

From this inequality, we find in turn that, for $n = 1, \ldots, N$,

$$
\begin{aligned}
& \|\boldsymbol{e}_n\|_h^2 - \|\boldsymbol{e}_{n-1}\|_h^2 \\
& \quad \le 2\, A_h(\boldsymbol{e}_n - \boldsymbol{e}_{n-1}, E_{n,\theta}(\boldsymbol{\Sigma})) \\
& \qquad + c\, k \left(\|\delta \boldsymbol{\Sigma}(t_n) - \dot{\boldsymbol{\Sigma}}(t_{n-1+\theta})\|_{\mathcal{T}} + c^h_{n,\theta}(\boldsymbol{w}) + c^h(\boldsymbol{C}, \boldsymbol{H})\, \|\dot{\boldsymbol{\Sigma}}\|_{L^\infty(0,T;\mathcal{T})} \right) \\
& \qquad\quad \times (\|E_{n,\theta}(\boldsymbol{\Sigma})\|_{\mathcal{T}} + M).
\end{aligned}
$$

Hence, recalling that $\boldsymbol{e}_0 = \boldsymbol{0}$, an induction on n yields

$$\begin{aligned}
\|\boldsymbol{e}_n\|_h^2 \leq 2 \sum_{j=1}^{n} A_h(\boldsymbol{e}_j - \boldsymbol{e}_{j-1}, E_{j,\theta}(\boldsymbol{\Sigma})) \\
+ ck \sum_{j=1}^{n} \Big(\|\delta \boldsymbol{\Sigma}(t_j) - \dot{\boldsymbol{\Sigma}}(t_{j-1+\theta})\|_{\mathcal{T}} + c_{j,\theta}^h(\boldsymbol{w}) \\
+ c^h(\boldsymbol{C}, \boldsymbol{H}) \, \|\dot{\boldsymbol{\Sigma}}\|_{L^\infty(0,T;\mathcal{T})} \Big) M \\
+ ck \sum_{j=1}^{n} \|E_{j,\theta}(\boldsymbol{\Sigma})\|_{\mathcal{T}} \Big(\|\delta \boldsymbol{\Sigma}(t_j) - \dot{\boldsymbol{\Sigma}}(t_{j-1+\theta})\|_{\mathcal{T}} \\
+ c_{j,\theta}^h(\boldsymbol{w}) + c^h(\boldsymbol{C}, \boldsymbol{H}) \, \|\dot{\boldsymbol{\Sigma}}\|_{L^\infty(0,T;\mathcal{T})} \Big).
\end{aligned}$$

We then use the identity

$$\begin{aligned}
\sum_{j=1}^{n} A_h(\boldsymbol{e}_j - \boldsymbol{e}_{j-1}, E_{j,\theta}(\boldsymbol{\Sigma})) \\
= \sum_{j=1}^{n-1} A_h(\boldsymbol{e}_j, E_{j,\theta}(\boldsymbol{\Sigma}) - E_{j+1,\theta}(\boldsymbol{\Sigma})) + 2 \, A_h(\boldsymbol{e}_n, E_{n,\theta}(\boldsymbol{\Sigma}))
\end{aligned}$$

to find that

$$\begin{aligned}
M^2 \leq c \left(\sum_{j=1}^{N-1} \|E_{j,\theta}(\boldsymbol{\Sigma}) - E_{j+1,\theta}(\boldsymbol{\Sigma})\|_{\mathcal{T}} + \|E_{n,\theta}(\boldsymbol{\Sigma})\|_{\mathcal{T}} \right) M \\
+ ck \sum_{j=1}^{N} \big(\|\delta \boldsymbol{\Sigma}(t_j) - \dot{\boldsymbol{\Sigma}}(t_{j-1+\theta})\|_{\mathcal{T}} + c_{j,\theta}^h(\boldsymbol{w}) \\
+ c^h(\boldsymbol{C}, \boldsymbol{H}) \, \|\dot{\boldsymbol{\Sigma}}\|_{L^\infty(0,T;\mathcal{T})} \big) M \\
+ ck \sum_{j=1}^{N} \|E_{j,\theta}(\boldsymbol{\Sigma})\|_{\mathcal{T}} \Big(\|\delta \boldsymbol{\Sigma}(t_j) - \dot{\boldsymbol{\Sigma}}(t_{j-1+\theta})\|_{\mathcal{T}} \\
+ c_{j,\theta}^h(\boldsymbol{w}) + c^h(\boldsymbol{C}, \boldsymbol{H}) \, \|\dot{\boldsymbol{\Sigma}}\|_{L^\infty(0,T;\mathcal{T})} \Big).
\end{aligned}$$

Applying the inequality (11.3) and recalling the definition of M, we get

$$
\begin{aligned}
\max_{0\le n\le N} &\|\boldsymbol{\Sigma}(t_n) - \boldsymbol{\Sigma}_n^{hk}\|_{\mathcal{T}} \\
&\le c\Big(\sum_{j=1}^{N-1} \|E_{j,\theta}(\boldsymbol{\Sigma}) - E_{j+1,\theta}(\boldsymbol{\Sigma})\|_{\mathcal{T}} + \|E_{N,\theta}(\boldsymbol{\Sigma})\|_{\mathcal{T}} \\
&\qquad + k \sum_{j=1}^{N} (\|\delta\boldsymbol{\Sigma}(t_j) - \dot{\boldsymbol{\Sigma}}(t_{j-1+\theta})\|_{\mathcal{T}} + c_{j,\theta}^h(\boldsymbol{w}))\Big) \\
&\quad + c\, c^h(\boldsymbol{C}, \boldsymbol{H})\, \|\dot{\boldsymbol{\Sigma}}\|_{L^\infty(0,T;\mathcal{T})} \\
&\quad + c\left\{ k \sum_{j=1}^{N} \|E_{j,\theta}(\boldsymbol{\Sigma})\|_{\mathcal{T}} \Big(\|\delta\boldsymbol{\Sigma}(t_j) - \dot{\boldsymbol{\Sigma}}(t_{j-1+\theta})\|_{\mathcal{T}} \right. \\
&\qquad\qquad \left. + c_{j,\theta}^h(\boldsymbol{w}) + c^h(\boldsymbol{C}, \boldsymbol{H})\, \|\dot{\boldsymbol{\Sigma}}\|_{L^\infty(0,T;\mathcal{T})} \Big) \right\}^{\frac{1}{2}}.
\end{aligned}
$$

Now assume that $\boldsymbol{\Sigma} \in W^{2,1}(0,T;\mathcal{T})$ if $\theta \in (\frac{1}{2}, 1]$ and $\boldsymbol{\Sigma} \in W^{3,1}(0,T;\mathcal{T})$ if $\theta = \frac{1}{2}$. Applying Lemmas 11.2, 11.3, and 11.4, we conclude that if $\theta \in (\frac{1}{2}, 1]$,

$$
\begin{aligned}
\max_{0\le n\le N} &\|\boldsymbol{\Sigma}(t_n) - \boldsymbol{\Sigma}_n^{hk}\|_{\mathcal{T}} \\
&\le c\, k\, \|\boldsymbol{\Sigma}\|_{W^{2,1}(0,T;\mathcal{T})} + c\, k \sum_{j=1}^{N} c_{j,\theta}^h(\boldsymbol{w}) \\
&\quad + c\, c^h(\boldsymbol{C}, \boldsymbol{H})\, \|\dot{\boldsymbol{\Sigma}}\|_{L^\infty(0,T;\mathcal{T})} \\
&\quad + c\left\{ k \sum_{j=1}^{N} k\, \|\ddot{\boldsymbol{\Sigma}}\|_{L^1(t_{j-1},t_j;\mathcal{T})} \big(\|\ddot{\boldsymbol{\Sigma}}\|_{L^1(t_{j-1},t_j;\mathcal{T})} \right. \\
&\qquad\qquad \left. + c^h(\boldsymbol{C}, \boldsymbol{H})\, \|\dot{\boldsymbol{\Sigma}}\|_{L^\infty(0,T;\mathcal{T})} + c_{j,\theta}^h(\boldsymbol{w}) \big) \right\}^{1/2} \\
&\le c\, k\, \|\boldsymbol{\Sigma}\|_{W^{2,1}(0,T;\mathcal{T})} + c\, k \sum_{j=1}^{N} c_{j,\theta}^h(\boldsymbol{w}) \\
&\quad + c\, c^h(\boldsymbol{C}, \boldsymbol{H})\, \|\dot{\boldsymbol{\Sigma}}\|_{L^\infty(0,T;\mathcal{T})} \\
&\quad + c\, k \left\{ \|\dot{\boldsymbol{\Sigma}}\|_{L^\infty(0,T;\mathcal{T})} \|\ddot{\boldsymbol{\Sigma}}\|_{L^1(0,T;\mathcal{T})} c^h(\boldsymbol{C}, \boldsymbol{H}) \right\}^{1/2} \\
&\quad + c\, k \left\{ \sum_{j=1}^{N} \|\ddot{\boldsymbol{\Sigma}}\|_{L^1(t_{j-1},t_j;\mathcal{T})} c_{j,\theta}^h(\boldsymbol{w}) \right\}^{1/2},
\end{aligned}
\tag{13.52}
$$

and if $\theta = \frac{1}{2}$,

$$\begin{aligned}
\max_{0 \le n \le N} & \|\boldsymbol{\Sigma}(t_n) - \boldsymbol{\Sigma}_n^{hk}\|_{\mathcal{T}} \\
& \le c\,k^2 \left(\|\boldsymbol{\Sigma}\|_{W^{3,1}(0,T;\mathcal{T})} + \|\ddot{\boldsymbol{\Sigma}}\|_{L^\infty(0,T;\mathcal{T})}\right) \\
& \quad + c\,k \sum_{j=1}^{N} c_{j,\theta}^h(\boldsymbol{w}) + c\,c^h(\boldsymbol{C},\boldsymbol{H})\,\|\dot{\boldsymbol{\Sigma}}\|_{L^\infty(0,T;\mathcal{T})} \\
& \quad + c \left\{ k \sum_{j=1}^{N} k^2 \|\ddot{\boldsymbol{\Sigma}}\|_{L^\infty(0,T;\mathcal{T})} \big(k\,\|\boldsymbol{\Sigma}^{(3)}\|_{L^1(t_{j-1},t_j;\mathcal{T})} \right. \\
& \qquad \left. + c_{j,\theta}^h(\boldsymbol{w}) + c^h(\boldsymbol{C},\boldsymbol{H})\|\dot{\boldsymbol{\Sigma}}\|_{L^\infty(0,T;\mathcal{T})}\big) \right\}^{1/2} \\
& \le c\,k^2 \left(\|\boldsymbol{\Sigma}\|_{W^{3,1}(0,T;\mathcal{T})} + \|\ddot{\boldsymbol{\Sigma}}\|_{L^\infty(0,T;\mathcal{T})}\right) \\
& \quad + c\,k \sum_{j=1}^{N} c_{j,\theta}^h(\boldsymbol{w}) + c\,c^h(\boldsymbol{C},\boldsymbol{H})\,\|\dot{\boldsymbol{\Sigma}}\|_{L^\infty(0,T;\mathcal{T})} \\
& \quad + c\,k^2 \left\{ \|\ddot{\boldsymbol{\Sigma}}\|_{L^\infty(0,T;\mathcal{T})} \|\boldsymbol{\Sigma}^{(3)}\|_{L^1(0,T;\mathcal{T})} \right\}^{1/2} \\
& \quad + c\,k\,\|\ddot{\boldsymbol{\Sigma}}\|^{1/2}_{L^\infty(0,T;\mathcal{T})} \left\{ k \sum_{j=1}^{N} c_{j,\theta}^h(\boldsymbol{w}) \right\}^{1/2} \\
& \quad + c\,k \left\{ c^h(\boldsymbol{C},\boldsymbol{H}) \|\dot{\boldsymbol{\Sigma}}\|_{L^\infty(0,T;\mathcal{T})} \right\}^{1/2},
\end{aligned} \tag{13.53}$$

where $c_{j,\theta}^h(\boldsymbol{w})$, $j - 1, \ldots, N$, are defined in (13.50), and $c^h(\boldsymbol{C},\boldsymbol{H})$ is defined in (13.51).

The final error estimates in terms of powers of k and h are derived from (13.52) and (13.53), and are dependent on the regularity of the Lagrangian multiplier $\boldsymbol{w}$ and that of $\boldsymbol{C}$ and $\boldsymbol{H}$. If we assume that

$$\boldsymbol{w} \in L^\infty(0,T;(H^2(\Omega))^3),$$

then from (13.49) we have

$$c_{n,\theta}^h(\boldsymbol{w}) \le c\,h\,\|\boldsymbol{w}\|_{L^\infty(0,T;(H^2(\Omega))^3)}. \tag{13.54}$$

And if we assume that

$$C_{ijkl} \in W^{1,\infty}(\Omega), \quad H_{ij} \in W^{1,\infty}(\Omega),$$

and that $\boldsymbol{C}_h$ and $\boldsymbol{H}_h$ are obtained from $\boldsymbol{C}$ and $\boldsymbol{H}$ through piecewise averaging, then

$$c^h(\boldsymbol{C},\boldsymbol{H}) \le c\,h. \tag{13.55}$$

We are now ready to state the results on the order error estimates for the fully discrete approximations.

THEOREM 13.7. *Assume* $\boldsymbol{\Sigma} \in W^{2,1}(0,T;\mathcal{T})$, $\boldsymbol{w} \in L^\infty(0,T;(H^2(\Omega))^3)$, $C_{ijkl} \in W^{1,\infty}(\Omega)$, *and* $H_{ij} \in W^{1,\infty}(\Omega)$. *Then for the fully discrete solutions defined in Problem* DUAL^{hk}, *we have the estimate*

$$\max_{0\le n\le N} \|\boldsymbol{\Sigma}(t_n) - \boldsymbol{\Sigma}_n^{hk}\|_{\mathcal{T}} = O(h+k).$$

In the case $\theta = \frac{1}{2}$, *if additionally* $\boldsymbol{\Sigma} \in W^{3,1}(0,T;\mathcal{T})$, *we have*

$$\max_{0\le n\le N} \|\boldsymbol{\Sigma}(t_n) - \boldsymbol{\Sigma}_n^{hk}\|_{\mathcal{T}} = O(h+k^2).$$

Convergence analysis under minimal regularity conditions. Now we prove the convergence of the fully discrete solutions under the basic regularity condition $(\boldsymbol{u},\boldsymbol{\Sigma}) \in H^1(0,T;V\times\mathcal{T})$. Again we denote $\boldsymbol{e}_n = \boldsymbol{\Sigma}(t_n) - \boldsymbol{\Sigma}_n^{hk}$, $n = 0,1,\ldots,N$, $\boldsymbol{e}_0 = \boldsymbol{0}$. We still have (13.46) and (13.47). Hence

$$\begin{aligned}&\frac{1}{2k}\left(\|\boldsymbol{e}_n\|_h^2 - \|\boldsymbol{e}_{n-1}\|_h^2\right)\\ &\qquad \le A_h(\delta\boldsymbol{e}_n, E_{n,\theta}(\boldsymbol{\Sigma})) + A_h(\delta\boldsymbol{e}_n, \boldsymbol{\Sigma}(t_{n-1+\theta}) - \boldsymbol{\Sigma}_{n-1+\theta}^{hk}). \qquad (13.56)\end{aligned}$$

We examine the second term on the right-hand side of (13.56):

$$\begin{aligned}&A_h(\delta\boldsymbol{e}_n, \boldsymbol{\Sigma}(t_{n-1+\theta}) - \boldsymbol{\Sigma}_{n-1+\theta}^{hk})\\ &\quad = A_h(\delta\boldsymbol{\Sigma}_n, -E_{n,\theta}(\boldsymbol{\Sigma})) + A_h(\delta\boldsymbol{\Sigma}_n, \theta\,\boldsymbol{e}_n + (1-\theta)\,\boldsymbol{e}_{n-1})\\ &\quad\quad - A_h(\delta\boldsymbol{\Sigma}_n^{hk}, \Pi^h\boldsymbol{\Sigma}(t_{n-1+\theta}) - \boldsymbol{\Sigma}_{n-1+\theta}^{hk})\\ &\quad\quad - A_h(\delta\boldsymbol{\Sigma}_n^{hk}, \boldsymbol{\Sigma}(t_{n-1+\theta}) - \Pi^h\boldsymbol{\Sigma}(t_{n-1+\theta})) \qquad (13.57)\end{aligned}$$

We take $\boldsymbol{T}^h = \Pi^h\boldsymbol{\Sigma}(t_{n-1+\theta}) \in \mathcal{P}^h$ in (13.41) to obtain

$$\begin{aligned}&-A_h(\delta\boldsymbol{\Sigma}_n^{hk}, \Pi^h\boldsymbol{\Sigma}(t_{n-1+\theta}) - \boldsymbol{\Sigma}_{n-1+\theta}^{hk})\\ &\qquad \le b(\boldsymbol{w}_{n-1+\theta}^{hk}, \Pi^h\boldsymbol{\sigma}(t_{n-1+\theta}) - \boldsymbol{\sigma}_{n-1+\theta}^{hk}).\end{aligned}$$

By (13.49),

$$b(\boldsymbol{w}_{n-1+\theta}^{hk}, \Pi^h\boldsymbol{\sigma}(t_{n-1+\theta}) - \boldsymbol{\sigma}_{n-1+\theta}^{hk}) = 0.$$

Therefore,

$$-A_h(\delta\boldsymbol{\Sigma}_n^{hk}, \Pi^h\boldsymbol{\Sigma}(t_{n-1+\theta}) - \boldsymbol{\Sigma}_{n-1+\theta}^{hk}) \le 0. \qquad (13.58)$$

Because Π^h is the orthogonal projection onto $\mathcal{T}^h$ in the inner product induced by the bilinear form $A_h(\cdot,\cdot)$, we have

$$A_h(\delta\boldsymbol{\Sigma}_n^{hk}, \boldsymbol{\Sigma}(t_{n-1+\theta}) - \Pi^h\boldsymbol{\Sigma}(t_{n-1+\theta})) = 0. \qquad (13.59)$$

Using (13.58) and (13.59) in (13.57), we see that

$$
\begin{aligned}
&A_h(\delta \boldsymbol{e}_n, \boldsymbol{\Sigma}(t_{n-1+\theta}) - \boldsymbol{\Sigma}^{hk}_{n-1+\theta}) \\
&\quad \le A_h(\delta \boldsymbol{\Sigma}_n, -E_{n,\theta}(\boldsymbol{\Sigma})) + A_h(\delta \boldsymbol{\Sigma}_n, \theta\, \boldsymbol{e}_n + (1-\theta)\, \boldsymbol{e}_{n-1}). \qquad (13.60)
\end{aligned}
$$

Now we take $\boldsymbol{T} = \boldsymbol{\Sigma}^{hk}_{n-1+\theta} \in \mathcal{P}$ in (13.2) and integrate the inequality over $I_n = [t_{n-1}, t_n]$:

$$
\int_{I_n} A(\dot{\boldsymbol{\Sigma}}(t), \boldsymbol{\Sigma}^{hk}_{n-1+\theta} - \boldsymbol{\Sigma}(t))\, dt + \int_{I_n} b(\boldsymbol{w}(t), \boldsymbol{\sigma}^{hk}_{n-1+\theta} - \boldsymbol{\sigma}(t))\, dt \ge 0,
$$

which after being divided by k can be rewritten as

$$
\begin{aligned}
&A(\delta \boldsymbol{\Sigma}_n, \theta\, \boldsymbol{e}_n + (1-\theta)\, \boldsymbol{e}_{n-1}) \\
&\quad \le \frac{1}{k} \int_{I_n} A(\dot{\boldsymbol{\Sigma}}(t), \theta\, \boldsymbol{\Sigma}(t_n) + (1-\theta)\, \boldsymbol{\Sigma}(t_{n-1}) - \boldsymbol{\Sigma}(t))\, dt \\
&\qquad + \frac{1}{k} \int_{I_n} b(\boldsymbol{w}(t), \boldsymbol{\sigma}^{hk}_{n-1+\theta} - \boldsymbol{\sigma}(t))\, dt. \qquad (13.61)
\end{aligned}
$$

Combining (13.56), (13.60), and (13.61) and rearranging some terms, we obtain

$$
\begin{aligned}
&\frac{1}{2} \left(\|\boldsymbol{e}_n\|_h^2 - \|\boldsymbol{e}_{n-1}\|_h^2 \right) \\
&\quad \le A_h(\boldsymbol{e}_n - \boldsymbol{e}_{n-1}, E_{n,\theta}(\boldsymbol{\Sigma})) \\
&\qquad + \int_{I_n} A(\dot{\boldsymbol{\Sigma}}(t), E_{n,\theta}(\boldsymbol{\Sigma}))\, dt - \int_{I_n} A_h(\dot{\boldsymbol{\Sigma}}(t), E_{n,\theta}(\boldsymbol{\Sigma}))\, dt \\
&\qquad + \int_{I_n} A_h(\dot{\boldsymbol{\Sigma}}(t), \theta\, \boldsymbol{e}_n + (1-\theta)\, \boldsymbol{e}_{n-1})\, dt \\
&\qquad - \int_{I_n} A(\dot{\boldsymbol{\Sigma}}(t), \theta\, \boldsymbol{e}_n + (1-\theta)\, \boldsymbol{e}_{n-1})\, dt \\
&\qquad + \int_{I_n} A(\dot{\boldsymbol{\Sigma}}(t), \boldsymbol{\Sigma}(t_{n-1+\theta}) - \boldsymbol{\Sigma}(t))\, dt \\
&\qquad + \int_{I_n} b(\boldsymbol{w}(t), \boldsymbol{\sigma}^{hk}_{n-1+\theta} - \boldsymbol{\sigma}(t))\, dt.
\end{aligned}
$$

Using the quantities $c^h(\boldsymbol{C}, \boldsymbol{H})$ and $\omega_k(\boldsymbol{\Sigma})$ defined in (13.51) and (13.15), we then derive from the above inequality

$$
\begin{aligned}
&\|\boldsymbol{e}_n\|_h^2 - \|\boldsymbol{e}_{n-1}\|_h^2 \\
&\quad \le 2\, A_h(\boldsymbol{e}_n - \boldsymbol{e}_{n-1}, E_{n,\theta}(\boldsymbol{\Sigma})) \\
&\qquad + c\, c^h(\boldsymbol{C}, \boldsymbol{H}) \int_{I_n} \|\dot{\boldsymbol{\Sigma}}(t)\|_{\mathcal{T}} dt\, \left(\|E_{n,\theta}(\boldsymbol{\Sigma})\|_{\mathcal{T}} + M \right) \\
&\qquad + c\, \omega_k(\boldsymbol{\Sigma}) \int_{I_n} \|\dot{\boldsymbol{\Sigma}}(t)\|_{\mathcal{T}} dt \\
&\qquad + \int_{I_n} b(\boldsymbol{w}(t), \boldsymbol{\sigma}^{hk}_{n-1+\theta} - \boldsymbol{\sigma}(t))\, dt, \qquad (13.62)
\end{aligned}
$$

where $M = \max_n \|\boldsymbol{e}_n\|_{\mathcal{T}}$. Now we estimate the last term in (13.62):

$$\begin{aligned}\int_{I_n} & b(\boldsymbol{w}(t), \boldsymbol{\sigma}^{hk}_{n-1+\theta} - \boldsymbol{\sigma}(t))\, dt \\ &= \int_{I_n} \big[b(\boldsymbol{w}(t), \boldsymbol{\sigma}^{hk}_{n-1+\theta} - \boldsymbol{\sigma}(t_{n-1+\theta})) \\ &\qquad + b(\boldsymbol{w}(t), \boldsymbol{\sigma}(t_{n-1+\theta}) - \boldsymbol{\sigma}(t))\big]\, dt.\end{aligned}$$

From (13.48) with an arbitrary $\boldsymbol{v}^h = \boldsymbol{v}^h(t) \in V^h$, we have

$$b(\boldsymbol{v}^h(t), \boldsymbol{\sigma}(t_{n-1+\theta}) - \boldsymbol{\sigma}^{hk}_{n-1+\theta}) = 0.$$

Then

$$\begin{aligned}\int_{I_n} & b(\boldsymbol{w}(t), \boldsymbol{\sigma}^{hk}_{n-1+\theta} - \boldsymbol{\sigma}(t))\, dt \\ &\le \int_{I_n} b(\boldsymbol{w}(t) - \boldsymbol{v}^h(t), \boldsymbol{\sigma}^{hk}_{n-1+\theta} - \boldsymbol{\sigma}(t_{n-1+\theta}))\, dt \\ &\quad + c\,\omega_k(\boldsymbol{\sigma}) \int_{I_n} \|\boldsymbol{w}(t)\|_V dt \\ &= \int_{I_n} b(\boldsymbol{w}(t) - \boldsymbol{v}^h(t), \theta\,(\boldsymbol{\sigma}_n - \boldsymbol{\sigma}(t_n)) + (1-\theta)\,(\boldsymbol{\sigma}_{n-1} - \boldsymbol{\sigma}(t_{n-1}))\, dt \\ &\quad + \int_{I_n} b(\boldsymbol{w}(t) - \boldsymbol{v}^h(t), E_{n,\theta}(\boldsymbol{\sigma}))\, dt + c\,\omega_k(\boldsymbol{\sigma}) \int_{I_n} \|\boldsymbol{w}(t)\|_V dt \\ &\le c \int_{I_n} \|\boldsymbol{w}(t) - \boldsymbol{v}^h(t)\|_V dt\; (M + \|E_{n,\theta}(\boldsymbol{\Sigma})\|_{\mathcal{T}}) \\ &\quad + c\,\omega_k(\boldsymbol{\sigma}) \int_{I_n} \|\boldsymbol{w}(t)\|_V dt.\end{aligned}$$

To simplify the notation, we define

$$E(\boldsymbol{\Sigma}) = \max_n \|E_{n,\theta}(\boldsymbol{\Sigma})\|_{\mathcal{T}}.$$

Hence, from (13.62), we have

$$\begin{aligned}\|\boldsymbol{e}_n\|_h^2 & - \|\boldsymbol{e}_{n-1}\|_h^2 \\ &\le 2\, A_h(\boldsymbol{e}_n - \boldsymbol{e}_{n-1}, E_{n,\theta}(\boldsymbol{\Sigma})) \\ &\quad + c\left(c^h(\boldsymbol{C}, \boldsymbol{H}) \int_{I_n} \|\dot{\boldsymbol{\Sigma}}(t)\|_{\mathcal{T}} dt + \int_{I_n} \|\boldsymbol{w}(t) - \boldsymbol{v}^h(t)\|_V\, dt\right) \\ &\qquad \times (E(\boldsymbol{\Sigma}) + M) + c\,\omega_k(\boldsymbol{\Sigma}) \int_{I_n} (\|\dot{\boldsymbol{\Sigma}}(t)\|_{\mathcal{T}} + \|\boldsymbol{w}(t)\|_V)\, dt.\end{aligned}$$

Apply the inequality recursively and notice that $\boldsymbol{e}_0 = \boldsymbol{0}$:

$$
\begin{aligned}
\|\boldsymbol{e}_n\|_h^2 \le 2 \sum_{j=1}^{n} A_h(\boldsymbol{e}_j - \boldsymbol{e}_{j-1}, E_{j,\theta}(\boldsymbol{\Sigma})) \\
+ c\left(c^h(\boldsymbol{C},\boldsymbol{H}) \int_0^{t_n} \|\dot{\boldsymbol{\Sigma}}(t)\|_{\mathcal{T}} dt + \int_0^{t_n} \|\boldsymbol{w}(t) - \boldsymbol{v}^h(t)\|_V dt\right) \\
\times (E(\boldsymbol{\Sigma}) + M) + c\,\omega_k(\boldsymbol{\Sigma}) \int_0^{t_n} (\|\dot{\boldsymbol{\Sigma}}(t)\|_{\mathcal{T}} + \|\boldsymbol{w}(t)\|_V)\, dt.
\end{aligned}
$$

We use the identity

$$
\begin{aligned}
&\sum_{j=1}^{n} A_h(\boldsymbol{e}_j - \boldsymbol{e}_{j-1}, E_{j,\theta}(\boldsymbol{\Sigma})) \\
&\quad = A_h(\boldsymbol{e}_n, E_{n,\theta}(\boldsymbol{\Sigma})) + \sum_{j=1}^{n-1} A_h(\boldsymbol{e}_j, E_{j,\theta}(\boldsymbol{\Sigma}) - E_{j+1,\theta}(\boldsymbol{\Sigma}))
\end{aligned}
$$

to obtain

$$
\begin{aligned}
M^2 \le c\,M \Big(\|E_{N,\theta}(\boldsymbol{\Sigma})\|_{\mathcal{T}} + \sum_{n=1}^{N-1} \|E_{n,\theta}(\boldsymbol{\Sigma}) - E_{n+1,\theta}(\boldsymbol{\Sigma})\|_{\mathcal{T}} \\
+ c^h(\boldsymbol{C},\boldsymbol{H})\, \|\dot{\boldsymbol{\Sigma}}\|_{L^1(0,T;\mathcal{T})} + \|\boldsymbol{w} - \boldsymbol{v}^h\|_{L^1(0,T;V)}\Big) \\
+ c\left(c^h(\boldsymbol{C},\boldsymbol{H})\, \|\dot{\boldsymbol{\Sigma}}\|_{L^1(0,T;\mathcal{T})} + \|\boldsymbol{w} - \boldsymbol{v}^h\|_{L^1(0,T;V)}\right) E(\boldsymbol{\Sigma}) \\
+ c\,\omega_k(\boldsymbol{\Sigma}) \left(\|\dot{\boldsymbol{\Sigma}}\|_{L^1(0,T;\mathcal{T})} + \|\boldsymbol{w}\|_{L^1(0,T;V)}\right).
\end{aligned}
$$

Now we apply the inequality (11.3) to obtain

$$
\begin{aligned}
M^2 \le c \Big(\|E_{N,\theta}(\boldsymbol{\Sigma})\|_{\mathcal{T}} + \sum_{n=1}^{N-1} \|E_{n,\theta}(\boldsymbol{\Sigma}) - E_{n+1,\theta}(\boldsymbol{\Sigma})\|_{\mathcal{T}} \\
+ c^h(\boldsymbol{C},\boldsymbol{H})\, \|\dot{\boldsymbol{\Sigma}}\|_{L^1(0,T;\mathcal{T})} + \|\boldsymbol{w} - \boldsymbol{v}^h\|_{L^1(0,T;V)}\Big) \\
+ c\Big\{\Big(c^h(\boldsymbol{C},\boldsymbol{H})\, \|\dot{\boldsymbol{\Sigma}}\|_{L^1(0,T;\mathcal{T})} + \|\boldsymbol{w} - \boldsymbol{v}^h\|_{L^1(0,T;V)}\Big) E(\boldsymbol{\Sigma}) \\
+ \omega_k(\boldsymbol{\Sigma}) \left(\|\dot{\boldsymbol{\Sigma}}\|_{L^1(0,T;\mathcal{T})} + \|\boldsymbol{w}\|_{L^1(0,T;V)}\right)\Big\}^{1/2}.
\end{aligned}
$$

Since $\boldsymbol{v}^h \in L^1(0,T;V^h)$ is arbitrary, we then get the estimate

$$
\begin{aligned}
&\max_{0\le n\le N} \|\boldsymbol{\Sigma}(t_n) - \boldsymbol{\Sigma}_n^{hk}\|_h^2 \\
&\quad \le c\Big(\|E_{N,\theta}(\boldsymbol{\Sigma})\|_{\mathcal{T}} + \sum_{n=1}^{N-1} \|E_{n,\theta}(\boldsymbol{\Sigma}) - E_{n+1,\theta}(\boldsymbol{\Sigma})\|_{\mathcal{T}} \\
&\quad\quad + c^h(\boldsymbol{C},\boldsymbol{H})\,\|\dot{\boldsymbol{\Sigma}}\|_{L^1(0,T;\mathcal{T})} + \inf_{\boldsymbol{v}^h\in L^1(0,T;V^h)} \|\boldsymbol{w} - \boldsymbol{v}^h\|_{L^1(0,T;V)}\Big) \\
&\quad\quad + c\Big\{\Big(c^h(\boldsymbol{C},\boldsymbol{H})\,\|\dot{\boldsymbol{\Sigma}}\|_{L^1(0,T;\mathcal{T})} + \inf_{\boldsymbol{v}^h\in L^1(0,T;V^h)} \|\boldsymbol{w} - \boldsymbol{v}^h\|_{L^1(0,T;V)}\Big)\, E(\boldsymbol{\Sigma}) \\
&\quad\quad\quad + \max_{0\le n\le N} \|E_{n,\theta}(\boldsymbol{\Sigma})\|_{\mathcal{T}} \inf_{\boldsymbol{v}^h\in L^1(0,T;V^h)} \|\boldsymbol{w} - \boldsymbol{v}^h\|_{L^1(0,T;V)} \\
&\quad\quad\quad + \omega_k(\boldsymbol{\Sigma})\,\big(\|\dot{\boldsymbol{\Sigma}}\|_{L^1(0,T;\mathcal{T})} + \|\boldsymbol{w}\|_{L^1(0,T;V)}\big)\Big\}^{1/2}. \qquad (13.63)
\end{aligned}
$$

Now for any $\varepsilon > 0$, there exists $\overline{\boldsymbol{w}} \in L^2(0,T;(H^2(\Omega))^3)$ such that

$$\|\boldsymbol{w} - \overline{\boldsymbol{w}}\|_{L^2(0,T;V)} < \varepsilon.$$

By the finite element interpolation error estimates, for a.a. $t \in (0,T)$,

$$\inf_{\boldsymbol{v}^h(t)\in V^h} \|\overline{\boldsymbol{w}}(t) - \boldsymbol{v}^h(t)\|_V \le c\,h\,\|\overline{\boldsymbol{w}}(t)\|_{(H^2(\Omega))^3}.$$

Then

$$
\begin{aligned}
&\inf_{\boldsymbol{v}^h\in L^1(0,T;V^h)} \|\boldsymbol{w} - \boldsymbol{v}^h\|_{L^1(0,T;V)} \\
&\quad \le \|\boldsymbol{w} - \overline{\boldsymbol{w}}\|_{L^1(0,T;V)} + \inf_{\boldsymbol{v}^h\in L^1(0,T;V^h)} \|\overline{\boldsymbol{w}} - \boldsymbol{v}^h\|_{L^1(0,T;V)} \\
&\quad \le c\,\varepsilon + c\,h\,\|\overline{\boldsymbol{w}}\|_{L^1(0,T;(H^2(\Omega))^3)}.
\end{aligned}
$$

The other terms on the right-hand side of (13.63) can be estimated as in Section 13.1 for the time-discrete schemes. Recall that $\|\cdot\|_h$ is a norm equivalent to $\|\cdot\|_{\mathcal{T}}$ with the equivalence constants independent of h. Therefore, we have proved the following convergence result.

THEOREM 13.8. *Under the basic regularity condition* $(\boldsymbol{u},\boldsymbol{\Sigma}) \in H^1(0,T;V\times \mathcal{T})$*, the fully discrete solution defined in the problem* DUAL^{hk} *converges:*

$$\max_{0\le n\le N} \|\boldsymbol{\Sigma}(t_n) - \boldsymbol{\Sigma}_n^{hk}\|_{\mathcal{T}} \to 0 \quad \text{as } k,h\to 0.$$

13.4 Predictor–Corrector Iterations

In actual computations, usually the discrete schemes discussed in Section 13.3 are not implemented directly, because of the large size of the discrete problems. What is done in practice is to use an iteration procedure to split the task of computing the generalized stress and the displacement. The iteration procedures used are all of the predictor–corrector type. In this section we formulate and analyze some predictor–corrector algorithms for solving the discrete problems discussed in the last section. We focus on algorithms that are in common use in current computational practice (see, for example, [114]). The presentation will be given for solving one step in the backward Euler time-discrete approximation of the dual problem, cf. the scheme DUAL_1^k; the treatment of other time-discrete and fully discrete approximations can be discussed similarly.

For convenience of discussion, we first formulate the dual problem DUAL in an equivalent form. We set

$$\boldsymbol{E}(\boldsymbol{u}) = (\boldsymbol{\epsilon}(\boldsymbol{u}), \boldsymbol{0}) \quad \text{and} \quad \boldsymbol{\Sigma}^e = (\boldsymbol{C}\boldsymbol{\epsilon}(\boldsymbol{u}), \boldsymbol{0}) = \boldsymbol{G}\,\boldsymbol{E}(\boldsymbol{u}),$$

where $\boldsymbol{G} = \operatorname{diag}[\boldsymbol{C}, \boldsymbol{H}]$. Then the variational inequality (13.2) can be rewritten as

$$A(\dot{\boldsymbol{\Sigma}}^e(t) - \dot{\boldsymbol{\Sigma}}(t), \boldsymbol{T} - \boldsymbol{\Sigma}(t)) \le 0 \quad \forall \boldsymbol{T} = (\boldsymbol{\tau}, \boldsymbol{\mu}) \in \mathcal{P}. \tag{13.64}$$

It is easy to see that the scheme DUAL_1^k can be rewritten as the following scheme DUAL^k, with $\boldsymbol{\ell}_n = \boldsymbol{\ell}(t_n)$ and the relation between $\boldsymbol{w}_n^k$ and $\boldsymbol{u}_n^k$ defined by

$$\boldsymbol{u}_n^k = k \sum_{j=1}^n \boldsymbol{w}_j^k, \quad n = 1, \ldots, N.$$

PROBLEM DUAL^k. Find $\{(\boldsymbol{u}_n^k, \boldsymbol{\Sigma}_n^k) = (\boldsymbol{u}_n^k, \boldsymbol{\sigma}_n^k, \boldsymbol{\chi}_n^k)\}_{n=0}^N \subset V \times \mathcal{P}$, with $(\boldsymbol{u}_0^k, \boldsymbol{\Sigma}_0^k) = \boldsymbol{0}$, such that for $n = 1, \ldots, N$,

$$b(\boldsymbol{v}, \boldsymbol{\sigma}_n^k) = \langle \boldsymbol{\ell}_n, \boldsymbol{v} \rangle \quad \forall \boldsymbol{v} \in V, \tag{13.65}$$

$$A(\boldsymbol{\Sigma}_n^{tr,k} - \boldsymbol{\Sigma}_n^k, \boldsymbol{T} - \boldsymbol{\Sigma}_n^k) \le 0 \quad \forall \boldsymbol{T} \in \mathcal{P}, \tag{13.66}$$

in which

$$\boldsymbol{\Sigma}_n^{tr,k} = \boldsymbol{\Sigma}_{n-1}^k + \boldsymbol{G}\,\Delta \boldsymbol{E}_n^k \tag{13.67}$$

and

$$\Delta \boldsymbol{E}_n^k = \boldsymbol{E}(\boldsymbol{u}_n^k) - \boldsymbol{E}(\boldsymbol{u}_{n-1}^k) = (\boldsymbol{C}\,\boldsymbol{\epsilon}(\boldsymbol{u}_n^k - \boldsymbol{u}_{n-1}^k), \boldsymbol{0}).$$

From the proof of Theorem 8.12 we see that under the assumptions of Theorem 8.12, the problem DUAL^k has a solution, and one can show that $\{\boldsymbol{\Sigma}_n^k\}_{n=1}^N \subset \mathcal{P}$ is unique. It seems difficult to prove the uniqueness of the sequence $\{\boldsymbol{u}_n^k\}_{n=1}^N$ directly, although we have seen in Theorem 8.12 that the limit of the sequence as $k \to 0$ is unique.

We note that once $\boldsymbol{\Sigma}_n^{tr,k}$ is known, the variational inequality (13.66) is equivalent to the minimization problem

$$J(\boldsymbol{T}) \equiv \tfrac{1}{2}\,\|\boldsymbol{\Sigma}_n^{tr,k} - \boldsymbol{T}\|_A^2 \to \inf, \quad \boldsymbol{T} \in \mathcal{P}, \tag{13.68}$$

where $\|\cdot\|_A$ is the norm induced by the bilinear form A, as well as to the projection problem

$$\boldsymbol{\Sigma}_n^k = \Pi_{\mathcal{P},A}\boldsymbol{\Sigma}_n^{tr,k}, \tag{13.69}$$

where, $\Pi_{\mathcal{P},A}$ denotes the projection operator onto $\mathcal{P}$ with respect to the inner product $(\cdot,\cdot)_A$.

The algorithms to be discussed here for solving (13.65)–(13.66) are all of the predictor–corrector type. Each iteration consists of a predictor step and a corrector step. In the predictor step, we update the quantity $\boldsymbol{u}_n^k$ by using the equation (13.65). Then we compute an updated value for $\boldsymbol{\Sigma}_n^{tr,k}$. In the corrector step we solve (13.66) (equivalently, (13.68) or (13.69)) to get a new iterate for $\boldsymbol{\Sigma}_n^k$. As with the primal problem, a variety of solution algorithms can be developed by using different schemes to update $\boldsymbol{u}_n^k$ in the predictor step. We will consider two types of predictors: the elastic predictor and a consistent tangent predictor.

In the literature (for example, [114]), an implementation of the corrector step is usually called a return map algorithm. Using an updated value for $\boldsymbol{u}_n^k$ from the predictor step, one computes the corresponding updated strain increment $\Delta\boldsymbol{E}_n^k$. Then an updated trial state $\boldsymbol{\Sigma}_n^{tr,k}$ for the generalized stress is calculated by the formula (13.67). If the trial state $\boldsymbol{\Sigma}_n^{tr,k}$ belongs to $\mathcal{P}$, then $\boldsymbol{\Sigma}_n^k = \boldsymbol{\Sigma}_n^{tr,k}$ is the solution, and we can move on to solve (13.65)–(13.66) for the next time level $n+1$. In general, however, the updated trial state lies outside the admissible region $\mathcal{P}$. The purpose of a corrector step is then to find a point in $\mathcal{P}$ that is close to the trial state in some sense; that is, the corrector step *returns* the iteration to some point in $\mathcal{P}$. This is also evident from the form of the projection problem (13.69).

The elastic predictor. For brevity in exposition, we drop the superscript k. A superscript i will be used later as the iteration counter in the algorithm.

We begin by returning to (13.1). Since the stress $\boldsymbol{\sigma}$ is implicitly a function of the displacement $\boldsymbol{u}$, we replace $\boldsymbol{\sigma}$ in (13.1) by $\boldsymbol{\sigma}^i$, the ith iterate, which is defined by

$$\boldsymbol{\sigma}^i \equiv \boldsymbol{\sigma}(\boldsymbol{\epsilon}(\boldsymbol{u}^i)) \approx \boldsymbol{\sigma}(\boldsymbol{\epsilon}(\boldsymbol{u}^{i-1})) + \boldsymbol{D}\boldsymbol{\epsilon}(\boldsymbol{u}^i - \boldsymbol{u}^{i-1});$$

the predictor step will be referred to as an elastic predictor by virtue of the fact that the modulus $\boldsymbol{D}$ will be assumed to be time-independent and to be related to the elastic modulus $\boldsymbol{C}$ in a definite way. Later, we will provide conditions on $\boldsymbol{D}$ that are sufficient to guarantee the convergence of the predictor–corrector algorithm.

From a known iterate $(\boldsymbol{u}_n^{i-1}, \boldsymbol{\Sigma}_n^{i-1})$, we therefore update $\boldsymbol{u}_n$ by solving the equation

$$\int_\Omega \boldsymbol{D}\boldsymbol{\epsilon}(\boldsymbol{u}_n^i - \boldsymbol{u}_n^{i-1}) : \boldsymbol{\epsilon}(\boldsymbol{v})\,dx = b(\boldsymbol{v}, \boldsymbol{\sigma}_n^{i-1}) - \langle \boldsymbol{\ell}_n, \boldsymbol{v}\rangle \quad \forall \boldsymbol{v} \in V$$

for $\boldsymbol{u}_n^i$. This equation may be regarded as an approximation to (13.1) at $t = t_n$, in which $\boldsymbol{\sigma}(t_n)$ is replaced by a first-order approximation $\boldsymbol{\sigma}_n^i \equiv \boldsymbol{\sigma}(\boldsymbol{\epsilon}(\boldsymbol{u}_n^i)) \approx \boldsymbol{\sigma}(\boldsymbol{\epsilon}(\boldsymbol{u}_n^{i-1})) + \boldsymbol{D}\boldsymbol{\epsilon}(\boldsymbol{u}_n^i - \boldsymbol{u}_n^{i-1})$. Thus, once $(\boldsymbol{u}_{n-1}, \boldsymbol{\Sigma}_{n-1})$ is known, a predictor–corrector algorithm for computing $(\boldsymbol{u}_n, \boldsymbol{\Sigma}_n)$ is the following procedure.

Initialization: $\boldsymbol{u}_n^0 = \boldsymbol{u}_{n-1}$, $\boldsymbol{\Sigma}_n^0 = \boldsymbol{\Sigma}_{n-1}$.

Iteration: For $i = 1, 2, \ldots,$

Predictor: Compute $\boldsymbol{u}_n^i \in V$ satisfying

$$\int_\Omega \boldsymbol{D}\boldsymbol{\epsilon}(\boldsymbol{u}_n^i - \boldsymbol{u}_n^{i-1}) : \boldsymbol{\epsilon}(\boldsymbol{v})\,dx = b(\boldsymbol{v}, \boldsymbol{\sigma}_n^{i-1}) - \langle \boldsymbol{\ell}_n, \boldsymbol{v}\rangle \quad \forall \boldsymbol{v} \in V, \tag{13.70}$$

and a trial state

$$\boldsymbol{\Sigma}_n^{tr,i} = \boldsymbol{\Sigma}_{n-1} + (\boldsymbol{C}\,\boldsymbol{\epsilon}(\boldsymbol{u}_n^i - \boldsymbol{u}_{n-1}), \boldsymbol{0}). \tag{13.71}$$

Corrector: Find $\boldsymbol{\Sigma}_n^i \in \mathcal{P}$ such that

$$A(\boldsymbol{\Sigma}_n^{tr,i} - \boldsymbol{\Sigma}_n^i, \boldsymbol{T} - \boldsymbol{\Sigma}_n^i) \le 0 \quad \forall \boldsymbol{T} \in \mathcal{P}. \tag{13.72}$$

Convergence of the elastic predictor. We first note that (13.66) can be rewritten in the alternative form (with the superscript k omitted)

$$A(\boldsymbol{\Sigma}_n, \boldsymbol{T} - \boldsymbol{\Sigma}_n) + b(\boldsymbol{u}_n, \boldsymbol{\tau} - \boldsymbol{\sigma}_n) \ge \langle \boldsymbol{L}_n, \boldsymbol{T} - \boldsymbol{\Sigma}_n\rangle \quad \forall \boldsymbol{T} \in \mathcal{P},$$

where $\boldsymbol{L}_n$ is a continuous linear form on $\mathcal{T}$ defined by

$$\langle \boldsymbol{L}_n, \boldsymbol{T}\rangle = A(\boldsymbol{\Sigma}_{n-1}, \boldsymbol{T}) + b(\boldsymbol{u}_{n-1}, \boldsymbol{\tau}).$$

To further simplify the notation, we will also drop the subscript n in the convergence analysis. Thus, given the continuous linear forms $\boldsymbol{\ell}$ and $\boldsymbol{L}$, the problem is to find $\boldsymbol{u} \in V$ and $\boldsymbol{\Sigma} = (\boldsymbol{\sigma}, \boldsymbol{\chi}) \in \mathcal{P}$ such that

$$b(\boldsymbol{v}, \boldsymbol{\sigma}) = \langle \boldsymbol{\ell}, \boldsymbol{v}\rangle \quad \forall \boldsymbol{v} \in V, \tag{13.73}$$

$$A(\boldsymbol{\Sigma}, \boldsymbol{T} - \boldsymbol{\Sigma}) + b(\boldsymbol{u}, \boldsymbol{\tau} - \boldsymbol{\sigma}) \ge \langle \boldsymbol{L}, \boldsymbol{T} - \boldsymbol{\Sigma}\rangle \quad \forall \boldsymbol{T} \in \mathcal{P}. \tag{13.74}$$

We assume that the problem has a solution $(\boldsymbol{u}, \boldsymbol{\Sigma})$, and that $\boldsymbol{\Sigma}$ is unique.

Given a modulus $\boldsymbol{D}$, independent of time, and an initial guess $(\boldsymbol{u}^0, \boldsymbol{\Sigma}^0) \in V \times \mathcal{P}$, the predictor–corrector algorithm for solving the problem (13.73)–(13.74) is this: For $i = 1, 2, \ldots,$ compute $\boldsymbol{u}^i \in V$ such that

$$\int_\Omega \boldsymbol{D}\boldsymbol{\epsilon}(\boldsymbol{u}^i - \boldsymbol{u}^{i-1}) : \boldsymbol{\epsilon}(\boldsymbol{v})\,dx = b(\boldsymbol{v}, \boldsymbol{\sigma}^{i-1}) - \langle \boldsymbol{\ell}, \boldsymbol{v}\rangle \quad \forall \boldsymbol{v} \in V, \tag{13.75}$$

and then compute $\mathbf{\Sigma}^i \in \mathcal{P}$ such that

$$A(\mathbf{\Sigma}^i, \boldsymbol{T} - \mathbf{\Sigma}^i) + b(\boldsymbol{u}^i, \boldsymbol{\tau} - \boldsymbol{\sigma}^i) \geq \langle \boldsymbol{L}, \boldsymbol{T} - \mathbf{\Sigma}^i \rangle \quad \forall \boldsymbol{T} = (\boldsymbol{\tau}, \boldsymbol{\mu}) \in \mathcal{P}. \tag{13.76}$$

We have the following result on convergence of the algorithm.

THEOREM 13.9. *Assume that the modulus $\boldsymbol{D}$ is chosen in such a way that it is symmetric, uniformly bounded, pointwise stable in the sense that for some constant $c > 0$,*

$$\boldsymbol{D}(\boldsymbol{x})\boldsymbol{\xi} : \boldsymbol{\xi} \geq c\, |\boldsymbol{\xi}|^2 \quad \forall \boldsymbol{\xi} \in M^3, \text{ a.e. } \boldsymbol{x} \in \Omega, \tag{13.77}$$

and such that its inverse $\boldsymbol{D}^{-1}$ is uniformly dominated by $\boldsymbol{C}^{-1}$ (or equivalently, $\boldsymbol{C}$ is uniformly dominated by $\boldsymbol{D}$) in the sense that for some constant $\alpha > 0$,

$$\boldsymbol{\xi} : \left(\boldsymbol{C}^{-1}(\boldsymbol{x}) - \boldsymbol{D}^{-1}(\boldsymbol{x})\right) \boldsymbol{\xi} \geq \alpha\, |\boldsymbol{\xi}|^2 \quad \forall \boldsymbol{\xi} \in M^3, \text{ a.e. } \boldsymbol{x} \in \Omega. \tag{13.78}$$

Then

$$\mathbf{\Sigma}^i \to \mathbf{\Sigma} \quad \text{as } i \to \infty,$$

and for some subsequence $\{\boldsymbol{u}^{i_j}\}_j$ of $\{\boldsymbol{u}^i\}_i$ and some element $\tilde{\boldsymbol{u}} \in V$,

$$\boldsymbol{u}^{i_j} \rightharpoonup \tilde{\boldsymbol{u}} \quad \text{as } j \to \infty.$$

The limits $\tilde{\boldsymbol{u}} \in V$ and $\mathbf{\Sigma} \in \mathcal{P}$ together solve the problem (13.73)–(13.74).

PROOF. Under the given assumptions, the iterative procedure given by (13.75) and (13.76) is well-defined. Set

$$\overline{\mathbf{\Sigma}}^i = \mathbf{\Sigma}^i - \mathbf{\Sigma} \quad \text{and} \quad \overline{\boldsymbol{u}}^i = \boldsymbol{u}^i - \boldsymbol{u}.$$

From (13.75) and (13.73) we find that

$$\int_\Omega \boldsymbol{D}\boldsymbol{\epsilon}(\overline{\boldsymbol{u}}^i - \overline{\boldsymbol{u}}^{i-1}) : \boldsymbol{\epsilon}(\boldsymbol{v})\, dx = b(\boldsymbol{v}, \overline{\boldsymbol{\sigma}}^{i-1}) = -\int_\Omega \overline{\boldsymbol{\sigma}}^{i-1} : \boldsymbol{\epsilon}(\boldsymbol{v})\, dx \quad \forall \boldsymbol{v} \in V.$$

Thus we get an important relation

$$\boldsymbol{D}\boldsymbol{\epsilon}(\overline{\boldsymbol{u}}^i - \overline{\boldsymbol{u}}^{i-1}) = -\overline{\boldsymbol{\sigma}}^{i-1}. \tag{13.79}$$

Since $\boldsymbol{D}$ is a symmetric positive definite operator, we can define its square root operator $\boldsymbol{D}^{1/2}$, which is also symmetric and positive definite. In particular, from (13.79) we have

$$\boldsymbol{D}^{1/2}\boldsymbol{\epsilon}(\overline{\boldsymbol{u}}^i) = \boldsymbol{D}^{1/2}\boldsymbol{\epsilon}(\overline{\boldsymbol{u}}^{i-1}) - \boldsymbol{D}^{-1/2}\overline{\boldsymbol{\sigma}}^{i-1}.$$

Taking the inner product of the relation with itself and integrating over Ω, we obtain

$$\int_\Omega \boldsymbol{D}\boldsymbol{\epsilon}(\overline{\boldsymbol{u}}^i) : \boldsymbol{\epsilon}(\overline{\boldsymbol{u}}^i)\,dx - \int_\Omega \boldsymbol{D}\boldsymbol{\epsilon}(\overline{\boldsymbol{u}}^{i-1}) : \boldsymbol{\epsilon}(\overline{\boldsymbol{u}}^{i-1})\,dx$$
$$= \int_\Omega \boldsymbol{D}^{-1}\overline{\boldsymbol{\sigma}}^{i-1} : \overline{\boldsymbol{\sigma}}^{i-1}\,dx - \int_\Omega \boldsymbol{\epsilon}(\overline{\boldsymbol{u}}^{i-1}) : \overline{\boldsymbol{\sigma}}^{i-1}\,dx. \quad (13.80)$$

Now take $\boldsymbol{T} = \boldsymbol{\Sigma}^i$ in (13.74) to obtain

$$A(\boldsymbol{\Sigma}, \boldsymbol{\Sigma}^i - \boldsymbol{\Sigma}) + b(\boldsymbol{u}, \boldsymbol{\sigma}^i - \boldsymbol{\sigma}) \geq \langle \boldsymbol{L}, \boldsymbol{\Sigma}^i - \boldsymbol{\Sigma}\rangle,$$

and take $\boldsymbol{T} = \boldsymbol{\Sigma}$ in (13.76) to obtain

$$A(\boldsymbol{\Sigma}^i, \boldsymbol{\Sigma} - \boldsymbol{\Sigma}^i) + b(\boldsymbol{u}^i, \boldsymbol{\sigma} - \boldsymbol{\sigma}^i) \geq \langle \boldsymbol{L}, \boldsymbol{\Sigma} - \boldsymbol{\Sigma}^i\rangle.$$

Adding these two inequalities we find that

$$A(\overline{\boldsymbol{\Sigma}}^i, \overline{\boldsymbol{\Sigma}}^i) \leq -b(\overline{\boldsymbol{u}}^i, \overline{\boldsymbol{\sigma}}^i) = \int_\Omega \boldsymbol{\epsilon}(\overline{\boldsymbol{u}}^i) : \overline{\boldsymbol{\sigma}}^i\,dx. \quad (13.81)$$

Combining (13.80) and (13.81), we get

$$\int_\Omega \boldsymbol{D}\boldsymbol{\epsilon}(\overline{\boldsymbol{u}}^i) : \boldsymbol{\epsilon}(\overline{\boldsymbol{u}}^i)\,dx - \int_\Omega \boldsymbol{D}\boldsymbol{\epsilon}(\overline{\boldsymbol{u}}^{i-1}) : \boldsymbol{\epsilon}(\overline{\boldsymbol{u}}^{i-1})\,dx$$
$$\leq \int_\Omega \boldsymbol{D}^{-1}\overline{\boldsymbol{\sigma}}^{i-1} : \overline{\boldsymbol{\sigma}}^{i-1}\,dx - A(\overline{\boldsymbol{\Sigma}}^{i-1}, \overline{\boldsymbol{\Sigma}}^{i-1}).$$

By the assumption that $\boldsymbol{H}$ is positive definite and (13.78), we then have, for some constant $c > 0$,

$$\int_\Omega \boldsymbol{D}\boldsymbol{\epsilon}(\overline{\boldsymbol{u}}^i) : \boldsymbol{\epsilon}(\overline{\boldsymbol{u}}^i)\,dx - \int_\Omega \boldsymbol{D}\boldsymbol{\epsilon}(\overline{\boldsymbol{u}}^{i-1}) : \boldsymbol{\epsilon}(\overline{\boldsymbol{u}}^{i-1})\,dx \leq -c\,\|\overline{\boldsymbol{\Sigma}}^{i-1}\|_{\mathcal{T}}^2. \quad (13.82)$$

The first consequence drawn from (13.82) is that the nonnegative sequence $\{\int_\Omega \boldsymbol{D}\boldsymbol{\epsilon}(\overline{\boldsymbol{u}}^i) : \boldsymbol{\epsilon}(\overline{\boldsymbol{u}}^i)\,dx\}_i$ is nonincreasing, and thus has a limit. Then again from (13.82), we see that

$$\|\overline{\boldsymbol{\Sigma}}^{i-1}\|_{\mathcal{T}} \to 0 \quad \text{as } i \to \infty,$$

that is, we have proved that

$$\boldsymbol{\Sigma}^i \to \boldsymbol{\Sigma} \quad \text{as } i \to \infty.$$

Moreover, since $\{\int_\Omega \boldsymbol{D}\boldsymbol{\epsilon}(\overline{\boldsymbol{u}}^i) : \boldsymbol{\epsilon}(\overline{\boldsymbol{u}}^i)\,dx\}_i$ is a nonincreasing sequence, and since $\boldsymbol{D}$ satisfies (13.77), we see that $\{\overline{\boldsymbol{u}}^i\}_i$ is a bounded sequence in V. Equivalently, the sequence $\{\boldsymbol{u}^i\}_i \subset V$ is bounded. Recall that V is a Hilbert space and is hence reflexive. Thus, we can find a subsequence $\{\boldsymbol{u}^{i_j}\}_j \subset \{\boldsymbol{u}^i\}_i$ and an element $\tilde{\boldsymbol{u}} \in V$, such that

$$\boldsymbol{u}^{i_j} \rightharpoonup \tilde{\boldsymbol{u}} \quad \text{in } V \text{ as } j \to \infty.$$

By (13.79),

$$\boldsymbol{u}^{i_j} - \boldsymbol{u}^{i_j - 1} \to 0 \quad \text{in } V \text{ as } j \to \infty.$$

Thus if we take the limit along the subsequence $\{i_j\}_j$ in (13.75) and (13.76), we find that the limit $(\tilde{\boldsymbol{u}}, \boldsymbol{\Sigma})$ satisfies (13.73) and (13.74). □

We observe that if a solution of the problem (13.73)–(13.74) is unique, then the whole sequence $\{\boldsymbol{u}^i\}_i$ converges weakly to $\boldsymbol{u}$.

We remark that it is not difficult to choose $\boldsymbol{D}$ such that both (13.77) and (13.78) are satisfied. For example, we may take $\boldsymbol{D} = \kappa\, \boldsymbol{C}$ for some $\kappa > 1$, or we may take $\boldsymbol{D} = \kappa\, \boldsymbol{I}$ with $\kappa > 1/C_0'$. Here C_0' is the constant in the inequality for the positive definiteness of the tensor $\boldsymbol{C}^{-1}$ (cf. Section 7.1).

Tangent predictor. This predictor takes as a starting point a first-order Taylor expansion of $\boldsymbol{\sigma}$, in which the modulus $\boldsymbol{D}$ introduced earlier is replaced by an appropriate tangent modulus. We follow the derivation of the symmetric consistent tangent modulus in Simo and Govindjee [115] and Simo [114] to obtain a formula for the tangent predictor. By the expression (13.69) and the formula (13.67), we see that $\boldsymbol{\Sigma}_n$ is a nonlinear function of $\boldsymbol{u}_n$. Here and below, we once again omit the superscript k. And we use i for the iteration index.

Assume that the $(i-1)$th iterate $(\boldsymbol{u}_n^{i-1}, \boldsymbol{\Sigma}_n^{i-1})$ is known. We will use (13.73) to update $\boldsymbol{u}_n$. By Taylor's expansion, we have the relation

$$\boldsymbol{\sigma}(\boldsymbol{\epsilon}(\boldsymbol{u}_n^i)) \approx \boldsymbol{\sigma}(\boldsymbol{\epsilon}(\boldsymbol{u}_n^{i-1})) + \frac{\partial \boldsymbol{\sigma}}{\partial \boldsymbol{\epsilon}}(\boldsymbol{\epsilon}(\boldsymbol{u}_n^{i-1})) : \boldsymbol{\epsilon}(\boldsymbol{u}_n^i - \boldsymbol{u}_n^{i-1}). \tag{13.83}$$

Now the question is how to find (an approximate value of) the quantity $\dfrac{\partial \boldsymbol{\sigma}}{\partial \boldsymbol{\epsilon}}(\boldsymbol{\epsilon}(\boldsymbol{u}_n^{i-1}))$. We start with the relation

$$\boldsymbol{\Sigma}_n = \boldsymbol{G}(\boldsymbol{E}_n - \boldsymbol{P}_n),$$

which comes from

$$\boldsymbol{\sigma} = \boldsymbol{C}(\boldsymbol{\epsilon}(\boldsymbol{u}) - \boldsymbol{p}) \quad \text{and} \quad \boldsymbol{\chi} = -\boldsymbol{H}\boldsymbol{\xi},$$

and the notation $\boldsymbol{P} = (\boldsymbol{p}, \boldsymbol{\xi})$. Here and below, we use the short-hand notation $\boldsymbol{\epsilon}_n = \boldsymbol{\epsilon}(\boldsymbol{u}_n)$, $\boldsymbol{E}_n = \boldsymbol{E}(\boldsymbol{u}_n)$, $\boldsymbol{\Sigma}_n = \boldsymbol{\Sigma}(\boldsymbol{\epsilon}_n)$, and $\boldsymbol{P}_n = \boldsymbol{P}(\boldsymbol{u}_n)$. We take the differentials of both sides of the relation to obtain

$$d\boldsymbol{\Sigma}_n = \boldsymbol{G}(d\boldsymbol{E}_n - d\boldsymbol{P}_n). \tag{13.84}$$

Thus we need to find an (approximate) expression of $d\boldsymbol{P}_n$ in terms of $d\boldsymbol{E}_n$. Using the relation

$$\dot{\boldsymbol{P}} = \lambda\, \nabla\phi(\boldsymbol{\Sigma}),$$

we get

$$\dot{\boldsymbol{P}}_n = \lambda_n\, \nabla\phi(\boldsymbol{\Sigma}_n).$$

Hence

$$\boldsymbol{P}_n - \boldsymbol{P}_{n-1} \approx \Delta\lambda_n\, \nabla\phi(\boldsymbol{\Sigma}_n),$$

where $\Delta\lambda_n = k\,\lambda_n$. The quantity $\boldsymbol{P}_{n-1}$ is assumed given. So, taking the differential of the above relation, we have

$$d\boldsymbol{P}_n \approx d(\Delta\lambda_n)\, \nabla\phi(\boldsymbol{\Sigma}_n) + \Delta\lambda_n\, \nabla^2\phi(\boldsymbol{\Sigma}_n)\, d\boldsymbol{\Sigma}_n.$$

Substitution of this relation into (13.84) yields

$$d\boldsymbol{\Sigma}_n \approx \boldsymbol{G}(d\boldsymbol{E}_n - d(\Delta\lambda_n)\, \nabla\phi(\boldsymbol{\Sigma}_n) - \Delta\lambda_n\, \nabla^2\phi(\boldsymbol{\Sigma}_n)\, d\boldsymbol{\Sigma}_n).$$

Hence

$$d\boldsymbol{\Sigma}_n \approx \mathcal{G}_n \left[d\boldsymbol{E}_n - d(\Delta\lambda_n)\, \nabla\phi(\boldsymbol{\Sigma}_n)\right], \tag{13.85}$$

where

$$\mathcal{G}_n = [\boldsymbol{G}^{-1} + \Delta\lambda_n\, \nabla^2\phi(\boldsymbol{\Sigma}_n)]^{-1}. \tag{13.86}$$

Thus the problem is reduced to one of finding an (approximate) expression for $d(\Delta\lambda_n)$ in terms of $d\boldsymbol{E}_n$. As in [115], we determine $d(\Delta\lambda_n)$ by enforcing the condition

$$d\phi(\boldsymbol{\Sigma}_n) = \nabla\phi(\boldsymbol{\Sigma}_n) : d\boldsymbol{\Sigma}_n = 0. \tag{13.87}$$

From (13.85) and (13.87) we find that

$$d(\Delta\lambda_n) \approx \frac{\nabla\phi(\boldsymbol{\Sigma}_n) : \mathcal{G}_n d\boldsymbol{E}_n}{\nabla\phi(\boldsymbol{\Sigma}_n) : \mathcal{G}_n \nabla\phi(\boldsymbol{\Sigma}_n)}. \tag{13.88}$$

Combining (13.85) and (13.88), we obtain the formula

$$d\boldsymbol{\Sigma}_n \approx [\mathcal{G}_n - \boldsymbol{N}_n \otimes \boldsymbol{N}_n]\, d\boldsymbol{E}_n, \tag{13.89}$$

where

$$\boldsymbol{N}_n = \frac{\mathcal{G}_n \nabla\phi(\boldsymbol{\Sigma}_n)}{\sqrt{\nabla\phi(\boldsymbol{\Sigma}_n) : \mathcal{G}_n \nabla\phi(\boldsymbol{\Sigma}_n)}}. \tag{13.90}$$

The relation (13.89) provides the formula

$$d\boldsymbol{\sigma}_n \approx \boldsymbol{C}_n \, d\boldsymbol{\epsilon}_n, \tag{13.91}$$

in which $\boldsymbol{C}_n$ can be viewed as an approximation of $\partial\boldsymbol{\sigma}/\partial\boldsymbol{\epsilon}(\boldsymbol{\epsilon}(\boldsymbol{u}_n))$.

Thus the tangent predictor step is constructed as follows. Once an iterate $(\boldsymbol{u}_n^{i-1}, \boldsymbol{\Sigma}_n^{i-1})$ is known, we find $\mathcal{G}_n^i$ from (13.86), with $\boldsymbol{\Sigma}_n$ there being replaced by $\boldsymbol{\Sigma}_n^{i-1}$. Then we find $\boldsymbol{N}_n^i$ from its definition (13.90), again with $\boldsymbol{\Sigma}_n$ there being replaced by $\boldsymbol{\Sigma}_n^{i-1}$. Now we have a relation between $d\boldsymbol{\Sigma}_n^i$ and $d\boldsymbol{E}_n^i$ from (13.89), which provides us the tangent modulus $\boldsymbol{C}_n^i$ from (13.91). After the predictor step is completed, we can then apply the corrector step (13.72) to update $\boldsymbol{\Sigma}_n$.

We note that $\mathcal{G}_n$ in (13.86) contains an undetermined scalar $\Delta\lambda_n$. In [115] this scalar is chosen in such a way that the computed value $\boldsymbol{\Sigma}_n$ belongs to $\mathcal{P}$.

As in the case of the primal problem, it is not clear how to prove convergence for the tangent predictor constructed above. In practice, however, it is known that the tangent predictor performs far more efficiently than the elastic predictor, particularly if a line search is incorporated.

Solution algorithms for the generalized midpoint discretization. The solution algorithms discussed so far are in the context of the backward Euler time-discrete approximation of the dual problem DUAL. We may as well consider the more general time-discrete approximations based on the generalized midpoint rule. Let $\theta \in \left[\frac{1}{2}, 1\right]$ be a parameter. With the same uniform partition of the time interval $[0, T]$ given as before, a typical step in solving the problem (13.1)–(13.2) is to find $\boldsymbol{u}_n \in V$ and $\boldsymbol{\Sigma}_n \in \mathcal{T}$ such that $\boldsymbol{\Sigma}_{n-1+\theta} \equiv \theta\,\boldsymbol{\Sigma}_n + (1-\theta)\,\boldsymbol{\Sigma}_{n-1} \in \mathcal{P}$, and

$$b(\boldsymbol{v}, \boldsymbol{\sigma}_n) = \langle \boldsymbol{\ell}_n, \boldsymbol{v}\rangle \quad \forall\, \boldsymbol{v} \in V, \tag{13.92}$$

$$\begin{aligned} &A(\boldsymbol{\Sigma}_n - \boldsymbol{\Sigma}_{n-1}, \boldsymbol{T} - \boldsymbol{\Sigma}_{n-1+\theta}) \\ &\quad + b(\boldsymbol{u}_n - \boldsymbol{u}_{n-1}, \boldsymbol{\tau} - \boldsymbol{\sigma}_{n-1+\theta}) \ge 0 \quad \forall\, \boldsymbol{T} = (\boldsymbol{\tau}, \boldsymbol{\mu}) \in \mathcal{P}. \end{aligned} \tag{13.93}$$

We can rewrite (13.92) and (13.93) as relations in terms of $\boldsymbol{u}_n$ and $\boldsymbol{\Sigma}_{n-1+\theta}$ (rather than $\boldsymbol{\Sigma}_n$). We have

$$\begin{aligned} b(\boldsymbol{v}, \boldsymbol{\sigma}_{n-1+\theta}) &= \theta\,\langle \boldsymbol{\ell}_n, \boldsymbol{v}\rangle + (1-\theta)\, b(\boldsymbol{v}, \boldsymbol{\sigma}_{n-1}) \\ &\quad \forall\, \boldsymbol{v} \in V, \end{aligned} \tag{13.94}$$

$$\begin{aligned} &A(\boldsymbol{\Sigma}_{n-1+\theta} - \boldsymbol{\Sigma}_{n-1}, \boldsymbol{T} - \boldsymbol{\Sigma}_{n-1+\theta}) \\ &\quad + \theta\, b(\boldsymbol{u}_n - \boldsymbol{u}_{n-1}, \boldsymbol{\tau} - \boldsymbol{\sigma}_{n-1+\theta}) \ge 0 \quad \forall\, \boldsymbol{T} = (\boldsymbol{\tau}, \boldsymbol{\mu}) \in \mathcal{P}. \end{aligned} \tag{13.95}$$

The problem (13.94)–(13.95) assumes exactly the form of (13.73) and (13.74). Thus the elastic predictor–corrector algorithm (13.75)–(13.76) can be applied directly to solve the problem (13.94)–(13.95). For the study of convergence, we can apply Theorem 13.6. It is also straightforward to derive a formula for a tangent predictor to solve the problem (13.94)–(13.95).

13.5 Computation of the Closest Point Projections

We observe that the core of a corrector step in the predictor–corrector algorithms discussed above is the solution of a variational inequality of the form

$$\boldsymbol{\Sigma} \in \mathcal{P}, \quad A(\boldsymbol{\Sigma}^{tr} - \boldsymbol{\Sigma}, \boldsymbol{T} - \boldsymbol{\Sigma}) \leq 0 \quad \forall \boldsymbol{T} \in \mathcal{P}. \tag{13.96}$$

This is an elliptic variational inequality of the first kind. Equivalently, the solution $\boldsymbol{\Sigma}$ is the closest-point projection of the trial generalized stress $\boldsymbol{\Sigma}^{tr}$ onto the admissible convex set $\mathcal{P}$. This is a standard problem in convex optimization. Thus, a prototype problem can be described as follows.

PROBLEM. Let $\mathcal{P}$ be a nonempty, closed, convex subset of a Hilbert space $\mathcal{T}$, and $\boldsymbol{G}^{-1}$ a symmetric, positive definite metric on $\mathcal{T}$. Given $\boldsymbol{\Sigma}^{tr} \in \mathcal{T}$, solve the problem

$$\min \left\{ \tfrac{1}{2} \left(\boldsymbol{\Sigma}^{tr} - \boldsymbol{\Sigma}\right) : \boldsymbol{G}^{-1} \left(\boldsymbol{\Sigma}^{tr} - \boldsymbol{\Sigma}\right) : \boldsymbol{\Sigma} \in \mathcal{P} \right\}. \tag{13.97}$$

We discuss a possible solution algorithm for solving the constrained minimization problem. To do this, we need the following equivalence result.

THEOREM 13.10. *$\boldsymbol{\Sigma} \in \mathcal{P}$ is the solution of the problem* (13.97) *if and only if there exists a scalar γ such that*

$$\begin{aligned} \boldsymbol{\Sigma} &= \boldsymbol{\Sigma}^{tr} - \gamma\, \boldsymbol{G}\, \nabla\phi(\boldsymbol{\Sigma}), \\ \phi(\boldsymbol{\Sigma}) \leq 0, &\quad \gamma \geq 0, \quad \gamma\, \phi(\boldsymbol{\Sigma}) = 0. \end{aligned} \tag{13.98}$$

PROOF. Let $\boldsymbol{\Sigma} \in \mathcal{P}$ be the solution of the problem (13.97). The Lagrangian associated with the constrained minimization problem is

$$\mathcal{L}(\boldsymbol{\Sigma}, \gamma) = \tfrac{1}{2} \left(\boldsymbol{\Sigma}^{tr} - \boldsymbol{\Sigma}\right) : \boldsymbol{G}^{-1} \left(\boldsymbol{\Sigma}^{tr} - \boldsymbol{\Sigma}\right) + \gamma\, f(\boldsymbol{\Sigma}).$$

By the Kuhn–Tucker optimality condition, we get (13.98).

Conversely, assume that (13.98) is satisfied for some $\boldsymbol{\Sigma} \in \mathcal{P}$ and some scalar γ. By the second-order sufficiency conditions [81], $\boldsymbol{\Sigma} \in \mathcal{P}$ is the solution of the constrained minimization problem. □

Following [114], an application of Newton's method to solve the system (13.98) results in a solution algorithm.

STEP 1. *Initialization.* Let $\delta > 0$ be a given error tolerance. Set $k = 0$, $\boldsymbol{\Sigma}^{(0)} = \boldsymbol{\Sigma}^{tr}$, and $\gamma^{(0)} = 0$.

STEP 2. *Convergence test and residual evaluation.* For current values $\boldsymbol{\Sigma}^{(k)}$ and $\gamma^{(k)}$, compute the yield function

$$\phi^{(k)} = \phi(\boldsymbol{\Sigma}^{(k)}).$$

If $\phi^{(k)} \leq \delta$, then $(\boldsymbol{\Sigma}, \gamma) = (\boldsymbol{\Sigma}^{(k)}, \gamma^{(k)})$, and the computation is completed. Otherwise, compute the residual

$$\boldsymbol{R}^{(k)} = \boldsymbol{G}^{-1}\left[\boldsymbol{\Sigma}^{tr} - \boldsymbol{\Sigma}^{(k)}\right] - \gamma^{(k)} \nabla\phi(\boldsymbol{\Sigma}^{(k)}).$$

STEP 3. *Linearization.* If $\phi^{(k)} > \delta$, then linearize the residual about the current iterate $(\boldsymbol{\Sigma}^{(k)}, \gamma^{(k)})$ and obtain a linear system for the increment $(\Delta\boldsymbol{\Sigma}^{(k)}, \Delta\gamma^{(k)})$,

$$\begin{aligned} -\Delta\boldsymbol{\Sigma}^{(k)} + \bar{\boldsymbol{G}}^{(k)}\left[\boldsymbol{R}^{(k)} - \Delta\gamma^{(k)}\, \nabla\phi(\boldsymbol{\Sigma}^{(k)})\right] &= 0, \\ \nabla\phi(\boldsymbol{\Sigma}^{(k)})\, \Delta\boldsymbol{\Sigma}^{(k)} + \phi^{(k)} &= 0, \end{aligned}$$

where,

$$\bar{\boldsymbol{G}}^{(k)} = \left[\boldsymbol{G}^{-1} + \gamma^{(k)}\, \nabla^2\phi(\boldsymbol{\Sigma}^{(k)})\right]^{-1}$$

is the tensor of algorithmic moduli.

STEP 4. *Solution of the linear system and update.* Solving the linear system in Step 3, we obtain

$$\Delta\gamma^{(k)} = \frac{\phi^{(k)} + \nabla\phi(\boldsymbol{\Sigma}^{(k)}) : \bar{\boldsymbol{G}}^{(k)} \boldsymbol{R}^{(k)}}{\nabla\phi(\boldsymbol{\Sigma}^{(k)}) : \bar{\boldsymbol{G}}^{(k)} \nabla\phi(\boldsymbol{\Sigma}^{(k)})}$$

and

$$\Delta\boldsymbol{\Sigma}^{(k)} = \bar{\boldsymbol{G}}^{(k)}\left[\boldsymbol{R}^{(k)} - \Delta\gamma^{(k)} \nabla\phi(\boldsymbol{\Sigma}^{(k)})\right].$$

Then set $k := k + 1$, $\boldsymbol{\Sigma}^{(k+1)} = \boldsymbol{\Sigma}^{(k)} + \Delta\boldsymbol{\Sigma}^{(k)}$, $\gamma^{(k+1)} = \gamma^{(k)} + \Delta\gamma^{(k)}$, and return to Step 2.

Although this algorithm performs well on some numerical examples in Simo [114], theoretically it is not guaranteed that $\gamma^{(k)} \geq 0$. Also, it is an open problem to prove the convergence of the algorithm rigorously.

Bibliography

[1] R.A. Adams, *Sobolev Spaces*, Academic Press, New York, 1975.

[2] S.S. Antman, *Nonlinear Problems of Elasticity*, Springer-Verlag, New York, 1995.

[3] S.S. Antman and J.E. Osborn, The principle of virtual work and integral laws of motion, *Archive for Rational Mechanics and Analysis* **69** (1979), 231–262.

[4] I. Babuška, The finite element method with Lagrangian multipliers, *Numer. Math.* **20** (1973), 179–192.

[5] I. Babuška and A.K. Aziz, Survey lectures on the mathematical foundations of the finite element method, in A.K. Aziz, ed., *The Mathematical Foundations of the Finite Element Method with Applications to Partial Differential Equations*, Academic Press, New York, 1972, 3–359.

[6] J. Barré de Saint Venant, Mémoire sur l'établissement des équations différentielles des mouvements intérieurs opérés dans les corps solides ductiles ..., *J. Math. Pures et Appl.* **16** (1871), 308–316.

[7] K.-J. Bathe, *Finite Element Procedures*, Englewood Cliffs, NJ, 1996.

[8] J. Bauschinger, Yearly report, Mitt. Mech. Lab. Munich, 1886.

[9] J. Bergh and J. Löfström, *Interpolation Spaces, An Introduction*, Springer-Verlag, Berlin, 1976.

[10] W.W. Bird and J.B. Martin, A secant approximation for holonomic elastic-plastic incremental analysis with a von Mises yield condition, *Eng. Comp.* **3** (1986), 192–201.

[11] W.W. Bird and J.B. Martin, Consistent predictors and the solution of the piecewise holonomic incremental problem in elasto-plasticity, *Eng. Structs.* **12** (1990), 9–14.

[12] G. Birkhoff, M.H. Schultz, and R.S. Varga, Piecewise Hermite interpolation in one and two variables with applications to partial differential equations, *Numer. Math.* **11** (1968), 232–256.

[13] E. Bonnetier, Mathematical treatment of the uncertainties appearing in the formulation of some models of plasticity, Ph.D. thesis, University of Maryland, 1988.

[14] D. Braess, *Finite Elements: Theory, Fast Solvers, and Applications in Solid Mechanics*, Cambridge University Press, Cambridge, 1997.

[15] S.C. Brenner and L.R. Scott, *The Mathematical Theory of Finite Element Methods*, Springer-Verlag, New York, 1994.

[16] H. Brezis, Problèmes unilateraux, *J. Math. Pures et Appl.* **51** (1972), 1–168.

[17] F. Brezzi, On the existence, uniqueness and approximation of saddle-point problems arising from Lagrange multipliers, *RAIRO Anal. Numér.* **8** (1974), 129–151.

[18] F. Brezzi and M. Fortin, *Mixed and Hybrid Finite Element Methods*, Springer-Verlag, Berlin, 1991.

[19] P. Carter and J.B. Martin, Weak bounding functions for plastic materials, *J. Appl. Mech.* **43** (1976), 434–438.

[20] J. Céa, Approximation variationnelle des problèmes aux limites, *Ann. Inst. Fourier (Grenoble)* **14** (1964), 345–444.

[21] P. Chadwick, *Continuum Mechanics: Concise Theory and Problems.* George Allen & Unwin, London, 1976.

[22] W.F. Chen and D.J. Han, *Plasticity for Structural Engineers*, Springer-Verlag, New York, 1988.

[23] P.G. Ciarlet, *The Finite Element Method for Elliptic Problems*, North Holland, Amsterdam, 1978.

[24] P.G. Ciarlet, *Mathematical Elasticity, Vol. I: Three-Dimensional Elasticity*, North Holland, Amsterdam, 1988.

[25] P.G. Ciarlet, Basic error estimates for elliptic problems, in P.G. Ciarlet and J.-L. Lions, eds., *Handbook of Numerical Analysis*, Vol. II, North-Holland, Amsterdam, 1991, 17–351.

[26] P. Clément, Approximation by finite element functions using local regularization, *RAIRO Anal. Numer.* **9R2** (1975), 77–84.

[27] B. Coleman and M. Gurtin, Thermodynamics with internal state variables, *J. Chem. Phys.* **47** (1967), 597–613.

[28] C. Comi and G. Maier, On the convergence of a backward difference iterative procedure in elastoplasticity with nonlinear kinematic and isotropic hardening, in D.R.J. Owen, E. Hinton, and E. Oñate, eds., *Computational Plasticity: Models, Software and Applications*, Pineridge Press, Swansea, 1989, 323–334.

[29] M.A. Crisfield, Accelerating and damping the modified Newton–Raphson method, *Computers and Structures* **18** (1984), 395–407.

[30] R. Dautray and J.L. Lions, *Mathematical Analysis and Numerical Methods for Science and Technology, Vol. II, Functional and Variational Methods*, Springer-Verlag, New York, 1988.

[31] J. Douglas Jr. and T. Dupont, Galerkin methods for parabolic equations, *SIAM J. Numer. Anal.* **7** (1970), 575–626.

[32] D.C. Drucker, A more fundamental approach to plastic stress–strain relations, in *Proc. 1st US National Congress of Applied Mechanics*, ASME, New York, 1951, 487–491.

[33] G. Duvaut and J.-L. Lions, *Inequalities in Mechanics and Physics*, Springer-Verlag, Berlin, 1976.

[34] I. Ekeland and R. Temam, *Convex Analysis and Variational Problems*, North-Holland, Amsterdam, 1976.

[35] L.C. Evans, *Partial Differential Equations*, Berkeley Mathematics Lecture Notes, Volume 3, 1994.

[36] R.A. Eve, T. Gültop, and B.D. Reddy, An internal variable finite-strain theory of elastoplasticity within the framework of convex analysis, *Quart. Appl. Math.* **48** (1990), 625–643.

[37] R.A. Eve and B.D. Reddy, The variational formulation and solution of problems of finite-strain elastoplasticity based on the use of a dissipation function, *Int. J. Num. Meths. Eng.* **37** (1994), 1673–1695.

[38] R.A. Eve, B.D. Reddy, and R.T. Rockafellar, An internal variable theory of elastoplasticity based on the maximum plastic work inequality, *Quart. Appl. Math.* **48** (1990), 59–83.

[39] R.S. Falk, Error estimates for the approximation of a class of variational inequalities, *Math. Comp.* **28** (1974), 963–971.

[40] G. Fichera, Problemi elastostatici con vincoli unilaterali: il problema di Signorini con ambigue condizioni al contorno, *Mem. Accad. Naz. Lincei* **8** (7) (1964), 91–140.

[41] R. Fletcher, *Practical Methods of Optimization. Volume 1: Unconstrained Problems; Volume 2: Constrained Problems*, Wiley, New York, 1980.

[42] A. Friedman, *Variational Principles and Free-Boundary Problems*, John Wiley & Sons, New York, 1982.

[43] V. Girault and P.-A. Raviart, *Finite Element Methods for Navier–Stokes Equations, Theory and Algorithms*, Springer-Verlag, Berlin, 1986.

[44] R. Glowinski, *Numerical Methods for Nonlinear Variational Problems*, Springer-Verlag, New York, 1984.

[45] R. Glowinski, J.-L. Lions, and R. Trémolières, *Numerical Analysis of Variational Inequalities*, North-Holland, Amsterdam, 1981.

[46] P. Grisvard, *Elliptic Problems in Nonsmooth Domains*, Pitman, Boston, 1985.

[47] M.E. Gurtin, Modern continuum thermodynamics, in S. Nemat-Nasser, ed., *Mechanics Today*, Pergamon, Oxford, 1974, 168–210.

[48] M.E. Gurtin, *An Introduction to Continuum Mechanics*, Academic Press, New York, 1981.

[49] B. Halphen and Q.S. Nguyen, Sur les matériaux standards généralisés, *J. Méc.* **14** (1975), 39–63.

[50] W. Han, *A posteriori* error analysis for linearizations of nonlinear elliptic problems and their discretizations, *Math. Meths. Appl. Sci.* **17** (1994), 487–508.

[51] W. Han, Quantitative error estimates for idealizations in linear elliptic problems, *Math. Meths. Appl. Sci.* **17** (1994), 971–987.

[52] W. Han, S. Jensen, and B.D. Reddy, Numerical approximations of internal variable problems in plasticity: error analysis and solution algorithms, *Numerical Linear Algebra with Applications* **4** (1997), 191–204.

[53] W. Han, S. Jensen, and I. Shimansky, The Kačanov method for some nonlinear problems, *Applied Numerical Analysis* **24** (1997), 57–79.

[54] W. Han and B.D. Reddy, On the finite element method for mixed variational inequalities arising in elastoplasticity, *SIAM J. Numer. Anal.* **32** (1995), 1778–1807.

[55] W. Han and B.D. Reddy, Computational plasticity: the variational basis and numerical analysis, *Computational Mechanics Advances* **2** (1995), 283–400.

[56] W. Han, B.D. Reddy, and G.C. Schroeder, Qualitative and numerical analysis of quasistatic problems in elastoplasticity, *SIAM J. Numer. Anal.* **34** (1997), 143–177.

[57] J. Haslinger, I. Hlaváček, and J. Nečas, Numerical methods for unilateral problems in solid mechanics, in P.G. Ciarlet and J.-L. Lions, eds., *Handbook of Numerical Analysis*, Vol. IV, North-Holland, Amsterdam, 1996, 313–485.

[58] R. Hill, The essential structure of constitutive laws for metal composites and polycrystals, *J. Mech. Phys. Solids* **15** (1967), 79–95.

[59] R. Hill, Constitutive dual potentials in cassical plasticity, *J. Mech. Phys. Solids* **35** (1987), 39–63.

[60] I. Hlaváček, A finite element solution for plasticity with strain-hardening, *RAIRO Anal. Numér.* **14** (1980), 347–368.

[61] I. Hlaváček, J. Haslinger, J. Nečas, and J. Lovíšek, *Solution of Variational Inequalities in Mechanics*, Springer-Verlag, New York, 1988.

[62] H. Huang, W. Han, and J. Zhou, The regularization method for an obstacle problem, *Numer. Math.* **69** (1994), 155–166.

[63] T.J.R. Hughes, *The Finite Element Method: Linear Static and Dynamic Finite Element Analysis*, Prentice-Hall, Englewood Cliffs, NJ, 1987.

[64] C. Johnson, Existence theorems for plasticity problems, *J. Math. Pures Appl.* **55** (1976), 431–444.

[65] C. Johnson, On finite element methods for plasticity problems, *Numer. Math.* **26** (1976), 79–84.

[66] C. Johnson, On plasticity with hardening, *J. Math. Anal. Appls.* **62** (1978), 325–336.

[67] C. Johnson, A mixed finite element method for plasticity problems with hardening, *SIAM J. Numer. Anal.* **14** (1977), 575–583.

[68] C. Johnson, A convergence estimate for an approximation of a parabolic variational inequality, *SIAM J. Numer. Anal.* **13** (1977), 599–606.

[69] C. Johnson, *Numerical Solutions of Partial Differential Equations by the Finite Element Method*, Cambridge University Press, Cambridge, 1987.

[70] N. Kikuchi and J.T. Oden, *Contact Problems in Elasticity: A Study of Variational Inequalities and Finite Element Methods*, SIAM, Philadelphia, 1988.

[71] D. Kinderlehrer and G. Stampacchia, *An Introduction to Variational Inequalities and Their Applications*, Academic Press, New York, 1980.

[72] W.T. Koiter, Stress-strain relations, uniqueness and variational theorems for elastic-plastic material with a singular yield surface, *Quart. Appl. Math.* **11** (1953), 29–53.

[73] W.T. Koiter, General theorems for elastic-plastic solids, in I.N. Sneddon and R. Hill, eds., *Progress in Solid Mechanics*, North-Holland, Amsterdam, 1960, 167–221.

[74] V.G. Korneev and U. Langer, *Approximate Solution of Plastic Flow Theory Problems*, Teubner-Texte **65**, Leibniz, 1984.

[75] J. Lemaitre and J.-L. Chaboche, *Mechanics of Solid Materials*, Cambridge University Press, Cambridge, 1990.

[76] M. Lévy, Extrait du mémoire sur les équations générales des mouvements intérieurs des corps solids ductiles au delà des limites où l'élasticité pourrait les ramener à leur premier état, *J. Math. Pures Appl.* **16** (1871), 369–372.

[77] Y. Li and I. Babuška, A convergence analysis of a p-version finite element method for one-dimensional elastoplasticity problem with constitutive laws based on the gauge function method, *SIAM J. Numer. Anal.* **33** (1996), 809–842.

[78] Y. Li and I. Babuška, A convergence analysis of an h-version finite element method with high order elements for two dimensional elastoplasticity problems, *SIAM J. Numer. Anal.* **34** (1997), 998–1036.

[79] J.-L. Lions and G. Stampacchia, Variational inequalities, *Comm. Pure Appl. Math.* **20** (1967), 493–519.

[80] J. Lubliner, On the thermodynamics foundations of non-linear solid mechanics, *Int. J. Nonl. Mech.* **7** (1972), 237–254.

[81] D.G. Luenberger, *Linear and Nonlinear Programming*, 2nd ed., Addison-Wesley, Reading, Mass., 1984.

[82] J.E. Marsden and T.J.R. Hughes, *Mathematical Foundations of Elasticity*, Prentice-Hall, New Jersey, 1983.

[83] J.B. Martin, *Plasticity: Fundamentals and General Results*, MIT Press, Cambridge, Mass., 1975.

[84] J.B. Martin, An internal variable approach to the formulation of finite element problems in plasticity, in J. Hult and J. Lemaitre, eds., *Physical Nonlinearities in Structural Analysis*, Springer-Verlag, Berlin, 1981, 165–176.

[85] J.B. Martin and S. Caddemi, Sufficient conditions for the convergence of the Newton-Raphson iterative algorithm in incremental elastic-plastic analysis, *Euro. J. Mech. A/Solids* **13** (1994), 351–365.

[86] J.B. Martin and A. Nappi, An internal variable formulation for perfectly plastic and linear hardening relations in plasticity. *Euro. J. Mech. A/Solids* **9** (1990), 107–131.

[87] J.B. Martin and B.D. Reddy, Variational principles and solution algorithms for internal variable formulations of problems in plasticity, in U. Andeaus et al., ed., *Omaggio a Giulio Ceradini*, Università di Roma "La Sapienza," Roma, 1988, 465–477.

[88] H. Matthies, Existence theorems in thermoplasticity, *J. Méc.* **18** (1979), 695–711.

[89] H. Matthies and G. Strang, The solution of nonlinear finite element equations, *Int. J. Numer. Meths. Eng.* **14** (1979), 1613–1626.

[90] H. Matthies, G. Strang, and E. Christiansen, The saddle point of a differential program, in R. Glowinski, E. Rodin, and O.C. Zienkiewicz, eds., *Energy Methods in Finite Element Analysis*, Wiley, New York, 1979.

[91] E. Melan, Zur Plastizität des räumlichen Kontinuums, *Ing. Arch.* **9** (1938), 116–125.

[92] R. von Mises, Mechanik der festen Körper im plastisch deformablen Zustand, *Gött. Nach. Math. Phys. Kl.* (1913), 582–592.

[93] R. von Mises, Mechanik der plastischen Formänderung von Kristallen, *ZAMM* **8** (1928), 161–185.

[94] J.J. Moreau, Sur les lois de frottement, de viscosité et plasticité, *C. R. Acad. Sc.* **271** (1970), 608–611.

[95] J.J. Moreau, Application of convex analysis to the treatment of elastoplastic systems, in P. Germain and B. Nayroles, eds., *Applications of Methods of Functional Analysis to Problems in Mechanics*, Springer-Verlag, Berlin, 1976.

[96] J.J. Moreau, Evolution problem associated with a moving convex set in a Hilbert space, *J. Diff. Eqns* **26** (1977), 347–374.

[97] J.T. Oden, Finite elements: an introduction, in P.G. Ciarlet and J.-L. Lions, eds., *Handbook of Numerical Analysis*, Vol. II, North-Holland, Amsterdam, 1991, 3–15.

[98] J.T. Oden and J.N. Reddy, *An Introduction to the Mathematical Theory of Finite Elements*, Wiley-Interscience, New York, 1976.

[99] P.D. Panagiotopoulos, *Inequality Problems in Mechanics and Applications*, Birkhäuser, Boston, 1985.

[100] W. Prager, Recent developments in the mathematical theory of plasticity, *J. Appl. Phys.* **20** (1949), 235–241.

[101] L.T. Prandtl, Spannungsverteilung in plastischen Körpern, in *Proc. 1st Intern. Congr. Mechanics*, Delft, 1924, 43–54.

[102] L.T. Prandtl, Ein Gedankenmodell zur kinetischen Theorie der festen Körper, *ZAMM* **8** (1928), 85–106.

[103] A. Quarteroni and A. Valli, *Numerical Approximation of Partial Differential Equations*, Springer-Verlag, Berlin, 1994.

[104] B.D. Reddy, Existence of solutions to a quasistatic problem in elastoplasticity, in C. Bandle et al., eds., *Progress in Partial Differential Equations: Calculus of Variations, Applications*, Pitman Research Notes in Mathematics **267**, Longman, London, 1992, 233–259.

[105] B.D. Reddy, Mixed variational inequalities arising in elastoplasticity, *Nonl. Anal. Theory, Meths. and Appls.* **19** (1992), 1071–1089.

[106] B.D. Reddy, *Introductory Functional Analysis with Applications to Boundary Value Problems and Finite Elements*, Springer-Verlag, New York, 1998.

[107] B.D. Reddy and J.B. Martin, Algorithms for the solution of internal variable problems in plasticity, *Comp. Meth. Appl. Mech. Engng.* **93** (1991), 253–273.

[108] B.D. Reddy and J.B. Martin, Internal variable formulations of problems in elastoplasticity: constitutive and algorithmic aspects, *Adv. Appl. Mech.* **47** (1994), 429–456.

[109] M. Renardy and R.C. Rogers, *An Introduction to Partial Differential Equations*, Springer-Verlag, New York, 1993.

[110] L.J. Rencontré, W.W. Bird, and J.B. Martin, Internal variable formulation of a backward difference corrector algorithm for piecewise linear yield surfaces, *Meccanica* **27** (1992), 13–24.

[111] A. Reuss, Berücksichtigung der elastischen Formänderung in der Plastizitätstheorie, *Zeit. Angew. Math. und Mech.* **10** (1939), 26–274.

[112] J.R. Rice, On the structure of stress-strain relations for time-dependent plastic deformation in metals, *J. Appl. Mech.* **137** (1970), 728–737.

[113] R.T. Rockafellar, *Convex Analysis*, Princeton University Press, Princeton, New Jersey, 1970.

[114] J.C. Simo, Topics on the numerical analysis and simulation of plasticity, in P.G. Ciarlet and J.-L. Lions, eds., *Handbook of Numerical Analysis* Vol. VI, North-Holland, Amsterdam, 1998, 183–499.

[115] J.C. Simo and S. Govindjee, Nonlinear B-stability and symmetry preserving return mapping algorithms for plasticity and viscoplasticity, *Int. J. Numer. Meths. Engng.* **31** (1991), 151–176.

[116] J.C. Simo and T.J.R. Hughes, *Computational Inelasticity*, Springer-Verlag, New York, 1998.

[117] J.C. Simo, J.G. Kennedy, and S. Govindjee, Non-smooth multisurface plasticity and viscoplasticity. Loading/unloading conditions and numerical algorithms, *Int. J. Numer. Meths. Engng.* **26** (1988), 2161–2185.

[118] J.C. Simo and R.L. Taylor, Consistent tangent operators for rate-independent elasto-plasticity, *Comp. Meth. Appl. Mech. Engng.* **48** (1985), 101–118

[119] G. Strang and G. Fix, *An Analysis of the Finite Element Method*, Prentice-Hall, Englewood Cliffs, NJ, 1973.

[120] A.M. Stuart and A.R. Humphries, Model problems in numerical stability theory for initial value problems, *SIAM Rev.* **36** (1994), 226–257.

[121] B. Szabó and I. Babuška, *Finite Element Analysis*, John Wiley & Sons, Inc., New York, 1991.

[122] R. Temam, *Mathematical Problems in Plasticity*, Gauthier-Villlars, Paris, 1985.

[123] R. Temam and G. Strang, Functions of bounded deformation, *Arch. Rational Mech. Anal.* **75** (1980), 7–21.

[124] R. Temam and G. Strang, Duality and relaxation in the variational problems of plasticity, *J. Méc.* **19** (1980), 493–527.

[125] H.E. Tresca, Mémoire sur l'écoulement des corps solids, *Mémoire Présentés par Divers Savants, Acad. Sci. Paris* **20** (1872), 75–135.

[126] C. Vuik, An L^2-error estimate for an approximation of the solution of a parabolic variational inequality, *Numer. Math.* **57** (1990), 453–471.

[127] C.C. Wang and C. Truesdell, *Introduction to Rational Elasticity*, Noordhoff, Leyden, 1973.

[128] E. Zeidler, *Nonlinear Functional Analysis and its Applications. I: Fixed-point Theorems*, Springer-Verlag, New York, 1985.

[129] E. Zeidler, *Nonlinear Functional Analysis and its Applications. IIA: Linear Monotone Operators*, Springer-Verlag, New York, 1990.

[130] E. Zeidler, *Nonlinear Functional Analysis and its Applications. III: Variational Methods and Optimization*, Springer-Verlag, New York, 1986.

[131] E. Zeidler, *Applied Functional Analysis: Applications of Mathematical Physics*, Springer-Verlag, New York, 1995.

[132] E. Zeidler, *Applied Functional Analysis: Main Principles and Their Applications*, Springer-Verlag, New York, 1995.

[133] O.C. Zienkiewicz, Origins, milestones and directions of the finite element method—a personal view, in P.G. Ciarlet and J.-L. Lions, eds., *Handbook of Numerical Analysis*, Vol. IV, North-Holland, Amsterdam, 1996, 3–67.

[134] O.C. Zienkiewicz and R.L. Taylor, *The Finite Element Method*, Vol. I (Basic Formulation and Linear Problems), McGraw-Hill, New York, 1989.

[135] O.C. Zienkiewicz and R.L. Taylor, *The Finite Element Method*, Vol. II (Solid and Fluid Mechanics, Dynamics and Nonlinearity), McGraw-Hill, New York, 1991.

[136] M. Życzkowski, *Combined Loadings in the Theory of Plasticity*, P.W.N. Polish Scientific Publishers, Warsaw, 1981.

Index

Interdisciplinary Applied Mathematics

---

1. *Gutzwiller:* Chaos in Classical and Quantum Mechanics
2. *Wiggins:* Chaotic Transport in Dynamical Systems
3. *Joseph/Renardy:* Fundamentals of Two-Fluid Dynamics:
   Part I: Mathematical Theory and Applications
4. *Joseph/Renardy:* Fundamentals of Two-Fluid Dynamics:
   Part II: Lubricated Transport, Drops and Miscible Liquids
5. *Seydel:* Practical Bifurcation and Stability Analysis:
   From Equilibrium to Chaos
6. *Hornung:* Homogenization and Porous Media
7. *Simo/Hughes:* Computational Inelasticity
8. *Keener/Sneyd:* Mathematical Physiology
9. *Han/Reddy:* Plasticity: Mathematical Theory and Numerical Analysis